Maths for Chemistry

A chemist's toolkit of calculations

Second Edition

Paul Monk[1] and Lindsey J. Munro[2]

[1] *Team Vicar in the Medlock Head Team Ministry, Oldham,*
and formerly Senior Lecturer in Physical Chemistry,
School of Biology, Chemistry, and Health Science,
Manchester Metropolitan University

[2] *Senior Lecturer, School of Biology, Chemistry, and Health Science,*
Department of Chemistry and Materials,
Manchester Metropolitan University

OXFORD
UNIVERSITY PRESS

OXFORD

UNIVERSITY PRESS

Great Clarendon Street, Oxford ox2 6DP

Oxford University Press is a department of the University of Oxford.
It furthers the University's objective of excellence in research, scholarship,
and education by publishing worldwide in

Oxford New York

Auckland Cape Town Dar es Salaam Hong Kong Karachi
Kuala Lumpur Madrid Melbourne Mexico City Nairobi
New Delhi Shanghai Taipei Toronto

With offices in

Argentina Austria Brazil Chile Czech Republic France Greece
Guatemala Hungary Italy Japan Poland Portugal Singapore
South Korea Switzerland Thailand Turkey Ukraine Vietnam

Oxford is a registered trade mark of Oxford University Press
in the UK and in certain other countries

Published in the United States
by Oxford University Press Inc., New York

First edition published 2006
This edition published 2010

British Library Cataloguing in Publication Data

Data available

Library of Congress Cataloging in Publication Data

Data available

Typeset by MPS Limited, A Macmillan Company
Printed in Italy on acid-free paper by
L.E.G.O. S.p.A.

ISBN 978–0–19–954129–4

1 3 5 7 9 10 8 6 4 2

Contents

Preface to the second edition

The rationale behind the first edition of *Maths for Chemistry* was fivefold: to produce a text that

1. Demonstrated at every stage how chemical calculations should be performed properly.
2. Illustrated each mathematical point using copious examples from chemistry, rather than use abstract mathematics on its own.
3. As well as mathematical notation, explained what was happening throughout a calculation using straightforward English prose.
4. Had additional help in the form of margin notes, a comprehensive bibliography, and an accompanying website.
5. Had fully worked model answers to all self-test questions.

The first edition of this book appeared in 2006, and was immediately greeted with near unanimous warmth and unexpectedly gracious praise. Clearly, the book had succeeded even better than planned. It has already gone through three printings.

But even as the ink of the first edition was drying, the book's deficiencies became ever clearer. Firstly, some reviewers wondered aloud why the book had not covered more introductory material. Students without an A-level in a mathematical discipline struggle to make the leap to degree-level chemistry if they have only GCSE-level mathematics. That insight alone explains why this second edition contains wholly new chapters on fractions, and dimensional analysis. Some of the other chapters are heavily reworked, particularly Chapters 1–3, 10, 12, 14–16 and 24.

Secondly, other reviewers asked the complementary question: 'why did the book stop too abruptly, before considering the mathematics necessary for courses in quantum chemistry?' For this reason, this edition includes wholly new chapters on partial differentials; integration by parts, by substitution, and integration tables; integration to yield areas and volumes; matrices and determinants; vectors; and complex numbers.

Having decided to produce a second edition of *Maths for Chemistry,* it was immediately apparent how the *whole* of the first edition book needed careful revision. It therefore contains proportionally more worked examples and self-test questions. The amount of prose in the answers is generally slightly greater. We have also subtly altered the order of the chapters. For example, in the first edition the two statistics chapters were appendices, but these have been assimilated fully into the text of the second edition. The book is also more thoroughly cross-referenced than the first edition, and also includes a substantial glossary.

But we have retained the original purpose and rationale: it is a book for chemistry students who want to learn mathematics as painlessly as possible.

PMSM and LJM: 1 August 2009

Preface to the first edition

Many people, when they first start learning advanced and degree-level chemistry, are dismayed when told they must also learn mathematics. They say, 'But I want to be a chemist and *not* a mathematician!' In fact, all chemists use mathematics each day of their working lives. It's one of our most useful tools, and we need to remember that mathematics is only a tool and never a master.

This book is intended for people who want to learn chemistry but either find mathematics very hard, or who have never really studied it and therefore need to master its rules. Most people, when they start to learn a musical instrument or a foreign language, initially find great difficulty in remembering the 'mechanics' of their new art. For example, they need to remember how to position their fingers every time they see a b-flat note from a sheet of music; again, the trainee linguist needs to remember how to conjugate a verb ending each time they encounter a new word. But after a remarkably short period of time, the music student's fingers will automatically reach for the right note, and the student learning a language can conjugate verbs seemingly without effort. Having attained this stage, the musician can concentrate on what the music actually sounds like; the linguist can concentrate on what the passage to be translated actually means.

Many people reading this book are similar to the trainee musician: we may know *how* to finger some of the notes, but we are not yet proficient enough to take the fingering for granted; and we need to learn more notes and develop our technique. We routinely forget to concentrate on the melody: the point of playing is to make a beautiful sound, not to move the fingers. Similarly, we will be better chemists after we master the simple rules of mathematics, because we will know what the equations actually *mean*.

Our first task, then, is to learn the grammar of mathematics.

Acknowledgement for the second edition

We would like to thank Dr Chris Rego, Principal Lecturer in Physical Chemistry at the School of Biology, Chemistry and Health Science, MMU, for reading three chapters and making helpful suggestions. We also thank Mr Richard Mundy for helpful discussions on stylistic points and pedagogical etiquette.

We thank the eight reviewers who OUP commissioned to read draft chapters from the second edition. We have not always incorporated their recommendations, but they will immediately discern where their comments have caused us to change or further develop the text.

We also thank Professor Pau Atela and Christophe Gole of the Mathematics Department, Smith College, USA, for permission to reproduce Fig. 7.1 (Fig. 6.1 in the first edition), and the Vicar and Wardens of Oldham Parish Church for permission to reproduce Fig. 23.1.

We would like to thank Dewi Jackson, Jonathan Crowe, and Ruth Hughes, of *Oxford University Press*, who have ensured the project's smooth running. We also thank Dr Peter Capper for his proof-reading skills, and Chantal Peacock and Keira Dickinson who were the production editors at OUP.

Clearly, any errors that remain are ours alone.

PMSM and LJM

Acknowledgement for the first edition

One of the greater joys in writing a book is the opportunity afforded me to thank the many colleagues who have made it possible. I therefore wish to thank Michael Rycraft (formerly of the Department of Mathematics, MMU) who read the entire manuscript, and made a great many clever comments. Also, Dr Chris Rego and Dr David Johnson, both of whom work in my own department, who each read several chapters. Again, their expertise is evident in the final version of the text. Similarly, I would like to thank the 2004–5 cohort of first-year students at MMU, who acted as 'guinea pigs' in our 'Mathematics for Chemists' unit, in which chapters of this book were issued in the form of weekly hand outs. Their comments on the first draft were frank but fair, and eventually appreciative. Any errors and obscurities remaining are of course entirely my own.

I wish to thank Professor Pau Atela and Christophe Gole of the Mathematics Department, Smith College, USA, for permission to reproduce Fig. 6.1 (7.1 in the second edition).

I would also like to thank Jonathan Crowe and Ruth Hughes, both of *Oxford University Press*, who have ensured the project's smooth running.

PMSM: October 2005

Instructions for the tutor

We intend this book to represent a course lasting a single academic year. Accordingly, each chapter will represent about one week's worth of work if lectures occur through three terms. One lecture and one tutorial per chapter will probably be an appropriate amount of time. When developing this course, we were adamant that early chapters should be taught without requiring the material from later chapters. In each of the (relatively rare) cases where later material has been mentioned, we include a cross-reference.

Each chapter is clearly too long for its entire content to be covered in depth during a single lecture. We suggest that tutors familiarize themselves with the text, and then offer the key points, perhaps using a few worked examples to act as ramification. Students will then be required to read the whole chapter properly in their own time before the respective tutorial.

Many students will have studied mathematics to an intermediate level, and feel the material in the early chapters is far too easy. Please recommend that they try the questions in the **ADDITIONAL PROBLEMS** section at the end of each chapter. These examples are intentionally much harder than those in the main body of the chapter. Only if the student can indeed perform these ADDITIONAL PROBLEMS should the student skip a chapter. Model answers to all the ADDITIONAL PROBLEMS as well as a set of Student Multiple Choice Questions for each chapter are available on the Internet at http://www.oxfordtextbooks.co.uk/orc/monk2e.

The contextual approach adopted here is intended to simplify the course, since most students learn material more readily if they already know a 'skeleton' and can add 'flesh' to it, rather than learning from scratch. Occasionally, students will find the chemistry more difficult than the mathematics; unfortunately, the topics deemed difficult will differ from student to student. We suggest that tutors regularly reassure nervous students, saying 'You only need to understand the examples; it is the mathematics you need to learn.' Only the numbered equations need to be learnt. The text contains about 130 such equations.

The answers to all the SELF-TEST QUESTIONS may be found in the appendices. Most are given in the form of complete model answers.

Instructions for the student

It is tempting to start a course in mathematics with the words, 'But I want to study *chemistry*!' We wrote this course with such a mindset in view. We intend the contextual approach here to emphasize how mathematics is a key tool in the chemist's toolchest, and not an 'add-on extra.' Occasionally the chemistry will obscure rather than enlighten. If you find this to be the case, simply jettison the chemistry and concentrate on the maths, or go on to the next worked example. You will never be asked in a maths course to remember the chemistry examples. The only equations you need to memorize for this course are those with a number, e.g. the symbol Δ is defined in eqn (2.1). The key equations are usefully summarized at the end of each chapter. On the average, there are around 5–10 such equations per chapter. Most are very easy; in fact, you will probably have met some of them before starting this course.

We recommend that you read a chapter through from end to end before starting any of the self-test questions. This allows for an overview and, hopefully, a clearer vision of what is involved. Then reread the relevant section before starting the Self-test questions. The answers to all these questions may be found in the appendices. Almost all are given as full model answers. You will find a set of Multiple Choice Questions to test yourself at the end of each chapter on the Internet at http://www.oxfordtextbooks.co.uk/orc/monk2e.

Many of you will have studied mathematics to an intermediate level before starting your chemistry course, and therefore feel the material in our early chapters is far too easy. Please try the questions in the ADDITIONAL PROBLEMS section at the end of each chapter, which are intentionally much more involved than in the body of the chapter. If you can indeed perform these ADDITIONAL PROBLEMS, then you should indeed skip the chapter. The model answers to these are available on the Internet at www.oxfordtextbooks.co.uk/orc/monk2e.

Symbols

Algebraic symbols

a acceleration; constant; van der Waals constant

a_0 atomic diameter

A   ampère

A area; Arrhenius pre-exponential factor; Debye–Hückel 'A' factor

Abs absorbance

b constant; van der Waals constant

c   centi

c concentration; intercept on a graph; speed of light; general constant; constant of integration; number of components

C   coulomb; Celsius

C heat capacity

C_p heat capacity at constant pressure

C_V heat capacity at constant volume

d   differential operator; deci

d inter-plane distance in a regular crystal; distance

∂ partial differential

D   dimension(s), e.g. 3D means three-dimensions or 3-dimensional

D diffusion coefficient

Da   Dalton

e   exponential

e^- electron (with charge indicated)

E energy

$\mathcal{E}$ electric field

E_a activation energy

$E_{O,R}$ electrode potential for the redox couple, $O + ne^- = R$

$E_{O,R}^{\ominus}$ standard electrode potential for the redox couple, $O + ne^- = R$

exp   exponential (i.e. $(2.178 \ldots)^x$)

f   femto

f generalized function; number of degrees of freedom

$f(x)$ function of x; distribution of x

F   Faraday constant; temperature in Fahrenheit

F function

g   gramme; acceleration due to gravity

G   giga

G Gibbs function

h height; Planck constant; Miller index

$\hbar$ $h \div 2\pi$, where h = Planck constant

hr   hour

H enthalpy

H_c enthalpy of combustion

H_f enthalpy of formation

H_r enthalpy of reaction

Hz   Hertz

i   indicates that a number is complex i.e. multiple of $\sqrt{-1}$

i item number in a list, current density

$\hat{\imath}$	unit vector in the x-direction	N_A	Avogadro number
$\mathbf{I}$	the identity matrix	O	oxidized form of a redox couple
I	ionic strength; current	p	pressure; number of phases, momentum
I_t	time-dependent current	$p^{\ominus}$	standard pressure
j	item number in a list, other than i	$p_{(x)}^{\ominus}$	partial pressure of pure compound x
$\hat{\jmath}$	unit vector in the y-direction	p	pico; the operator $-\log_{10}(\ldots)$, e.g. pH $= -\log_{10}[H^+]$
J	Joule	p	pressure; momentum
k	kilo	ph	phenyl
$\hat{k}$	unit vector in the z-direction	pH	$= -\log[H^+]$
k	rate constant; bond force constant; abbreviation for (RT/F); Miller index	P	probability
k_B	Boltzmann constant	P	peta
k_n	rate constant of the nth step in a reaction series	q	charge; heat
k_{-n}	rate constant of the reverse of the nth step in a reaction series	Q	quinone
		Q	charge density; result of a Q-test
K	kelvin	$Q_{(exp)}$	calculated Q-test quotient
K	equilibrium constant; thermal conductivity	r	correlation coefficient; radius of a circle or sphere; radial distance of a vector
K_{sp}	equilibrium constant of solubility ('solubility product')	R	reduced form of a redox couple
l	Miller index	R	gas constant; resistance; alkyl group
l	length; optical path length	$\mathfrak{R}$	Rydberg constant
ln	logarithm to base e (i.e. natural logarithm)	s	second
log	logarithm in base 10	s	sample standard deviation
$\log_x$	logarithm in base x	S	entropy
m	metre; milli	t	time; temperature in Celsius
m	mass; gradient of a graph	$t_{1/2}$	half-life
M	mega	T	absolute temperature
M	molar mass	u	top function within a quotient; first function in a product; substituent in a chain-rule problem; initial velocity
mol	mole		
n	amount of material; number; number of electrons transferred in a redox reaction; number of X-ray reflections	U	internal energy
		v	bottom function within a quotient; second function in a product; velocity; final velocity
n_0	amount of material at the start of a process		
N	number; number of members in a series or data set; number of monomer units in a polymer; number of π-electrons	V	volt
		V	volume
		V_m	molar volume $= V \div n$

x	variable; the controlled variable; the horizontal axis on a graph ('abscissa'); mole fraction; displacement from equilibrium position	ν	wave number; scan rate in cyclic voltammetry
X	argument of an algebraic function	λ	wavelength; eigenvalue of a matrix
y	variable; the observed variable; the vertical axis on a graph ('ordinate')	Λ	ionic conductivity
		Λ°	ionic conductivity at infinite dilution
z	charge on an ion	Λ^2	the Legendrian operator
Z	atomic number	π	the ratio of a circle's diameter and circumference
Z_{eff}	effective charge on an atomic nucleus	Π	Pi product operator, as defined in eqn (2.3)
		ρ	density
γ	mean ionic activity coefficient	σ	population standard deviation
δ	small increment; thickness of the Nernst diffusion layer	σ_x	innate error associated with the variable x
Δ	Delta operator, as defined in eqn (2.1)	Σ	Sigma operator, as defined in eqns (2.2)
Δ	matrix determinant	υ	kinematic viscosity
ε	molar extinction coefficient; permittivity	ψ	atomic wavefunction
ε_0	permittivity of free space	ψ_{1s}	wavefunction of 1s atomic orbital
η	overpotential	Ψ	molecular wavefunction
κ	ionic conductivity	ϕ	electric potential; azimuth angle in spherical polar coordinates
θ	angle of focus in a triangle; vector inclination angle; elevation angle in cylindrical and spherical polar coordinates	ω	angular rotation speed (e.g. of a RDE); frequency of a sinusoidal wave
ζ	extent of reaction	Ω	ohm
μ	reduced mass; ionic or electronic mobility	∇	del or Nabla
		∇	**curl** of a vector

Additional subscripts, superscripts, and typographic notations

$\ominus$	standard state	r	reaction
#	activation parameter	T	the transpose of a matrix
c	combustion	$\bar{x}$	an 'overbar' indicates a *mean* value of x; e.g. see eqn (14.1)
i	the ith member of a series or group	$\wedge$	the adjoint of a matrix
m	molar		

Common standard prefixes

P   peta ($\times 10^{12}$)

G   giga ($\times 10^{9}$)

M   mega ($\times 10^{6}$)

k   kilo ($\times 10^{3}$)

d   deci ($\times 10^{-1}$)

c   centi ($\times 10^{-2}$)

m   milli ($\times 10^{-3}$)

μ micro ($\times 10^{-6}$)

n   nano ($\times 10^{-9}$)

p   pico ($\times 10^{-12}$)

f   femto ($\times 10^{-15}$)

Acronyms and abbreviations

ac   alternating current

BODMAS   acronym (p. 35): brackets, of, division, multiplication, addition, subtraction

CFC   chlorofluorocarbon

cos   cosine

DDT   dichlorodipheny ltrichloroethane

d.p.   decimal places

DRG   degree, radians, grads (a button on a calculator)

DVM   digital voltmeter

emf   electromotive force

H_2Q hydroquinone

IUPAC   International Union of Pure and Applied Chemistry

IQ   intelligence quotient

KE   kinetic energy

LCAO   linear combination of atomic orbitals

MV   methyl viologen (1,1'-dimethyl-4,4'-bipyridilium)

NMR   nuclear magnetic resonance

ppb   parts per billion

ppm   parts per million

RDE   rotated-disc electrode

RRDE   rotated ring-disc electrode

rms   root mean square

SCE   saturated calomel electrode

s.f.   significant figure

sin   sine

SLS   sodium lauryl sulphate

SN   signal-to-noise ratio

sohcahtoh   acronym to use sin, cos, and tan (see p. 249)

tan   tangent

Irrational numbers

π (pi)   3.142 …

τ (tau)   1.618 …

e   2. 178 …

1

The display of numbers

Standard factors, scientific notation, significant figures and decimal places

Introducing algebraic phrases

By the end of this section, you will know:

- An algebraic phrase includes a variable, a number and (usually in chemistry) units.
- One or more of the terms in the phrase may also be multiplied by a factor.
- We give the name compound variable to the product of two or more variables.

When chemists make a chemical compound, for example precipitating a salt by metathesis, they only require a certain number of building 'blocks.' The blocks in this case are anions and cations. Similarly, a surveyor tells the builder how many bricks and window frames are needed to build a house, and will write a quantity beside each on his order form: ten thousand bricks, twenty window frames, etc.

When we have a different **variable** such as velocity, we require the units of 'm' and 's^{-1}', and then quantify it, saying something like, 'He ran fast, covering a distance of 10 metres per second'. By this means, any variable is defined both in terms of a **number** but also its **units**. In chemistry, we generally write the variable with a symbol of some kind, enabling us to write an **algebraic phrase** such as energy $E = 12$ kJ mol^{-1}.

We write a phrase to describe the mass, length, velocity, etc. using a standard format:

$$\text{variable} = \text{number} \times \text{factor} \times \text{units} \tag{1.1}$$

> We give the name **compound unit** to several units written together. We need to leave a space between each constituent unit when writing such a compound, so kJ mol^{-1} is correct but kJmol^{-1} is not.

■ **Worked Example 1.1**

Look at the algebraic phrase, 'energy = 12 kJ mol^{-1}': identify the variable, number, factor and the unit.

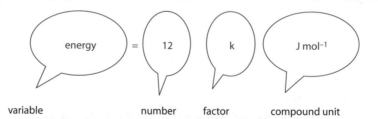

variable   number   factor   compound unit

Reasoning:

- **Variable:** In simple mathematical *phrases* such as this, we almost always write the variable on the left. A variable is a quantity, the value of which can be altered.
  - The variable on the left might be multiplied by other variables, in which case we have a *compound variable*.
- **Number:** The easy part! It will be made up of numbers 0, 1, 2, 3, . . . 9.
  - The number could be written as a decimal or as a fraction.
- **Factor:** A factor is something we multiply by. If we need a factor, we always write it between the number and the unit(s).
  - Table 1.1 is a comprehensive list of the standard factors, including micro μ, centi c, deci d, kilo k, mega M, etc. Alternatively, the factor could be written in terms of scientific notation (as below).

Table 1.1 Powers of ten: energy in joules. Anyone requiring help when manipulating these powers of ten should go to Chapter 11.

10^{-18}	a	atto	0.000 000 000 000 000 001	1 aJ = energy of a single photon of blue light ($\lambda = 10^{-7}$ m)
10^{-15}	f	femto	0.000 000 000 000 001	1 fJ = energy of a single high-energy photon (γ-ray of $\lambda = 10^{-10}$ m)
10^{-12}	p	pico	0.000 000 000 001	1 pJ = energy consumption of a single nerve impulse
10^{-9}	n	nano	0.000 000 001	1 nJ = energy per beat of a fly's wing
10^{-6}	μ	micro	0.000 001	1 μJ = energy released per second by a single phosphor on a TV screen
10^{-3}	m	milli	0.001	1 mJ = energy consumption per second of an LCD watch display
10^{-2}	c	centi	0.01	1 cJ = energy per mole of low-energy photons (radio wave of $\lambda = 10$ m)
10^{-1}	d	deci	0.1	1 dJ = energy released by passing 1 mA across 1 V for 100 s (energy = Vit)
$10^{0} = 1$		1	1	1 J = the kinetic energy (via $\frac{1}{2}mv^2$) of 2 kg travelling at a velocity of 1 m s^{-1}
10^{3}	k	kilo	1000	1 kJ = half the energy of room temperature (E at RT of 298 K = 2.5 kJ)
10^{6}	M	mega	1 000 000	1 MJ = energy of burning $\frac{1}{3}$ mole of glucose (ca. a chocolate bar)
10^{9}	G	giga	1 000 000 000	1 GJ = energy of 1 mole of γ-ray photons ($\lambda = 10^{-10}$ m)
10^{12}	T	tera	1 000 000 000 000	1 TJ = energy (via $E = mc^2$) held in a mass of 10 μg
10^{15}	P	peta	1 000 000 000 000 000	1 PJ = energy released by detonating a very small nuclear bomb

- **Units:** The units are always written on the far right of these phrases. In the example above, there are two units: joules (J) and moles (as 'mol^{-1}', in this case), so we call it a *compound unit*.
  - The way we write the unit(s) ought to follow the IUPAC scheme.
  - When there are several units, we separate each by a single space.

A *factor* is simply a form of shorthand, so we could dispense with it by writing the *number* differently, saying energy = 12 000 J mol^{-1}, as below. By contrast, the units are *not* dispensable.

In the next sections, we will learn first how to write factors, then how to write the number in a meaningful way. ■

Self-test 1.1

In each of the following algebraic phrases, identify the variable, number, unit and (where appropriate) the factor:

1.1.1 mass = 2.65 kg

1.1.2 frequency = 94.5 MHz

 Continued . . .

> **Self-test 1.1 continued…**
>
> **1.1.3**   wavelength = 500 nm
>
> **1.1.4**   current = 0.3 mA
>
> **1.1.5**   mass of beaker = 340 g
>
> **1.1.6**   mass of large car = 1.4 Mg
>
> **1.1.7**   energy liberated = 34 MJ mol^{-1}
>
> **1.1.8**   length of bond = 130 pm
>
> **1.1.9**   potential difference = 550 mV
>
> **1.1.10**   speed = 34 m s^{-1}

Writing an algebraic phrase with a standard factor

By the end of this section, you will know:

- That the standard factor is merely a shorthand.
- The names, values and symbols of the common IUPAC factors.
- How to write an algebraic phrase in terms of standard factors.

In chemistry, we give the name **factor** to any number by which we multiply the numerical value of a variable. Factors are usually expressed in shorthand notation.

We get the names of the factor **pico** from the Latin *pica*, meaning 'tiny'. This root also helps explain why a very small flute is a 'piccolo.'

A typical chemical bond has a length l of about 0.000 000 000 1 m (where m here is the IUPAC symbol for 'metre', *not* milli). This number is cited in a ridiculous way—instinctively, we know that writing so many zeros is not only a waste of time and effort, but also increases the risk of error. For example, we could misread it and write either too many or too few. Surely there is a shorter way of relaying this information?

In this section, we look at a form of shorthand that is popular with chemists. Some numbers are huge and others tiny, but we never need to write out so many zeros, either before or after a decimal point.

For example, we could either say the length of a carbon–carbon double bond is l = 0.000 000 000 13 m, or abbreviate by writing l = 130 pm. Here, the small 'p' means '**pico**', where, 1 pm = 0.000 000 000 001 m. We have not changed the length l in any way by saying it is 130 pm; we have only made it much easier to read or write.

The symbol 'p' is called a **standard factor** (or **prefix**). Table 1.1 lists the IUPAC standard factors, together with a few examples. It is worth memorizing the factors in this table, because we will need them constantly. We never print these factors in italic type. Note how some letters are capital and others are lower case (only factors greater than 1 appear in upper case).

Many chemists also use an additional, non standard, factor—the angström, Å, where 1 Å = 10^{-10} m. The length is named after the Swedish physicist Anders Jonas Ångström (1814–1874), one of the founders of modern spectroscopy. This length is particularly convenient when considering crystallography and chemical structures, because bond lengths are typically 1–3 Å. Many old-fashioned books express bond lengths in terms of 'Ångström units'. More modern (but still non-SI) texts prefer the simpler 'angstroms'.

■ Worked Example 1.2

Burning one mole of *n*-octane (**I**), e.g. from petrol, liberates 5 471 000 J of energy. Write this energy in terms of standard factors.

I

In words, the energy released is about five millions joules. The standard factor meaning 'million' is '**mega**' (symbol M). We could write the energy as $5.471 \times 1\,000\,000$ J. By substituting for the factor of 'million' with the shorthand M, we obtain:

energy released $= 5.471$ MJ ■

The name of the factor **mega** comes from the Greek *megas*, which means 'great.' Other common words coming from the same root include 'megaphone' (a device to make the voice bigger, i.e. louder) and 'megalith' (a huge stone such as those at Stonehenge).

Self-test 1.2

Rewrite the following using standard factors:

1.2.1   energy $= 12\,300$ J mol^{-1}

1.2.2   frequency $= 500\,000\,000$ Hz

1.2.3   length of road $= 3400$ m

1.2.4   voltage $= 30\,000$ V

1.2.5   mass of truck $= 36\,000\,000$ g

1.2.6   amount of material $= 1\,200\,000$ mol

1.2.7   energy evolved $= 2\,034\,000$ J mol^{-1}

1.2.8   wavelength $\lambda = 0.000\,000\,98$ m

1.2.9   bond length $= 0.000\,000\,000\,156$ m

1.2.10 diameter of a fly's eye $= 0.000\,01$ m

Writing algebraic phrases using scientific notation

By the end of this section, you will know:

- That scientific notation is a different form of shorthand from the standard factors in the previous section, but fulfils the same function.

- How to write a number with standard factors as, $a.b \times 10^c$, where a is an integer and has a value between 1 and 9.

The approximate speed of light in a vacuum c is 300 000 000 m s^{-1}. We could write this speed with a standard factor, e.g. as $c = 0.3$ Gm s^{-1}, where G stands for Giga (10^9). Many people not only find it difficult to write a large number such as this, but also do not like writing the standard factors. They prefer using **scientific notation**. In this form of mathematical shorthand, we normally split the number into a simple number

The name **Giga** comes from the Latin *gigans*, meaning 'a giant.' We get the English word 'gigantic' from this same root.

Whenever we read aloud, the correct wording would be, 'The speed of light is 3 point nought times ten to the power of eight metres per second.'

(generally, although not always, a two-or-three digit with a decimal point), followed by a factor. In this case, we avoid the standard factors in the previous section, but write the factor as ten raised to an appropriate power. In the case of c above, we write the speed of light *in vacuo* as 3.0×10^8 m s^{-1}.

■ Worked Example 1.3

Write the Boltzmann constant k_B in scientific notation. The value of k_B is 0.000 000 000 000 000 000 000 013 8 J K^{-1}.

Aim: To write a number having the form $a.b \times 10^c$.

Strategy:

(i)   We write the number $a.b$ before the factor of 10^c. Here, a and b are both integers and a lies in the range 1–9. We write only one digit *before* the decimal point. In this case, the number is simply '1.38' because these are the only actual numbers in the string of digits in the question.

(ii)   We decide the number of tens, i.e. the decimal place. In this case, the decimal point is 23 places to the *left* of the '1', which is another way of saying 10^{-23}.

(iii)   We **MULTIPLY** the answers from parts (i) and (ii), obtaining 1.38×10^{-23}.

(iv)   Finally, we usually include the units, which in this example are J K^{-1}.

Therefore, $k_B = 1.38 \times 10^{-23}$ J K^{-1} ■

Aside

Using scientific notation on a calculator:
We need to mention a common error experienced when typing numbers into a pocket calculator. The correct procedure when typing the Boltzmann constant would be:

(i) Type '1.38'

(ii) Press the button marked $\times 10^x$, *EXP* or *EE* (which is usually positioned at the bottom of the number pad)

(iii) Type in '−23'.

It is *incorrect* to enter the number as:

(a) '1.38'

(b) $\times 10$

(c) Press $\times 10^x$, *EXP* or *EE*

(d) Type in '−23'.

The calculator is programmed to think of '1.38 × 10^{-23}' as '1.38 × 1e^{-23}', so typing an extra '10' in step (b) here will introduce an additional factor of ten.

Using scientific notation on a computer spreadsheet:

Using *Excel*™, we type '1.38e−23', where the small 'e' stands for 'exponent' (as explained in Chapter 11).

Self-test 1.3

1.3.1–10   Rewrite each of the algebraic phrases in Self-test 1.2 above, this time in terms of scientific notation.

1.3.11   The Faraday constant has a value of 96 500 C mol^{-1}. Rewrite this number in scientific notation.

1.3.12   The charge q on an electron is 0.000 000 000 000 000 000 16 C (where 'C' is the unit called the coulomb). Rewrite this charge in scientific notation.

Typing '10' instead of '1' in step (b) here causes the answer to be 10 times too large

The use of significant figures

By the end of this section, you will:

- Know what is meant by the term 'significant figures' (s.f.).
- Appreciate that the number of significant figures is important.

» Continued . . .

>> *Continued...*

- Appreciate how the number of significant figures makes a statement concerning the precision of our data: either we do not have the information available to allow greater precision, or the context means that more data are not useful.
- Know how to determine the correct number of significant figures.

Imagine a trip to a well-stocked History Museum, to look at the fossils of a great dinosaur. Beneath the bones is a small card saying '100 million years old' (that is, an age of 100 000 000 years).

Next, imagine going back again the following year and on the same date, and looking at the same fossil. This second time, we notice how the display card has been amended slightly to read, '100 million and 1 years'. We would surely wonder 'how meaningful was the additional 1 year?'

The value of 100 million years is huge and therefore worth our attention; by contrast, citing an age as '100 million and 1' is silly: the 'and 1' tells us nothing when compared to the vastness of 100 million, because it is so small. The 100 million is significant; the 1 is simply irrelevant by comparison. In fact, we cannot know the age with any precision, so we signal that lack of knowledge by citing the age in an imprecise manner.

In a different way, imagine two children describing their ages: the first says he is five years of age, but the second says her age is 6 years, 143 days and 3 hours. The first is giving an appropriate amount of information; the second gives us so much data that we would normally want to ignore most of it. It is foolish in this context to cite an age to this precision, because there is no need to know so much information.

Rewriting an answer to a prescribed number of significant figures is easy.

- When we say *n* significant figures, we display the number using only the first *n* digits.
- The answer may require rounding up or down: an enthalpy of 124.6 kJ, expressed to 3 s.f., is 125 kJ.
- For numbers smaller than 1, we start counting the number of significant figures from the first non-zero digit. For example, if we express an *emf* of 0.00234 V to 2 s.f., we write 0.0023 V or 2.3×10^{-3} V, not 0.0 V.

> Stating an age to this level of precision is doubly unwise:
>
> (i) It causes information overload, and makes the girl look silly.
> (ii) Within a few minutes, the answer is wrong, for the girl has aged (slightly).

■ **Worked Example 1.4**

The value of π is irrational. To 12 s.f., its value is 3.141 592 653 59. Express this value to different number of significant figures.

- The value of π is 3 to **one** s.f.
- The value of π is 3.1 to **two** s.f.
- The value of π is 3.14 to **three** s.f.
- The value of π is 3.142 to **four** s.f.
- The value of π is 3.141 6 to **five** s.f.
- The value of π is 3.141 59 to **six** s.f.
- The value of π is 3.141 593 to **seven** s.f.
- The value of π is 3.141 592 7 to **eight** s.f.
- The value of π is 3.141 592 65 to **nine** s.f.
- The value of π is 3.141 592 654 to **ten** s.f.

> We say a number is **irrational** if we cannot express its value as a *simple* decimal. The decimal of an irrational number is infinitely long, and has no pattern to the number.

Notice how we often need to round up or down so, for example, while $\pi = 3.141\,59$ to 6 s.f., its value is 3.141 6 to 5 s.f.

In fact, the first few digits (3.142) are generally useful, and tell us the value of pi. We are unlikely to need more s.f. if we wanted to calculate the area of a circle or the volume of a sphere or cone. The next three digits (592) merely 'tweak' the final answer; and chemists rarely cite the three digits after these (654). Indeed, they are useless to just about everyone. They tell us nothing useful at all.

Finally, a simple way to remember the value of pi is to recite the sentence 'How I wish I could calculate pi': the number of letters in each word gives the value of each successive digit, 3.141 592. ∎

> A number can be cited to any number of significant figures. In thermochemistry, for example, it is rarely useful to cite more than 3 s.f. as a result of experimental imprecision during measurement.

Self-test 1.4

A student's age is 18 years, 178 days, 5 hours and fifteen minutes. Calculate the student's age:

1.4.1   To 1 s.f.

1.4.2   To 2 s.f.

1.4.3   To 3 s.f.

1.4.4   To 4 s.f.

1.4.5   To 5 s.f.

By part 1.4.5, you will probably find the answer has changed between starting and finishing the Self-test problem!

> We often abbreviate the term **significant figures** with the letters **s.f.**

> We say a number is **irrational** if it goes on forever (is 'infinite') and does not repeat itself.

In both these examples, dinosaurs and children, we see how the quantity of information says a lot about the precision of measurement. Often, particularly in everyday life, the precision with which a number is cited is erroneous, because the inferred precision is greater than the actual data allow. We need to therefore consider the correct way to cite numbers.

The number of **significant figures** is a commonly adopted measure of how precise or imprecise a number is. If we think it wise to express a lack of precision, we cite to only one significant figure. If we have reason to feel more confident, we will use two s.f. And if we know our data are good, we may dare to cite to three or even four s.f.

In practice, we usually calculate any parameter with slightly too many significant figures and then, as a final step, simplify by appropriately rounding up or down, e.g. if we know our final answer should be expressed to 3 s.f., we perform the calculation using 4 s.f.

It is important to recognize that we cannot cite a quantitative answer without using significant figures. The number of s.f. makes a powerful statement about the precision of our method or experiment. For example, if we say $\Delta H = 12$ kJ mol^{-1}, we are in fact saying the technique used to measure ΔH was incapable of distinguishing between the enthalpy changes 11.5 and 12.5 kJ mol^{-1}. By contrast, if we say $\Delta H = 12.58$ kJ mol^{-1}, we imply the actual value lies between 12.575 and 12.585 kJ mol^{-1}. The latter experiment is 100 times more precise than the former.

Self-test 1.5

Re-express each of the following with a smaller number of significant figures. In each case, the required number of s.f. is given after the algebraic phrase in square brackets.

❯❯ Continued . . .

> **Self-test 1.5 continued...**
>
> **1.5.1** energy change $= -134.99$ kJ [3]
>
> **1.5.2** $emf = 1.4324$ V [4]
>
> **1.5.3** volume $= 1.986$ m^3 [2]
>
> **1.5.4** amount of material $= 3.221$ mol [3]
>
> **1.5.5** mass $= 1\,002\,010$ g [1]
>
> **1.5.6** In the SI system, a length of 1 metre $= 1,650,763.73$ wavelengths of the light emitted *in vacuo* by krypton-86 [7]

The use of significant figures when performing MULTIPLYING or DIVIDING:

We often have to multiply or divide in situations when the data are expressed with differing numbers of significant figures. So we need to decide the number of s.f. of the final answer. The rule is simple: look at each number to determine their number of s.f. Express the answer to the same number of s.f. as the value with the least number of s.f.

■ **Worked Example 1.5**

We perform a kinetic experiment, reacting methanol (**III**) and salicylic acid (**II**) to form the sweet-smelling ester methyl salicylate (**IV**).

The rate of formation of (**II**) is given by the expression:

rate $= k$ [methanol] [salicylic acid]

where k is a rate constant. What is the value of the rate when $k = 3.06 \times 10^{-4}$ (mol dm^{-3})$^{-1}$ s^{-1}, [methanol] $= 5$ mol dm^{-3} and [salicylic acid] $= 0.45$ mol dm^{-3}?

These numbers are clearly expressed in terms of different numbers of significant figures. If we insert the numbers into the rate equation above, we obtain:

rate $= 3.06 \times 10^{-4} \times 5 \times 0.45$

so   rate $= 6.885 \times 10^{-4}$ mol dm^{-3} s^{-1}

The variable with the limiting precision was the concentration of methanol (which was expressed to 1 s.f.). Accordingly, we must express this rate to 1 s.f.:

rate $= 7 \times 10^{-4}$ mol dm^{-3} s^{-1} ■

....
The old-fashioned name for methyl salicylate (**IV**) is 'Oil of Wintergreen.' This ester is commonly added to medications designed to alleviate muscular pain.
....

Significant figures when using *Excel*™:

Chemists often need to display their data on computer spreadsheets. As professionals, we would not want to be seen citing a number to a silly number of s.f.

To obtain a *sensible* number of significant figures, we need to know how to display our data. The procedure below relates to *Excel*™:

(i)   Highlight cells.

(ii)   Right click in cells, when highlighted.

(iii) Select the FORMAT CELLS page.

(iv)   The box that appears has 'FORMAT CELLS' along its top. Click on the tab saying SCIENTIFIC:

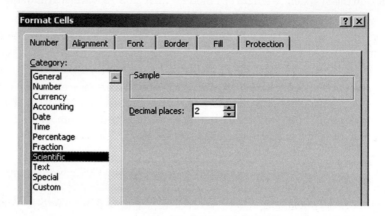

(v)   Having clicked on SCIENTIFIC with the mouse, a small box will allow us to choose the number of decimal places. Into this box, we select the number of s.f. we want (by typing in a number that is $(n_{SF} - 1)$, or use the small up and down arrows), then press RETURN.

Self-test 1.6

Calculate the following multiplication and division problems, citing each answer to the correct number of significant figures:

1.6.1 Volume of a unit cell, $V = $ side $a \times$ side $b \times$ side c. What is V if $a = 120$ pm, $b = 151$ pm and $c = 146.5$ pm?

1.6.2 Amount of substance, $n = $ volume $\times$ concentration. What is n if volume $= 0.250$ dm^3 and concentration $= 0.05$ mol dm^{-3}?

1.6.3 Rate $= k \times$ [concentration]. Calculate the rate if $k = 9.3 \times 10^{-2}$ s^{-1} and [concentration] $= 0.3$ mol dm^{-3}.

1.6.4 Concentration, $c = $ amount of material $\div$ volume. Calculate c when the amount of material $= 3.2 \times 10^{-4}$ mol and the volume $= 0.5$ dm^3.

1.6.5 Amount of substance, $n = $ mass $\div$ molar mass. Calculate n when mass $= 4$ g and molar mass $= 422$ g mol^{-1}.

1.6.6 Charge density, $Q = $ charge $\div$ area. What is Q when charge $= 87.3$ mC and area $= 0.32$ cm^2?

1.6.7 From the ideal gas equation, $p = \dfrac{nRT}{V}$. What is p if $n = 0.13$ mol, $R = 8.314$ J K^{-1} mol^{-1}, $T = 298$ K and $V = 14.2$ m^3?

1.6.8 From the van't Hoff isotherm $\Delta G^{\ominus} = -RT \ln K$. What is the value of $\Delta G^{\ominus}$ if $(\ln K) = 4.0$, $T = 298$ K and $R = 8.314$ J K^{-1} mol^{-1}?

The use of decimal places

By the end of this section, you will:

- Know what is meant by the term 'decimal places'.
- Know how many decimal places to use.
- Know the difference between decimal places and significant figures.

When we add or subtract numbers, we tend not to use significant figures. The results can be unhelpful or even ambiguous. Instead, we use **decimal places**. For example, the number 2.0 is expressed to 1 d.p. because there is a single digit after the decimal point (the dot). Conversely, the enthalpy change $\Delta H_c = 23.76$ kJ mol^{-1} is expressed to two d.p.

We use the correct number of decimal places to indicate the correct precision when adding or subtracting. The procedure is two-fold:

(1) We look at each component term and note the number of decimal places in each, and note what is the minimum number of d.p.

(2) We add or subtract as normal, including *all* the decimal places from all the component terms. We will call this the 'preliminary' or 'uncorrected answer.'

(3) We round the preliminary answer up or down to decrease the number of decimal places until it is the same as the value with the minimum number of d.p.

> We often abbreviate the term **decimal places** with the letters **d.p.**

■ **Worked Example 1.6**

Generally, an electrochemical current I comprises several component parts. A current is measured and comprises the following three components: $I_{(analyte)} = 5.37$ mA, $I_{(double\text{-}layer\ charging)} = 43\ \mu A$, and $I_{(side\ reactions)} = 1$ mA. What is the total current, expressed to an appropriate number of d.p.?

We start by writing a sum to enable the calculation of a total current:

$$I_{(total)} = I_{(analyte)} + I_{(double\text{-}layer\ charging)} + I_{(side\ reactions)}$$

where

- $I_{(side\ reactions)}$ is expressed to 0 d.p
- $I_{(analyte)}$ is expressed to 2 d.p.
- $I_{(double\text{-}layer\ charging)}$ is expressed to 3 d.p. (that is, after conversion from μA to mA i.e. $43\ \mu A = 0.043$ mA.)

The smallest number of d.p. is 0, so we therefore express the final answer to 0 d.p.

Before we calculate the final sum, it is usually helpful to express the numbers in a consistent way, with each component having the same common factor:

i.e. $I_{(total)} = 5.37$ mA $+ 0.043$ mA $+ 1$ mA

so $I_{(total)} = 6.413$ mA

When expressed with a precision of 0 decimal places, this current is 6 mA. ■

> We tend to 'exhaust the data' when we subtract two nearly equal numbers, so it may be wise to perform the calculation with too many d.p., then redisplay the answer.

Occasionally, when the numbers are *not* comparable in magnitude, we will sometimes exclude one or more component parts because they are too small. We then express the final answer to the minimum number of d.p. of the remaining components.

■ **Worked Example 1.7**

An industrial chemist wishes to make a batch of paint. The paint is to be made in a vat. To this vat, the chemist adds 12.0 tonnes of titanium dioxide (as a 'filler'), 5.0 tonnes of polyurethane monomer, and 15 kg of pigment. Cite the overall mass to an *appropriate* number of decimal places.

To ensure consistency, it is convenient to first express each mass with the same units. In this case, expressed in tonnes, the mass of pigment is 0.015 tonnes. We obtain:

$$
\begin{array}{r}
12.0 \\
+\quad 5.0 \\
0.015 \\
\hline
17.015
\end{array}
$$

The third mass is clearly inconsequential when compared with the others, so we simply omit it. Such an omission is the same as citing the answer to 1 d.p.:

overall mass = 12.0 + 5.0 tonnes = 17.0 tonnes

It would not have made sense to say '17.015 tonnes' (which involves 3 d.p.s) because the two larger masses are each expressed to only 1 d.p. In other words, the spread of values expressed by the next smallest mass '5.0 tonnes' is itself greater than the mass of 15.0 kg. ■

Self-test 1.7

Calculate the following addition and subtraction problems, citing each answer to the correct number of decimal places:

1.7.1 mass = 12.0 g + 1.001 kg − 130.62 g

1.7.2 charge = 96 000 C − 67.27 C − 1096.3 C

1.7.3 amount of material = 12.1 mol − 2.754 mol + 0.5419 mol

1.7.4 time = 60.4 s + 12 μs + 33.96 ms + 4.0 s

Additional problems

1.1 Rank the following masses in order of increasing size: 3×10^{-3} g; 500 μg; 0.6 mg; 110 cg; and 4 000 000 ng.

1.2 A protein has a molar mass of 1.302×10^5 g. Rewrite this mass in the form of a straightforward number.

1.3 Cobalt chloride hexahydrate has a molar mass is 238 g mol^{-1}. A sample of mass 1.5 g is dissolved in water (140 cm^3). What is the concentration of the solution, cited to 1, 2, and 3 s.f.?

1.4 A current I of 23.4 μA passes through an electrode. The area A of the electrode is 4.1×10^{-4} m^2. Calculate the current density i, where i is defined as $= I \div A$.

$$i = \frac{I}{A} = \frac{23.4 \times 10^{-6}\,\text{A}}{4.1 \times 10^{-4}\,\text{m}^2}$$

1.5 The closest distance between two chloride nuclei in a crystal of sodium chloride is 0.000 000 000 174 m. Write this length in terms of scientific notation.

1.6 The Faraday constant F is obtained as the product of the Avogadro number $N_A = 6.022 \times 10^{23}$ mol^{-1} and the charge on a single electron $q = 1.602 \times 10^{-19}$ C, i.e. $F = N_A \times q$. Calculate the value of F.

1.7 The velocity of light in vacuo c is 2.9980×10^8 m s^{-1}. Write this velocity to 2 s.f.

1.8 The gradient of a graph is defined as the change in the vertical axis Δy divided by the change in the horizontal axis Δx, i.e. gradient $= \Delta y \div \Delta x$. Calculate the gradient if $\Delta y = 0.000324$ and $\Delta x = 41$.

1.9 The number of transistors produced worldwide each year is 2×10^{19}. Write this number with a standard factor.

1.10 A chemist makes 0.37 g of potassium chloride. The molar mass of KCl is 74.5 g mol^{-1}. If 'amount = mass ÷ molar mass', what is the amount of KCl made?

2

Algebra I

Introducing notation, nomenclature, symbols and operators

Mathematical shorthand: symbols

By the end of this section, you will know that:

- Mathematical symbols are merely shorthand, each telling us to *do* something.
- All mathematical expressions contain operators.

Our word 'mathematics' comes from the Greek word *mathatās*, which means literally a 'disciple.' A mathematician is therefore someone who is a disciple of a master, and closely follows the master's rules. In this case, the 'master' we follow is the simple rules of how mathematical symbols behave.

We already know many mathematical symbols, such as +, −, ×, ÷, %, and so on. We must first appreciate how these symbols are merely a form of shorthand: they are intended to save us time and ink. They are never more than symbols, and are never magical. And there is no way someone can look at one of these symbols and decide from its appearance alone what it means. In fact, a symbol means what we want it to mean: we *define* its meaning. In other words, the only reason Δ means what it means is because we say so.

The simplest symbol is the plus sign, '+'. Each plus sign we encounter operates in the same way and tells us to do something: in effect, it says, 'look at the numbers either side of me and add them together.' For this reason, we call the plus sign an **operator**, because it tells us what *operation* to perform. In the same way, a multiplication sign '×' is also an operator, because it tells us to multiply together whatever is positioned to its immediate left and immediate right.

While these examples are utterly trivial, they tell us a lot about the way that even a simple operator works, and demonstrate how they are merely one form of shorthand notation. In each case we could have used words. We have a choice, and could say either 'the number obtained when we assemble eight groups of five' or '8 × 5.' We obtain the same answer in both cases: the only difference is the *notation*.

In summary, we use mathematical operators to save time and space, not to make life more complicated. The rest of this book merely explores the mathematics of operators, and how chemists can master them.

We call symbols, such as the plus sign, **operators**, because they tell us what operation to perform.

An **operator** is merely a shorthand notation.

Some mathematicians reserve the term 'operator' **for more complicated** operations such as those involving differentials and integrals. They call simpler operations 'functions'.

Especially in music, linguistics or mathematics, the word **notation** means the way we write something by using symbols.

The Delta notation Δ

By the end of this section, you will know that the Delta symbol Δ:

- Is a shorthand notation for an operator.
- Is meaningless when written on its own: we have to write it with an argument.
- Tells us to subtract the initial value of the argument from its final value.

As chemists, we often see the term ΔH, particularly when we study thermodynamics and calorimetry. As we learn more chemistry, we will also encounter ΔS, ΔU and ΔG. In each case, the symbol Δ is the Greek letter **Delta**. (We write 'Delta' with upper-case letters rather than 'delta,' because a lower-case delta looks like δ.)

Table 2.1 List of symbols having more than one meaning.

Symbol	Meaning
Δ	difference, according to eqn (2.1); crystal-field splitting energy
σ	standard deviation; electronic conductivity*, symmetry number
c	concentration; constant; intercept on a graph
I	electrical current; intensity; ionic strength; iodine
k	rate constant; kilo; constant
K	equilibrium constant; potassium; Kelvin
T	transmittance; tritium
U	internal energy; uranium
x	horizontal axis on a graph; unknown; mole fraction

* To ensure consistency with ionic conduction and the conduction of heat, many scientists use the symbol κ (kappa) for electronic conductivity. In electrochemistry and physics, σ is the preferred choice.

H is the thermodynamic quantity, enthalpy. The collection of symbols ΔH is a **phrase**, and means the change in enthalpy occurring during a reaction or process. Therefore, the symbol Δ in the phrase ΔH must be some form of shorthand for 'change in enthalpy.' The expression ΔH does not mean 'Delta **MULTIPLIED** by H'; in fact, if we want to gain the full significance of Δ, we ought perhaps to say 'Delta of H'.

The symbol Δ in the phrase ΔH tells us that *something happens* to H. Because Δ does something to H, we say that Δ **operates** on H. In more technical language, we go further by saying Δ is an **operator**.

If we see the symbol Δ on its own, it will not be an operator because an operator must always operate *on* something. Table 2.1 lists a few other symbols, some of which are also operators. We will need to learn the whole of this list because it does not display any inner *logic*—for example, there is nothing about a capital H that makes us say 'enthalpy.' Rather, we have assigned the letter H to enthalpy.

The previous paragraph said the operator Δ must operate *on* something. We call that 'something' the **argument**. In the case of the phrase ΔH, the operator is Δ and the argument is H. In words, we say that Δ operates on H. Δ does something to the H. We will investigate the scope of an argument as we progress through this chapter.

Now to specifics. We call the things that Δ does its **operation**. The Δ operation tells us to subtract the *initial* value of the argument from its *final* value. For an argument in general (call it X):

$$\Delta X = X_{final} - X_{initial} \tag{2.1}$$

so when seeing the phrase ΔH, we know we are to subtract the initial value of enthalpy H from the final value of enthalpy. This operation is shown diagrammatically in Fig. 2.1.

It should be clear from Fig. 2.1 that the value of ΔH has not only a magnitude—which relates to the vertical separation between the two bold lines—but also a sign. If the arrow points upwards, then ΔH is positive, and if it points downwards, then ΔH is negative.

The expression ΔH does not mean 'Delta multiplied by H'

It is vitally important to distinguish between Δ and H : Δ is an **operator** because it *does something* (according to eqn (2.1)). H is a **symbol** because it is *shorthand* for enthalpy.

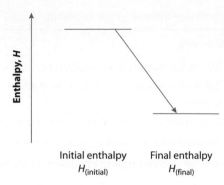

Figure 2.1 In an exothermic process, the final product has *less* energy than the initial starting materials. Energy has been given out.

■ **Worked Example 2.1**

In thermodynamics, the letter S indicates *entropy*. Therefore, using the definition of Δ above, the phrase ΔS means 'a change in entropy.' If $S_{final} = 12$ J K^{-1} mol^{-1} and $S_{initial} = 25$ J K^{-1} mol^{-1}, what is the value of ΔS?

Inserting values into eqn (2.1),

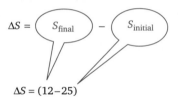

$$\Delta S = (12 - 25)$$

so $\Delta S = -13$ J K^{-1} mol^{-1}

We calculate the value of ΔS as -13 J K^{-1} mol^{-1}.

The change in S we calculated is negative. We need to note how the definition of Δ allows answers that are either negative or positive, depending on the *relative* values of the final and initial terms. ■

Self-test 2.1

2.1.1   If the final value of X is 10 and the initial value is 22, what is ΔX?

2.1.2   If the final value of X is 33 and the initial value is 5.2, what is ΔX?

2.1.3   If the initial and final values of X are 9.34 and 9.37, respectively, what is ΔX?

2.1.4   If the final value of the Gibbs function G is 8.0 kJ mol^{-1} and the initial value of G is 3.6 kJ mol^{-1}, what is ΔG?

2.1.5   If the optical absorbance Abs decreases from 1.1 to 0.6 during a reaction, what is the value of $\Delta(Abs)$?

2.1.6   If the electromotive force emf increases from 1.45 V to 1.50 V during a process, what is the value of $\Delta(emf)$?

The Sigma notation Σ

By the end of this section, you will know that:

- The sigma notation Σ is a shorthand symbol for an operator.
- The sigma notation Σ tells us to add up a series of terms.
- The terms to be summed can themselves be sums of smaller terms.

Another simple operator that occurs often in chemistry is **Sigma** Σ, which is defined by eqn (2.2):

$$\sum_{i=1}^{n} X_i = X_1 + X_2 + X_3 + \cdots + X_n \tag{2.2}$$

We write Sigma with a big (Greek) S because a lower-case sigma looks different: σ.

In eqn (2.2), i takes the value 1, 2, 3, 4 all the way up to n. Each of these values of i is then operated upon, so X_i here is a function of the variable i. In fact, the mathematical form of the function X is the same for each term in the sum in eqn (2.2). For example, if the function is 'reciprocal,' so X means $1/i$, then the sum is

$$\sum_{i=1}^{n} \frac{1}{i} = \frac{1}{1} + \frac{1}{2} + \frac{1}{3} + \cdots + \frac{1}{n}$$

We should beware of looking at the Σ sign and somehow guessing that it means 'sum everything to my right.' There is nothing intrinsic in the Σ symbol to indicate this operation. Rather, we have *defined* the symbol to mean a simple summation operation.

We know that Σ is an operator because it is telling us to *do* something. In this case, the symbol tells us to sum everything to the right of the Σ sign. In this, the general case, we are to add together n terms. We write a small n above the large Σ sign to tell us that this number of terms is the extent of the summation operation. Additionally, we write a small $i = 1$ directly below the large Σ to tell us to start the summation operation with the first term. We call the terms above and below the Σ sign its **limits**.

This description seems abstract, so we will now do an actual example.

■ **Worked Example 2.2**

What is the value of $\displaystyle\sum_{i=3}^{6} i^2$?

Before we start the actual operation, we will 'read' the equation:

- The Σ sign tells us the operation is a summation process.
- The terms to be summed are squares because the symbol straight after the Σ sign is i^2. In terms of the definition in eqn (2.2), $X = i^2$.
- The first variable to be squared in this sum is 3, because '$i = 3$' is written in small print beneath the Σ sign.
- The last term to be squared in this sum is 6, because '6' is written in small print above the Σ sign.
- While not stated explicitly, it is assumed that each of the i terms is an integer, i.e. whole numbers.

We can now do the sum: $\displaystyle\sum_{i=3}^{6} i^2 = 3^2 + 4^2 + 5^2 + 6^2 = 9 + 16 + 25 + 36 = 86$.

We could have done this sum without any additional notation, but $\sum\limits_{i=3}^{6} i^2$ manages to compress all this information into a very small space. So the reason why we use the summation operator Σ is because it represents a more efficient form of shorthand than writing a string of plus signs. ∎

■ Worked Example 2.3

Write an expression starting with Σ for the following sum: $\dfrac{1}{2} + \dfrac{1}{3} + \dfrac{1}{4} + \dfrac{1}{5} + \dfrac{1}{6}$

Strategy:

(i)   We decide which mathematical function is displayed by the data in the series. Write a general form of this function in terms of the general integer i.

(ii)   Write a capital Σ followed by the general form of the expression, from part (i) immediately above.

(iii) Decide which are the first and the last values of the variable displayed by the data in the series.

(iv)   Below the Σ sign, write '$i = \ldots$' followed by the first value of the variable displayed by the data in the series.

(v)   Above the Σ, write the last value of the variable displayed by the data in the series.

Solution:

(i)   The function is '1/variable.' As we write the variable as i, the function is $\dfrac{1}{i}$.
       In terms of the definition of Σ in eqn (2.2), $X = \dfrac{1}{i}$.

(ii)   The first and last values of the variable are 2 and 6, respectively.

(iii) $\sum \dfrac{1}{i}$

(iv) $\sum\limits_{i=2} \dfrac{1}{i}$

(v) $\sum\limits_{i=2}^{6} \dfrac{1}{i}$. ∎

It is very common to see a summation with one of the limits being infinity ∞. Such situations are particularly common in situations involving statistical mechanics.

■ Worked Example 2.4

What is the value of $\sum\limits_{i=1}^{\infty} \dfrac{1}{i^3}$?

In this example, the function X of i is $1/i^3$.

We could write a sum for all values from $i = 4$ through to $i = \infty$, but it would take a long time! But we don't have to. Look at the way the values of each term changes:

$1/1^3 = 1$

$1/2^3 = \dfrac{1}{8} = 0.125$

The word **strategy** comes from the Greek word *stratos* meaning 'army'. A strategy was originally the procedure(s) involved in planning a military campaign or manoeuvre.

$$1/3^3 = \tfrac{1}{27} = 0.037\,037$$

$$1/4^3 = \tfrac{1}{64} = 0.015\,625$$

$$1/5^3 = \tfrac{1}{125} = 0.008\,000$$

$$1/6^3 = \tfrac{1}{216} = 0.004\,630$$

This trend shows how the reciprocal cube terms are getting progressively smaller, and rapidly so. Indeed, after about ten terms (i.e. after $i = 10$) the values of $1/i^3$ are so small that they make no material difference to the overall sum. Any additional terms in the sum would be so very small as to be utterly meaningless.

In summary, if we see a summation in which the terms get progressively smaller, and with an upper limit of infinity, we usually need to sum only about 10 terms ... and sometimes we need to sum even fewer. ■

The three Worked Examples above all involve simple functions of i. In such cases, the Σ operator is merely a novel way of rewriting a series of terms without using a string of plus signs.

A great many examples with the Σ sign involve far more complicated functions, such as when we want to add up several smaller sums. In this way, we manage to compress even more information into a small space. In other words, the shorthand method becomes even more efficient.

....

If the terms in a sum become progressively smaller, we say the series **converges**.

We say the series **diverges** if the terms get progressively larger. We cannot ignore terms in this way if a series diverges.

....

■ **Worked Example 2.5**

Calculate the mass per mole of the amino acid, L-proline (**I**). Use an approach based on the Σ operator.

I

Strategy:

(i) Devise an expression for the molar mass in terms of the Σ operator. This will look a little different from Worked Example 2.2.

(ii) Determine the elemental formula of L-proline.

(iii) Use the expression from part (i) to calculate the mass of one mole.

Solution:

(i) We define the molar mass as the sum of 'the mass per atom $\times$ the number of such atoms.' In terms of the Σ operator, we write

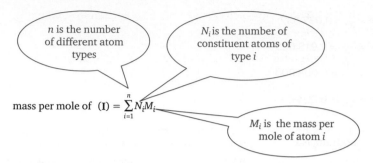

(ii)   The elemental formula is $C_5H_9O_2N_1$.

(iii)   We insert the data:

mass per mole of $(\mathbf{I}) = \displaystyle\sum_{i=1}^{4} N_i\, M_i$ where again N_i is the number of different atoms of type i and M_i is the corresponding molar mass.

$$\text{so} = \sum_{i=1}^{4} N_i\, M_i = N_C\, M_C + N_H\, M_H + N_O\, M_O + N_N\, M_N$$

$$\sum_{i=1}^{4} N_i\, M_i = \boxed{5 \times 12} + \boxed{9 \times 1} + \boxed{2 \times 16} + \boxed{1 \times 14}$$

$$\qquad\quad\;\; carbon \qquad\quad hydrogen \qquad\quad oxygen \qquad\quad nitrogen$$

mass per mole of $(\mathbf{I}) = 60 + 9 + 32 + 14$

mass per mole of $(\mathbf{I}) = 115 \text{ g mol}^{-1}.\ \blacksquare$

The units 'g mol^{-1}' (the SI units of molar mass) were omitted during the calculation for the sake of clarity.

■ **Worked Example 2.6**

When exploring the chemistry of solutes in aqueous solution, one of the most useful parameters is the ionic strength I, which is defined as:

$$I = \tfrac{1}{2}\sum_{i=1}^{n} c_i\, z_i^2$$

where c is concentration of the ions in solution i, and z is the charge per ion. Calculate the ionic strength I of a solution that contains chloride ions (0.3 mol dm^{-3}), sodium ions (0.4 mol dm^{-3}), sulphate ions (0.1 mol dm^{-3}) and calcium ions (0.05 mol dm^{-3})?

The solution contains four ions, so the resultant sum will involve four terms: $c\, z^2$ for the chloride ions, $c\, z^2$ for sodium ions, $c\, z^2$ for sulphate ions and $c\, z^2$ for calcium ions. We tabulate the necessary data in Table 2.2.

Inserting the data from Table 2.2 into the equation defining ionic strength I yields:

$$I = \tfrac{1}{2} \times \left(\boxed{0.05 \times 2^2} + \boxed{0.3 \times (-1)^2} + \boxed{0.4 \times 1^2} + \boxed{0.1 \times (-2)^2} \right)$$

$$\qquad\quad Calcium \qquad\qquad Chloride \qquad\qquad Sodium \qquad\qquad Sulphate$$

So $I = \dfrac{1}{2}\ \{0.2 + 0.3 + 0.4 + 0.4\} = \dfrac{1}{2} \times 1.3$ and $I = 0.65$ mol dm^{-3}.

Table 2.2 Data for inclusion within the expression defining ionic strength I.

Ion in solution	Concentration/ mol dm^{-3}	Charge per ion, z	z^2
Calcium, Ca^{2+}	0.05	+2	4
Chloride, Cl^-	0.3	−1	1
Sodium, Na^+	0.4	+1	1
Sulphate, SO_4^{2-}	0.1	−2	4

Ionic strength I has the same units as concentration. This follows straightforwardly from the definition of I since z has no units and c is a concentration. ■

It is quite common for an expression to require more than one Σ term. The mathematics of First-Law thermodynamics is a good example. In such calculations, the Σ terms are used to calculate enthalpies from Hess-law cycles.

■ **Worked Example 2.7**

We wish to calculate a standard enthalpy change of reaction $\Delta H_r^{\ominus}$. To this end, we sum the molar enthalpies of formation $\Delta H_{f,i}^{\ominus}$ for each chemical participating in a reaction:

$$\Delta H_r^{\ominus} = \sum_{i=\text{products}} \left(v_i \Delta H_{f,i}^{\ominus} \right) - \sum_{j=\text{reactants}} \left(v_j \Delta H_{f,j}^{\ominus} \right)$$

where v_i is the stoichiometric number of species i, i.e. the number printed immediately before the participating species in a fully balanced chemical reaction. We use a subscripted j rather than i to avoid confusing the terms. Consider the simple oxidation process that occurs in a catalytic converter:

$$2CO(g) + O_2(g) \rightarrow 2CO_2(g)$$

Calculate $\Delta H_r^{\ominus}$ for this process using the data in the table below.

	$\Delta H_f^{\ominus}$ / kJ mol^{-1}	Stoichiometric number, v
CO	−110.5	2
O_2	0	1
CO_2	−393.0	2

We first write an expression based on the expression in the question:

$$\Delta H_r^{\ominus} = \underbrace{\left(2 \times \Delta H_r^{\ominus}(CO_2) \right)}_{products} - \underbrace{\left([2 \times \Delta H_r^{\ominus}(CO)] + [1 \times \Delta H_r^{\ominus}(O_2)] \right)}_{reactants}$$

The small r in $\Delta H_r^{\ominus}$ refers to *reaction*. The small f in $\Delta H_f^{\ominus}$ stands for formation. In this particular example, we could have said $\Delta H_c^{\ominus}$, where the c stands for combustion.

Inserting values yields:

$$\Delta H_r^{\ominus} = (2 \times [-393.0]) - (2 \times [-110.5] + 1 \times [0])$$

so $\Delta H_r^{\ominus} = (-786.0) - (-221.0)$

and $\Delta H_r^{\ominus} = -565 \text{ kJ mol}^{-1}$

We deduce that burning carbon monoxide is very exothermic: the catalytic converter will get very hot. ∎

Self-test 2.2

For each of the following, starting with an expression involving a summation sign Σ, calculate the molecular mass by adding up the relative atomic masses:

2.2.1   Potassium hexacyanoferrate(II), $K_4[Fe(CN)_6].3H_2O$

2.2.2   Anhydrous copper sulphate, $CuSO_4$

2.2.3   Hydrogen peroxide, H_2O_2

2.2.4   Benzoic acid, $C_7H_6O_2$, (**II**)

2.2.5   Pyrrole, $C_5N_1H_5$ (**III**)

2.2.6   Aspirin, $C_9O_4H_8$, (**IV**)

II   III   IV

The relative atomic masses are: H=1, C=12, N=14, O=16, S=32, K=39, Fe=56, and Cu=64.

The large brackets here are not strictly necessary, but help clarify the expression.

Calculations with Σ can be more complicated still, which illustrates the power of this method. For example, we sometimes need to calculate **the sum of a sum**. We can write an expression of this type

$$\text{mass of product} = \sum_{i=1}^{n} \left(\sum_{j=1}^{n'} X_{ij} \right)$$

We next look closely beneath each sigma sign, we see one has a small i and the other a small j. This notation merely emphasizes that each summation adds up a different set of numbers. The n and n' terms are likely to differ.

In the next Worked Example, we will calculate the molar masses of a product, starting from the molar masses of reactants.

■ **Worked Example 2.8**

Methyl viologen (MV^{2+}) (**VI**) and phenol (PhOH) (**VII**) combine to form a molecular complex. Molecules of the methanol solvent (**V**) are also incorporated, according to the stoichiometry:

$$2CH_3OH + MV^{2+} + 2PhOH \rightarrow \text{complex}$$

V   VI   VII

Calculate the mass per mole of the complex formed. To this end, use the expression

$$\sum_{i=1}^{n} v_i \left(\sum_{j=1}^{m_i} N_{ij} M_j \right),$$ where N_{ij} is the number of atoms of type j in the molecule i, M is the

molar mass of atom j, and v_i is the stoichiometric number for molecule i.

We first perform three sums to calculate the mass per mole of the reactants, (V), (VI), and (VII):

m_i is the number of different atom types in molecule i.

Molecule **V**:   mass $= \displaystyle\sum_{j=1}^{3} N_{1j} M_j = N_C M_C + N_H M_H + N_O M_O = (1 \times 12) + (4 \times 1) + (1 \times 16)$
$= 32 \text{ g mol}^{-1}$

Molecule **VI**:   mass $= \displaystyle\sum_{j=1}^{4} N_{2j} M_j = N_C M_C + N_H M_H + N_N M_N + N_{Cl} M_{Cl} = (12 \times 12) + (14 \times 1) +$
$(2 \times 14) + (2 \times 35.5) = 257 \text{ g mol}^{-1}$

Molecule **VII**: mass $= \displaystyle\sum_{j=1}^{3} N_{3j} M_j = N_C M_C + N_H M_H + N_O M_O = (6 \times 12) + (6 \times 1) +$
$(1 \times 16) = 94 \text{ g mol}^{-1}$

We next calculate the mass of complex produced, saying:

$$\text{molar mass of complex} = \sum_{i=1}^{n} v_i \left(\sum_{j=1}^{m_i} N_{ij} M_j \right)$$

so   molar mass of complex $= (2 \times \text{mass of } (\mathbf{V})) + (1 \times \text{mass of } (\mathbf{VI})) + (2 \times \text{mass of } \mathbf{VII}))$

and   molar mass of complex $= (2 \times 32) + (1 \times 257) + (2 \times 94)$

so the mass per molar of the complex is 509 g mol^{-1}. ∎

Self-test 2.3

2.3.1   Consider the formation of a cyclic acetal formed by reaction between acetone (**IX**) and 1,2-dihydroxyethane (**X**):

VIII   IX

Scheme 2.1

Assuming the reaction is quantitative, what is the mass of product if we react 0.35 mol each of (**IX**) and (**X**)?

Formulate the answer in the form of eqn (2.2).

 Continued . . .

> **Self-test 2.3 continued...**

2.3.2 A student's overall mark depends on the percentage of each constituent units studied. Calculate the student's overall mark if:

(i) The maths unit is worth 10 credits, and the student has a percentage of 45.

(ii) The inorganic unit is worth 20 credits, and the student has a percentage of 56.

(iii) The organic unit is worth 20 credits, and the student has a percentage of 70.

(iv) The physical unit is worth 20 credits, and the student has a percentage of 62.

(v) The laboratory unit is worth 30 credits, and the student has a percentage of 40.

Formulate the answer in the form of eqn (2.2).

The Pi notation $\prod$

By the end of this section, you will know that the Pi notation $\prod$:

- Is another shorthand symbol for an operator.
- Tells us to multiply together a series of terms.
- A special form of the Pi operator is called a factorial, *n!*

In everyday life, we could either say '*multiplying* the numbers 4 and 5 yields the number 20' or 'the *product* of 4 and 5 is 20.' Both sentences are correct, but the latter is shorter.

In mathematics, the word **product** always bears a sense of 'multiply together.' Using the word 'product' in this mathematical context, a **Pi product** is another operator, this time involving multiplication. Its name derives from the way this new operator is written with a capital Greek letter Pi (Π). Each time we see a Π, we are to recognize that it is a shorthand way of saying, 'multiply together series of terms.'

We call it Pi with a *capital* letter, because a lower case pi looks slightly different: π.

■ **Worked Example 2.9**

What is the Pi product of the numbers 4, 6, and 9?
The value of the Pi product is, $4 \times 6 \times 9 = 216$.

When we write a Pi product, we follow the Π symbol with a phrase or bracket. In mathematical notation, we would write the numbers after the Π symbol, each number or item separated by a comma:

$$\prod(4, 6, 9) = 216. \quad ■$$

The word **generic** here means 'concerning origin' and comes from the Greek word *genesis*, which means 'in the beginning.' The English *genetics* comes from this same root.

Occasionally, we need to multiply together a series of generic terms. In which case, we tend to write the Pi product in a slightly more complicated-looking way:

$$\prod_{i=1}^{n} X_i$$

where the term X_i means a mathematical function involving the general species *i*.

We know that Π is an operator because it is telling us to *do* something. In this case, the symbol tells us to multiply everything to the right of the Π sign. In this, the general case, we are to multiply together n terms. We write a small n above the large Π sign to tell us that this number of terms is the extent of the summation operation. Additionally, we write a small $i = 1$ directly below the large Π to tell us to start the summation operation with the first term. As with Sigma, the terms above and below the Π sign are called the *limits* of the product.

The definition above could have written:

$$\prod_{i=1}^{n} X_i = X_1 \times X_2 \times X_3 \times \ldots \times X_n \tag{2.3}$$

■ **Worked Example 2.10**

Consider following series of numbers:

$$1 \times 4 \times 9 \times 16 \times 25 \times 36$$

Write a Pi product expression in the style of eqn (2.3) to describe them.

Analysing these numbers tells us we have a series of *square numbers*. They range from 1^2 through to 6^2.

Strategy:

(i)   We write the Pi symbol, because we must *multiply* the members of the series: Π.

(ii)   The variable is i. It has been squared, so we write against the Π symbol 'i^2': $\Pi\, i^2$.

(iii) The lowest value of i in the series is 1, so we write '$i = 1$' beneath the Π symbol:
$$\prod_{i=1} i^2$$

(iv)   Finally, the highest value of i in the series is 6, so we write a small '6' above the Π symbol: $\displaystyle\prod_{i=1}^{6} i^2$ ■

> In writing this expression, it is assumed that all values of i are **integers**.

■ **Worked Example 2.11**

A common way of defining an equilibrium constant K is to write:

$$K = \frac{\displaystyle\prod_{i=1}^{n} [\text{product } i]^{\upsilon_i}}{\displaystyle\prod_{j=1}^{m} [\text{reactant } j]^{\upsilon_j}}$$

where the square brackets denote concentrations, and the υ terms relate to stoichiometric numbers.

Starting with Pi product notation, write an expression for the equilibrium constant K for the following oxidation reaction:

$$Ce^{IV} + Fe^{II} \rightarrow Ce^{III} + Fe^{III}$$

Each stoichiometric number υ_i is 1. The products are written on the right-hand side and the reactants on the left. We write eqn (2.3) as:

$$K = \frac{\prod \left([Ce^{III}], [Fe^{III}] \right)}{\prod \left([Ce^{IV}], [Fe^{II}] \right)}$$

For many people, this notation will look somewhat odd—artificial even. They will prefer simple multiplication signs:

$$K = \frac{[Ce^{III}] \times [Fe^{III}]}{[Ce^{IV}] \times [Fe^{II}]}$$

But most chemists will choose to omit the multiplication signs altogether, and write:

$$K = \frac{[Ce^{III}][Fe^{III}]}{[Ce^{IV}][Fe^{II}]}$$

This example illustrates how it is possible to have different (but equally valid) notations.

All three notations are correct and entirely valid, but the third is probably the easiest to read and therefore the most common. ∎

Sometimes we write equilibrium constants in such a way that we retain the word 'product'.

■ **Worked Example 2.12**

Solid limestone ($CaCO_3$) is almost water insoluble. The process of dissolving limestone is:

$$CaCO_3 \, (s) \rightarrow Ca^{2+} \, (aq) + CO_3^{2-}(aq)$$

Write the equilibrium constant K for this dissolution process and hence work out the relationship between the equilibrium constant and its Pi product. (Do not include the $CaCO_3$ in the equilibrium constant K, because it remains an undissolved solid.)

We write the equilibrium constant K for this process saying,

$$K = [Ca^{2+} \, (aq)] \, [CO_3^{2-}(aq)]$$

The right-hand side is a Pi product, so we write this simplified equilibrium constant as

$$K = \prod (\text{concentrations of ions in solution})_i \quad ∎$$

We often call equations of this sort 'the **solubility product**' of limestone, and give K the additional subscript 'sp,' as K_{sp}. The word 'product' here alerts us to the need to multiply together the ionic concentrations.

The Pi product is also useful in other areas of chemistry.

■ **Worked Example 2.13**

A chemist wishes to make the complexing ligand 4-(4-bromophenyl)-2,2′-bipyridine (**XI**) in three steps, according to Scheme 2.2. The yield of step **1** is 66%, the yield of step **2** is 92% but the yield of step **3** is only 41%. What is the overall yield of (**XI**)?

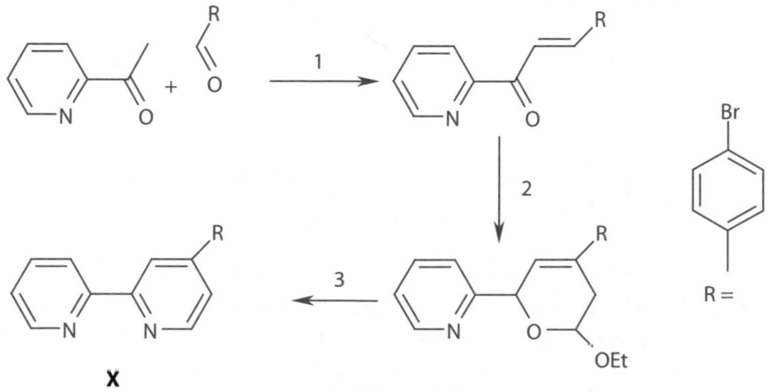

Scheme 2.2

We calculate the overall yield of (**XI**) as a Pi product in the style of eqn (2.3), saying:

$$\text{overall yield} = \prod_{i=1}^{3} \left(\text{yield of step } i\right)$$

so we compute by multiplying together the individual yields:

overall yield of (**XI**) = (yield of step 1) × (yield of step 2) × (yield of step 3)

so   overall yield of (**XI**) = (0.66) × (0.92) × (0.41)

and   overall yield of (**XI**) = 0.249 or 24.9%

For help with percentages, see Chapter 5.

The rather low overall yield illustrates why, when performing any multistep reaction, we need to maximize the yield of each step. Otherwise, we will never generate an economically viable amount of product. ■

Sometimes we wish to write a Pi product expression to represent the multiplication of a series of numbers. When such a Pi product operates starts with the number 1, we call it a **factorial**. We give it the symbol $n!$

■ **Worked Example 2.14**

Write the factorial 7! as a Pi product.

By definition, the factorial $7! = 1 \times 2 \times 3 \times 4 \times 5 \times 6 \times 7$. This is clearly a Pi product starting with $i = 1$ and ending with 7. Therefore, we write, $\prod_{i=1}^{7} i$. ■

This result gives us the formal definition of a factorial, in eqn (2.4):

$$n! = \prod_{i=1}^{n} i \qquad (2.4)$$

Factorials occur relatively infrequently in the mathematics we encounter as chemists, but often enough to be useful.

Self-test 2.4

Calculate the values of the following Pi products in the style of eqn (2.3):

2.4.1 $\prod_{i=1}^{5} \text{number, } i$

2.4.2 $\prod_{i=2}^{4} \left(\dfrac{1}{i^2}\right)$

2.4.3 $\prod_{i=5}^{7} i^3$

2.4.4 $\prod_{i=1}^{3} \sqrt{i}$

Rewrite each of the following as a Pi product, e.g. $1 \times 2 \times 3 = \prod_{i=1}^{3} i$

2.4.5 $5 \times 6 \times 7 \times 8 \times 9 \times 10$

2.4.6 $\dfrac{1}{9} \times \dfrac{1}{10} \times \dfrac{1}{11} \times \dfrac{1}{12}$

2.4.7 $3^2 \times 4^2 \times 5^2 \times 6^2$

2.4.8 $1 \times 4 \times 9 \times 16 \times 25$

2.4.9 $6!$

2.4.10 $14!$

Help with multiplying fractions can be found in Chapter 5.

Other operators commonly encountered in chemistry

We have looked at three of the chemist's most common operators, Delta, Sigma and Pi. They have much in common: firstly, we write a symbol for each using a capital Greek letter. More importantly, they save us a lot of time because they allow us to represent a mathematical reality using a form of shorthand notation.

Table 2.3 Operators commonly encountered in chemistry. In each case here, the letter X represents the argument.

Function	Symbol	Usual notation	Simple example
square	2	X^2	$3^2 = 3 \times 3 = 9$
cube	3	X^3	$4^3 = 4 \times 4 \times 4 = 64$
powers in general	n	X^n	e.g. if $n = 4$, $3^4 =$ $3 \times 3 \times 3 \times 3 = 81$
square root	$\sqrt{}$	$\sqrt{X}$ or $\sqrt{}X$	$\sqrt{4} = 2$
cube root	$\sqrt[3]{}$	$\sqrt[3]{X}$ or $\sqrt[3]{}X$	$\sqrt[3]{8} = 2$
nth root	$\sqrt[n]{}$	$\sqrt[n]{X}$ or $\sqrt[n]{}X$	$\sqrt[5]{32} = 2$
percentage	%	$X\%$	5% of $120 = \frac{5}{100} \times 120 = 6$
sine	sin	$\sin X$ or $\sin (X)$	
cosine	cos	$\cos X$ or $\sin (X)$	
logarithm in base 10	log	$\log X$ or $\log_{10} X$ or $\log_{10} (X)$	
logarithm in base e	ln	$\ln X$ or $\ln (X)$ or $\log_e X$ or $\log_e (X)$	
differential	$\dfrac{dy}{dx}$	if $y = 2x^3$, $\dfrac{dy}{dx} = 6x^2$, see especially Chapter 17	
partial differential	$\dfrac{\partial}{\partial x}$	see Chapter 22	
integral	$\int$	see Chapters 23–26	

Table 2.3 lists a few of other more common operators, together with their usual notations.

We need to note the following:

- Operators such as those representing power, logarithm and trigonometric operations are often referred to simply as 'functions'. The term 'operator' tends to be reserved for more complicated operations, such as those involving differentials and integrals.

- We generally write the operator on the *left* of the argument, e.g. we write $\sin x$ rather than $x \sin$. Power operators such as square X^2, cube X^3, etc., are the only common exceptions to this rule. (We describe powers in Chapter 11.)

- Some operators have more than one correct notation. For example, $\sqrt{X}$ or $\sqrt{}X$ or $X^{1/2}$ are equally valid; and $\exp(X)$ and e^X are also equally valid. (We describe exponentials in Chapter 12.)

- Sometimes the function looks like a number. For example, the square root operator $\sqrt{X}$ can also be written as $X^{1/2}$: X here is the argument and the superscripted '½' is the operator.

- Each time we write an argument X, we must appreciate that it could be a constant number (whole or partial), or a variable of some kind. The variable can be simple such as a single term (velocity, mass, current, height), or it can be a series of variables arranged in an equation. Therefore, if we write an expression such as X^2, the argument X could be 2, 3.373, velocity, or a complicated string of terms such as

$$\left(x^2 - \sqrt{x} + \frac{2}{x} \right).$$

- If we write $\sqrt{X}$ (with an overbar positioned over the X), we are being told to take the root of *everything* enclosed within the root symbol, $\sqrt{}$. By contrast, writing, $\sqrt{X}$ could mean the root of everything positioned to the right of the $\sqrt{}$ symbol, or merely the first term to its right. Clearly, a good chemist will not tolerate any ambiguity.

- Therefore, we should enclose the argument within brackets if it comprises more than a single term. As a simple example, we write $\sin(x^3)$ rather than $\sin x^3$, because '$\sin x^3$' could be mistaken for the different function $(\sin x)^3$. Similarly, we write $(2\pi f)^2$ if the argument is '$2\pi f$' because '$2\pi f^2$' means $2 \times \pi \times f \times f$, i.e. the '2' and the '$\pi$' terms have *not* been squared. (We describe trigonometric functions such as 'sin' and 'cos' in Chapter 16.)

- In this context, we can consider a root sign to be a bracket, so writing $\sqrt{(mgh)}$ is exactly the same as writing $\sqrt{mgh}$.

- When performing the operations in calculus, we usually need to read the operator with care: while a simple integral sign $\int$ looks fairly straightforward, it may have additional component parts that 'refine' the way we actually perform the whole operation, e.g. $\int_2^4$; see Chapters 23–26.

- In differential calculus, the operators can look more complicated still: $\dfrac{dy}{dx}$, $\dfrac{\partial^2 \Omega}{\partial u^2}$ and $\dfrac{\partial}{\partial u}\left(\dfrac{\partial \Omega}{\partial v} \right)$ are all operators, as described in Chapters 17–22.

Summary of key equations in the text

Definition of Δ: $\quad \Delta X = X_{\text{final}} - X_{\text{initial}}$ (2.1)

Definition of Σ: $\quad \displaystyle\sum_{i=1}^{n} X_i = X_1 + X_2 + X_3 + \ldots + X_n$ (2.2)

Definition of Π: $\quad \displaystyle\prod_{i=1}^{n} X_i = X_1 \times X_2 \times X_3 \times \ldots \times X_n$ (2.3)

Definition of factorials, $n!$: $\quad n! = \displaystyle\prod_{i=1}^{n} i$ (2.4)

Additional problems

2.1 What is the value of $\displaystyle\prod_{i=1}^{\infty} \frac{2}{i^2}$ to two decimal places?

2.2 Consider the expression, 'current $= 35\ \mu A$'. Identify each part of this algebraic phrase.

2.3 When plotting a graph, the vertical axis (the *y-axis*) increases from 12 cm to 32 cm. What is the value of Δy?

2.4 Determine the value of the sum, $\sum\limits_{i=4}^{12} 3i^2$

2.5 To obtain the standard enthalpy change on combustion $\Delta H_c^{\ominus}$, we sum the molar enthalpies of each chemical participating in the reaction:

$$\Delta H_c^{\ominus} = \sum_{\text{products}} v\, H_f^{\ominus} - \sum_{\text{reactants}} v\, H_f^{\ominus}$$

Calculate the value of $\Delta H_c^{\ominus}$ for methane at 25 °C using the molar enthalpies of formation, $\Delta H_f^{\ominus}$. The necessary data are:

Species (all as gases)	CH_4	O_2	CO_2	H_2O
$\Delta H_f^{\ominus}$ / kJ mol^{-1}	−74.81	0	−393.51	−285.83
v	1	2	1	2

The stoichiometry of the reaction is: $CH_4 + 2O_2 \rightarrow CO_2 + 2H_2O$.

2.6 One definition of an **equilibrium constant**, is $K = \dfrac{\prod\limits_{i=1}^{n}[\text{products}]^{v_i}}{\prod\limits_{j=1}^{m}[\text{reactants}]^{v_j}}$.

Now consider the following redox reaction:

$$Cr^{VI} + 2Co^{II} \rightarrow Cr^{IV} + 2Co^{III}$$

Calculate a value of K for this reaction using the following equilibrium concentrations: $[Cr^{VI}] = 3.22 \times 10^{-9}$, $[Co^{II}] = 2.9 \times 10^{-8}$, $[Cr^{IV}] = 0.09$, and $[Co^{III}] = 0.4361$, all expressed in the same units of 'mol dm^{-3}'.

2.7 Determine a value of the product, $\prod(1.2, 4, 5.5, 7, 9)$

2.8 Determine a value of the product, $\prod\limits_{i=3}^{5} \dfrac{i^2}{3}$

2.9 During the course of a reaction, the entropy S changes from −15.1 J K^{-1} mol^{-1} to −32.5 J K^{-1} mol^{-1}. Calculate a value of ΔS.

2.10 Determine a value of the sum, $\sum\limits_{i=2}^{5} i^2$

3

Algebra II

The correct order to perform a series of operations: BODMAS

The order in which we perform a calculation

By the end of this section, you will know:

- When a calculation involves more than one operator, the resultant value depends on the order in which we employ the various operators.
- Only if we perform the operations in the correct order will the answer be correct.

When we write a sentence, we need to be careful that the words say what we want them to say. For example, the words in two sentences, 'The cat sat on the man' and 'The man sat on the cat' are identical, yet the meaning of the sentences differ as a result of scrambling the order. In the same way, we need to exercise care when composing a mathematical sentence because we otherwise 'misread' it, and obtain an incorrect answer.

■ **Worked Example 3.1**

A man walks into a shop and asks for a bottle of milk and two packets of crisps. Each bottle of milk costs £1 and each packet of crisps costs 40p. How much does the man need to pay?

We could just add up this bill in our heads and say, correctly, £1.80. And we would be quite correct. But a formal **methodology** would say:

Methodology means the way we choose to perform a calculation or procedure.

Total = (price per bottle of milk × number of bottles)

+ (price per packet of crisps × number of packets)

The methodology for this simple example is to perform two multiplication calculations:

- First (price per bottle of milk × number of bottles)
- Next, (price per packet of crisps × number of packets).

We obtain the final answer by adding up the two sums:

Total = (£1 × 1) + (40p × 2)
Total = £1.80 ■

This example may be simple, but it illustrates the way that most calculations are made up of 'building blocks.' Each building block is itself a simple calculation.

To answer the sum 4×3, we could say either 'four times three' or 'three times four'. The answer is the same either way, if we count properly. But in any sum involving *two* operators, inevitably one will have priority over the other. If we ignore the correct order of priorities, we may obtain the wrong answer. For example, notice how we performed the multiplication steps first in Worked Example 3.1 and only then did we add up the individual components. We say that multiplication, as an operator, has priority over addition.

In the language of mathematics, if the answer does not depend on the order in which we do it, we say a calculation is **associative**. For example, the calculation '$4 \times 3 \times 2$' is associative, because we obtain the same answer of 24 if we first multiply '4×3', then multiply its answer of 12 by 2, or if we multiply '4×2' by 3 or '3×2' by 4. We could write this, saying:

$$(4 \times 3) \times 2 = 4 \times (3 \times 2) = 3 \times (4 \times 2) = 24$$

In a similar way, addition is associative: $3 + 7 + 1 = (3 + 7) + 1 = 3 + (7 + 1) = (3 + 1) + 7$.

Sums involving only addition or only multiplication are always associative, but most sums are not. In fact, the majority of operators do not operate in an associative sense. Accordingly, we need to formulate simple rules to ensure we perform a sum in the correct order.

> We say a calculation is **associative** if the answer does not depend on the order in which we do it.

■ **Worked Example 3.2**

Show that the problem $3 + 4 \times 5$ is *not* associative.

First, we consider the case in which we perform the ADDITION step first: $3 + 4$. The result of this problem is then multiplied by 5:

$$3 + 4 = 7$$

then
$$7 \times 5 = 35$$

Second, we perform the MULTIPLICATION step first: 4×5. We then add 3 to the result of this problem:

$$4 \times 5 = 20$$

then
$$20 + 3 = 23$$

The results from these two approaches clearly differ, although the initial data were identical. Of these two answers, the correct one is 23. ■

Worked Example 3.2 helps illustrate the need for a series of rules when performing even simple problems. We know some of these rules already. For example, look at the use of the minus sign, which is a more complicated operator than +. The complexity arises because the operation it describes depends on the positions of the numbers or symbols: if we see the sum '4 − 3', we automatically subtract the number positioned on the *right* of the operator from the number on the *left*. Swapping the 3 and 4 yields the wrong answer.

Deciding the correct order when performing calculations

By the end of this section, you will know:

- The BODMAS scheme is a simplest way to remember the correct order in which to perform a calculation comprising several operations.

We saw in Worked Example 3.2 that multiplication has priority over addition. When assigning priorities, we often start with the acronym **BODMAS**. In priority order, the letters stand for BRACKETS, OF, DIVISION, MULTIPLICATION, ADDITION, SUBTRACTION. The word 'OF' here is not a normal grammatical preposition, but implies more complicated operators, such as powers, roots, and those we meet in subsequent chapters. For example, we might says, 'The square *of ...*' a number.

Before we start a calculation, write 'BODMAS' in large letters across the page:

$$B \quad O \quad D \quad M \quad A \quad S$$

← *Highest priority during a calculation*

→ *Lowest priority during a calculation*

■ Worked Example 3.3

A chemist has to perform a thermodynamic calculation, using the equation:

$$\Delta G^{\ominus} = \Delta H^{\ominus} - T \times \Delta S^{\ominus}$$

where $\Delta G^{\ominus}$ is the change in the Gibbs function during a reaction, $\Delta H^{\ominus}$ is the change in enthalpy during the same reaction, $\Delta S^{\ominus}$ is the entropy change and T the temperature at which the reaction is performed.

Consider the Haber process to produce ammonia:

$$N_2(g) + 3H_2(g) \leftrightharpoons 2NH_3(g)$$

The word **data** is plural. The singular is *datum* and never *datas*.

At 300 K, the relevant thermodynamic data are $\Delta H^{\ominus} = -92$ kJ mol^{-1} and $\Delta S^{\ominus} = -200$ J K^{-1} mol^{-1}. What is the value of $\Delta G^{\ominus}$?

The equation above contains two operators, both '+' and '×'. Before we perform the calculation, we must first decide whether to perform the addition first, or the multiplication first. We will obtain the wrong answer for $\Delta G^{\ominus}$ if we use the operators in the wrong order.

Which operator has priority? When we look at the line saying 'BODMAS', we see that the left-hand side is marked as highest priority and the right-hand end has a lower priority. 'M' for 'multiplication' is nearer to the left than is 'A' for 'addition.'

Remember from Chapter 1 how a 'kilo joule' kJ is shorthand for 1000 J.

From now on, each time we meet one of these operators, we will print it in SMALL CAPITALS. We therefore perform the MULTIPLICATION step *first*. Inserting values into the equation:

$$\Delta G^{\ominus} = (-92 \text{ kJ mol}^{-1}) - (300 \text{ K} \times -200 \text{ J K}^{-1} \text{ mol}^{-1})$$

The brackets around −200 are to prevent confusion: the minus sign is not an operator but merely indicates a negative number.

It is a common error to forget to convert the standard factors. In effect, the sum is:

$$\Delta G^{\ominus} = -92 \times 1000 \text{ J mol}^{-1} - 300 \text{ K} \times (-200) \text{ J K}^{-1} \text{ mol}^{-1}$$

If we had forgotten to multiply by 1000, the answer would have been wrong.

Using the rules of BODMAS gives:

$$\Delta G^{\ominus} = (-92000 \text{ J mol}^{-1}) - (300 \text{ K} \times -200 \text{ J K}^{-1} \text{ mol}^{-1})$$

So, $\Delta G^{\ominus} = -32\,000$ J mol^{-1}

As chemists, we might like to write this answer as −32 kJ mol^{-1}. ■

■ Worked Example 3.4

A charged ion will accelerate when it enters a magnetic field. Its acceleration is a. Its initial velocity is u. The distance travelled by the particle is s. The variables s, u and a are related according to Newton's Law of motion, where t is the time after entering the field:

$$s = ut + \tfrac{1}{2}at^2$$

In what order should we perform the calculation if we wish to calculate s?

Remember from Chapter 1 how a 'kilo joule' kJ is shorthand for 1000 J.

Before we start, we notice how u and t are written together, but without an operator. The convention says we must assume variables are MULTIPLIED if written together without an apparent operator. We see how 'ut' is really $u \times t$, and '$\frac{1}{2} at^2$' is really $\frac{1}{2} \times a \times t^2$.

We now analyse the equation to see which operators it contains, and find ADDITION, MULTIPLICATION (twice) and a SQUARE function. In sequence:

(1) The operator with the highest priority is the square function (within BODMAS, we call it OF, because 't^2' is a function OF t), so we would first calculate t^2.

(2) The operator with the next highest priority is MULTIPLICATION, so we perform the two MULTIPLICATION steps: $u \times t$ (i.e. $u\,t$) and $\frac{1}{2} \times a \times t^2$ (i.e. $\frac{1}{2} at^2$).

(3) The last operator is ADDITION, so we add together the compound terms 'ut' and '$\frac{1}{2} at^2$' terms, and obtain: $s = ut + \frac{1}{2} at^2$.

Some people find it easier to understand the order of priorities by drawing a series of self-contained boxes. The correct order of performing the operations involves working from the inside and going outwards:

$$S = \boxed{ut} + \frac{1}{2}a\boxed{t^2} \rightarrow S = \boxed{ut} + \boxed{\frac{1}{2}a\boxed{t^2}} \rightarrow S = \boxed{\boxed{ut} + \boxed{\frac{1}{2}a\boxed{t^2}}} \quad \blacksquare$$

■ Worked Example 3.5

We can express the heat capacity C of a substance in terms of a power series:

$$C = a + bT + cT^2$$

where T is the temperature, and a, b, and c are constants obtained experimentally. In what order should we calculate C?

Equations of this type are sometimes called virial equations. As before, we first look to see which functions are involved. We see ADDITION, MULTIPLICATION (because bT means '$b \times T$' and cT^2 means '$c \times T^2$'), and a function OF, because T^2 is a function OF T.

(1) In the BODMAS scheme, the operator with the highest priority is OF, so we first calculate T^2.

(2) The operator with the next highest priority is MULTIPLICATION, so we calculate $b \times T$ and $c \times T^2$.

(3) The final operator is ADDITION. We finally add together the individual terms, as $C = a + bT + cT^2$. ■

■ Worked Example 3.6

We sometimes find it helpful to think of a chemical bond as behaving like a spring (see Fig. 3.1). The equation below describes the frequency in wave numbers at which the bond vibrates $\bar{v}$, and relates the force constant k with the so-called 'reduced mass' of the vibrating species μ and the speed of light c:

$$\bar{v} = \frac{1}{2\pi c} \sqrt{\frac{k}{\mu}}$$

If we know c, k and μ, in what order should we perform the calculation of $\bar{v}$?

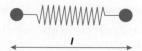

Figure 3.1 The vibration of a chemical bond has many analogies with two solid objects separated by a straightforward elastic spring. The length l extends from the centre of mass of one object to the centre of mass of the other.

Before we start this example, we notice the fraction immediately after the equals sign. We call fractions of the form '1 ÷ anything' a **reciprocal**.

We now look at the contents of this bracket. The bottom line clearly contains three terms, yet no operators are printed. It is most *un*likely that '$2\,\pi c$' is a single variable. Again, we are to read '$2\,\pi c$' as though it was '$2 \times \pi \times c$'. We've discovered yet another shorthand notation. In fact, we could have written the equation as:

$$\bar{v} = \frac{1}{(2 \times \pi \times c)} \times \sqrt{\frac{k}{\mu}}$$

We can now start planning the calculation.

1. From the laws of BODMAS, we give the highest priority to BRACKETS, because 'B' for BRACKET lies nearest the high priority end. We must first calculate $(2 \times \pi \times c)$, because it is bracketed.

2. The next operator after BRACKETS comes OF, which means 'a function of.' The simplest functions are powers and roots, so we next calculate a value of the term $\sqrt{\frac{k}{\mu}}$, i.e. calculate k/μ and then take the square root of its value.

3. Next comes DIVISION, so we divide, 1 by the bracket $(2\,\pi c)$.

4. Finally, the last task is to MULTIPLY together the two terms $\frac{1}{(2\,\pi c)}$ and $\sqrt{\frac{k}{\mu}}$

If we were to indicate the correct order in which to perform this calculation using concentric boxes, as on p. 37 above, we would draw the following:

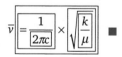

■ **Worked Example 3.7**

The ideal-gas equation, $pV = nRT$ is rarely realistic because of strong inter-particle forces. The van der Waals equation is one of the better ways to take account of such forces:

$$\left(p + \frac{n^2}{V^2}\,a\right)(V - n\,b) = n\,R\,T$$

where the constants a and b are called the van der Waals constants, the values of which depend on the gas. The term n is the amount of gas.

To calculate a value of the compound variable '$n\,R\,T$', what would be the correct order in which to calculate the left-hand side of the equation?

There are six different operators on the left-hand side of the equation: ADDITION, SUBTRACTION, MULTIPLICATION, DIVISION, BRACKETS and OF. Before we start, we need to mention that the two terms V^2 and n^2 terms are best treated as functions OF V and n, each of which we consider within the BODMAS scheme under OF.

The a term in this equation reflects the strength of the interaction between gas particles, and the b term reflects the size of a particle.

1. Within the BODMAS scheme, the BRACKETS have the highest priority, so we calculate their contents first. The right-hand bracket is easy: we first multiply n with b (MULTIPLICATION) and only then SUBTRACT this compound term from V (i.e. saying '$V - nb$').

2. We next consider the left-hand bracket. We notice within this bracket both V^2 and n^2. These squares are functions OF V and n, so we square both terms before any further calculations.

3. Staying with the left-hand bracket, we next consider DIVISION, dividing the n^2 term by V^2 to yield, $\dfrac{n^2}{V^2}$.

4. Next, again staying with this bracket, we consider MULTIPLICATION: We multiply a with $\dfrac{n^2}{V^2}$ to yield, $\dfrac{n^2}{V^2} \times a$, which we generally choose to write without the MULTIPLICATION sign, as $\dfrac{n^2}{V^2} \times a$.

5. To complete the left-hand bracket, we ADD the p term to $\dfrac{n^2}{V^2}a$, which completes the equation's left-hand side.

6. Finally, we MULTIPLY together the two brackets.

$$\left(p + \boxed{\dfrac{n^2}{V^2}}\, a \right)\left(\left(V - \boxed{nb} \right) \right) = n\,R\,T \quad \blacksquare$$

■ Worked Example 3.8

We now return to Worked Example 3.5. It is sometimes appropriate to model the heat capacity C_p of a substance in terms of a power series (also called a *virial equation*). For chloroform (**I**), the appropriate series if

$$C_p = 91.47 + 7.5 \times 10^{-2}\,T - 10^{-6}\,T^2$$

where T is the thermodynamic temperature, and the factors relate to values of C_p expressed in units of $J\,K^{-1}\,mol^{-1}$. What is the value of C_p at 400 K?

I

The equation may confuse us because it contains so much detail. Accordingly, we will clarify its appearance by rewriting it slightly, with all the operators written in **bold** print:

$$C_p = (91.47) + (7.5 \times 10^{-2})\ T - 10^{-6} \times \boxed{T^2}$$

These two pairs of brackets are written purely for clarity.

1. While there are two pairs of brackets, they are present purely to clarify the equation's appearance. So the operator of highest priority is the square (i.e. it's a function OF T). Our first step is to calculate T^2, which equals $400 \times 400 = 160\,000$. The equation becomes:

$$C_p = (91.47) + (7.5 \times 10^{-2}) \times T - 10^{-6} \times 160\,000$$

2. Next in priority are the MULTIPLICATION steps. There are three. In order to empha-sise the portions to be MULTIPLIED, we draw square brackets around the relevant portions:

$$C_p = (91.47) + \boxed{(7.5 \times 10^{-2}) \times 100} - \boxed{10^{-6} \times 160\ 000}$$

which becomes

$$C_p = 91.47 + 30 - 0.16$$

3. Finally, we calculate C_p by ADDING and SUBTRACTING the resultant terms, to yield

$$C_p = 121.31\ \mathrm{J\,K^{-1}\,mol^{-1}}. \blacksquare$$

Self-test 3.1

In each of the following, calculate the value of x using the BODMAS rule:

3.1.1 $x = 3 \times 2 \times 6$

3.1.2 $x = 6 \times 7 - 8 \times 2$

3.1.3 $x = (2 + 3) \times 5$

3.1.4 $x = 4 + 5^2 - 3^4$

3.1.5 $x = \dfrac{4 \times 6}{7 \times 2}$

3.1.6 $x = \dfrac{6 \times 9 - 2}{3 + 5 \times 2}$

3.1.7 $x = 5 \times \sqrt{4 - 6 \times 2 + 44}$

3.1.8 $x = \left(\dfrac{5+7}{6}\right)^2 - 56$

Summary of key equation in the text

B = brackets
O = 'of'
D = division
M = multiplication
A = addition
S = subtraction

Additional problems

3.1 The molar mass of sulphuric acid is 98 g mol⁻¹, and the molar mass of water is 18 g mol⁻¹. What is the mass of 2 moles of sulphuric acid dissolved in 12 moles of water?

3.2 Calculate the molar mass of ferric nitrate nonahydrate, $Fe(NO_3)_3 \cdot 9(H_2O)$. The relevant atomic masses are Fe = 56, N = 14, O = 16 and H = 1 g mol⁻¹.

3.3 A chemist orders three bottles of chemicals: twelve of A and seven of B. The mass of a single bottle of A is 500 g, while each bottle of B has a mass of 250 g. What is the overall mass of the bottles ordered?

3.4 The acceleration a of a proton inside a mass spectrometer may be calculated using the expression

$$a = \frac{v - u}{t}$$

where the velocity changes from an initial value u to a different final value v during a time t. Calculate a if $u = 30$ m s^{-1}, $v = 650$ m s^{-1} and $t = 3.9$ s.

3.5 We define the gradient of a graph as

$$\text{gradient} = \frac{y_2 - y_1}{x_2 - x_1}$$

where y values relate to vertical distances and x relates to horizontal distances. Calculate the gradient when $y_1 = 3.0$, $y_2 = 12.0$, $x_1 = 4.1$ and $x_2 = 5.5$.

3.6 The **entropy change** $\Delta S_{(cell)}$ for an electrochemical cell as it discharges depends on temperature, according to the following equation:

$$\Delta S_{(cell)} = nF \left(\frac{E_{(cell)2} - E_{(cell)1}}{T_2 - T_1} \right)$$

where n is the number of electrons transferred in the balanced redox reaction, F is the Faraday constant, $E_{(cell)}$ is the voltage of the cell, and T is temperature. Calculate a value of $\Delta S_{(cell)}$ if $n = 2$, $F = 96\,485$ C mol^{-1}, $E_{(cell)}$ is 1.440 V at 298 K (i.e. at T_1) and 1.436 V at T_2 of 330 K.

3.7 A student sits four exams. The physical chemistry exam is worth 20 credits, and has a mark of 70%; the inorganic chemistry exam is worth 20 credits and has a mark of 63%, the organic exam is again worth 20 credits and has a mark of 59%, and the analytical exam is worth 40 credits, and has a mark of 50%. The student therefore acquires 100 credits. What is the *overall* mark?

3.8 The **Kirchhoff equation** expresses the way an enthalpy change ΔH alters with temperature T:

$$\Delta H_2 = \Delta H_1 + C_p (T_2 - T_1)$$

where C_p is a heat capacity. If $\Delta H_1 = 12$ kJ mol^{-1} at a T_1 of 298 K, and $C_p = 31.2$ J K^{-1} mol^{-1}, calculate the value of ΔH_2 when $T_2 = 330$ K.

3.9 Consider the **Einstein equation**

$$E = mc^2$$

that relates the energy released E when a mass m is converted entirely to energy, for example in a bomb; c is the speed OF light. Calculate the energy released when 0.11 kg of uranium is converted. Take $c = 3.0 \times 10^8$ m s^{-1}.

3.10 An **electrode potential** E varies with temperature T. For the silver bromide–silver redox couple, the value OF $E_{AgBr,Ag}$ varies according to a virial expression:

$$E_{AgBr,Ag} = 0.07131 - 4.99 \times 10^{-6}\, T - 3.45 \times 10^{-8}\, T^2$$

Calculate the value of $E_{AgBr,Ag}$ when $T = 312$ K.

Algebra III

Simplification and elementary rearrangements

Balancing the variables: why we need the equals sign, =

By the end of this section, you will know:

- The equals sign '=' is not an operator, so it does not *do* anything to the variables.
- An equation is only complete when it contains an equals sign to relate the expressions on either side of the equation.
- An equals sign tells us that the magnitudes of the two sides of the equation are identical in every way, but it tells us nothing about the content of either side.

It may seem trite to begin saying an equals sign '=' lies at the heart of our mathematics. The equals sign acts like the verb at the heart of a sentence. For example, the two phrases 'the cat' and 'the mat' have nothing in common and are wholly unrelated until we place some form of 'doing word' between them, saying something like 'the cat *sat* on the mat' or 'the cat *liked* the mat.' The subject and the object are completely unrelated without the italicized verbs.

In just the same way, a variable such as 'pH' is not particularly interesting until we say something *about* it. Similarly, a number such as '3.2' is only a free-standing number until we attach it to something, using it to describe a variable. But when we say, 'pH = 3.2', we suddenly learn something. By relating a variable (pH) to a number (here, '3.2') we, as chemists, suddenly know quite a lot.

The equals sign in an equation acts in just the same way as does the verb 'is.' In words, we could have said, 'The pH *is* 3.2,' and we would have said neither more nor less than when writing, 'pH = 3.2.' They are the same.

The equals sign is not an operator. It does not do something to the variables in the way an operator such as '+' or '×' does. It merely gives us information.

Introducing the concept of simplification: collecting like terms

By the end of this section, you will know:

- Collecting together similar terms is one of the easiest ways of simplifying an equation.
- By 'similar terms' we mean algebraic portions that have the same form or content.

We collect terms together in everyday life. It simplifies otherwise difficult calculations. For example, we ask the grocer for a 'dozen eggs', rather than requesting them one egg at a time. We say 'six buses' drove past, rather than listing each one.

Collecting terms in this way saves us time and ink. It simplifies our lives, and we can simplify our maths in just the same way.

■ **Worked Example 4.1**

After analysing the composition of a reaction, the product contains the following atoms: C, C, C, C, C, H, H, H, H, H, C, C, O, O, H, and H. Simplify the way we express the composition by collecting together like terms.

I

The product contains three different elements: carbon, hydrogen and oxygen. Adding up the 'C' terms, we see the product contains seven carbons, call them 7C. It also contains two oxygens, 2O, and six hydrogens, 6H. We have simplified the maths by writing the composition as $7C + 6H + 2O$. In fact, the product is benzoic acid, **(I)**. ■

The plural of **formula** is formulae (which we pronounce as, *for-mew-lee*). The plural is never *formulas*.

■ **Worked Example 4.2**

A chemist obtains a mass spectrum, which reveals that the molecule responsible for the flavour of blackcurrants **(II)** contains the following fragments: CH_3, CH_3, CH_3, SH, CO, CH, CH_2, CH_2, CH_2, CH and C.

What is its empirical formula?

II

Adding together the terms, we have 10C, 18H, 1O and 1S. We can write the formula as $C10 + H18 + 1O + 1S$, although most organic chemists prefer to write the formula of **(II)** in the form, $C_{10}H_{18}OS$. ■

Simplifying data in this way usually saves us a lot of time: calculating the molar mass of compound **(I)** would have taken much, much longer, and would have been more likely to have been wrong, if we had inserted numbers *before* simplifying.

Self-test 4.1

Collect together the like terms, and thereby simplify the expression:

4.1.1 $a + 2a + 3a + a$

4.1.2 $b + 4b + 6b + 22b$

4.1.3 $c + 2c + c + 4c + 4c + 4c + 7c$

4.1.4 $g + h + i + 5g + 6g + 2h + 6i$

4.1.5 $12g + 5g - 2g + 4g - 3g$

4.1.6 $6f - 5e + 6e - 7d + 4e + 3e + 8f$

4.1.7 $h + 9f - 3f - h + 5h - 6h + 4f$

4.1.8 $CH_3 + CH + SH + OH + CH_2 + CH_2$

Further simplifications: introducing the concept of balancing

By the end of this section, you will know:

- We say an equation is *balanced* when the collection of variables and numbers written to the left of the equals sign has the same value as the collection of variables and numbers written to its right, even if they look different.

- We can change the magnitude of either the variables or the numbers on the left-hand side of an equation provided we do *exactly* the same to the right-hand side.

- The equation is no longer balanced if we change only one side of the equation and retain the other. We say, *left-hand side ≠ right-hand side.*

Care: Weight and mass are not identical. A mass represents an absolute gauge of how heavy an item is, while the weight depends on the extent of the Earth's gravitational pull.

Equations often *look* difficult– usually more difficult than they really are. It is wise therefore to make the equation look less intimidating. This section is designed to help us when we want to simplify an equation. The simplest two ways of simplifying an equation are factorizing (see Chapter 7) and cancelling, which is the subject of this chapter.

Before we investigate any form of simplification, we need to learn a few 'ground rules.' First, we need to see how the equals sign '=' lies at the very heart of an equation. An equation is not a real equation without one. We can think of the equals sign as acting much like an old-fashioned balance like that in Fig. 4.1.

Just as the balance beam needs to be horizontal if the two pans contain the same mass, so the maths written on the left of the equals sign needs to balance the maths written to its right. If the two pans bear a different mass, the arms will tilt and we experience **inequality**. In a similar way, if we do something to one side of an equation, the two sides are not equal, meaning that we cannot write an equals sign between them.

Mathematically, we say,

$$mass_{\text{left-hand side}} = mass_{\text{right-hand side}}$$

if the two sides bear the same mass, and

The symbol ≠ means 'is *not* equal to.'

$$mass_{\text{left-hand side}} \neq mass_{\text{right-hand side}}$$

if they differ.

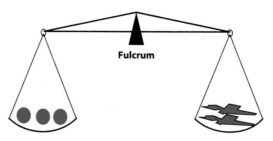

Figure 4.1 An old-fashioned balance comprises a beam with pans attached at equal distances from the central *fulcrum* (or *pivot*). The left- and right-hand pans remain balanced only if they bear the same mass. Provided we do the same to both pans, the balance remains equilibrated as demonstrated by the beam being horizontal.

The analogy of the arm balance is perfect. The two sides will always balance provided the two sides bear the same mass. The equality will continue to hold:

- Regardless of the actual magnitudes of the individual masses.
- Regardless of the identity of the materials on the balances.

We know the equality holds because the beam remains horizontal. In just the same way, an equals sign is a valid way of expressing the truth: 'the value of the left-hand side equals the value of the right-hand side.' This is true:

- Regardless of the actual magnitudes of the left- and right-hand sides.
- Regardless of the algebraic terms written on the equation's two sides.

The similarity extends. The balance beam remains horizontal, meaning the two pans bear the same mass, even if we change the contents of one pan, provided we change the contents of the other pan *by the same amount*. We can change the amounts on the two pans in one of two ways:

- We can add the same mass to both sides (we use the ADDITION operator)
- We can change the proportion on both sides (we use the MULTIPLICATION operator)

We can change the value of the right-hand side provided we perform the same operation to both sides. For example, we can ADD the same to both sides:

> ■ **Worked Example 4.3**
>
> Think of the old-fashioned balance in Fig 4.1. The balance beam is horizontal, therefore the contents of the two pans must have an equal mass. If we ADD a mass of 25 g to the left-hand side, what must we do to make it balance again?
>
> To make the balance beam horizontal, we must do the same to both sides. As 25 g was ADDED to the left-hand side, we must also ADD 25 g to the right-hand side. ■

We can also simplify an expression by SUBTRACTING from *both* sides.

> ■ **Worked Example 4.4**
>
> After a reaction, the mass recorded was 10.4 g. Incorporated in that mass is the mass of the weighing boat, of mass 0.3 g. What is the mass of product?
>
> We start with the equation,
>
> $$m_{total} = m_{product} + m_{weighing\ boat}$$
>
> where m represents mass. The equation will continue to balance provided we do the same to both sides. We want to isolate the term $m_{product}$. Unfortunately, $m_{product}$ does not appear on its own, but in conjunction with another mass, $m_{weighing\ boat}$. Since this additional term has been ADDED to the quantity we want, we perform the opposite operation, and SUBTRACT: we SUBTRACT the term $m_{weighing\ boat}$ from *both* sides:
>
> $$m_{total} - m_{weighing\ boat} = m_{product} + (m_{weighing\ boat} - m_{weighing\ boat})$$
>
> The equation is still balanced. The last two terms on the right-hand side (which, for clarity, are written within round brackets) are the same. Any mathematical term minus itself is zero, so the bracket is zero. We write:
>
> $$m_{total} - m_{weighing\ boat} = m_{product} + 0$$
>
> In fact, there is no need to write the last term. We see how the mass of the product is
>
> $$m_{total} - m_{weighing\ boat}$$
>
> So, inserting values, this yields,
>
> $$m_{product} = 10.4\ g - 0.3\ g = 10.1\ g.\ ■$$

Instead of saying 'opposite,' we usually say we perform the **inverse** operation.

■ Worked Example 4.5

We wish to scale up a reaction involving thionyl chloride (**III**) with water:

$$Cl-\underset{\underset{O}{\|}}{\overset{\overset{O}{\|}}{S}}-Cl \;+\; 2\,H_2O \longrightarrow HO-\underset{\underset{O}{\|}}{\overset{\overset{O}{\|}}{S}}-OH \;+\; 2\,HCl$$

III

Before scaling up, we reacted 2 moles of water with one mole of (**III**). We want to react 30 mol of water. How much (**III**), n_{III}, do we need?

We start by writing an equation to summarize the information in the question. We say:

$$2n_{III} = n_{H_2O}$$

where n is amount of material i involved in the reaction.

Halving both sides gives, $\quad \frac{1}{2} \times 2n_{III} = \frac{1}{2} n_{H_2O}$

when $n_{H_2O} = 30$ mol, n_{III} is half this amount, so $n_{III} = \frac{1}{2} \times 30 = 15$ mol. ■

It is usual practice to omit the multiplication signs, and write $n_{III} = \frac{1}{2} n_{H_2O}$

■ Worked Example 4.6

We wish to scale up a reaction. In our first attempt, we made 1 mol of nicotine (**IV**), which had a mass of 165 g. We now want to make 4.2 mol of (**IV**). How much nicotine will we make?

IV

We start by writing the equality,

$$1 \text{ mol of } (\textbf{IV}) = 165 \text{ g}$$

We can MULTIPLY either side of the equation by any **factor** we like, provided we MULTIPLY *both* sides by the same factor. In this case, we want 4.2 moles of (**IV**), so here we MULTIPLY both sides by 4.2. We write:

$$4.2 \times 1 \text{ mol of } (\textbf{IV}) = 4.2 \times 165 \text{ g}$$

So the mass of 4.2 mol of (**IV**) will be,

$$4.2 \times 165 \text{ g} = 693 \text{ g}. \blacksquare$$

A **factor** is either one of the components within a number or algebraic expression, or a term having a constant value by which we multiply a variable.

■ Worked Example 4.7

We need some of the artificial food flavouring, ethyl vanillin (3-ethoxy-hydroxy-benzaldehyde, **V**) to impart a strong vanilla flavour. We wish to scale down the reaction: we started by making 1 mol of (**V**) and made 166 g but now we want to decrease the amount thirty-fold. How much (**V**) do we make?

V

We start with the equation,

$$1 \, mol = 166 \, g.$$

We then DIVIDE both sides by thirty:

$$\frac{1 \, mol}{30} = \frac{166 \, g}{30}$$

so $\dfrac{1}{30}$ mol of (**V**) has a mass of 5.53 g. ∎

We see how treating the equals sign in an equation as a balance is a powerful way of scaling up or down, as achieved by MULTIPLYING or DIVIDING.

Self-test 4.2

Transform each of the following by ADDING or SUBTRACTING from *both* sides of the equation:

4.2.1   Rearrange $x + 9 = 12$, to obtain x on its own.

4.2.2   Rearrange $2x + 3x - 4x - 4v = 4v$, to obtain x on its own.

4.2.3   Rearrange $a = x - 12$ to make the term on the right, x.

4.2.4   Rearrange $a = x + 4c$ to make the term on the right, x.

4.2.5   Rearrange $4p = m_{one} - m_{two}$ to make the term on the right, m_{one}.

4.2.6   Rearrange $p = 4x - 3x + 5b$ to make the term on the right, x.

Simplification by cancelling

By the end of this section, you will know:

- Cancelling is a powerful way of simplifying equations.
- When we cancel, we look for like terms on the top and bottom of a fraction.
- If the like terms are multiples of each other, then they do not cancel completely, since a factor will remain.

The easiest way to simplify an equation is to **cancel**. We often find the first effect of rearranging an equation is to make it look more complicated.

Before we start, we will explore the statement, 'A number divided by itself is 1.' A couple of examples readily demonstrate the statement's validity:

- One mole of sulphur contains one mole of sulphur.
- Four divides into four only once (in other words, '1×4' has a value of 4).

These statements are obviously true if somewhat trivial. We would not usually state these truths in such words, though; rather, we generally employ the symbols learnt from the previous chapter. We can state such relationships in a mathematical way, saying that a thing (any thing) divided by itself is **unity**, so '$x \div x = 1$'.

Anything divided by itself equals '1'.

■ **Worked Example 4.8**

Fractions are described in Chapter 5.

Simplify the equation $y = \dfrac{6cd}{2}$.

We start by rewriting the equation by separating the numbers from the algebraic letters:

$$y = \frac{6}{2} \times cd$$

In words, the fraction on the left is 'two into six,' which clearly has a value of '3'.

Note how the '6' on the top could be written as '3×2.' We **cancel** the 2 on the bottom with the 2 s on the top (which leaves the 3 untouched). The bracket term '$2 \div 2$' has a value of 1. We then write:

$$y = \left(\frac{\cancel{2}}{\cancel{2}}\right) \times 3 \times cd$$

We sometimes call the diagonal lines **cancelling lines**. We normally omit the multiplication sign $\times$, and write just $3cd$.

Alternatively, if we can see intuitively how 2 straightforwardly goes into 6, we can cancel without first splitting the number '6':

The '3' on the top row is the result of cancelling (i.e. $6 \div 2 = 3$).

$$y = \frac{\overset{3}{\cancel{6}}}{\cancel{2}} \times cd$$

When we collect together the remaining terms, we are left with: $3cd$. ■

We need to note:

• Unless told otherwise, we write the letters in alphabetical order.

• By convention, we write the numbers *before* the letters: $3cd$ is acceptable, but neither of the permutations '$c3d$' or '$cd3$' are regarded as standard, and should never be used.

• The cancelling lines are optional. Use them only if they help.

■ **Worked Example 4.9**

A reaction proceeds in three steps:

$$A \xrightarrow{k_1} B \xrightarrow{k_2} C \xrightarrow{k_3} D$$

A is the initial reactant and D is the final product. B and C are reactive intermediates. Each of the three processes proceeds at a characteristic speed, so each is defined by is respective rate constant, k_i. Having performed a kinetics analysis, the equation describing the overall rate of reaction is:

$$\text{rate} = \frac{3k_1 \, k_2^2 \, k_3}{6k_1 k_2}$$

Simplify this equation by cancelling.

We need to note:

- Each k_i term is a rate constant, where i here means 'generic'.

- Sometimes, we need to distinguish between similar-looking terms. One of the simplest ways is to employ subscripts, which we write to the right of the symbol.

- The subscripted numbers are only included to help us distinguish between the three different rate processes.

- k_2 means k of the second process. If it meant 'two times k', we would have written the number *before* the k and would have written the two on the *same* level as the k, rather than subscripted.

- The term k_2^2 means that the rate constant k_2 has been squared, i.e. $k_2^2 = k_2 \times k_2$

Before we start, we will rewrite the expression. We first include *all* the multiplication signs, then collect the numbers together. To ease the analysis, we group the like terms, one above another:

$$\text{rate} = \frac{3}{6} \times \frac{\cancel{k_1} \times k_2 \times k_2 \times k_3}{\cancel{k_1} \times k_2}$$

We next cancel the '$k_1 \div k_1$' term, writing in its place '1'.

$$\text{rate} = \frac{3}{6} \times 1 \times \frac{\cancel{k_2} \times k_2 \times k_3}{\cancel{k_2}}$$

We next cancel the '$k_2 \div k_2$' term, again writing in its place '1'. This leaves:

$$\text{rate} = \frac{3}{6} \times 1 \times 1 \times k_2 \times k_3$$

The 1×1 in the middle is a bit silly, so we'll ignore it from now on. Finally, we want to simplify the fraction of '3 ÷ 6'. To this end, we note how '6' on the bottom is the same as '3×2', which allows us to rewrite the fraction, using a different notation. We say:

$$\text{rate} = \left(\frac{\cancel{3}}{\cancel{3} \times 2} \right) \times k_2 \times k_3$$

The 3 on top and bottom will cancel, leaving a factor of ½. There is now no need to retain the brackets, so we obtain:

$$\text{rate} = \frac{1}{2} \times k_2 \times k_3$$

Accordingly, our expression collapses down to a manageable size. Omitting both multiplication signs, we write:

$$\text{rate} = \frac{k_2 k_3}{2} \ \blacksquare$$

Self-test 4.3

By use of cancelling, simplify the following expressions:

4.3.1 $\quad y = \dfrac{18}{6} c$

4.3.2 $\quad y = \dfrac{2d}{4}$

4.3.3 $\quad y = \dfrac{d}{4d}$

4.3.4 $\quad y = \dfrac{bcd}{cde}$

4.3.5 $\quad y = \dfrac{4}{3d} \times \dfrac{6}{20}$

4.3.6 $\quad y = \dfrac{6a^2 b}{6a}$

Introducing rearrangements: reversing the MULTIPLICATION and DIVISION operators

By the end of this section, you will know:

- As well as MULTIPLYING both sides of an equation, we can also DIVIDE.
- Before rearranging, we look for the operator. We then perform the reverse operation.
- The description 'cross-multiplying' recognizes the way that terms swap from the top to the bottom when MULTIPLICATION or DIVISION occurs.

In the previous sections, we have been *rearranging* equations, by terms (numbers or letters) from one part of an equation to another. Rearrangement is one of the most useful, and powerful, of all the mathematical techniques we need to learn. In this section, we consider rearrangement more formally.

Rearranging equations is easy, provided we follow these two simple rules:

- Whatever we do to one side of an equation, we must do exactly the same to the other.

- To reverse an operation, we identify the operator and perform the inverse operation, again to both sides of the equation.

The list of operations is simply vast. The key is to identify the operation, and do the opposite. Table 4.1 lists a few simple inverses.

So, for example,

- If the operation is to MULTIPLY by 4 the inverse is to DIVIDE, again by 4.
- If the operation is to DIVIDE by a^2, the inverse is to MULTIPLY by a^2.
- If the operation is SQUARE ROOT $\sqrt{}$ of y, the inverse is to SQUARE it: $\left(\sqrt{y}\right)^2$.
- If the operation raises $5d$ to its fifth POWER, the inverse takes the fifth ROOT of $5d$.

Table 4.1 Some functions and their inverses.

Function	Inverse
ADD a	SUBTRACT a
SUBTRACT a	ADD a
MULTIPLY by a	DIVIDE by a
DIVIDE by a	MULTIPLY by a
SQUARE	SQUARE ROOT or POWER ½
SQUARE ROOT or POWER ½	SQUARE
nth POWER	nth ROOT
nth ROOT	nth POWER

See Chapter 11 for an explanation of why a power of ½ is the same as a square root.

■ Worked Example 4.10

The ideal-gas equation

$$pV = nRT$$

relates the pressure and temperature T of n moles of gas housed in a volume V. We call the term R the gas constant. By use of cancelling, obtain an equation in which p is the **subject**.

Before we start, we need to define the new term above. We write the **subject** of an equation as, 'A solitary algebraic term = ...' Its **object** is then the remainder of the equation, positioned to the right of the = sign. This method of labelling an equation might remind us of English grammar, and emphasizes how we can think of an equals sign as behaving a bit like the *verb*, 'to be.'

We want the symbol for pressure written on its own, in the form '$p = ...$' To this end, we note how the symbol pressure p is presently MULTIPLIED by the symbol for volume V. We need to remove the V term to ensure p appears on its own, so we cancel the V term.

The operator doing the opposite function to MULTIPLY is divide. To remove the V term multiplied to the p term, we must divide the **compound variable** pV by V.

Before we proceed further, we need to retain the equals sign '='. We must therefore DIVIDE *both* sides of the ideal-gas equation by V. If we DIVIDED only the left-hand side by V, the two sides would no longer have the same value, so we DIVIDE both sides by V:

$$\frac{pV}{V} = \frac{nRT}{V}$$

Look at the left-hand side. It contains two V terms: one each on the top and the bottom. We can therefore rewrite the equation slightly:

$$p \times \left(\frac{\cancel{V}}{\cancel{V}} \right) = \frac{nRT}{V}$$

This version is identical to the one above except we have collected together the two V terms. Since 'V DIVIDED by V' is a number DIVIDED by itself, the bracket has a value of '1'. We therefore rewrite the equation as

$$p \times 1 = \frac{nRT}{V}$$

This MULTIPLYING by '1' is silly. Clearly, $1 \times p$ is the same as p on its own, so we write the left-hand side as 'p' on its own. Therefore, when p is the subject, we obtain:

$$p = \frac{nRT}{V} \quad \blacksquare$$

> ■ **Worked Example 4.11**
>
> A chemist has a stock solution of sodium thiosulphate solution (0.2 mol dm^{-3}). How many moles of thiosulphate does 25 cm^3 contain?

We start by writing the definition: one litre (1 dm^3) of a solution having a concentration 0.2 mol dm^{-3} contains 0.2 mol. We then turn these words into a form of equation:

$$1 \text{ dm}^3 \times 0.2 \text{ mol dm}^{-3} = 0.2 \text{ mol}$$

One litre contains 1000 cm^3, so to determine how much thiosulfate is contained in *one* cm^3, we divide by 1000:

$$\frac{1 \text{ dm}^3 \times 0.2 \text{ mol dm}^{-3}}{1000} = \frac{0.2 \text{ mol}}{1000}$$

But we don't want 1 cm^3; we want 25, so we multiply both sides by 25:

$$25 \times \frac{1 \text{ dm}^3 \times 0.2 \text{ mol dm}^{-3}}{1000} = 25 \times \frac{0.2 \text{ mol}}{1000}$$

so 25 cm^3 of solution contains $(25 \times 0.2)/1000$ mol = 0.005 mol, or 5 mmol. ■

Two variables multiplied together are called a **compound variable**.

We retain the units to prove the equation is valid.

Cross-multiplying Notice how, in this example, we started with V on the top left and end with V positioned on the bottom right. This is a general result: when we start with a fraction, and rearrange by MULTIPLYING or DIVIDING in this way, the variable we move changes from one side of the equation to the other, and also moves from top to bottom or *vice versa*. Look what happens when we MULTIPLY both sides by $\boldsymbol{b}$:

$$\frac{a}{\boldsymbol{b}} \nearrow \frac{c}{d} \;\rightarrow\; \frac{a}{\boldsymbol{1}} = \frac{b \times c}{d} \tag{4.1}$$

The direction of the arrow helps explain why we use the phrase, **cross-multiply**.

Similarly, when we DIVIDE, the variable moves from one side of the equation to the other. Look what happens when we divide both sides by $\boldsymbol{c}$:

We need the '1' on the top of the right-hand fraction here, because a fraction must have *something* its numerator. The '1' here comes from rewriting the top line as '1 × c' before we divided by c.

$$\frac{a}{b} \swarrow \frac{\boldsymbol{c}}{d} \;\rightarrow\; \frac{a}{b \times \boldsymbol{c}} = \frac{1}{d} \tag{4.2}$$

■ Worked Example 4.12

n moles of solute are dissolved in a volume V of solvent to make a solution of concentration c. n, V and c are related by the equation:

$$c = \frac{n}{V}$$

By use of rearranging, obtain an equation in which n is the subject.

We first analyse the equation, and see the symbol for the amount of substance n has been DIVIDED by the symbol for volume, V. To isolate the n term, we must get rid of the V term. We reverse this operation, and MULTIPLY *both* sides of the equation by V:

$$c \times V = \frac{n \times V}{V}$$

As before, we see two V terms in the equation, both on the top and bottom of the right-hand side. We group them together:

$$c \times V = n \times \left(\frac{\cancel{V}}{\cancel{V}} \right)$$

The two V terms cancel, because anything DIVIDED by itself equals 1, so $V \div V = 1$:

$$c \times V = n \times 1$$

We again follow the convention that tells us to write the subject on the *left* of the algebraic expression.

Omitting the multiplication sign yields, $n = cV$. ■

We need to note how in both these examples:

- We looked at the equation before starting.

- By inspection, we saw which mathematical operations were involved i.e. the process by which the variable of interest transmutes into the answer. In each case, it had either been MULTIPLIED or DIVIDED by something else.

- We then applied the *inverse* operator, in order to remove the other variables. In this way, we make the variable of interest the subject of the equation.

- Sometimes we cancel in order to obtain a number, rather than to rearrange an equation.

Self-test 4.4

Rearrange the following equations, in each case to make x the subject:

4.4.1 $y = mx$ **4.4.2** $24x = 5y$

4.4.3 $y^2x = 3z$ **4.4.4** $abx = 12$

4.4.5 $mgh = 55x$ **4.4.6** $mgx = 55h$

Collect together terms, then rearrange, again to make x the subject:

4.4.7 $5x + 6x = g$ **4.4.8** $6x + 4x + x - 3x = 34$

4.4.9 $5f + 3f - 2f = 4x + 9x$ **4.4.10** $p + q + 4q - 7q = 6x$

When we cross-multiply, *everything* on the other side of the equation is affected—it is a very common error to assume it involves only the first term.

■ Worked Example 4.13

The value of the Gibbs function $\Delta G^{\ominus}$ varies quite strongly with temperature T, according to the **Gibbs–Helmholtz equation**:

$$\frac{\Delta G^{\ominus}}{T} = \frac{\Delta H^{\ominus}}{T} + c$$

where $\Delta H^{\ominus}$ is the change in enthalpy of the same reaction and c is a constant. Write an expression in which $\Delta G^{\ominus}$ is the only term on the left-hand side.

> By writing only $\Delta G^{\ominus}$ on the left-hand side of an equation, we make it the **subject** of the equation.

We want to make $\Delta G^{\ominus}$ the subject of the equation. Before attempting an answer, we analyse the equation to see how the value of ΔG on the left-hand side has been changed. In this example, $\Delta G^{\ominus}$ has been DIVIDED by the temperature T. We therefore rearrange the equation in such a way that the T term disappears from the left-hand side. The opposite function to DIVIDE is MULTIPLY, so we MULTIPLY both sides of the equation by T:

$$T \times \frac{\Delta G^{\ominus}}{T} = T\left(\frac{\Delta H^{\ominus}}{T} + c\right)$$

We need the brackets around the right-hand term '$\Delta H^{\ominus} \div T + c$' because we have MULTIPLIED *everything* on that side by T. In other words, we must treat $\Delta H^{\ominus} \div T + c$ as an inseparable whole. It is a **compound variable**.

We now return to the left-hand side, we can rewrite the equation slightly:

$$\left(\frac{\cancel{T}}{\cancel{T}}\right) \times \Delta G^{\ominus} = T\left(\frac{\Delta H^{\ominus}}{T} + c\right)$$

> We could multiply out the right-hand side, and write '$\Delta H^{\ominus} + Tc$'. The expression here is just as valid, although the version with brackets looks a little neater.

The term in brackets on the left-hand side equate to '1' because any number DIVIDED by itself is unity. The resultant expression for $\Delta G^{\ominus}$ therefore becomes:

$$\Delta G^{\ominus} = T\left(\frac{\Delta H^{\ominus}}{T} + c\right)$$

■ Worked Example 4.14

The pressure inside an autoclave is 1 300 000 Pa, which is 13 times atmospheric pressure, $p^{\ominus}$. What is atmospheric pressure?

We first write the data in the form of an equation. We write, 1 300 000 Pa = 13 $p^{\ominus}$.

Because we want $p^{\ominus}$, but the equation so far contains 13 MULTIPLIED by $p^{\ominus}$, we perform the opposite operation and DIVIDE both sides by 13. We write:

$$\frac{1\,300\,000\ \text{Pa}}{13} = \frac{13 \times p^{\ominus}}{13}$$

On the right-hand side, the two factors of '13' cancel, leaving us with,

$$\frac{1\,300\,000\ \text{Pa}}{13} = p^{\ominus}$$

so $\qquad p^{\ominus} = 100\,000\ \text{Pa} = 10^5\ \text{Pa}$ ■

Rearrangements involving simple powers and roots

By the end of this section, you will know:

- We reverse a POWER by taking the appropriate ROOT.
- We reverse a ROOT by taking the appropriate POWER.

Aside:

To perform this cube ROOT on pocket calculator:

1. Type in '1.728 × 10⁻³⁰'
2. Press the $x^{1/y}$ button.*
3. Type in '3'
4. Press '='
5. Most cube ROOTs are irrational numbers, so remember to cite the answer to a sensible number of significant figures.

* Some calculators, such as the popular *Casio fx-85ES* do not have this button. If your calculator is similar, look out instead for buttons saying , or $x^{\square}$.

When we have an operator such as a power (which is treated within BODMAS as a function 'OF'), we first identify the nature of the power, then perform the reverse operation. For example, we reverse a simple SQUARE by take the SQUARE ROOT, reverse a CUBE by taking the CUBE ROOT, etc.

■ Worked Example 4.15

Sodium chloride crystallizes in a simple cubic structure. Each side of the cube—we call it a **unit cell**—has a length l. If the unit cell's volume is 1.728×10^{-30} m³, calculate the length of each side.

In this example, we first rewrite the data in a form of an equation. Because the volume is that of a cube, $l^3 = 1.728 \times 10^{-30}$. To reverse this operation, we identify the operator and perform the reverse operation: the operator is 'cube of', so we take the cube root. Clearly, we must take the cube root of *both* sides:

$$l = \sqrt[3]{1.728 \times 10^{-30}\ \text{m}^3}$$

where the small '3' to the left of the root sign means 'cube root.'
We find, $\qquad l = 1.2 \times 10^{-10}$ m ■

This strategy works because a cube ROOT is the same as having a power of ⅓. For a detailed argument to support this idea, see Chapter 11.

Self-test 4.5

Rearrange each of the following, in each case making x the subject:

4.5.1 $\quad y = x^2$ **4.5.2** $\quad 4y = x^3$

4.5.3 $\quad by = x^4$ **4.5.4** $\quad y = \sqrt[3]{x}$

4.5.5 $\quad ay = \sqrt{x}$ **4.5.6** $\quad 4y = \sqrt[4]{x}$

Summary of key equations in the text

Cross-multiplying:

$$\frac{a}{b} \diagup \frac{c}{d} \rightarrow \frac{a}{1} = \frac{b \times c}{d} \tag{4.1}$$

$$\frac{a}{b} \diagdown \frac{c}{d} \rightarrow \frac{a}{b \times c} = \frac{1}{d} \tag{4.2}$$

Additional problems

The answers to some of these questions will be obvious using intuition alone. Nevertheless, determine the answer using the methodology outlined in this chapter.

4.1 The mass of 4 moles of aspirin is 1424 g. Calculate the mass of one mole.

4.2 The volume of one mole of ethanol is 45 cm³. Calculate the volume occupied by 6.2 mol of ethanol.

4.3 The ideal-gas equation

$$pV = nRT$$

relates the volume V of an ideal gas to the temperature T, the amount of gas n and the pressure p. R is the gas constant. Rearrange this equation to make T the subject.

4.4 The Beer–Lambert law

$$Abs = \varepsilon c\, l$$

relates the optical absorbance Abs of a coloured sample to its concentration c, the thickness of the sample l, and the molar absorptivity ε. Rearrange the equation to make c the subject.

4.5 The equation relating temperatures in Kelvin T and Celsius T' is

$$T = T' + 273.15$$

Rearrange this expression to make T' the subject.

4.6 A chemist weighs some compound onto a balance, and records an overall mass of 12.443 g. If the weighting boat has a mass of 0.250 g, what is the mass of sample?

4.7 Consider the simple second-order esterification reaction: acid + alcohol → ester + water. The rate equation for reaction is

$$\text{rate} = k_2\, [\text{acid}]\, [\text{alcohol}]$$

Rearrange this rate equation to make the second-order rate constant k_2 the subject.

4.8 The area A of a SQUARE electrode is 7.2 cm². Calculate the length of each side of the square l (remember $A = l^2$).

4.9 The equation relating the changes in Gibbs function $\Delta G^{\ominus}$, enthalpy $\Delta H^{\ominus}$ and entropy $\Delta S^{\ominus}$ is:

$$\Delta G^{\ominus} = \Delta H^{\ominus} - T\Delta S^{\ominus}$$

where T is the temperature. Rearrange the equation to make $\Delta H^{\ominus}$ the subject.

4.10 The Clausius equality relates minuscule changes in energy dq with temperature T and the change in entropy ΔS:

$$\Delta S = \frac{\mathrm{d}q}{T}$$

Rearrange this expression to make dq the subject.

5

Algebra IV
Fractions and percentages

Concepts and nomenclature

By the end of this section, you will know:

- A fraction involves one quantity divided by another.
- The numerator of the fraction is above the line and the denominator is below it.

The word fraction comes from the Latin *fractus*, which means 'broken' or 'a discontinuity.' We see this root when we say that certain bone breakages are 'fractures'.

The *Concise encyclopedia of mathematics* defines a fraction as an example of a specific type of ratio, in which the two numbers are related in a part-to-whole relationship, rather than as a comparative relation between two separate quantities. This complicated sentence means that a fraction, like a broken bone, has two parts, one either side of a discontinuity.

Fractions are written with a straight line separating the two component parts, like the crack separating the two halves of a broken bone. If the line is not horizontal, it is called a **solidus** (or, more rarely in mathematics, a **virgule**). While the solidus in fractions like $\frac{2}{5}$ is identical to a forward slash so treasured by Internet users, we will not call it a 'forward slash.' Rather, in this text, we will call the line separating the two halves of the fraction a solidus whether it's horizontal or not, because its angle does not affect the meaning of the fraction: $1/2 = \frac{1}{2} = \frac{1}{2}$. We *never* write a fraction with a *back* slash, as $\frac{2}{1}$.

Occasionally, writing a fraction as '1/2' with a sloping solidus can lead to ambiguity: in some fonts and with some handwriting styles, $1/2a$ can mean either $1/(2a)$ or $\frac{1}{2} \times a$. For this reason, we generally use an upright fraction like $\frac{1}{2}$.

We call the part of the fraction above the solidus line the **numerator**, and the part below the **denominator**. If the solidus is printed at an angle, the numerator is the part written to the left of the solidus, and the denominator is to the right.

We generally obtain a fraction by dividing, for example using a pocket calculator.

> A simple way to remember which part of the fraction is the numerator and which the denominator, we say:
> nUmerator = Up
> Denominator = Down

■ Worked Example 5.1

A chemist takes 200 ml of solution from a litre. Express the proportion removed as a fraction.

Strategy:

(i) We make sure all the data are expressed in the same way.

(ii) We next decide which number is the numerator to place on top and which the denominator to write beneath the solidus. Generally, the denominator represents the total number possible, and the numerator represents the proportion under scrutiny.

Solution:

(i) In this example, it is convenient to note how the removed portion is expressed in terms of ml, so we need to express the original volume (a 'litre') in the same units. A litre = 1000 ml.

(ii) In this case, 200 ml is the numerator and the denominator is 1000 ml. Inserting values yields:

$$\frac{200 \text{ ml}}{1000 \text{ ml}}$$

The units of ml cancel (see Chapter 4). Further cancelling simplifies the numerical answer: $\frac{200}{1000}$ is the same as $\frac{1}{5}$. In other words, for every five ml of the original solution, one is removed. Accordingly, the fraction is one fifth. ■

The simplification of fractions

By the end of this section, you will know that fractions can be simplified by:

- Cancelling portions of the numerator and denominator.
- Converting into a decimal.

Cancelling a fraction:

Many fractions can be simplified. The most common way of simplifying a fraction is to **cancel**, as explored in Chapter 4. For example, it should be obvious that the fraction $\frac{2}{4}$ ('two quarters') is the same as a half. It is better to write the fraction as $\frac{1}{2}$, which is obtained by cancelling the factor of '2' on top and bottom of the fraction.

Sometimes it is obvious that cancelling is possible, particularly if the numbers involved are quite small. For larger numbers, however, the following four rules may ease the cancelling process:

- If the numbers in the numerator and denominator both end with an even number, then both top and bottom are divisible by 2.
- If the numbers in the numerator and denominator both end with 5, then both top and bottom are divisible by 5.
- If the numbers in the numerator and denominator both end with 0, then both top and bottom are divisible by 10.
- If one of the numbers in the numerator and denominator ends with 5 and the other with 0, then both top and bottom are divisible by 5.

Trial and error is advisable if these four rules do not apply, and intuition and common sense are not available. Whatever the method employed, we must always remember the golden rule: do not cancel if in doubt. It is better to have a correct fraction that looks complicated rather than an incorrect but more visually appealing fraction.

Self-test 5.1

Simplify the following fractions by cancelling:

5.1.1 $\dfrac{3}{9}$ 5.1.2 $\dfrac{3}{81}$

5.1.3 $\dfrac{25}{200}$ 5.1.4 $\dfrac{252}{48}$

5.1.5 $\dfrac{12abc}{3acd}$ 5.1.6 $\dfrac{6a}{3a^2}$

Converting a fraction into a decimal:

The second way to simplify a fraction is to convert it into a decimal. We can only simplify the numerical parts of a fraction. We cannot, for example, simplify the fraction $\frac{a}{b}$ before we have substituted straightforward numbers for both a and for b.

Before we convert a fraction into a decimal, it is sensible to look carefully, deciding whether the numerator is larger or smaller than the denominator. The decimal will be less than 1 if the numerator is *smaller* than the denominator. In the jargon, such a fraction is called a **proper fraction**. Examples include $\frac{3}{4}$, $\frac{1}{2}$ or $\frac{5}{102}$, for which the decimals are 0.75, 0.5 and 0.0490 (to 3 s.f.), respectively.

Conversely, if the numerator is *greater* than the denominator, the value of the resultant decimal must be greater than 1. Such fractions are called **improper**. Examples include $\frac{3}{2}$, $\frac{8}{5}$ and $\frac{123}{50}$. In fact, many people think that improper fractions should be rewritten with whole numbers followed by a proper fraction. For example, in this mindset we express $\frac{3}{2}$ as $1\frac{1}{2}$; $\frac{8}{5}$ should be $1\frac{3}{5}$, and $\frac{123}{50}$ is best rewritten as $2\frac{23}{50}$. Regardless of how we write these improper fractions, their respective decimals are clearly all greater than 1: 1.5, 1.6, and 2.46.

Most students convert a fraction into a decimal using a pocket calculator (many then forget to check the answer …). It is always wise to estimate the answer, so the fraction $\frac{3}{7}$ will lie in the range 0 to 0.5, because the numerator is less than half the denominator.

We can do better than estimating, though, by checking a calculation by hand, using long division.

■ Worked Example 5.2

A chemist takes 11 g of water (which has a molar mass of 18 g mol⁻¹). As a decimal, what amount of water has been weighed out?

The fraction is $\dfrac{11}{18}$. We perform the division problem using a special *bracket*, as:

$$\text{fraction} = 18\overline{)11}$$

The numerator 10 is less than the denominator 18, so 18 cannot go directly into 11. In other words, the decimal is less than 1. We amend the bracket, top and bottom:

$$\begin{array}{r} 0. \\ 18\overline{)11.0} \end{array}$$

18 *does* go into 110, six times in fact: $6 \times 18 = 108$. Because $110 - 108 = 2$, we have a 'remainder' of 2. Again, we rewrite the bracket:

$$\begin{array}{r} 0.6... \\ 18\overline{)11.0} \\ \underline{108} \\ 02 \end{array}$$

Yet again, we ask the question: concerning the remainder of 2, how many times does 18 go into 2? Clearly, not at all. We need to amend the bracket once more:

$$\begin{array}{r} 0.61 \\ 18\overline{)11.00} \\ \underline{108} \\ 020 \\ \underline{18} \\ 2 \end{array}$$

18 goes into 20 once, so another '1' appears on top of the bracket. The remainder is yet again 2, and so on. In this way, we arrive at the decimal, $\dfrac{11}{18} = 0.6\dot{1}$. ■

We need to learn a third term while thinking about converting fractions into decimals. **Vulgar fractions** cannot be expressed *exactly* in decimal notation. To a first approximation, $\frac{1}{7} = 0.142\,857\,142$, but the value is not exact until expressed to an infinite number of terms. The fraction $\frac{1}{7}$ is therefore vulgar because its decimal is never exact. The fraction $\frac{1}{3}$ is also vulgar: although its decimal value can be written as a recurring decimal ($\frac{1}{3} = 0.\dot{3}$), its decimal value is never exact. By contrast, the fraction $\frac{2}{5}$ is not vulgar because it equates *exactly* with the decimal 0.4.

Care: This argument is only truly correct if we say the *modulus* of the decimal is smaller or greater than 1. The modulus of a number or algebraic term has magnitude only but not a sign. The modulus is always expressed as a positive number.

Self-test 5.2

Convert the following fractions into decimals. Comment whether the fraction is proper, improper or vulgar.

5.2.1 $\dfrac{3}{9}$

5.2.2 $\dfrac{6}{24}$

5.2.3 $\dfrac{5}{21}$

5.2.4 $\dfrac{6}{13}$

5.2.5 $\dfrac{125}{24}$

5.2.6 $\dfrac{16}{5}$

Aside

The simplest way to convert a fraction into a decimal is to use a pocket calculator.

- On straightforward calculators, use the ÷ key: type in the fraction $\frac{3}{8}$ as 3 ÷ 8.
- On more esoteric machines, it may be necessary to use the solidus key, which may look like a back-to-front '**L**'.
- In a computer spreadsheet such as *Excel*™, we use the same key as the forward slash '/' when typing an Internet URL. For example, we would type in the fraction 3/8 exactly as it is written.

The multiplication of fractions

By the end of this section, you will know:

- When multiplying fractions, the new numerator is the product of the numerators in the constituent fractions.
- When multiplying fractions, the new denominator is the product of the denominators in the constituent fractions.

It is easy to multiply simple fractions together. We can often say (in common parlance) 'of' instead of 'multiplied.' For example, it is easy to deduce the answer for $\frac{1}{2} \times \frac{1}{2}$ by saying 'a half of a half.' Clearly, the answer is a quarter.

For more complicated examples, we first place the two fractions side by side, then multiply the two numerators together and then multiply the two denominators:

$$\frac{a}{b} \times \frac{x}{y} = \frac{a \times x}{b \times y} \tag{5.1}$$

For the simple example of $\frac{1}{2} \times \frac{1}{2}$, we obtain the same answer using eqn (5.1) as using simple logic.

In the same way, if we have more than two fractions, we multiply together *all* the numerators and multiply together *all* the denominators. Using the notation of a Π-product from chapter 2, we say:

$$\frac{a}{b} \times \frac{c}{d} \times \cdots \times \frac{y}{z} = \frac{\prod_{a}^{y} \text{numerators}}{\prod_{b}^{z} \text{denominators}} \tag{5.2}$$

■ Worked Example 5.3

Barium carbonate is a partially soluble solid. A sample of $BaCO_3$ is immersed in a beaker of water, and left to reach equilibrium:

$$BaCO_3(s) \rightarrow Ba^{2+}(aq) + CO_3^{2-}(aq)$$

The equilibrium constant of dissolution is often termed a solubility product, K_{sp} and has the form $[Ba^{2+}]\,[CO_3^{2-}]$. (Being a solid, the $BaCO_3$ does not appear in the expression for K_{sp}.)

At the end of the reaction, the concentration of Ba^{2+} is $\frac{5}{36}$ mol dm^{-3} and the concentration of CO_3^{2-} is $\frac{2}{70}$ mol dm^{-3}. What is the value of the equilibrium constant, K_{sp}?

Following eqn (5.1), $K_{sp} = \dfrac{5}{36} \times \dfrac{2}{70} = \dfrac{10}{2520}$

> Notice how a fraction can have units. Fractions are not always dimensionless.

This answer can be simplified by cancelling, to $\frac{1}{252}$. It can also be expressed as a decimal as 3.97×10^{-3} (mol dm^{-3})^2. ■

■ Worked Example 5.4

A chemist prepares a solution containing $\frac{1}{40}$ mole of benzoic acid (**I**) in one litre of water. How many moles of (**I**) do we remove by taking an aliquot of volume 200 ml?

I

Before we start, we recognize from Worked Example 5.1 that 200 ml is one fifth of a litre.

> We discuss more fully the way units cancel in Chapter 30.

The number of moles $= \dfrac{1}{40} \times \dfrac{1}{5} = \dfrac{1 \times 1}{40 \times 5} = \dfrac{1}{200}$.

There is one two hundredth of a mole of (**I**). Notice how this problem is dimensionally correct: the units of the first fraction is $\frac{1}{40}$ mol litre^{-1}, and the units of the second is litre. The units of 'litre' and 'litre^{-1}' cancel, leaving 'mol.' ■

Self-test 5.3

Determine the value of the following fraction problems. It usually helps to cancel terms before starting.

5.3.1 $\quad \dfrac{1}{2} \times \dfrac{1}{3}$
5.3.2 $\quad \dfrac{3}{22} \times \dfrac{14}{3}$

5.3.3 $\quad \dfrac{1}{2} \times \dfrac{1}{3} \times \dfrac{1}{4} \times \dfrac{1}{5}$
5.3.4 $\quad \dfrac{1}{2} \times \dfrac{11}{10} \times \dfrac{2}{7} \times \dfrac{3}{31}$

5.3.5 $\quad \dfrac{1}{a} \times \dfrac{1}{3b} \times \dfrac{1}{2c}$
5.3.6 $\quad \dfrac{1}{b} \times \dfrac{c}{12a} \times \dfrac{2d}{a}$

The addition of fractions

By the end of this section, you will know:

- To add fractions, we use a common denominator.
- The numerator of the fraction contains a plus sign.

Simple sums can be done mentally: a half plus another half equals one. But we very soon find that adding together fractions using mental arithmetic can be quite difficult without a conceptual framework.

■ **Worked Example 5.5**

A chemist mixes two solutions of phenol (**II**), which was often called 'carbolic acid.' The first contains one third a mole of (**II**), and the second contains two fifths of a mole. How many moles of (**II**) does the combined mixture contain?

II

Strategy:

(i) We first write the two fractions side by side: $\dfrac{1}{3} + \dfrac{2}{5}$

(ii) We write a new horizontal solidus. It needs to be long. Because it's an addition problem, there is an addition sign in the centre of the *numerator*. In the *denominator* of the new fraction, we multiply together the denominators of the constituent fractions: $\dfrac{1}{3} + \dfrac{2}{5} = \dfrac{\cdots + \cdots}{3 \times 5} = \dfrac{\cdots + \cdots}{15}$

We call this procedure, finding a **common denominator**.

(iii) In this example, the first fraction is *one* third, so the first part of the new numerator is '1'. We multiply this '1' by '5', because 3 goes into 15 (in the denominator) five times: $\dfrac{1}{3} + \dfrac{2}{5} = \dfrac{(1 \times 5) + \cdots}{15}$.

(iv) The second fraction has a numerator of 2. For this reason, the second term in the numerator starts with 2. We multiply this two by '3', because 5 goes into 15 three times: $\dfrac{1}{3} + \dfrac{2}{5} = \dfrac{5 + (2 \times 3)}{15}$

The sum becomes $\dfrac{1}{3} + \dfrac{2}{5} = \dfrac{5 + 6}{15} = \dfrac{11}{15}$. There are $\dfrac{11}{15}$ moles of phenol (**II**). ■

> **Aside**
> We often call multiplication of this sort **cross-multiplying**. The following diagram illustrates the reason why:
>
> $\dfrac{1}{3} \diagup\!\!\!\!\!\diagdown \dfrac{2}{5} = \dfrac{(1 \times 5) + (2 \times 3)}{15}$

The addition of simple fractions is entirely commutative, so it does not matter which fraction is written first.

In general algebraic terms, we say,

$$\frac{a}{b} + \frac{c}{d} = \frac{ad + bc}{bd} \qquad (5.3)$$

We will need the rule of BODMAS from Chapter 3 if we are to obtain the correct answer.

■ **Worked Example 5.6**

A chemist prepares two batches of the weed-killer paraquat (1,1′-dimethyl-4,4′-bipyridilium dichloride) (**III**). The first batch yields $\frac{1}{4}$ mole of (**III**) and the second batch yields $\frac{2}{7}$ mole. How much paraquat is produced in total?

III

We insert numbers into the template in eqn (5.3): $\frac{1}{4}+\frac{2}{7}=\frac{(1\times7)+(2\times4)}{4\times7}=\frac{7+8}{28}=\frac{15}{28}$.

This result is then made complete by adding the unit: the yield is $\frac{15}{28}$ mol. Most chemists would then present the answer as a decimal: yield = 0.536 mol. ■

Self-test 5.4

Determine the value of the following fraction problems.

5.4.1 $\frac{1}{2}+\frac{1}{2}$ 5.4.2 $\frac{1}{22}+\frac{4}{3}$

5.4.3 $\frac{2}{5}+\frac{1}{15}$ 5.4.4 $\frac{1}{a}+\frac{1}{b}$

5.4.5 $\frac{1}{2}+\frac{1}{2}+\frac{1}{2}$ 5.4.6 $\frac{3}{22}+\frac{4}{3}+\frac{1}{5}$

The subtraction of fractions

By the end of this section, you will know:

- To subtract fractions, we use a common denominator.
- The numerator of the fraction contains a minus sign.

Conceptually, subtracting fractions is almost identical to adding them. The principal difference is the way the operation is not commutative: unlike addition, it *does* matter which fraction we write first.

To subtract two fractions, we require an equation template similar to eqn (5.3). We write:

$$\frac{a}{b}-\frac{c}{d}=\frac{ad-bc}{bd} \tag{5.4}$$

so we have replaced the plus sign in the numerator with a minus sign.

We call the ~ accent above the Greek v a **tilde**. We use it to follow the spectroscopist's convention that the quantity is expressed in wave numbers, (cm^{-1}).

■ **Worked Example 5.7**

The Balmer series describes an electron's energy transitions in the hydrogen atom:

$$\tilde{v} = \frac{1}{\lambda} = \Re_H \left(\frac{1}{2^2} - \frac{1}{n^2} \right)$$

where $\tilde{v}$ is just a single variable for the energy. $\Re_H$ is the Rydberg constant (the H subscript on $\Re_H$ has no mathematical meaning, and merely tells us we are looking at the hydrogen atom). n can have any value from 1 to ∞.

What is the value of the bracket when $n = 5$?

Ignoring the outer parts of the equation, we can write: $\dfrac{1}{2^2} - \dfrac{1}{5^2} = \dfrac{1}{4} - \dfrac{1}{25}$.

Inserting numbers into eqn (5.4) yields: $\dfrac{1}{4} - \dfrac{1}{25} = \dfrac{25 - 4}{100} = \dfrac{21}{100}$ ■

■ **Worked Example 5.8**

A solution of silver nitrate contains $\frac{2}{7}$ of a mole of silver ion. Potassium chloride is titrated into the solution, causing precipitation:

$$Ag^+(aq) + Cl^-(aq) \rightarrow AgCl(s)$$

The amount of Ag^+ ion lost is $\dfrac{4}{29}$ of a mole. How much silver ion remains?

In examples of this sort, it is often useful to first re-express the data in the form of an equation:

$$\text{fraction remaining} = \frac{2}{7} - \frac{4}{29}$$

Inserting values into eqn (5.4) yields $\dfrac{2}{7} - \dfrac{4}{29} = \dfrac{(2 \times 29) - (4 \times 7)}{7 \times 29} = \dfrac{30}{203}$ mol. ■

Self-test 5.5

Determine the value of the following fraction problems.

5.5.1 $\dfrac{1}{2} - \dfrac{1}{6}$ 5.5.2 $\dfrac{2}{3} - \dfrac{1}{6}$

5.5.3 $\dfrac{11}{12} - \dfrac{3}{8}$ 5.5.4 $\dfrac{4}{13} - \dfrac{1}{22}$

5.5.5 $\dfrac{1}{2a} - \dfrac{b}{3a}$ 5.5.6 $\dfrac{2}{a^2} - \dfrac{3}{a}$

Dividing by a fraction

By the end of this section, you will know:

- When dividing by a fraction, we multiply the numerator by the *inverse* of the fraction on the denominator, and multiply in the usual way.

In practice, we do not actually DIVIDE by a fraction. Rather, we MULTIPLY the numerator by the *inverse* of the fraction on the denominator. In this context, an 'inverse fraction' means a fraction after it has been turned upside down:

$$\text{If 'fraction'} = \frac{a}{b} \text{ then the 'inverse fraction'} = \frac{b}{a} \qquad (5.5)$$

Therefore, when we divide by a fraction, we employ the template in eqn (5.6):

$$\frac{a}{\left(\frac{c}{d}\right)} = a \times \frac{d}{c} = \frac{ad}{c} \qquad (5.6)$$

It is important to note how the numerator remains intact. Only the fraction in the denominator (the portion below and beneath the solidus) has changed.

For generations, school-children were taught to divide in this way using the simple rhyme, 'The number you're dividing by, turn upside down and multiply.'

■ **Worked Example 5.9**

IV

Powdered paracetamol (N-(4-hydroxyphenyl)acetamide, **IV**) is to be incorporated into tablets. A bottle contains $22\frac{1}{5}$ g of (**IV**). Each tablet needs to contain $\frac{1}{5}$ g (200 mg) of paracetamol. How many tablets can be obtained from this bottle?

Firstly, to keep fractions (and often to aid the process of cancelling), we convert the decimal 22.5 into a fraction: $22.5 = \frac{45}{2}$. Inserting numbers into the template in eqn (5.6) yields, $\frac{45}{2} \div \frac{1}{5} = \frac{45}{2} \times \frac{5}{1} = \frac{225}{2} = 112\frac{1}{2}$.

The sample of (**IV**) will form 112 tablets, with a small amount left over. ■

The numerator (the portion above and left of the solidus) can also be a fraction, in which case eqn (5.6) will look slightly different:

$$\frac{\left(\frac{a}{b}\right)}{\left(\frac{c}{d}\right)} = \frac{a}{b} \times \frac{d}{c} = \frac{ad}{bc} \qquad (5.7)$$

Logarithms (symbolized here as ln) can be ignored in this chapter. They are covered in chapter 14.

■ **Worked Example 5.10**

Protons H$^+$ are so small, they can readily move by diffusion or migration through a solid matrix. The chemical potential μ of the proton takes the form

$$\mu(\text{H}^+) = A + 2Bx + nRT \quad \ln\left(\frac{x}{1-x}\right)$$

where A and B are constants, n is the amount of material, R is the gas constant and T the thermodynamic temperature. The variable x is the so-called mole fraction. 'ln' is the *logarithm* operator that we describe later in Chapter 12. Re-express the bracket if $x = \frac{1}{4}$.

The denominator is clearly $\frac{3}{4}$, so the fraction in the bracket is $\frac{\left(\frac{1}{4}\right)}{\left(\frac{3}{4}\right)}$.

Inserting numbers into eqn (5.7) yields: $\frac{1}{4} \times \frac{4}{3}$, which is the same as $\frac{1}{3}$. ∎

Self-test 5.6

Determine the value of the following fraction problems.

5.6.1 $\dfrac{1}{2} \div \dfrac{1}{6}$ **5.6.2** $\dfrac{2}{5} \div \dfrac{2}{7}$

5.6.3 $\dfrac{3}{7} \div \dfrac{1}{12}$ **5.6.4** $\dfrac{1}{7} \div \dfrac{5}{12}$

5.6.5 $\dfrac{1}{a} \div \dfrac{1}{ab}$ **5.6.6** $\dfrac{b^2 c}{a^2} \div \dfrac{bc^2}{a}$

Percentages

By the end of this section, you will know:

- The concept of a percentage relates to the hundredth part of a thing.

The word **percentage** comes from the Latin 'for every hundred' (or 'hundredth'). We see the same linguistic root *cent* (meaning hundred) in many words such as 'century' and 'centipede' and in the standard prefix for 10^2, c (see p. 3). Also, possibly in a majority of currencies, the hundredth part of the currency unit is called a cent ($\frac{1}{100}$ of a dollar $ in the USA, Australia, Barbados, Belize, Canada, or Singapore, etc.), $\frac{1}{100}$ of a euro € in much of Europe, and $\frac{1}{100}$ part of a Franc (in Congo, France, Gabon or Haiti, etc.) is a centime. 'Cent' is also the French word for 100.

There is no difference between a number expressed as a *percent* and as *per cent*.

Percentages were once used quite freely when denoting concentration. For example, a bottle labelled 'nitric acid 15%' would indicate a concentration of 0.15 mol dm^{-3}, which is 15% of 1.00 mol dm^{-3}. Percentages used in this way were often ambiguous, and should not be used.

The symbol for percentage is %. One per cent of a thing is a way of expressing a number as a fraction of a hundred. Percentages are used to express how large one quantity is relative to another quantity.

To express a fraction as a percentage, we *multiply* by 100:

$$\text{percentage \%} = \text{fraction} \times 100 \tag{5.8}$$

■ **Worked Example 5.11**

A student obtains 11 marks in a maths class test. If all questions had been correct, the number of marks would have been 40. What is the student's test mark, expressed as a percentage?

We will answer this question in two stages although in simple examples such as this, the two steps could have been performed together.

Strategy:

(i)   Express the data as a simple fraction. Then convert to a decimal (see p. 62 above).

(ii)   Convert this answer into a percentage using eqn (5.8).

Solution:

(i)   The fraction of the marks achieved is $\frac{11}{40}$, $= 0.275$.

(ii)   The percentage is therefore $0.275 \times 100 = 27.5\%$.

The student either needs to work harder, or read this book! ∎

■ **Worked Example 5.12**

What is the elemental composition by mass of aspirin (**V**), expressed as a percentage?

V

Strategy:

(i)   Write the empirical formula.

(ii)   Calculate the mass of carbon, oxygen and hydrogen in compound (**V**), and hence calculate the molar mass of (**V**).

(iii)   The percentage of carbon is given in a variant of eqn (5.8) :

$$\% \text{ of element X} = \left(\frac{\text{mass of element X in compound in grammes}}{\text{molar mass in grammes}} \right) \times 100 \quad (5.9)$$

Solution:

(i)   The empirical formula of aspirin (**V**) is $C_9H_8O_4$.

(ii)   The mass of carbon is $9 \times 12 \text{ g} = 108 \text{ g}$.

The mass of hydrogen is $8 \times 1 \text{ g} = 8 \text{ g}$.

The mass of oxygen is $4 \times 16 \text{ g} = 64 \text{ g}$.

The molar mass is therefore $(108 + 8 + 64) \text{ g} = 180 \text{ g}$.

(iii)   Inserting numbers in eqn (5.9) and expressing the result to the nearest integer:

$$\% \text{ carbon} = \frac{108}{180} \times 100 = 60\%$$

$$\% \text{ hydrogen} = \frac{8}{180} \times 100 = 4\%$$

$$\% \text{ oxygen} = \frac{64}{180} \times 100 = 36\%$$

The elemental composition of aspirin (**V**) is 60% carbon, 4% hydrogen and 36% oxygen. This result looks likely because:

(a)   The percentage by mass of hydrogen is the smallest. We should expect this result because the hydrogen atom is so much lighter than carbon or oxygen.

(b) The sum of these percentages is 100%. In other words, the mass of the compound is entirely due to carbon, hydrogen and oxygen. No other elements are involved: all of the compound's mass comes from the three elements it comprises.

We need to take care with this second criterion. The sum of these elemental percentages is often very slightly more or very slightly less than 100%, e.g. 99.9%. This problem arises because we have to round up or round down the percentage of each composition. This problem is more likely to occur if we express the percentages to a certain number of significant figures or to the nearest integer rather than a certain number of decimal places. ■

Instead of calculating a result and expressing it as a percentage, we often want to work in reverse: we know the *total* value of a thing, and want to know the value of a proportion of it, that proportion being expressed as a percentage. In such situations, we employ a variant of eqn (5.8) :

$$x \text{ per cent of } y = x\% \times \frac{y}{100} \qquad (5.10)$$

This equation is often expressed in a slightly different format, as

$$x \text{ per cent of } y = \frac{x}{100} \times y \qquad (5.11)$$

This second usage emphasizes the term 'per *cent*' because the x is divided by 100.

This equation looks reasonable. The fraction $\frac{y}{100}$ gives us the value of 1% (remember, 'per cent' means one-hundredth part of …'). We then multiply by x, which scales the value of 1%.

We obtain eqns (5.8) or (5.11) by rearranging eqn (5.10) using the laws of algebra in Chapters 4 and 6.

■ **Worked Example 5.13**

This question concerns *Viagra* ™ (**VI**), which has the trade name *Sildenafil*. *Viagra* is 1-[4-ethoxy-3-(6,7-dihydro-1-methyl-7-oxo-3-propyl-1H-pyrazolo[4,3-d]pyrimidin-5-yl)phenylsulfonyl]-4-methylpiperazine as the citrate salt. Clinical trials suggest that tablets of (**VI**) are too potent and can cause unnecessary side effects. Other data suggest the dose of 250 mg should be cut by 70%. What is the dose of (**VI**) in the new tablet?

VI

If the dose is to be reduced by 70%, then the amount of (**VI**) remaining will be $(100 - 70)\% = 30\%$.

Inserting numbers into eqn (5.10) : $\text{dose} = 30\% \times \dfrac{250 \text{ mg}}{100} = 75 \text{ mg}$.

The tablets contain 75 mg of *Viagra* ™, (**VI**). The rest of the tablet will probably comprise inert 'fillers' such as silica, chalk or talc. ■

Self-test 5.7

5.7.1   A chemist starts an experiment with 0.6 mol of reagent. The final yield is 0.35 mol. What is the percentage yield?

5.7.2   A reagent's price increases from £12.60 to £14.00. What is the percentage increase?

5.7.3   A solution is diluted from 0.60 mol dm^{-3} to 0.23 mol dm^{-3}. What is the percentage dilution?

5.7.4   A chemist performs a reaction at two different temperatures. At 298 °C, the reaction rate is 3.10×10^4 (mol dm^{-3})$^{-1}$ s^{-1}, but at 288 °C the rate decreases to 8.54×10^3 (mol dm^{-3})$^{-1}$ s^{-1}. Express the decrease in rate as a percentage.

Express the following problems as a percentage.

5.7.5   What is 72% of 0.75 mol dm^{-3}?

5.7.6   What is 0.3% of 50 g?

5.7.7   What is 17.5% of £12 000?

5.7.8   What is 43% of 12 mol?

Summary of key equations in the text

Multiplying fractions:

for two terms : $\dfrac{a}{b} \times \dfrac{x}{y} = \dfrac{a \times x}{b \times y}$ (5.1)

for many terms: $\dfrac{a}{b} \times \dfrac{c}{d} \times \cdots \times \dfrac{y}{z} = \dfrac{\prod_{a}^{y} \text{numerators}}{\prod_{b}^{z} \text{denominators}}$ (5.2)

Adding fractions: $\dfrac{a}{b} + \dfrac{c}{d} = \dfrac{ad + bc}{bd}$ (5.3)

Subtracting fractions: $\dfrac{a}{b} - \dfrac{c}{d} = \dfrac{ad - bc}{bd}$ (5.4)

Reciprocal of a fraction: if 'fraction' $= \dfrac{a}{b}$ then the 'inverse fraction' $= \dfrac{b}{a}$ (5.5)

Dividing by a fraction: $\dfrac{a}{\left(\dfrac{c}{d}\right)} = a \times \dfrac{d}{c} = \dfrac{ad}{c}$ (5.6)

Dividing a fraction with a different fraction:

$$\dfrac{\left(\dfrac{a}{b}\right)}{\left(\dfrac{c}{d}\right)} = \dfrac{a}{b} \times \dfrac{d}{c} = \dfrac{ad}{bc}$$ (5.7)

Converting a fraction into a percentage:

percentage % = fraction × 100   (5.8)

Defining elemental composition in terms of percentages:

$$\% \text{ of element X} = \left(\frac{\text{mass of element X in compound in grammes}}{\text{molar mass in grammes}} \right) \times 100 \qquad (5.9)$$

Defining a proportion as a percentage: x per cent of $y = x\% \times \dfrac{y}{100}$ $\qquad (5.10)$

alternative notation x per cent of $y = \dfrac{x}{100} \times y$ $\qquad (5.11)$

Additional problems

5.1 Consider the following reaction:

4-bromo-benzaldehyde (0.105 mol) is the limiting reagent. The yield of product is 0.069 mol. What is the percentage yield?

5.2 The reaction in Additional problem 5.1 is then reacted further to make the complexing ligand 4-(4-bromophenyl)-2,2′-bipyridine. The overall reaction proceeds in three steps, according to the scheme below. The fractional yield of step **1** is $\frac{2}{3}$, the fractional yield of step **2** is $\frac{22}{33}$ but the fractional yield of step **3** is only $\frac{4}{10}$. What is the overall yield of the desired ligand?

5.3 The activation energy E_a for ionic movement has been modelled by Anderson and Stuart. A charge of M^{z+} (of charge z_+ and radius r_+) is transferred over a distance d from an oxygen ligand (of radius r_-), to a vacancy near a similar oxygen (each bearing a charge z_-). E_a is:

$$E_a = \frac{B\, z_+ z_- e^2}{\varepsilon(r_+ + r_-)} - \frac{2\, z_+ z_- e^2}{\frac{1}{2} d\varepsilon} + \frac{\Gamma \pi\, l(r_+ - r_d)^2}{2}$$

where z_+ and z_- are the respective charge numbers on the cation and the oxygen, and r_+ and r_- are the corresponding radii. ε here is the relative permittivity of the material, and B/ε is a form of effective Madelung constant representing the loss of lattice stabilization at the onset of the ionic 'jump' from its initially stable lattice site. The second term is the Coulombic stabilization acquired by interaction with two oxygens at the midpoint of the jump, i.e. to $\frac{1}{2}d$, midway between these oxygens.

Re-express the middle term on the right-hand side without a fraction.

5.4 A chemist has prepared $\frac{3}{4}$ mol of the anticancer drug, *cis*-platin:

From this batch, $\frac{1}{17}$ mol is removed for clinical trials. How much *cis*-platin remains?

5.5 This question concerns vitamin C (L-ascorbic acid). What is the elemental composition of this vitamin?

5.6 A chemical reaction occurs. The initial amount of the limiting reagent is 1.25 mol. After reaction and purification, the amount of product is 0.85 mol. What is the percentage yield?

5.7 The price of a reagent increases by a third. If the initial cost of the reagent is £12.50 per kg, what is the new cost?

5.8 The mole fraction x of a substance is defined as

$$x = \left(\frac{\text{amount of } x \text{ (in moles)}}{\text{total number of moles}} \right)$$

A mixture contains three components, A, B and C. The mixture comprises 4.5 mol of A, 3.2 mol of B and 11.6 mol of C. What is the mole fraction of B, x_B?

5.9 Hydrogen peroxide is a pale blue liquid. It is chemically unstable, and will spontaneously yield elemental oxygen, i.e. neat H_2O_2 is an explosion risk. To avoid explosion, the usual concentration for storage is 30%. All concentrations below this level are wholly stable.

What volume of hydrogen peroxide do we need if we wish to make a diluted solution of total volume 150ml?

5.10 *Abacavir* ™ is a nucleoside analogue reverse transcriptase inhibitor used in the treatment of AIDS. Its IUPAC name is [(1R)-4-[2-amino-6-(cyclopropylamino)purin-9-yl]-1-cyclopent-2-enyl]methanol. A chemist makes three batches of *Abacavir*: the yields are $\frac{1}{40}$ mol, $\frac{2}{55}$ mol and $\frac{3}{80}$ mol. What is the total amount of *Abacavir*?

6

Algebra V

Rearranging equations according to the rules of algebra

Rearranging equations of more than one operator: BODMAS

By the end of this section, you will know:

- When we need to rearrange complicated equations, we first identify the operators present.
- Secondly, we assign priorities according to the BODMAS scheme.
- Thirdly, starting with the operator of lowest priority, we perform the inverse operation to both sides of the equation.
- We invert each operation in turn until the rearrangement is complete.

The great mathematician and physicist Albert Einstein once said, 'Reading a mathematical equation is much like reading a detective novel: the symbols are the clues and we are the detectives who use them to solve the mystery.' We will learn in this chapter how to find the clues, and how to 'read' them. The 'mystery' is what happened to a variable *en route* to becoming the final equation.

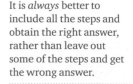

It is *always* better to include all the steps and obtain the right answer, rather than leave out some of the steps and get the wrong answer.

We can often perform a rearrangement in our head without needing the methods outlined below, that is, without formal logic. Many examples here fall into that category. But it is *always* better to include all of the steps in the algebra and get the right answer rather than leave them out, and get it wrong. The simple examples here are included to show how formal logic yields the same answer as that we would have reached without formal logic. We will also look at examples in which intuition is wholly inadequate.

The examples in previous chapters were all straightforward since each contained a single operator. Each time we rearranged an equation, we first identified the operator, and then performed its inverse. We will use a similar approach here. Unfortunately, however, we cannot do that straightaway because these equations are more complicated insofar as they each contain more than one operator.

Chapter 3 introduced us to the BODMAS rule. There, we learnt the correct order in which to perform a series of operations. For example, we could only use SUBTRACTION and ADDITION operators *after* we had MULTIPLIED or DIVIDED. In the same way, when we rearrange an equation involving different operators, we must learn the correct step-by-step sequence in which the rearrangement proceeds.

In this chapter, we are not performing a series of operations to obtain an answer; rather, we wish to **rearrange** the equation. To do so, we will employ the methodology introduced in the last chapter: first, we look at the equation to see what has happened to the variable of interest. The equations in this chapter have a minimum of two operators. Secondly, we will rank operators into a hierarchy. It is our job therefore to unpick the equation, bit by bit. The order of the operators in the BODMAS scheme is the same whether we perform a calculation of rearrangement. The sole difference is that we work *backwards* during rearrangement.

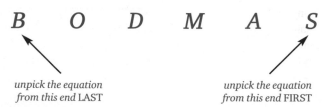

$$B \quad O \quad D \quad M \quad A \quad S$$

unpick the equation from this end LAST   *unpick the equation from this end* FIRST

■ Worked Example 6.1

The strong magnets inside a mass spectrometer accelerate a charged ion. After a time, t, the ion's velocity has increased from u to v. The acceleration is a. The variables are interrelated according to the equation:

$$v = u + a\,t$$

Rearrange the equation to yield an expression for the acceleration a, expressed in terms of the other variables, i.e. obtain a final equation of the form, '$a = ...$'

Strategy:

(i) We look at the equation to ascertain which operators are involved.

(ii) We decide which operators have priority, according to the BODMAS scheme.

(iii) We ascertain the order in which the equation will have been compiled, i.e. if we started with a, what was the sequence of steps that occurred in order to obtain the final equation?

(iv) We *reverse* each operation, starting with the operator of *lowest* priority.

Solution:

(i) There are two operators:

- MULTIPLICATION (yielding, $a \times t$, which we write as at)

- ADDITION (which combines the two terms 'u' and 'at', forming '$u + at$').

(ii) Within the BODMAS scheme, the ADDITION operator has a lower priority than the MULTIPLICATION operator.

(iii) The value of a was MULTIPLIED by t to form the compound variable at. To this compound variable, u was ADDED to yield the final velocity v.

(iv) We start to 'unpick' the equation in the order dictated by the BODMAS scheme. Since ADDITION has the lowest priority, we first remove the term ADDED to at by performing the inverse operation. Since u was ADDED to at, we SUBTRACT u from both sides the equation:

$$v - u = \boxed{(u - u)} + a\,t$$

$$\uparrow$$

The value is equal to zero

> The equation would not have balanced if we had not performed the same operation to *both* sides.

In writing a term 'u' on both sides, we demonstrate that we remembered how the equals sign '=' at the centre of the equation requires that we do the same to both sides. The value of '$u - u$' on the right-hand side is clearly zero (as we intended), and effectively removes u as a variable from that side of the equation.

The equation becomes:

$$v - u = at$$

To obtain a on its own, we next attack the MULTIPLICATION step, which is the only operator remaining. The a term that we want has been MULTIPLIED by t. To obtain a on its own, we must perform the inverse (i.e. opposite) operation to MULTIPLICATION, so we DIVIDE by t. The equals sign '=' tells us to do the same to both sides of the equation:

> **Remember:** We are to assume that two variables are multiplied together when we see them written side-by-side without a printed operator.

$$\frac{v - u}{t} = \frac{at}{t}$$

We must remove the two t terms on the right-hand side. To do so, we could simply say '$t \div t$' = '1'. More formally, we rewrite the equation in a slightly different way:

$$\frac{v-u}{t} = a \times \left(\frac{\cancel{t}}{\cancel{t}} \right)$$

Any number or algebraic term divided by itself has a value of '1'.

This equation is identical to that above except we have reformatted its right-hand side. The bracketed term has a value of 1. In effect, it says 't goes into t once.' The bracket may be substituted with '1'. We say their values **cancel**, as discussed more fully in Chapter 4.

Having removed the two t terms, we obtain:

$$\frac{v-u}{t} = a \times 1$$

We say the acceleration a is the **subject** of this equation. It would be ludicrous to write '$\times 1$', so we'll not mention the factor of 1 any further.

We might choose to write this answer with a bracket, as

Adding brackets causes no material difference to this equation, but some people think it looks neater.

$$a = \left(\frac{v-u}{t} \right) \blacksquare$$

We need to note:

- When we rearrange in order to make one term the subject, the usual practice is to write the final line with the subject on the *left*.

- The IUPAC convention says that we write time with a lower-case letter t and temperature with an upper-case letter, T. We write both in italic type because they are both variables.

■ **Worked Example 6.2**

The Hückel rule says that a fused-ring system is aromatic if the number N of its π-electrons obeys the equation:

$$N = 4n + 2$$

Where n is an **integer** such as 1, 2, 3, etc. For example, $n = 4$ for tetracene (**I**). Rearrange the equation to make n the subject.

Integer is the mathematical word meaning 'whole number.'

I

Two operators are apparent, both ADDITION and MULTIPLICATION. The BODMAS rule says that ADDITION has the lower priority, so we reverse the ADDITION step, and perform the inverse operation SUBTRACTION to both sides:

$$N - 2 = 4n + (2 - 2)$$

The bracket on the right-hand side disappears, as intended. We obtain:

$$N - 2 = 4n$$

One operator remains on the right-hand, since n has been MULTIPLIED by 4. We therefore perform the inverse operation of DIVIDE. We DIVIDE both sides by 4:

$$\frac{N-2}{4} = \frac{4n}{4}$$

The two factors of 4 on the right-hand side cancel, leaving n as the subject of the equation:

$$n = \frac{N-2}{4}$$

Many people are tempted to cancel the 2 and 4 on the right-hand side. In fact we cannot cancel them, because the 2 on the top is not a multiplication factor. It has merely been SUBTRACTED from N. ∎

■ **Worked Example 6.3**

Consider the fraction $n = \dfrac{10-2}{8-4}$. Show how inappropriate cancelling yields an incorrect result.

First, we evaluate the simple sums on top and bottom:

$$n = \frac{10-2}{8-4} = \frac{8}{4}$$

Clearly, $n = 2$.

Secondly, we cancel—but this time inappropriately. The most common way of doing this is to notice how 2 goes into 4 twice, and say:

$$n = \frac{10 - \overset{1}{\cancel{2}}}{8 - \underset{2}{\cancel{4}}}$$

so $\quad n = \dfrac{9}{8-2} = \dfrac{9}{6} = 1\tfrac{1}{2}$

This approach is wrong because we should have cancelled *all* of the top with *all* of the bottom, for example dividing each term by 2.

In this example, n this time has a value of 1.5, which clearly differs greatly from the correct answer above. ∎

Sometimes we want an equation in which the subject is not a single term but a phrase or compound variable.

■ **Worked Example 6.4**

The **Kirchhoff equation** tells us how the enthalpy change accompanying a reaction ΔH varies according to the temperature T at which we perform it. If the enthalpy change at a temperature T_1 is ΔH_1 and the enthalpy change at a temperature T_2 is ΔH_2, then:

$$\Delta H_2 = \Delta H_1 + C_p (T_2 - T_1)$$

where C_p is the heat capacity at constant pressure. Rearrange the Kirchhoff equation to make C_p the subject.

Before we start, we look critically at the equation to discern what operators influence C_p. We see it has been MULTIPLIED by the bracket $(T_2 - T_1)$ to form '$C_p \times (T_2 - T_1)$'. ΔH_1 was then ADDED to this resultant compound variable.

(i) When taking apart this equation, the operator of lowest priority is ADDITION (so ΔH_1 has been ADDED), so we perform the reverse operation of SUBTRACTION, and first SUBTRACT ΔH_1 from both sides to yield:

$$\Delta H_2 - \Delta H_1 = C_p \times (T_2 - T_1)$$

Before cancelling, the right-hand side of the equation read, '$(\Delta H_1 - \Delta H_1) + C_p \times (T_2 - T_1)$.'

We ignore the two operators subtraction and brackets because the subtraction only occurs within the bracket, and we move the bracket without affecting its contents in any way.

(ii) Having moved the ΔH_1 term from the right-hand side, we see how C_p was multiplied by the bracket $(T_2 - T_1)$. We now need to reverse this MULTIPLICATION step by performing the inverse operation of DIVISION. In this case, we must DIVIDE both sides by the *whole of the bracket*:

$$\frac{\Delta H_2 - \Delta H_1}{(T_2 - T_1)} = C_p \frac{\cancel{(T_2 - T_1)}}{\cancel{(T_2 - T_1)}}$$

The twin BRACKETS on the right-hand side cancelled (as intended) to leave:

$$C_p = \frac{\Delta H_2 - \Delta H_1}{(T_2 - T_1)} \quad \blacksquare$$

We need to note:

- The laws of algebra permit us to ADD, SUBTRACT, DIVIDE or MULTIPLY by a complete bracket, provided we do not alter the bracket's contents in any way.

- The bracket line on the bottom of the last line is superfluous, and can be omitted.

Rearranging equations with negative coefficients

By the end of this section, you will know:

- If a coefficient is negative, we can regard the term as being MULTIPLIED by −1.

We now consider what amounts to a special case, and learn how to proceed when rearranging equations that contain negative terms.

■ Worked Example 6.5

A reaction is thermodynamically feasible provided the value of the Gibbs function $\Delta G^{\ominus}$ is negative. The magnitude of $\Delta G^{\ominus}$ depends on the change in the enthalpy of reaction $\Delta H^{\ominus}$ and the entropy change $\Delta S^{\ominus}$, according to the equation:

$$\Delta G^{\ominus} = \Delta H^{\ominus} - T\Delta S^{\ominus}$$

where T is the thermodynamic temperature.

Rearrange this equation to make the change in entropy, $\Delta S^{\ominus}$, the subject.

As usual, before we start, we look to identify the operators present, and assess the order in which they are invoked. We see how $\Delta S^{\ominus}$ was first MULTIPLIED by the temperature T to form $T\Delta S^{\ominus}$, and next, this compound variable $T\Delta S^{\ominus}$ was SUBTRACTED from $\Delta H^{\ominus}$.

As usual, we will invert the roles of these operators, and deal with the operator of lowest priority first. In fact, we can rearrange this equation in two ways: one long and the other shorter but a little more involved.

The longer (but simpler) route:

We start by inverting the SUBTRACTION step by ADDING the compound variable $T\Delta S^{\ominus}$ to both sides:

$$\Delta G^{\ominus} + T\Delta S^{\ominus} = \Delta H^{\ominus} + (-T\Delta S^{\ominus} + T\Delta S^{\ominus})$$

The bracketed portion on the right-hand side clearly equates to zero and vanishes, which is what we wanted. We are left with the equation:

$$\Delta G^{\ominus} + T\Delta S^{\ominus} = \Delta H^{\ominus}$$

Next we SUBTRACT $\Delta G^{\ominus}$ from both sides in much the same way, to yield:

$$(\Delta\cancel{G}^{\ominus} - \cancel{\Delta G^{\ominus}}) + T\Delta S^{\ominus} = \Delta H^{\ominus} - \Delta G^{\ominus}$$

Again, the bracketed portion cancels, yielding:

$$T\Delta S^{\ominus} = \Delta H^{\ominus} - \Delta G^{\ominus}$$

Finally, to invert the MULTIPLICATION of $\Delta S^{\ominus}$ by T, by performing the opposite operation, and DIVIDE both sides by T:

$$\frac{T\Delta S^{\ominus}}{T} = \frac{\Delta H^{\ominus} - \Delta G^{\ominus}}{T}$$

The two T terms on the left-hand side cancel, to yield:

$$\Delta S^{\ominus} = \frac{\Delta H^{\ominus} - \Delta G^{\ominus}}{T}$$

The shorter but more involved route:

In the shorter route, as with the longer route, we start by identifying the operators present. Again, we start by noting how $\Delta S^{\ominus}$ is MULTIPLIED by a function of temperature, but this time we say $T\,\Delta S^{\ominus}$ is MULTIPLIED by '–1' to yield '$-T\,\Delta S^{\ominus}$'. $\Delta H^{\ominus}$ is then ADDED to the compound variable $T\,\Delta S^{\ominus}$.

To make ΔS the subject, we first subtract $\Delta H^{\ominus}$ as in the longer version, to obtain

$$\Delta G^{\ominus} - \Delta H^{\ominus} = -1 \times T\Delta S^{\ominus}$$

Note how the factor of -1 persists. The right-hand side has been MULTIPLIED by '$-1 \times T$', so we perform the opposite operation and DIVIDE both sides by $-1 \times T$:

$$\frac{\Delta G^{\ominus} - \Delta H^{\ominus}}{-1 \times T} = \frac{-1 \times T \times \Delta S^{\ominus}}{-1 \times T}$$

'$-1 \times T$' terms on the right-hand side cancel, to yield $\Delta S^{\ominus}$, so

$$\Delta S^{\ominus} = \frac{\Delta G^{\ominus} - \Delta H^{\ominus}}{-T}$$

This answer ought to be the same as that derived above using the longer route, because each derives from the same equation, '$\Delta G^{\ominus} = \Delta H^{\ominus} - T\Delta S^{\ominus}$', and each expresses $\Delta S^{\ominus}$ as their subject. While they look slightly different, in fact they are the same: the only difference is that the right-hand sides have been rearranged somewhat.

To demonstrate the equality of the two equations, we will take the version immediately above, and MULTIPLY both top and bottom by '–1'. We write:

$$\Delta S^{\ominus} = \left(\frac{-\cancel{1}}{-\cancel{1}}\right) \times \frac{\Delta G^{\ominus} - \Delta H^{\ominus}}{-T}$$

which is the same equation because the new bracketed term has a value of '1', and multiplying by '1' will not change the values. The value of '-1×-1' on the bottom equals '+1'.

Remember: '$-1 \times -1 = +1$, so '$-1 \times -T$' $= +T$.

We then MULTIPLY the top line to yield:

$$\Delta S^{\ominus} = \frac{-\Delta G^{\ominus} + \Delta H^{\ominus}}{T}$$

Rearranging the top line yields:

$$\Delta S^{\ominus} = \frac{\Delta H^{\ominus} - \Delta G^{\ominus}}{T}$$

so the two versions of the equation are the same.

This kind of rearrangement, multiplying by '–1', is simply a mathematical dodge. It only works because $-1 \times -1 = +1$. ■

Self-test 6.1

Rearrange each equation to make x the subject:

6.1.1 $y = 5x - 4x + 7x + 1$ **6.1.2** $y = 3x - 7$

6.1.3 $y = 5 - 4x$ **6.1.4** $y = \dfrac{x-4}{7}$

6.1.5 $y = 8\,(x-1)$ **6.1.6** $y = 8x\,(b-1)$

6.1.7 $y = \dfrac{d}{x}$ **6.1.8** $y = \dfrac{3}{x-2}$

6.1.9 $y = \dfrac{a-x}{4d}$ **6.1.10** $y = \dfrac{x-9}{c+d}$

6.1.11   A temperature can be expressed in either degrees Celsius (C) or in degrees Fahrenheit (F). These temperature notations are linked by the equation

$$C = \frac{5}{9} \times (F - 32)$$

Rearrange to make F the subject.

Rearrangements involving simple powers and roots

By the end of this section, you will know:

- The BODMAS scheme accommodates powers under the heading, OF.
- Accordingly, POWER operators have a high priority in the BODMAS scheme, and only BRACKETS are higher.
- To rearrange a POWER, we perform the inverse operation, which is generally a ROOT.

> We will discuss the mathematics of powers more fully in Chapters 11–13.

In this section, we merely continue our exploration of the trends in the previous section, and look at rearrangements involving power functions.

> When taking a multiple root, such as $\sqrt[3]{x}$, the small figure nestling before the root sign tells us the level of the **root**. We should assume a *square* root if we see only a root sign with no small figure.

The first power functions are simple SQUARES. Just looking at simple numbers ought to convince us that SQUARE and SQUARE ROOT are inverse functions. For example, $3^2 = 9$ and $\sqrt{9} = 3$. A little more thought reveals this rule to have a wider application, so $4^3 = 64$ and $\sqrt[3]{64} = 4$. In fact, we can deduce a more general rule still, which says that 'raising anything to its nth power' and 'taking the nth root' are inverse functions.

In general, we write:

$$a^n = b \quad \text{so} \quad a = \sqrt[n]{b} \tag{6.1}$$

■ **Worked Example 6.6**

The dithionite ion (**II**) is an extremely powerful reducing agent, and readily reduces organic aromatic species such as tetracene (**I**) from Worked Example 6.2.

II

The rate law when reducing with dithionite is:

$$\text{rate} = k\,[\text{aromatic}]\,\sqrt{[(\mathbf{II})]}$$

where k is a rate constant. Rearrange the equation to make [(II)] the subject.

Looking at the right-hand side of the equation shows how the concentration [(II)] was first SQUARE ROOTED to form $\sqrt{[(II)]}$, which was then multiplied by 'k [aromatic]'.

The first step when obtaining [(II)] is therefore to DIVIDE both sides of the equation by 'k [aromatic]'. We will treat 'k [aromatic]' as a *compound variable*. We therefore perform the inverse operation, and DIVIDE both sides in a single step:

$$\frac{\text{rate}}{k\,[\text{aromatic}]} = \sqrt{[(II)]}$$

Since the right-hand side is a SQUARE ROOT, to obtain [(II)] from $\sqrt{[(II)]}$ we perform the inverse operation and SQUARE both sides:

$$\left(\frac{\text{rate}}{k\,[\text{aromatic}]}\right)^2 = \left(\sqrt{[(II)]}\right)^2$$

The SQUARE of a SQUARE ROOT yields the thing itself. In this case, it yields '[(II)]', so

$$[(II)] = \left(\frac{\text{rate}}{k\,[\text{aromatic}]}\right)^2 \;\blacksquare$$

We can rearrange more complicated equations incorporating ROOT operators, provided we stick closely to the BODMAS rules.

■ **Worked Example 6.7**

The ionic conductivity Λ of ions moving through a solution relates to its ionic concentration c according to the **Onsager** equation:

$$\Lambda = \Lambda^\circ - b\sqrt{c}$$

where b is a constant, and Λ° is the ionic conductivity when the concentration of ions is zero. Rearrange this equation to make concentration c the subject.

The concentration term has been operated on in three separate, consecutive ways:
- This concentration c was SQUARE ROOTED (i.e. a function OF c) to yield $\sqrt{c}$.
- The root $\sqrt{c}$ was then MULTIPLIED by $-b$, to yield $-b\sqrt{c}$.
- Finally, Λ° was ADDED to the compound variable '$-b\sqrt{c}$' to yield, $\Lambda^\circ - b\sqrt{c}$

It often comes as something of a surprise to suggest that '$\sqrt{c}$' was multiplied by *minus b*; it would have been equally valid to suggest that '$\sqrt{c}$' was multiplied by $+b$ and only then was '$b\sqrt{c}$' subtracted from 'Λ°'. Our approach here is intended to be easy and to save time.

To make c the subject of the equation, we first SUBTRACT Λ° from both sides:

$$\Lambda - \Lambda^\circ = -b\sqrt{c}$$

Note how the minus sign persists on the right-hand side. Next, we note how $\sqrt{c}$ has been MULTIPLIED by a factor of $-b$, so we DIVIDE both sides by '$-b$':

$$\frac{\Lambda - \Lambda^\circ}{-b} = \sqrt{c}$$

Finally we perform the inverse operation to SQUARE ROOT, and square both sides:

$$c = \left(\frac{\Lambda - \Lambda^\circ}{-b} \right)^2$$

We could rewrite this result, multiplying top and bottom within the bracket by '−1':

$$c = \left(\frac{\Lambda^\circ - \Lambda}{b} \right)^2 \blacksquare$$

■ **Worked Example 6.8**

Figure 6.1 shows a rotated-disc electrode (RDE). An electrochemist immerses a RDE in a solution of analyte at a concentration of c. The viscosity of the solution is v. The RDE rotates at a frequency of ω. The current at the RDE is given by the Levich equation:

$$I = \frac{0.62\, nFA\, c\, D^{2/3}\, \omega^{1/2}}{\sqrt[6]{v}}$$

> We discuss the physical meaning of powers, e.g. $D^{2/3}\ \omega^{1/2}$ in Chapter 11.

where A is the area of the disc, D is the speed at which the analyte diffuses through solution, n is the number of electrons transferred in the redox reaction, and F is the Faraday constant. Rearrange the Levich equation to make v the subject.

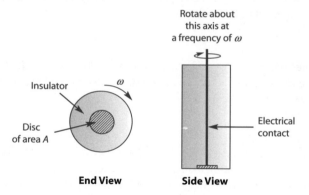

Rotate about this axis at a frequency of ω

Insulator

Disc of area A

ω

Electrical contact

End View   **Side View**

Figure 6.1 A rotated-disc electrode (RDE) of radius r. The electrode is spun about its long, cylindrical axis at a constant frequency of ω.

> Strictly, D is called the diffusion coefficient and v is called the kinematic viscosity.

As usual, before we start, we ask the question, what has been done to v.

1. The sixth ROOT has been taken, to form $\sqrt[6]{v}$.

2. $\sqrt[6]{v}$ has been DIVIDED into a collection of constants ' $0.62\, nFAc\, D^{2/3}\ \omega^{1/2}$. While this collection looks formidable, we do not really need to think about them. We merely note how $\sqrt[6]{v}$ has been DIVIDED into them *en masse*. If it makes the equation look less scary, rewrite it as $k \div \sqrt[6]{v}$, where k represents all these constants.

To rearrange the equation, we reverse these two operations:

1. To obtain $\sqrt[6]{v}$ on its own, we MULTIPLY both sides of the equation by $\sqrt[6]{v}$ and DIVIDE both sides by I. We obtain:

$$\sqrt[6]{v} = \frac{0.62\, nFA\, c\, D^{2/3}\omega^{1/2}}{I}$$

2. To reverse the sixth ROOT, we must take the sixth POWER. We obtain:

$$\left(\sqrt[6]{v}\right)^6 = \left(\frac{0.62\ nFA\ c\ D^{\frac{2}{3}}\omega^{\frac{1}{2}}}{I}\right)^6$$

The left-hand side then collapses to yield

$$v = \left(\frac{0.62\ nFA\ c\ D^{\frac{2}{3}}\omega^{\frac{1}{2}}}{I}\right)^6 \blacksquare$$

In general, we invert an nth root by raising to the nth power, just as we invert a square root by taking a square.

Self-test 6.2

In each of the following, rearrange to make x the subject. In each case where a root is part of the answer, assume only the positive root is required:

6.2.1 $y = x^2$

6.2.2 $y = -4x^2$

6.2.3 $y = x^2 + 7$

6.2.4 $y = 4\,(c - x^2)$

6.2.5 $y = c\,(x^2 + 1)$

6.2.6 $y = (x - a)^2$

6.2.7 $y = \sqrt{x - 9}$

6.2.8 $y = 5 \times \sqrt{x - v}$

6.2.9 $y = \sqrt{a - x}$

6.2.10 $y = \left(\dfrac{x + 1}{5}\right)^2$

6.2.11   When a rotating disc electrode (RDE) is immersed in a solution, electrolysis only occurs within the very thin layer of solution adjacent to electrode's surface. We call this layer, the **Nernst diffusion layer,** δ. The following equation

$$\delta = \frac{1.61\,v^{\frac{1}{6}}\,D^{\frac{1}{3}}}{\sqrt{\omega}}$$

defines the thickness of δ, where D is the diffusion coefficient, v the kinematic viscosity and, ω is the rotation speed. Rearrange the equation to make ω the subject.

Summary of key equation used in the text

$$a^n = b \quad \text{so} \quad a = \sqrt[n]{b} \tag{6.1}$$

Additional problems

The answers to some of these questions may be intuitively obvious. Nevertheless, determine the answer using the strategies outlined in this chapter.

6.1 Consider the right-angled triangle below:

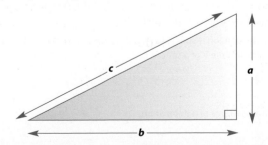

Pythagoras' theorem says the lengths of the sides follow the relationship

$$c^2 = a^2 + b^2$$

Rearrange Pythagoras' equation to make b the subject.

6.2 Consider the following rate equation:

$$\text{rate} = \frac{k_1 k_2^2}{k_1 k_2 k_3}$$

Simplify the equation and thence rearrange to make k_3 the subject.

6.3 The Rydberg equation

$$\bar{v} = \Re_H \left\{ \frac{1}{n_1^2} - \frac{1}{n_2^2} \right\}$$

describes the energies (as a wavenumber $\bar{v}$) of electronic transitions emitted in atoms or ions with only 1 electron, e.g. the hydrogen atom. $\Re_H$ is the Rydberg constant for the hydrogen atom, and n_1 and n_2 are integers. Rearrange the equation to make n_2 the subject.

6.4 Anything dissolved in a solution will move through the solvent by diffusion. The solution of **Fick's second law** (a diffusion equation) relates the distance a solvated species in solution diffuses, l, over a period of time, t, with the 'velocity of diffusion,' known as the diffusion coefficient, D:

$$l = \sqrt{2Dt}$$

Rearrange the expression to make 't' the subject.

6.5 The **Einstein equation**

$$E = mc^2$$

relates the energy released, E, when a mass, m, is converted entirely to energy. c is the speed of light *in vacuo*. Rearrange the equation to make, c, the subject.

6.6 Consider the reaction: ethane $\rightarrow$ ethene $+ H_2$ (a process known in the petrochemical industry as **cracking**).

The equilibrium constant for the cracking process is

$$K = \frac{[H_2][\text{ethene}]}{[\text{ethane}]}$$

Rearrange the expression for the equilibrium constant, to make $[H_2]$ the subject.

6.7 The acceleration a of a molecular fragment inside a mass spectrometer is determined via the expression:

$$a = \frac{v - u}{t}$$

where the velocity changes from an initial value u to a different final value v during a time t. Rearrange the equation to make v the subject.

6.8 Assume that an atom of neon is spherical with a volume, V, of 4.85×10^{-24} m^3. If the equation relating radius r and volume V of a sphere is

$$V = \frac{4}{3} \pi r^3$$

determine the atomic radius of a neon atom.

6.9 Particles experience a strong force when they approach. The magnitude of the resultant potential energy varies with their separation, r, according to the **Lennard-Jones equation**:

$$\text{energy} = 4\varepsilon \left\{ \left(\frac{\sigma}{r}\right)^{12} - \left(\frac{\sigma}{r}\right)^{6} \right\}$$

where ε is an energy and σ is the distance of closest approach of 2 atoms, i.e. σ is a constant. At most values of r, the first term in the bracket becomes vanishingly small, and can be ignored. We can therefore approximate this equation, as follows:

$$\text{energy} = -4\varepsilon \left(\frac{\sigma}{r}\right)^{6}$$

Rearrange this latter equation make r the subject.

6.10 The energy, E, of an electron in a one-dimensional box of length l is:

$$E = \frac{n^2 h^2}{8ml^2}$$

Rearrange this expression to make l the subject.

Algebra VI

Simplifying equations: brackets and factorizing

Multiplying with brackets

By the end of this section, you will know:

- The correct way to multiply with a bracket is to MULTIPLY each term within the bracket by the factor positioned outside it.
- That when two brackets are positioned side by side, we MULTIPLY them term by term.

We know from the BODMAS rule in Chapter 2 that brackets have the highest priority in a calculation. We now explore the best way to MULTIPLY them.

The simplest process is MULTIPLYING a single bracket with a factor or variable.

■ Worked Example 7.1

The amount of energy E needed to warm an amount of water n from an initial temperature $T_{initial}$ to a final temperature T_{final} is given by the expression:

$$E = nC\,[T_{final} - T_{initial}]$$

where C is the heat capacity of the water. Multiply out the bracket.

We will treat n and C as a compound variable, nC. We MULTIPLY nC by each temperature in turn:

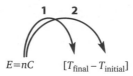

$$E = nC \qquad [T_{final} - T_{initial}]$$

The first multiplication step (**1**) yields $nC \times T_{final}$. The second multiplication step (**2**) yields $nC \times T_{initial}$. Combining the two terms:

$$E = nC\,T_{final} - nC\,T_{initial}$$
$$\quad\;\; \mathbf{1} \qquad\quad \mathbf{2}$$

....
We often say we **multiply out** the brackets.
....

Sometimes an equation incorporates several brackets, positioned together. In the absence of an operator between the brackets, we MULTIPLY them together. ■

■ Worked Example 7.2

The van der Waals equation is one of the better ways to take account of the forces that cause the ideal-gas equation to fail:

$$\left(p + \frac{n^2 a}{V^2}\right)(V - nb) = nRT$$

MULTIPLY together the two brackets on the left-hand side.

In a similar way to MULTIPLYING a single bracket with a factor, we here MULTIPLY term by term. In effect, we treat the first bracket as though it comprised two factors, 'p' and '$n^2\,a \div V^2$'. We first MULTIPLY both of the terms in the second bracket by p, then multiply both terms in the second bracket by $n^2 a \div V^2$. Finally, we add together the two results.

First we MULTIPLY, term by term, the whole of the right-hand bracket with p from the left-hand bracket:

$$\left(p + \frac{n^2 a}{V^2}\right)\ (V - nb) = nRT$$

The results of the two MULTIPLICATION steps are:

Step 1: pV and Step 2: $-pnb$

Then we MULTIPLY term by term, the right-hand bracket with, $n^2 a \div V^2$:

$$\left(p + \frac{n^2 a}{V^2}\right)\ (V - nb) = n\,R\,T$$

> Each bracket contains two terms. The final result will therefore comprise 4 terms, because $2 \times 2 = 4$

The results of the two MULTIPLICATION steps are:

Step 3: $\dfrac{n^2 a}{V^2} \times V$ Step 4: $\dfrac{n^2 a}{V^2} \times -nb$

Summing the terms: 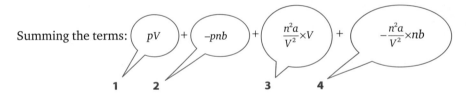 $\left(\ pV\ \right) + \left(\ -pnb\ \right) + \left(\ \dfrac{n^2 a}{V^2} \times V\ \right) + \left(\ -\dfrac{n^2 a}{V^2} \times nb\ \right)$

        1   2   3   4

We will learn later how to tidy up this answer. ■

Self-test 7.1

MULTIPLY out the following expressions:

7.1.1 $p\,(a + b)$ **7.1.2** $bp\,(x - y)$

7.1.3 $(a + 1)\,(c + 2)$ **7.1.4** $(a + b^2)\,(a^2 - b)$

7.1.5 $g\,(H + 2)\,(J + 3)$ **7.1.6** $(J + 2)\,(J - 2)$

7.1.7 $(\tfrac{1}{2} + 4v)\left(3 - \dfrac{a}{c}\right)$ **7.1.8** $\left(\dfrac{1}{c} - \dfrac{1}{d}\right)(e + 5)$

Squaring a bracket: the 'perfect square'

By the end of this section, you will know:

- When we square a bracket, we multiply it with itself to form a 'perfect square'.
- Squaring the bracket $(a + b)$ generates $a^2 + 2ab + b^2$.

We employ the same methodology when we SQUARE a bracket. For example, if we square the bracket $(a+b)$, i.e. $(a+b)^2$, we are asked to multiply $(a+b)$ with itself:

$$(a+b)^2 = (a+b) \times (a+b)$$

so

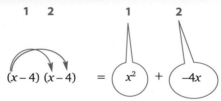

therefore

$$(a+b)^2 = a^2 + 2ab + b^2 \tag{7.1}$$

Multiplication is commutative, so ab is the same as ba.

We generally consider that squaring a bracket is a special case, because we can simply write the answer via eqn (7.1) and thereby obviate the necessity to multiply out each bracket separately.

■ **Worked Example 7.3**

Show that $(x-4)^2$ follows the general scheme in eqn (7.1).

First we consider the x in the first bracket:

Secondly, we consider the 4 in the first bracket:

so $(x-4)^2 = x^2 + (-4x) + (-4x) + 16$.
We then add together the terms in x, so $(x-4)^2 = x^2 - 8x + 16$. ■

Self-test 7.2

Multiply out each of the following squares, and check that the result is the same as that generated with eqn (7.1):

7.2.1	$(a+b)^2$	7.2.2	$(x-y)^2$
7.2.3	$(-x+c)^2$	7.2.4	$(-x-y)^2$
7.2.5	$(2x-y)^2$	7.2.6	$(ax-by)^2$
7.2.7	$(5-y^2)^2$	7.2.8	$(1-4y)^2$

Factorizing simple expressions

By the end of this section, you will know:

• Factorizing is one of the simplest ways to simplify an equation.

• Factorizing is often the inverse process of multiplying brackets.

When we **factorize**, we look at an equation and look for factors in common. For example, if we have '$ab + ac$', then both terms contain the common factor 'a'. Similarly, if we have '$4x + 8$', then '4' is a factor in each of the two terms, because 4 goes into both 4 and 8. The factors can be more complicated, so the expression '$4x^2 + 2x$' has '$2x$' as a factor.

A **factor** is a number or algebraic symbol with which we multiply something else.

■ **Worked Example 7.4**

The energy E of a gas particle has two components, both kinetic and potential energies. The potential energy is mgh, and the kinetic energy is $\frac{1}{2}mv^2$, where m is mass, g is the acceleration due to gravity, h is height and v is velocity. Factorize the expression:

$$E = mgh + \frac{1}{2}mv^2$$

Both terms contain m as a factor. In effect, then, we could write the expression as:

$$E = m \times (gh) + m \times (\frac{1}{2}v^2)$$

This expression is identical in every way to that above except our choice of notation style: we have separated the factor of m from each of the two terms. Since m is common to both, we then write

$$E = m\,(gh + \frac{1}{2}v^2)$$

This is the factorized expression. In other words, we have collected together the two common m factors. If we were to multiply out this bracket, we would regain the original expression. ■

■ **Worked Example 7.5**

Dalton's Law says the total pressure exerted by a mixture of gases $p_{(total)}$ equals the sum of the partial pressures of the gases. If a mixture comprises argon, nitrogen and hydrogen at pressure of $p(\mathrm{Ar}) = ap^{\ominus}$, $p(\mathrm{N_2}) = bp^{\ominus}$, and $p(\mathrm{H_2}) = cp^{\ominus}$, use Dalton's Law to add up these pressures, then factorize them. ($p^{\ominus}$ here is standard pressure.)

Dalton's Law says $\quad p_{(total)} = p(\mathrm{Ar}) + p(\mathrm{N_2}) + p(\mathrm{H_2})$

Substituting for the numbers: $\quad p_{(total)} = ap^{\ominus} + bp^{\ominus} + cp^{\ominus}$

Factorizing, we say, $\quad p_{(total)} = (a + b + c) \times p^{\ominus}$

It would have been equally correct to have written, $p_{(total)} = p^{\ominus}(a + b + c)$. ■

Self-test 7.3

Factorize each of the following:

7.3.1	$a + ab$	**7.3.2**	$2x - 2y$
7.3.3	$ab + 2ac + ad$	**7.3.4**	$4a + 2ac + 6ab$
7.3.5	$a^2 + ab$	**7.3.6**	$a^2 + 4a$
7.3.7	$a^2 + 3a + 6ab$	**7.3.8**	$xy - x^2 + 2x$
7.3.9	$x^3 + 6x^2 + 4x$	**7.3.10**	$(x + 4)^2 + (2x - 3)^2 - (25 + x)$

Remember: It is always easy to get mixed up with signs, so always check factorized equations by multiplying out, and make sure the result is the same as the original expression.

Factorizing the difference between two squares

By the end of this section, you will know:

- We call equations of the sort $x^2 - y^2$ 'the difference between two squares'.
- The difference between two squares, $x^2 - y^2$ factorizes as $(x - y)\,(x + y)$.

Possibly the simplest factorizations start with equations of the type $a^2 - b^2$, which factorizes in a characteristic way:

$$a^2 - b^2 = (a - b)\,(a + b) \tag{7.2}$$

We will call the right-hand side of eqn (7.2) the **difference between two squares**.

■ Worked Example 7.6

Consider the right-angled triangle below:

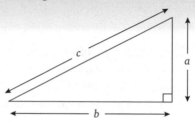

Pythagoras' theorem says that, $b^2 = c^2 - a^2$. Factorize this equation.

The right-hand side of the Pythagoras equation represents the difference between two squares, so

as $b^2 = c^2 - a^2 =$ the difference between two squares

so $b^2 = (c - a)\,(c + a)$ ■

Sometimes, when dealing with the difference between two squares, the second term in eqn (7.2) does not immediately *look* like a square. In which case, we rewrite the difference between the two squares somewhat differently, and say

$$(x^2 - a) = \left(x - \sqrt{a}\right)\left(x + \sqrt{a}\right) \tag{7.3}$$

■ Worked Example 7.7

Factorize the expression, $y = x^2 - 9$.

As the right-hand side represents the difference between two squares, the factors will be $\left(x - \sqrt{a}\right)$ and $\left(x + \sqrt{a}\right)$, where each $\sqrt{a}$ term here is the square root of 9. The equation factorizes as:

$y = (x - 3)\,(x + 3)$ ■

Self-test 7.4
Factorize the following differences between two squares, using eqn (7.3):

7.4.1 $x^2 - b^2$ 7.4.2 $c^2 - p$

7.4.3 $\alpha^2 - \beta^2$ 7.4.4 $\omega^2 - 16\pi$

7.4.5 $x^2 - 4$ 7.4.6 $a^2 - 81$

7.4.7 $d^2 - 10\,000$ 7.4.8 $e^2 - 5.32$

Completing a square

By the end of this section, you will know:

- It is possible to factorize an equation to conform to a perfect square, eqn (7.1).
- Completing a square is often a powerful way of factorizing an equation.

Occasionally, it is helpful to take an expression and make a perfect square from it, even though some of the necessary components are missing. We use the following equation

$$x^2 + 2ax + b^2 = (x+a)^2 + (b^2 - a^2) \tag{7.4}$$

■ **Worked Example 7.8**

Make a perfect square from the equation $x^2 + 10x + 20 = 0$, and hence determine the values of x that satisfy the equation.

Strategy:

(i) Equation (7.4) tells us the coefficient of x is twice the second term within the bracket of a perfect square – call it $2a$. So we divide this coefficient by 2, and write a perfect square: $(x+a)^2 = x^2 + 2ax + a^2$.

(ii) We add the constant term $(b^2 - a^2)$ onto this perfect square, as shown in eqn (7.4).

(iii) To determine the value of x, we rearrange the equation in the usual ways (see the previous chapters) by making x the subject.

Solution:

(i) The equation must take the form $(x+5)^2$, because the second term in the bracket is half the coefficient of x; and $10/2 = 5$.

(ii) To determine the coefficient outside the square, we use eqn (7.4), saying $(20 - 5^2) = -5$. The equation is, $(x+5)^2 - 5 = 0$. We have **completed the square**.

(iii) To determine the value of x that satisfies the equation is now straightforward: simple rearrangement yields, $(x+5)^2 = 5$, so $x+5 = \pm\sqrt{5}$, and hence $x = \pm\sqrt{5} - 5$. ■

Factorizing quadratic equations

By the end of this section, you will know:

- A quadratic equation has four terms, hence its name, and has the form $y = ax^2 + bx + c$, where a, b and c are factors (or 'coefficients').
- If $a = 1$, and b and c are whole numbers, we factorize the equation $y = x^2 + bx + c$ saying $(x+A)(x+B)$, where $b = (A+B)$ and $c = A \times B$.
- When $a \neq 1$, we may have to factorize by a process of trial and error.

The equations we factorized above were relatively simple. For example, a perfect square only has a single term containing x. We now look at more complicated equations, in which there is more than one term in x, and which cannot straightforwardly be factorized.

The Latin prefix **quad** always means 'four,' so a *quadrangle* is an area bounded by four walls, and a *quad bike* has four wheels.

■ **Worked Example 7.9**

What are the factors of the equation $y = x^2 + 3x + 2$?

Altogether, this equation contains four terms, so we call it **a quadratic equation**, following the Latin convention which says that *quad* always means '4'.

Before factorizing the expression, quite reasonably we assume it must have come from somewhere—it was formed when two brackets were MULTIPLIED together. Our task is to work backwards, asking the question, 'What was in those two brackets?'

Next, we consider the general case where we multiply together the two brackets and compare it with the equation in the question:

General case $y = (x + A) \times (x + B) = x^2 + (A + B) x + AB$

Specific case $y = \qquad\qquad = x^2 + \qquad 3x + 2$

By comparing the right-hand side of the equation in the question $(x^2 + 3x + 2)$ with the general equation $(x^2 + (A + B) x + AB)$, we see how $(A + B) = 3$ and $AB = 2$.

By inspection: $A = 1$ and $B = 2$

So: $y = x^2 + 3x + 2$ factorizes to, $y = (x + 1)(x + 2)$ ■

We could equally have said $A = 2$ and $B = 1$, but the answer would have been the same, because multiplying brackets is **associative** (see Chapter 3).

Occasionally, this exercise can be quite difficult:

■ **Worked Example 7.10**

What are the factors of the equation $y = x^2 + 22x + 40$?

Using the relationships immediately above, the coefficient '40' is the *product* of A and B, while the coefficient of '22' is the *sum* of A and B. In an example such as this, the identity of A and B may not be immediately apparent. We therefore need a method to determine their values.

Strategy:

(i) We 'brainstorm', looking for pairs of numbers whose *product* is 40. Obvious examples involving only whole numbers are: 1×40, 2×20, 4×10, and 5×8.

(ii) We look closely at the pairs of numbers generated in step (i). Specifically, we ask what is their *sum*. Remember: we are looking for a pair with a sum of '22.'

Product	Sum
1×40	41
2×20	22
4×10	14
5×8	13

At this stage, it should be clear that only '2' and '20' fulfil both criteria. Accordingly, these are the roots of this quadratic.

Therefore, $y = (x + 2)(x + 20)$ ■

Table 7.1 The various permutations of signs and coefficients encountered when solving quadratic equations by inspection. $(x+A)(x+B)$ and $A > B$ (which is important for the last two examples).

Sign of coefficient on x	Sign of the number	Sign of A	Sign of B	Example
Positive	positive	positive	positive	$x^2 + 5x + 6 = (x+3)(x+2)$
Negative	positive	negative	negative	$x^2 - 5x + 6 = (x-3)(x-2)$
Positive	negative	positive	negative	$x^2 + x - 6 = (x+3)(x-2)$
Negative	negative	negative	positive	$x^2 - x - 6 = (x-3)(x+2)$

Self-test 7.5

Factorize the following quadratic equations using the *inspection* above.

7.5.1 $x^2 + 4x + 3$ **7.5.2** $x^2 + 5x + 6$

7.5.3 $x^2 + 5x + 4$ **7.5.4** $x^2 + 7x + 6$

7.5.5 $x^2 + 9x + 20$ **7.5.6** $x^2 + 25x + 100$

7.5.7 $x^2 + 6x + 9$ **7.5.8** $x^2 + 15x + 56$

7.5.9 $x^2 + 7x + 10$ **7.5.10** $x^2 + 8x + 15$

Factorizing with 'the formula'

By the end of this section, you will know:

- 'The formula' will solve any quadratic equation, although sometimes the answer may be 'complex' (see Chapter 29).
- We factorize quadratic equations with 'the formula' when the coefficients a, b and c are not integers.
- To use 'the formula,' we sometimes need to first rearrange in order to start with an equation of the form '0 = …'

If we cannot factorize a quadratic equation using simple patterns (as above), we employ 'the formula.' Provided the quadratic has the form $0 = ax^2 + bx + c$, we say,

$$x = \frac{-b \pm \sqrt{b^2 - 4ac}}{2a} \tag{7.5}$$

where the symbol $\pm$ means '**either plus or minus**.' In effect, the $\pm$ sign tells us there are *two* possible answers that will correctly satisfy the equation. We might have expected this result because a quadratic equation should factorize to form *two* brackets.

It will prove very useful to memorize eqn (7.5).

■ **Worked Example 7.11**

Use the formula to factorize, $0 = 3x^2 + 14x + 8$.

Strategy:

(i) We identify the coefficients, a, b and c.

(ii) We use the formula in eqn (7.5) to determine the two values of x that satisfy the quadratic equation.

(iii) Knowing the values of x, we derive two brackets as the factors of the equation.

Solution:

(i) We start by identifying the factors: the number MULTIPLYING x^2 is a (in this case, $a = 3$). The number MULTIPLYING x is $b = 14$ and the free-standing number at the end is $c = 8$.

(ii) We then insert these coefficients into eqn (7.5):

$$x = \frac{-14 \pm \sqrt{14^2 - 4 \times 3 \times 8}}{2 \times 3}$$

which becomes

$$x = \frac{-14 \pm \sqrt{196 - 96}}{6}$$

The value of the square root is $\sqrt{100}$, which is clearly '10'. As a result of the $\pm$ sign, factorizing with the formula yields *two* answers:

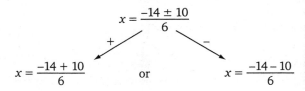

$$x = \frac{-14 \pm 10}{6}$$

$$x = \frac{-14 + 10}{6} \quad \text{or} \quad x = \frac{-14 - 10}{6}$$

so $\quad x = -\dfrac{4}{6} = -\dfrac{2}{3} \quad$ or $\quad x = -\dfrac{24}{6} = -4$

We call the two values of x that **satisfy** this quadratic its **roots**. The roots of the equation, $0 = 3x^2 + 14x + 8$ are $-\frac{2}{3}$ and -4.

Before progressing to the next stage of this problem, we should check that these solutions do actually solve the source equation. To check, we substitute the answers we obtained into the original equation: they are correct if the answer is indeed zero:

If $\quad x = -\dfrac{2}{3}, \quad 3 \times \left(-\dfrac{2}{3}\right)^2 + 14 \times \left(-\dfrac{2}{3}\right) + 8 = 3 \times \dfrac{4}{9} - \dfrac{28}{3} + 8 = 0 \checkmark$

If $\quad x = -4, \quad 3 \times (-4)^2 + 14 \times (-4) + 8 = 48 - 56 + 8 = 0 \checkmark$

So both roots are correct.

(iii) To obtain the equation's factors, we rearrange the formulae of these **roots**:

If $x = -4$ then adding 4 to both sides yields $0 = (x + 4)$

and $x = -\dfrac{2}{3}$ then adding $\dfrac{2}{3}$ to both sides yields $0 = \left(x + \dfrac{2}{3}\right)$

The second bracket contains a fraction, and therefore needs modification. We multiply throughout by '3', which yields: $(3x + 2) = 0$. Therefore, the equation $0 = 3x^2 + 14x + 8$ factorizes to: $0 = (x + 4)(3x + 2)$. ∎

■ Worked Example 7.12

The leaves of a plant never grow randomly around its stem, but grow in a regular spiral with successive leaves forming at fixed angles (see Fig. 7.1). This fraction of a circle between successive leaves represents the proportion of a circle that fulfils the equation, $x^2 = x + 1$. What is x?

Figure 7.1 The angle between successive leaves of *Fibonacci phyllotaxis* is always close to the Golden Ratio of about 137.5°. (Figure reproduced by kind permission of Professor Pau Atela, Mathematics Department, Smith College, USA.)

The angle 137.5° occurs throughout nature. When expressed as a proportion of a circle, it represents the so-called **Golden Ratio**.

We give it the Greek letter τ, (from the Greek word for 'the cut' or 'the section'). Some authors call it phi φ. The ratio of the circle sectors is, $1 : \tau$, where τ fulfils the equation $\tau^2 = \tau + 1$.

We first rearrange slightly so instead of saying $x^2 = x + 1$, we say $0 = x^2 - x - 1$. The coefficients are, $a = 1$, $b = -1$ and $c = -1$.

Next, we insert coefficients into the formula (eqn (7.5)):

$$\frac{1 \pm \sqrt{1^2 - (4 \times 1 \times -1)}}{2 \times 1}$$

so $\dfrac{1 \pm \sqrt{1 + 4}}{2}$

and $\dfrac{1 \pm \sqrt{5}}{2}$ where $\sqrt{5} = 2.236$.

Note: We often choose to leave these numbers as fractions and roots: $\dfrac{1 \pm \sqrt{5}}{2}$ is precise and $1.618\ldots$ and $-0.618\ldots$ are not.

There will be two roots to the equation:

$$x = \frac{1 \pm 2.236}{2}$$

$$x = \frac{1 + 2.236}{2} \quad \text{or} \quad x = \frac{1 - 2.236}{2}$$

so $x = \dfrac{3.236}{2} = 1.618$ or $x = -\dfrac{1.236}{2} = -0.618$

so the equation $0 = x^2 - x - 1$ has the roots $x = 1.618$ and $x = -0.618$.

The equation factorizes to, $(x - 1.618)(x + 0.618) = 0$.

We say that τ is the fraction of the circle such that it fulfils the equation, $\tau = (360 - \theta)/\theta$. Inserting $137.5°$ as θ yields $\tau = 1.618$; so this value of θ relates to the Golden Ratio. Being negative, the second root does not correspond to a real, physical situation: a negative angle is not useful. ∎

■ Worked Example 7.13

Triflouroethanoic acid (**I**) ionizes in water to form one anion and a single solvated proton:

The ionization reaction does not proceed to completion; rather, a proportion x ionizes while the remainder persists as a neutral, covalent molecule, according to the equilibrium constant:

$$K = \frac{[\text{H}^+][\text{anion}^-]}{[\text{un-ionized acid, } \textbf{I}]}$$

At equilibrium, the concentrations of $[\text{H}^+]$ and $[\text{anion}^-]$ are both cx, and the concentration of un-ionized (**I**) will be $c(1-x)$. To simplify, take the concentration $c = 1$ mol dm^{-3}.

What is the value of x if $K = 0.3$?

Strategy:

(i)   We insert numbers into the equation for K.

(ii)   We rearrange the equation to generate a quadratic equation.

(iii)   We solve the quadratic equation using the formula.

Solution:

(i) $$0.3 = \frac{cx \times cx}{c(1-x)}$$

Clearly, one of the c terms on the top row will cancel with the c term in the denominator, so, $0.3 = \dfrac{cx^2}{(1-x)}$.

(ii)   We first MULTIPLY both sides of the equation by $(1 - x)$, i.e. $0.3 (1 - x) = cx^2$. The value of c is 1, so the equation simplifies slightly to become $0.3 (1 - x) = x^2$. MULTIPLYING the bracket yields, $0.3 - 0.3x = x^2$. Slight rearranging yields, $0 = x^2 + 0.3x - 0.3$.

(iii)   The quadratic coefficients are $a = 1$, $b = 0.3$ and $c = -0.3$. Inserting these values into 'the formula' (eqn (7.5)) yields:

$$\frac{-0.3 \pm \sqrt{0.3^2 - (4 \times 1 \times -0.3)}}{2 \times 1}$$

which becomes

$$\frac{-0.3 \pm \sqrt{0.09 + 1.2}}{2}, \quad \text{where} \quad \sqrt{1.29} = 1.136.$$

The equation has two roots:

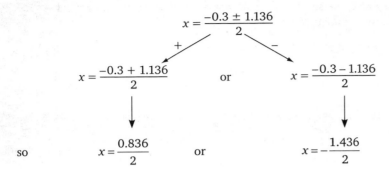

$$x = \frac{-0.3 \pm 1.136}{2}$$

$$x = \frac{-0.3 + 1.136}{2} \qquad \text{or} \qquad x = \frac{-0.3 - 1.136}{2}$$

so $\qquad x = \frac{0.836}{2} \qquad$ or $\qquad x = -\frac{1.436}{2}$

The roots of the equation are $x = 0.418$ or -0.718.

To suggest that the proportion of the acid that ionizes is negative is pure nonsense. In this context, therefore, while a negative root solves the mathematical equation, it cannot have any meaning in chemical fact.

So we calculate the proportion of the acid that dissociates as 0.418, (or 41.8%) which tells us that triflouroethanoic acid (**I**) is a strong acid.

Again, we need to take note that while the equation has two roots, only one actually has any physical significance. For example, it would be complete nonsense here to say the proportion of the acid dissociating is *minus* 71.8%! ■

> **Care:** It is a common error to suppose the word **strong** here means the acid solution is concentrated. The word 'strong' tells us the proportion of the acid that ionizes is high.

Self-test 7.6

Use the formula to solve and factorize the following quadratic equations:

7.6.1 $0 = 2x^2 + 5x - 12$ **7.6.2** $0 = 4x^2 + 10x + 6$

7.6.3 $0 = 10x^2 + 9x - 1$ **7.6.4** $0 = 5x^2 - 11x + 2$

7.6.5 $0 = 0.5x^2 + 1.75x - 1$ **7.6.6** $0 = 2x^2 - 1.6x - 0.4$

7.6.7 $0 = x^2 + 0.579x + 0.0561$ **7.6.8** $0 = x^2 - 0.25x - 0.5$

Summary of key equations in the text

The perfect square $\qquad\qquad (a+b)^2 = a^2 + 2ab + b^2$ (7.1)

The difference between two squares:

(a): $\qquad\qquad\qquad a^2 - b^2 = (a-b)\,(a+b)$ (7.2)

(b): $\qquad\qquad\qquad (x^2 - a) = \left(x - \sqrt{a}\right)\left(x + \sqrt{a}\right)$ (7.3)

Completing a square: $\quad x^2 + 2ax + b^2 = (x+a)^2 + (b^2 - a^2)$ (7.4)

The solution to a quadratic equation: $\quad x = \dfrac{-b \pm \sqrt{b^2 - 4ac}}{2a}$ (7.5)

Additional problems

The answers to some of these questions will be obvious through intuition alone. Nevertheless, determine the answer using the methodology outlined in this chapter, and thereby grow accustomed to the way the methodology works.

7.1 A complicated reaction occurs in two sequential steps:

$$A \underset{k_{-1}}{\overset{k_1}{\rightleftharpoons}} B \underset{k_{-2}}{\overset{k_2}{\rightleftharpoons}} C$$

The final **kinetic rate equation** defining the rate at which the concentration of B changes is:

$$\text{rate} = k_1 [A] - k_{-1} [B] - k_2 [B] + k_{-2} [C]$$

Factorize this expression.

7.2 When two atomic particles approach, they experience a potential energy. The magnitude of this energy varies with separation r according to the **Lennard-Jones equation**:

$$\text{energy} = 4\,\varepsilon \left\{ \left(\frac{\sigma}{r}\right)^{12} - \left(\frac{\sigma}{r}\right)^{6} \right\}$$

where ε is the depth of the potential well, and σ is the distance of closest approach. Factorize the right-hand side of this expression.

7.3 A chemical bond acts much like a spring, allowing the atoms at either end to vibrate. As the atoms vibrate, so the length of the bond between the atoms x deviates from its average length $\bar{x}$. The force between the atoms is a simple function of the difference $(x - \bar{x})$, according to the equation:

$$\text{force} = -k(x - \bar{x})$$

Square both sides of this equation, and multiply out the bracket.

7.4 The value of the Gibbs function $\Delta G^{\ominus}$ varies with temperature T. Sometimes the value of $\Delta G^{\ominus}$ can be approximated by a **virial equation**:

$$\Delta G^{\ominus} = 34000 - 1700\,T$$

Factorize this expression.

7.5 During the derivation of the **Gibbs' phase rule**, the number of degrees of freedom f is given by the expression:

$$f = cp + 2 - p - c(p - 1)$$

Multiply out the bracket and then simplify the expression by factorizing.

7.6 When calculating the line frequencies within a spectral band edge of an emission spectrum, the frequencies $\tilde{\nu}$ of individual lines are calculated using the Rydberg equation:

$$\tilde{\nu} = \Re_{\text{H}} \left(\frac{1}{n_1^2} - \frac{1}{n_2^2} \right)$$

where $\Re_{\text{H}}$ is the Rydberg constant for the hydrogen atom, and n_1 and n_2 are integers. Assuming the right-hand bracket represents the difference between two squares, factorize the right-hand side of the equation.

7.7 Use 'the formula' to solve the equation, $0 = 4x^2 - 3x - 2$.

7.8 Consider a right-angled triangle similar to that shown in Worked Example 7.6. The length of the longest side (the diagonal) is $2d$, the length of the shortest side is $\sqrt{d}$, and the length of the horizontal side is 3. Insert these lengths into the **Pythagoras equation** given in Worked Example 7.6, factorize the expression, and hence calculate the length d.

7.9 The **dissociation constant** K of ethanoic acid (**II**) is 1.75×10^{-5}. Using the same methodology as that in Worked Example 7.12, calculate the proportion x of the acid that is dissociated if the concentration of the acid is 0.1 mol dm^{-3}.

$$CH_3 - \overset{\displaystyle O}{\underset{\displaystyle O-H}{\diagdown\!\!\!\!\!\diagup}}$$

II

7.10 Complete the following square: $x^2 + 18x + 70 = 0$, and hence determine the values of x that satisfy the equation.

Graphs I

Pictorial representations of functions

The concepts underlying graphical representations

By the end of this section, you will know:

- A graph is a pictorial representation of data.
- The names of the x- and y-axis.
- The x-axis is the horizontal axis, and represents the controlled variable.
- The y-axis is the vertical axis, and represents the observed variable.

As the old proverb says it all, 'A picture is worth a thousand words.' So are graphs.

Scientists often want to know whether a relationship exists between sets of variables. It's a more advanced version of the child who says, 'What happens if I press this button?' Such scientists will probably call their investigations, 'an experiment.'

The scientist's investigation is more advanced than the child for several reasons. First, the scientist is more responsible, so he does not try sticking a finger into an electrical socket, but actually thinks about the experiment before starting it. Secondly, the scientist knows that variables come in two general types, both 'controlled' and 'observed,' as follows. Imagine that the scientist designs an experiment to see if a radio gets louder while adjusting the VOLUME control. The two variables are 'loudness' and 'the position of the VOLUME knob.' The scientist deliberately tweaks the volume control to see whether the radio output changes. We give the name **controlled variable** to the adjustments to the VOLUME knob, because we dictate its position. We give the name **observed variable** to the loudness resulting from adjusting the controlled variable (the VOLUME knob).

It's easy to turn the knob and measure how loud is the radio; it is more difficult to decide beforehand what loudness we want and then find the necessary position of the VOLUME knob: we would probably need to turn the knob back and forth until we obtain the loudness we want. This 'fine tuning' illustrates the way scientists should design their experiments: it is best when the magnitude of the controlled variable spans a wide range. It is generally more difficult if we want swap the variables, in this case deciding what the loudness should be and seeing where to position the VOLUME knob.

So the scientist performs a series of experiments, varying the controlled variable (say) a dozen times, and obtaining a dozen values of the observed variable. The first test is to repeat the experiment, merely switching the loudspeaker on and off, and seeing if the same response occurs each time. If 'yes,' the result is reproducible. If not, it is irreproducible. This topic is explored in greater depth in Chapters 14 and 15, which discuss the simple statistics of data collection.

Remember: The word **data** implies a plural; the singular of *data* is **datum**.

Secondly, the scientist varies the position of the VOLUME knob, to decide if a relationship does indeed connect the two variables, continually asking the question, 'Is there a relationship between the position of the VOLUME knob and loudness?'

Next, the scientist looks at the data. If changing one variable causes the other to change, then we discern the existence of a **relationship**. Some scientists might say a **correlation** exists, which means exactly the same thing.

Few scientists remain content with knowing a relationship exists: they usually want to know the *nature* of that relationship. While a table of data demonstrates the existence of a relationship, it is generally a lousy way of ascertaining the nature of the relationship.

Before we look at the ways of determining the mathematical nature of a relationship, we need to introduce two more terms. A **qualitative** measurement tells us the qualities of the things investigated. Looking at a table of data tells us the result that loudness and the position of a radio's volume knob are related. Other simple qualitative statements might be, 'This colour is brighter than that one,' 'This reaction gets hotter than that one,' or 'This student is the tallest in the class.' Each refers to a genuine difference in the quality of a thing, but none suggests a numerical value. If we want to put a number to the differences, perhaps saying, 'I travelled twice as fast as you,' we say more than a qualitative statement: we say something quantitative. Quoting the numerical value of an enthalpy, entropy or an *emf* is a quantitative statement.

Self-test 8.1

Decide which of the following statements are qualitative and which are quantitative:

8.1.1 $\Delta H_{\text{reaction}} = 12\,\text{kJ mol}^{-1}$

8.1.2   Your flask is bigger than my beaker

8.1.3   The temperature rose faster during the second experiment

8.1.4 $\lambda = 450\,\text{nm}$

8.1.5   We are late for the class

8.1.6   The temperature was raised by 2.3 °C

Quantitative measurements are generally more difficult to perform than qualitative measurements, so we generally start a chemical investigation by assessing qualitative relationships and only then do we make quantitative statements—it would be silly to try to quantify if no relationship exists. For example, we first perform a reaction and ascertain whether a reaction has occurred, and what is formed. Only then is it sensible to calculate the chemical yield. The yield of a reaction cannot be calculated if we don't even know what has been formed.

Drawing a graph is one of the best ways to discern the quantitative relationship between the observed and controlled variables. The most common graphical method is that attributed to the French philosopher and scientist René Descartes (1596–1650). His name explains why we say many parameters relating to graphs are Cartesian. Other types of graphical coordinate, such as polar coordinates, will be discussed in Chapter 16.

We always start a graph by drawing the controlled variable along the horizontal axis. We generally call this axis the *x*-axis. Its more scientific name is the abscissa. We draw the data concerning the observed variable along the vertical axis, which we call either the *y*-axis or the ordinate. We call the intersection of the two axes the origin, because the two axes originate from this point.

The *x*-axis is the horizontal axis while the *y*-axis is the vertical axis. A simple way to remember which axis is which is to say, 'an e**X**panse of road goes horizontally along the *x*-axis', and 'a **Y**o-Yo goes up and down the *y*-axis'

The word **abscissa** comes via the Latin *absindere* that means in this context 'to cut.' The word **ordinate** again comes from the Latin, and represents an abbreviation of *linea ordinate applicata*—'a line drawn parallel'.

Generating a graph: plotting

By the end of this section, you will know:

- The purpose of drawing a graph is to discern whether two variables are related—we look for a *correlation*.
- How to plot a graph, both manually and using computer programmes such as *Excel*™.
- How to label the graph's two axes.

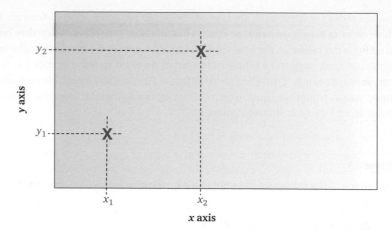

Figure 8.1 We draw a graph by drawing a vertical line through each value of x and a horizontal line through each value of y. We draw a small cross × to represent the point where the two lines intersect, and then draw the line of best fit through the crosses. We discuss the phrase 'best fit' in Chapter 12.

To **plot** a graph, we first obtain pairs of data, either by calculation or experiment. The data pairs must be *complementary*, by which we mean that each value of the controlled variable x must be associated with its own value of observed variable y. When citing the data, we always cite the controlled variable first: (x_n, y_n). The subscripted n here represents a convenient form of labelling: perhaps it tells us the order in which we obtained the data, or maybe it indicates their numerical order. But each pair will need a unique label.

Next, for each pair, we follow the vertical line through the value of x and the horizontal line through the value of y to the point of intersection. We indicate the point where these 'guidelines' intersect with a small cross ×, as represented in Fig. 8.1, a small circle as with many of the graphs here, or simply a well-defined dot. In general, a cross is better because its position is more precise; a circle can be used to indicate errors; see Chapter 15.

It does not matter how large or small we draw the crosses, provided they are legible, consistent, and neat, and it is easy to identify the specific location of the data point. We lightly draw these guidelines in pencil when we first attempt such graphs, but we no longer need to draw them physically after a little practise. We see them 'in our mind's eye' as a prelude to drawing a physical cross. Some people draw the guidelines, draw a cross, and then rub out the guidelines to make the graph neater.

Having plotted the data, we then label the resulting graph. Half-way along the x-axis scale we write a description (involving both words, symbols and units) to indicate the nature of the controlled variable. We position the label beneath the rows of numbers below the line. We position the label for the y-axis half-way up the axis and slightly to the left of the numbers making up the y-axis scale.

■ **Worked Example 8.1**

Hydrogen peroxide H_2O_2 decomposes in the presence of excess cerous ion Ce^{III} to form ceric ion Ce^{IV}:

$$2H_2O_2 + 2Ce^{III} \rightarrow 2H_2O + Ce^{IV} + O_2$$

The data below were obtained at 298 K, and show how the concentration of H_2O_2 decreases with time t. Plot the data as concentration $[H_2O_2]_t$ against time, t.

Time, t / s	2	4	6	8	10	12	14	16	18	20
$[H_2O_2]_t$ / mol dm^{-3}	6.23	4.84	3.76	3.20	2.60	2.16	1.85	1.49	1.27	1.01

The subscripted 't' printed on the outer right-hand side of the concentration bracket tells us this concentration changes with time, t, after the reaction starts. In this same nomenclature, the concentration at the start of the reaction is $[H_2O_2]_0$.

Strategy:

(i) We first decide which is the controlled and which the observed variables.

(ii) Having decided, we plot the controlled variable on the horizontal x-axis and the observed variable on the vertical y-axis.

Solution:

There are two ways we can decide which is the controlled and which the observed variables:

(i) (a) When compiling a table of data, it is common practice to write the controlled axis on the top row and the observed data as lower rows. This convention instructs us that time is the controlled variable. Concentration is therefore the observed variable. This method is not always safe—for example, the person compiling the table might not have known the convention.

(b) So the best way to decide which is the controlled and which the observed variable is to imagine ourselves performing the experiment ourselves. A moment's thought suggests it would be very difficult indeed to dictate the concentration of the peroxide, because it keeps changing. We can't wait until the concentration reaches a certain value, and then notice the time. It would be too fiddly. On the other hand, it's quite easy to imagine ourselves with a stop-clock in one hand, deciding when to take each reading.

As corroboration, we notice how the time increments are evenly spaced and the concentrations are not. Normally, only a controlled variable will be evenly spaced in this way.

(ii) Knowing that the time is the controlled variable, we plot times t along the horizontal x-axis. 'Concentration' represents the observed variable, so we plot $[H_2O_2]_t$ up the vertical y-axis; see Fig. 8.2.

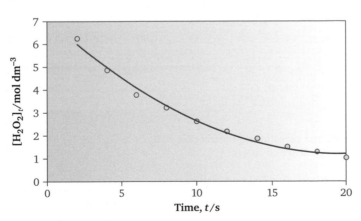

Figure 8.2 Graph of the concentration $[H_2O_2]_t$ as the observed variable, plotted on the y-axis, and time as the controlled variable, plotted on the x-axis. ■

Aside

Plotting a graph using *Excel* ™ 2007 is very straightforward:

1. Open a new *Excel* spreadsheet.

2. Type in the data. We can enter the data either horizontally or vertically:

- If entering the data vertically, type the *x*-values into a column positioned immediately to the left of the *y*-values.
- If horizontally, type the *x*-data into a column positioned immediately above the *y*-data.

3. Using the mouse, highlight the data, and select the 'Insert' tab at the top of the Worksheet.

4. When the next window opens, select the 'Scatter' option. The chart will appear in the main window.

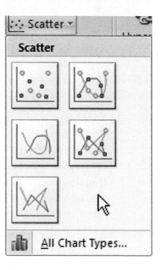

5. To label the chart title and axes, click on the 'Layout' tab.

6. To put the chart on a separate sheet, right click on it, then select 'Move Chart.'

Drawing graphs with *Excel 2003* is explained in the appendix to this chapter. We discuss other options with *Excel* graphs in the next chapter.

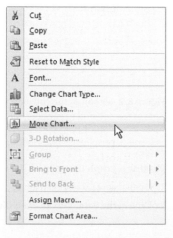

Self-test 8.2

Plot the following graphs. In each case, use values of x from -4 to $+4$.

8.2.1 $y = 5x + 3$ **8.2.2** $y = 12x - 2$

8.2.3 $y = -4x + 1$ **8.2.4** $y = 0.2x$

8.2.5 $y = 0.3x + 0.2$ **8.2.6** $y = x$

8.2.7 $y = 0$ **8.2.8** $x = 2$

The correct appearance of a graph

By the end of this section, you will know that a graph:

- Must always be labelled, and where to place the labels.

- Should always fill the page as far as possible.

- If linear, we should scale the axes to ensure the line is inclined at an angle close to 45° to both axes, therefore enhancing the accuracy when reading from the graph.

Position on a page:

If we want to interpret the graph, we should plot the data such that they *fill the page*. Graphs that occupy one small part of the page waste paper. Furthermore, if the graph is squeezed into a small space, we are likely to experience errors in reading from it, and in calculating the gradient.

Labelling:

Both axes on a graph must be labelled. Figures 8.1 and 8.2 both demonstrate the correct positions when positioning the labels:

- We position the x-axis label just below the x axis, and midway along its length.

- We position the y-axis label just to the left of the y-axis, half-way up its height.

We now look at each label in Fig. 8.2. The vertical scale says more than just 'concentration of analyte, c,' and the horizontal scale says more than 'time, t' because neither description is complete until we indicate the *units*. The label is incomplete without the appropriate units.

The straightforward way to include the units is to write the label as, 'quantity ÷ units', so a temperature axis might be labelled as, 'temperature T/K' and a concentration axis might be labelled 'concentration $c/\text{mol dm}^{-3}$'. We call this way of labelling axes **quantity calculus**, and discuss it in detail in Chapter 30.

The angle of the line:

If a graph is linear (see below), we should scale the axes in such a way that we ensure that the line is positioned at an angle close to 45° to either axis. If we plot the line such that it is nearly horizontal on the page, or nearly vertical, we will probably experience profound difficulties reading from it.

For example, look at the graph in Fig. 8.3. The spacing between the values of the controlled variable x_1 and x_2 are wide, and therefore easily read from the graph. Conversely,

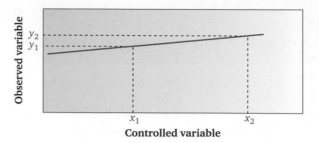

Figure 8.3 It is not possible to read from the y-axis with any accuracy if we draw the graph with a near-horizontal line.

because the line was drawn at such a shallow angle, the respective values of the observed variable y_1 and y_2 are so close together they cannot be easily differentiated. Accordingly, if we were to use Fig. 8.3 as a calibration curve, it is likely that the values we read from the graph would be in error.

In a similar way, if we draw the graph with a straight line that is nearly vertical, the errors associated with reading x_1 and x_2 are again high. For these reasons, we can minimize errors when drawing graphs by scaling the axes, so that we can draw the line close to an angle of 45°.

If a graph is curved, we should manipulate the values of x and y in such a way that the portion of interest is neither too shallow nor too steep. And, above all else, we should aim to fill the page as far as practicable.

Reading the different types of graph

By the end of this section, you will:

- Know that looking at a graph will often tell us the general type of correlation between the two variables.
- Know that one of the common causes of graphs showing scatter is the uncompensated existence of *compound variables*.

Next, we look at several types of graph, each demonstrating a different kind of relationship between the observed and controlled variables. In common parlance, we talk about plotting the observed variable 'against' the controlled variable. It is bad practice to cite the controlled variable first; most scientists have been taught to cite the *observed* variable first. If we plotted a graph to show the loudness of a radio as a function of the volume knob's position, we would say, 'A graph of loudness against knob position.' It would be wrong to say, 'A graph of knob position against loudness.'

Figure 8.4 shows a schematic of a straight-line graph, which passes through the origin. We would still call it a 'straight-line graph' even if the line had not passed through the origin. Figure 8.4 tells us that when we vary the controlled variable x, the observed variable y changes in direct proportion. A glass of orange cordial affords an obvious example of such a situation: the intensity of its colour increases in linear proportion to the concentration of the cordial concentrate, according to the Beer–Lambert law. The graph in Fig. 8.4 goes through the origin because no orange colour is discernible before we add

The word **schematic** means either to represent something with symbols or to draw a general example. The word comes from the Greek *schema,* meaning a 'form' or 'figure.'

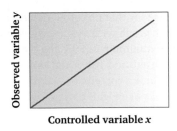

Figure 8.4 A graph of observed variable (along the y-axis) against controlled variable along the x-axis: a simple linear proportionality, so $y = \text{constant} \times x$.

Linear means line, from the Latin *linea*, meaning 'a line.' Scientists generally take the word 'linear' to mean a *straight* line.

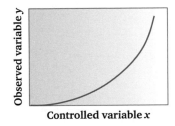

Figure 8.5 A graph depicting the situation whereby the observed variable y is not a simple function of x, although a relationship clearly does exist. A simple example might be an exponential graph (see Chapter 14).

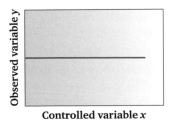

Figure 8.6 A graph depicting a situation when the variable x is independent of variable x: the yield of a chemical reaction (as y) against the time of day (as x).

cordial to the water (i.e. its concentration is zero before we add any coloured cordial). We draw a bold straight line through the data if they fall on a perfect straight line, or the best straight line through the data if they display any scatter.

Figure 8.5 shows a different kind of correlation. The data display a clear *trend* although the graph is clearly not a straight line. Again, we discern a relationship between the observed variable y and the controlled variable x. This time, the observed variable y increases at a faster rate than the controlled variable x. A simple example might be the rate at which a reaction proceeds (the observed variable, y) as we raise the reaction temperature (the controlled variable, x).

Figure 8.6 shows another straight-line graph, but this time it is horizontal. It tells us the observed variable y remains unchanged whatever we do to the controlled variable x. A simple example of such a graph is the yield of a chemical reaction (as y) against the time of day (the controlled variable, x): in the absence of other chemicals and variables, the yield will be same whether we react the chemicals in the morning, afternoon or evening. The graph shows no correlation because changing x does not cause y to change. We say, 'the variable y is not a function of x'.

We must be careful before we say that no correlation exists. It is easy to lose sight of a genuine correlation if we plot a graph with too insensitive a y-axis scale. For example, if the values of y vary between a minimum of 50.1 and a maximum of 50.2, then a graph plotted using y-scale limits such as 0 and 100 would look just like that in Fig. 8.6. Insensitive axis scales can completely hide a correlation—as discussed in the previous section.

If we plot a graph and it looks like Fig. 8.6, we should always replot it, using a more sensitive y-axis scale. When replotted, the data in Fig. 8.6 would generate a graph looking much like Fig. 8.4 or Fig. 8.5.

■ **Worked Example 8.2**

The data below relate to the thermal isomerization of 1-butene (**I**) to form *trans* 2-butene (**II**):

T/K	686	702	733	779	826
K	1.72	1.63	1.49	1.36	1.20

The equilibrium constants of reaction K clearly depend on the temperature. To demonstrate the importance of choosing the axis scales, we plot the data twice:

(a) Using a silly choice of axes: y-scale extends from 0.0 to 2 and the x-axis extending from 0 to 900 K.

(b) Using a sensible choice of axes: the y-scale extends from 1.1 to 1.8, and the x-axis extending between 650–850 K.

Look at the graphs in Fig. 8.7. The lower graph is vastly superior because:

• The data actually fill the page, so we waste very little of the paper.

• The data occupy the whole of both axes, so it is easier to read data from the graph and to discern the type of relationship that exists.

• Because the data is spread out, we can see the type of trend. It is very difficult to tell from Fig. 8.7(a) whether the graph is curved or linear. Figure 8.7(b) clearly demonstrates how the graph has a slight curvature. ■

We need to note:

• The maximum on the y-axis should be very similar to the largest value of y.

• The minimum on the y-axis should be very similar to the smallest value of y.

• We plot the data of a straight-line graph to *fill the page*.

• If at all possible, we should draw the straight line at an angle of about 45°.

The graph in Fig. 8.8 shows another common situation: the data seem to indicate no straightforward relationship whatsoever. Certainly, this graph might be telling us that no relationship exists at all because the magnitude of the controlled variable x does not have any bearing on the observed variable y. We say the observed variable y is *independent* of the controlled variable x. In other words, there is no correlation. Nevertheless, we do see a range of y values as x is varied.

Notice the two different uses of K here. The italic K indicates a variable (equilibrium constant) while the upright K is a unit (kelvin).

When plotting these data with *Excel*, the second set of axes is the default choice of scales.

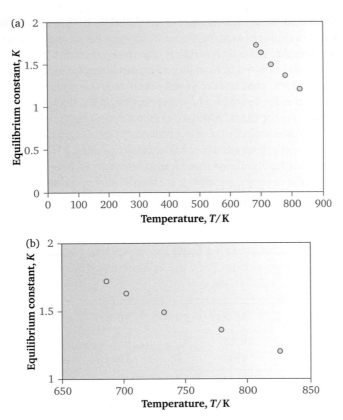

Figure 8.7 Graphs to demonstrate the need for a sensible scale on the y-axis: (a) foolish axes, and (b) sensible axes.

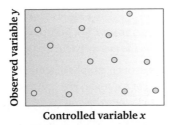

Controlled variable x

Figure 8.8 A graph depicting the situation in which no relationship is apparent between the observed variable y and the controlled variable x, although y does vary.

Several possibilities could explain the apparent lack of any correlation.

1. The value of y is indeed completely random. If this were the case, repeating the experiment using the same values of controlled variable x would generate wholly different values of the observed variable y. We would then conclude that our data are **irreproducible**. Faulty equipment represents one of the simplest causes of irreproducible data, and poor experimental technique is another.

2. Perhaps x represents a **compound variable** (see p. 2). We initially assumed the observed variable y responded to a *single* controlled variable x, but perhaps our original analysis was naïve, and in fact several variables dictate the observed value of y. In other words, we failed to design our experiment properly.

Perhaps we can call these various types of controlled variable x_1, x_2, etc. An everyday example might be a student's exam performance as y and his IQ as x_1. The lack of a clear correlation suggests that while IQ is certainly important, other variables also impinge on the magnitude of the exam result: examples might include commitment as a further controlled variable (call it x_2) and state of health—both mental and physical—as x_3. A more realistic choice of controlled variable x here therefore might be, 'How good is the student at chemistry on the day of the exam,' which is certainly a complicated function of x_1, x_2 and x_3.

That we did not discern more than one contributory factor suggests we failed to design our experiment properly. If we need to consider many variables, it is better to keep all but one of them constant during an experiment, and then ascertain how the *single* controlled variable x_1 affects the observed variable, y. We then modify x_2 and see its affect on y, and so on. In this way, we will obtain a more realistic idea of how the variables interrelate.

■ **Worked Example 8.2**

We inflate a car tyre. The pressure p inside the tyre changes (increases) concurrently with changes in the gas volume V and the internal temperature T, according to the data below:

Pressure p/Pa	10 000	20 000	30 000	40 000	50 000
Volume V/m³	0.4	0.1	0.5	0.3	0.05
Temperature T/K	481	240	1804	1443	301

The data refer to one mole of an ideal gas. Plot the following three graphs:

(a) Temperature T (as y) against pressure p (as x).

(b) Temperature T (as y) against volume V (as x).

(c) Temperature T (as y) against the *compound* variable pressure × volume, i.e. $p \times V$ (as x).

Pressure p, volume V and temperature T are interrelated, according to the ideal-gas equation, $pV = nRT$, where R is the gas constant and n is the amount of gas.

If all three variables change at once, graphs of temperature T as y against other variables as x will exhibit scatter. For example, Fig. 8.9(a) shows a graph of temperature T (as y) against pressure p (as x) when the volume was uncontrolled; and Fig. 8.9(b) shows a graph of temperature T (as y) against volume V (as x) when the pressure was uncontrolled. Both graphs clearly show appalling scatter.

To ensure the graph is linear, we must plot the compound variable '$p \times V$' as x. Figure 8.9(c) shows such a graph, with temperature T (as y) plotted against the *compound* variable 'pressure × volume', i.e. $p \times V$ (as x). This graph is suitably linear.

Incidentally, Worked Example 8.2 also demonstrates why chemists recommend that we obtain data using **thermostatted** apparatus. We eliminate much of the scatter on the graphs merely by ensuring the temperature T does not vary.

A **thermostat** is a device for keep a temperature constant. The word comes from the two Greek roots *thermo*, which means 'energy' or 'temperature,' and 'stat–', which derives from the *statikos*, meaning 'to stand', i.e. not move or alter.

$$T = \frac{V}{nR} \times p$$

$$T = \frac{p}{nR} \times V$$

$$T = \frac{1}{nR} \times pV$$

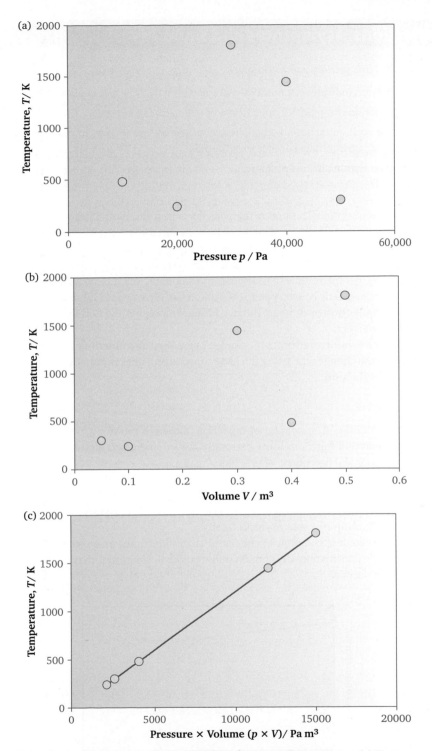

Figure 8.9 Data relating to one mole of an ideal gas inside an inflating car tyre. (a) Temperature T (as y) against 'pressure p' (as x) but the volume was uncontrolled; (b) Temperature T (as y) against 'volume V' (as x) but the pressure was uncontrolled; (c) Temperature T (as y) against the *compound* variable 'pressure × volume', i.e. $p \times V$ (as x). ■

Inflections, asymptotes and sigmoids

By the end of this section, you will know:

- An inflection indicates the point on a graph where two linear regions meet.
- How to obtain the 'point' of inflection when the graph is curved.
- An asymptotic curve never actually reaches the axis, but appears to skim alongside it, becoming essentially parallel with the axis.
- A sigmoidal graph has an 'S' shape.

Inflections: Many graphs show more than one linear portion. Typically, the graph is linear at low values of x and also linear at higher values of x, but the linear portions have different slopes. Usually, a smooth curve connects the two straight sections.

Figure 8.10 shows an example of an inflection. It shows the molar conductivity Λ of a solution (as y) against volume of sodium lauryl sulphate, SLS, (as x) during a titration. An inflection can clearly be seen where the curved line joins two straight lines. We obtain the **point of inflection** by drawing the two **blue** extrapolants and noting the values of x where they intersect.

We call such a sudden and dramatic change in gradient an **inflection**. Often, we locate the centre of the change (e.g. using the **blue** construction lines in Fig. 8.10) and call this the **point of inflection**.

Asymptotes:

An asymptote differs from a point of inflection, although the two are often confused. Some mathematical functions form a strongly curved line when plotted. For example, the simple reciprocal function $y = 1/x + 10$ in Fig. 8.11.

The curved line in Fig. 8.11 never leaves the first quadrant and therefore never crosses either axis. But it does look as though the line will actually touch the y-axis at really tiny values of x and huge values of y. We say the y-axis is an **asymptote**; and the line approaches the ordinate y-axis **asymptotically**.

Asymptotes are less common in chemistry than inflections, and (generally) convey extremely little information. In common undergraduate chemistry, they are only useful in adsorption science.

In fact, at $x = 0$, the line will in fact *just* touch the y-axis, but such a situation will probably not correspond to any actual physicochemical reality.

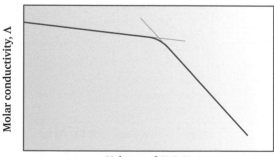

In this example, the value of x at the point of inflection relates to the endpoint of the titration.

Figure 8.10 When a solution of sodium lauryl sulphate (SLS) is added to a beaker of water, the molar conductivity Λ of the resultant solution undergoes a sudden change. Graph of molar conductivity Λ (as y) against volume of SLS (as x). The conductivity curve clearly shows an inflection.

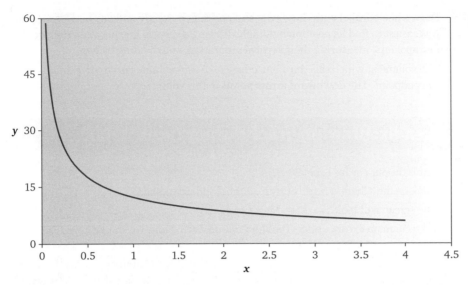

Figure 8.11 Graph of the function $y = 1/x + 10$. The graph approaches the y-axis asymptotically as x tends to 0.

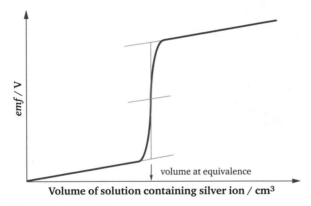

Figure 8.12 Schematic plot depicting a sigmoidal graph. The data were obtained during a potentiometric titration, measuring the *emf* of a simple cell (as 'y') while a solution of silver ion was added to a solution of KCl (as 'x').

Sigmoidal graphs:

The peculiar-looking word *sigmoid* simply means 'looks like an S,' so sigmoidal graphs have an S shape. They are fairly common in chemistry, especially as a result of a titration.

Figure 8.12 shows a sigmoidal graph, which we would normally obtain during a potentiometric titration. The graph differs from Fig. 8.10 in insofar as the second linear portion is parallel with the first. The curved portion in the middle of the 'S' may or may not appear linear—if it *is* linear, the fact of its being linear conveys no physicochemical meaning.

Having performed such a titration, we usually need to determine the endpoint. We do so by finding the graph's **point of inflection**, as follows:

- Extrapolate the two linear portions (in **blue**). These extrapolants will usually be parallel.

The purpose of the sigmoidal graph in Fig. 8.12 is to determine the solubility product of AgCl.

We can use the corresponding *emf* to determine the solubility product K_{sp} of a nearly insoluble salt.

- Draw a line parallel with these two, and positioned exactly midway between them. It is easiest to find its position by simply drawing a vertical line between the two extrapolants, measuring their vertical separation, and dividing by two.

- The volume at which this third line crosses the *emf*–volume curve corresponds to the endpoint. The downward arrow points to this volume.

Appendix: drawing graphs using Excel™ 2003

1. Enter the data as for Excel 2007.
2. Click on the 'Chart Wizard' button on the standard toolbar.
3. The screen will look like this—Step 1 of 4.

   It's a common error to select the third option down, 'line' we should select 'XY (scatter)', and press 'next.'

   The screen will change, giving Step 2 of 4. This screen is important insofar as it previews the final graph. If it shows appalling scatter, we can abort before wasting more time on it. If it shows mistyped data, we press 'back' and correct the problem.

   A common cause of error occurs when we accidentally swap the variables *x* and *y*. In addition, we must ensure the correct tab is selected, either Rows or Columns.

   If OK, press the 'Next' button. The screen will change, giving Step 3 of 4, and will look like this:

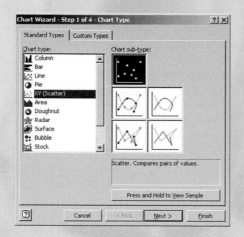

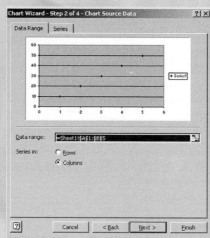

4. If we wish to label the axes, type them in the respective boxes on the left of the Chart Wizard. Note that such labels will not appear in either bold or italic type at this stage, but we can change them later, if needed. When satisfied, press the 'Next' button. The screen will change, showing Step 4 of 4:

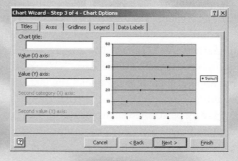

5. The last stage of the Wizard allows us to decide the format of the final graph:
    *   If we want the graph to appear on its own, as an entire page, click the top button.
    *   If we want the graph to appear next to the data, slick the bottom button. The graph will appear as an 'object'. This option is the default setting.
6. Click 'Finish.'

Additional problems

8.1–5 Decide which of the following statements are qualitative and which are quantitative:

8.1   (a) I performed more experiments than you.

8.2   (b) The spectrum contains a peak at 500 nm.

8.3   (c) The lecture lasted 50 minutes.

8.4   (d) The lecture was longer than last week's lecture.

8.5   (e) Five hundred people attended the lecture.

8.6   **The Beer–Lambert Law** says the optical absorbance A of a chromophore is directly proportional to its concentration c. The data below refer to the absorbances Abs of aqueous solutions of permanganate ion MnO_4^- of concentration c. (The optical path length, l, was 1 cm, and the wavelength of observation was 523 nm.)

$c/\text{mol dm}^{-3}$	0	0.0001	0.0002	0.0003	0.0004	0.0005	0.0006
Absorbance, Abs	0	0.2334	0.4668	0.7002	0.9336	1.167	1.4004

Note that optical absorbance Abs has no units.

Plot a graph of A (as 'y') against c (as 'x') to confirm Beer's law.

8.7   When we leave a mixture of volatile liquids to stand, the composition of the vapour above the liquid obeys **Raoult's law**: the amount of a material in the vapour mixture above the mixture relates to the product of the vapour pressure of the *pure* liquid $p^{\ominus}$ and the mole fraction x of the component in the liquid mixture. These data relate to benzene.

Mole fraction, x	0	0.2	0.4	0.6	0.8	1
Partial pressure, p/Pa	0	14 940	29 880	44 820	59 760	74 700

Plot a graph of pressure p (as 'y') against mole fraction of benzene x (as 'x') to show that Raoult's law is obeyed.

8.8 The **ideal-gas equation** suggests a correlation between the volume V of an ideal gas and its pressure p. Plot a graph of pressure p (as 'y') against volume V (as 'x') to ascertain whether such a relationship holds. The data refer to helium.

V / m³	0.01	0.02	0.025	0.03	0.04	0.05	0.06	0.07	0.08	0.1	0.2
p / Pa	100	50	40	33.3	25	20	16.7	14.3	12.5	10	5

8.9 The **ideal-gas equation** suggests the volume of an ideal gas should be directly proportional to the temperature. Using the data below, plot a graph of volume V (as 'y') against temperature T (as 'x').

T / K	280	300	320	340	360	380	400	420	440
V / m³	0.0233	0.0249	0.0266	0.0283	0.0299	0.0316	0.0333	0.0349	0.0366

8.10 Consider the data below, which relate to the second-order racemization of glucose in aqueous hydrochloric acid at 17 °C. The concentration of glucose is '[A]'.

Time, t / s	0	600	1200	1800	2400
[A] / mol dm⁻³	0.400	0.350	0.311	0.279	0.254

Plot a **second-order kinetic** plot, i.e. plot $1/[A]_t$ (as 'y') against time (as 'x').

Graphs II

The equation of a straight-line graph

The algebraic equation of a straight line is $y = mx + c$

By the end of this section, you will know that:

- We can write the mathematical form of a straight-line graph as $y = mx + c$.
- 'm' is the gradient of the graph.
- 'c' is the intercept of the graph on the y-axis.
- We may need to *reduce* the equation if the equation has a form $n \times y = mx + c$, by dividing throughout by 'n'.

Remember: The lack of a printed operator between m and x means they are multiplied together, so m and x are MULTIPLIED in eqn (9.1).

The equation of a straight line will always take the algebraic form in eqn (9.1):

$$y = mx + c \qquad (9.1)$$

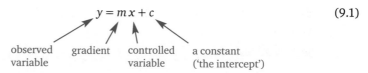

observed variable   gradient   controlled variable   a constant ('the intercept')

Self-test 9.1

For each of the following, what is the gradient m and what is the intercept c:

9.1.1 $y = 2x + 4$

9.1.2 $y = 3.5x - 2$

9.1.3 $y = 4x + 8.4$

9.1.4 $y = 4x + 3$

9.1.5 $y = x + 22$

9.1.6 $y = -4x - 4.3$

If we see an expression written to look like eqn (9.1), but with a factor in front of the y, then in such cases, it is common practice to divide throughout by that constant. We say we **reduce** the equation.

A **straight line** will always follow an equation of the type $y = mx + c$, where m is the gradient and c is the intercept on the y-axis (i.e. when the value of x is 0).

■ Worked Example 9.1

The relationship linking the electrochemical current I (in milliampères) through an electrode of area A (in cm²) follows the form:

$$5.2\,I = 3\,A + 12.2$$

Reduce the relationship to the form $y = mx + c$.

In words, the equation indicates that for every three-fold increase in area, the current increases by a factor of 5.2. We deduce that current I is the observed variable y and the electrode area A is the controlled variable x. The additional current of 12.2 milliampères is constant, whatever the size of the electrode, and probably represents a fault in the circuitry measuring the currents.

To reduce the equation, we divide throughout by the factor of 5.2:

$$\frac{5.2}{5.2}I = \frac{3.2A}{5.2} + \frac{12.2}{5.2}$$

and, because $\frac{5.2}{5.2} = 1$

$$I = \frac{3A}{5.2} + \frac{12.2}{5.2}$$

We might have written out the equation in this way:

$$
\begin{array}{c|c|c|c}
I & = & \dfrac{3}{5.2} & A & + & \dfrac{12.2}{5.2} \\
y & = & m & x & + & c
\end{array}
$$

It would be silly to leave the equation like this, so we would then perform the two divisions, to yield:

$$I = 0.58A + 2.35 \ \blacksquare$$

Self-test 9.2

Reduce the following equations to the form $y = mx + c$:

9.2.1 $2y = 4x + 10$

9.2.2 $3y = 5x - 30$

9.2.3 $-5y = 10x + 15$

9.2.4 $7.1y = 4x$

9.2.5 $30.2y = 19x + 10.4$

9.2.6 $-4y = -4x + 12$

9.2.7 $3.17y = -2x + 1.22$

9.2.8 $10^6 y = 10^4 x - 10^7$

'Satisfying' the equation of a line

By the end of this section, you will know that:

- Data satisfy the equation only if they lie on the straight line.
- A point not on the line does not satisfy the equation.

Consider a straight line of the form $y = 4x + 2$. If we know the value of x, then we can calculate the value of y. In this case, we say that y is a sum: in words, y equals four times the value of x plus an additional two. If $x = 0$, then $y = 2$. If $x = 1$, then $y = 6$, and so on. We could write this data in the form of a table:

x	−3	−2	−1	0	1	2	3
$4x$	−12	−8	−4	0	4	8	12
$4x + 2 = y$	−10	−6	−2	2	6	10	14

We could then construct a graph in just the same way we drew Fig. 8.1, with the values of x from the table plotted against the respective values of y we have just determined. It is convenient to group the data together: we often put them in brackets, with the value of x first, followed by a comma, then the value of y. In this example, we would write, $(-3, 10)$, $(-2, -6)$, $(-1, -2)$, and so on. We might call these **data pairs**, or merely the **points**.

The point $(0, 2)$ lies on this straight line, as does $(2, 10)$. We say these points **satisfy** the equation. The point $(2, 7)$ does not lie on the line, so we say it does *not* satisfy the equation. A moment's thought ought to persuade us that an infinite number of points will satisfy a single line, because a straight line can stretch from $x = -\infty$ through to $+\infty$.

The intercept c

By the end of this section, you will know that:

- 'c' is the intercept of the graph on the y-axis.
- The intercept on the y-axis is only the same as 'c' if the x-axis extends as far as $x = 0$.
- The intercept will be positive if it intercepts the y-axis above the x-axis and negative if it intercepts below the x-axis.
- The intercept is zero if the line goes through the origin.
- A straight line occupies at least two of a graph's quadrants.

We can tell the value of the constant c at a glance, because the straight line strikes the y-axis at a value of c. For this reason, it is quite common to see c called 'the intercept.' The intercept can have any value, and can be positive, negative or goes through the origin, in which case the intercept is zero.

If the straight line goes through the origin of the graph, the value of the constant c is zero (because the value of y at the origin is 0). Such a situation would arise if we had tared the weighing boat before weighing our sample on, say, a top-pan balance. If $c = 0$, we would then write either $y = m\,x + 0$ or even omit the constant, saying $y = m\,x$. These two equations mean the same.

If we have the equation of a line (and it has been *reduced*, as above), then we can read off the intercept as the final number, that is, the figure expressed without an accompanying x or y. For example, if the equation is $y = 4x + 2$, then the intercept must be '2'. It is *not* true, however, to say the intercept is the same as the constant if we forgot to first reduce the equation to the form $y = m\,x + c$, but had some multiple of y.

A **tare** is the allowance made for the weight of a weighting boat or the packaging encasing something. The law says we must tare the weight of the goods we buy in a shop to accommodate additional weights such as packaging. This explains why we see the words 'NET WEIGHT' printed on a packet.

■ Worked Example 9.2

Having dissolved varying amounts of the organic dye crystal violet (**I**) in water, we determine the optical absorbance of each solution. Figure 9.1 is a Beer's law plot constructed with the data.

What is the figure's intercept?

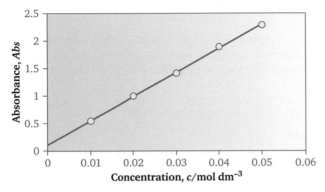

I

Figure 9.1 Graph of optical absorbance *Abs* (as '*y*') of Crystal Violet (I) in aqueous solution against concentration *c* of absorbing species in solution (as '*x*'): Beer's Law.

The bold line touches the *y*-axis at an absorbance of 0.1, so the intercept is 0.1. The equation of this straight line therefore has the algebraic form $y = mx + 0.1$. As yet, we do not need to know the magnitude of the gradient *m*. ■

Sometimes the intercept on a graph is negative.

Beer's Law suggests the line should pass through the origin. In this example, the cause of the graph's non-zero intercept suggests that we failed to 'blank' the spectrum.

■ **Worked Example 9.3**

A solution of the chiral alkyl halide (**II**) reacts with hydroxide ion to form a racemic mixture. When a beam of plane-polarized light is passed through the reaction mixture, its angle of twist θ changes steadily with time until it reaches zero, according to the graph in Fig. 9.2. What is the intercept?

II

The bold line in Fig. 9.2 intercepts both axes: it cuts the *x*-axis at $t = 35$ min and crosses the *y*-axis at $\theta = -15°$. We only need the intercept on the *y*-axis when deriving the equation of the straight line, so we can ignore where the line crosses the *x*-axis. Accordingly, the intercept is $-15°$.

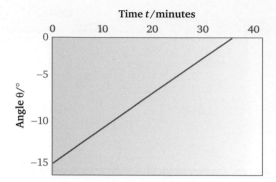

Figure 9.2 A beam of plane-polarized light twists through an angle of θ as it passes through a solution containing a chiral compound: graph of θ (as 'y') against time (as 'x') showing the way that θ increases to 0 with time if racemization occurs during the reaction.

We write the equation of this straight line as either

$$y = mx + (-15)$$

or as

$$y = mx - 15.$$

These equations mean the same, but we would usually use the latter style because it looks less complicated. Note that we could also say $\theta = mt - 15$, where θ is simply the variable on the ordinate axis and t the variable on the abscissa. ∎

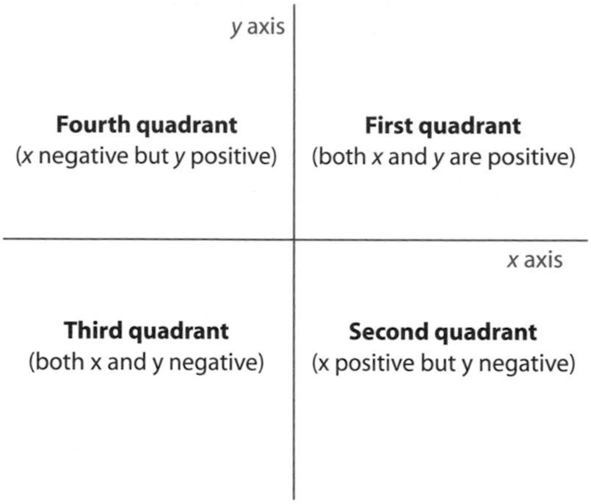

	y axis	
Fourth quadrant (*x* negative but *y* positive)	**First quadrant** (both *x* and *y* are positive)	
		x axis
Third quadrant (both x and y negative)	**Second quadrant** (x positive but y negative)	

Figure 9.3 Graph labelled with the names of the four quadrants.

A **quadrant** of a Cartesian graph is one of the four segments produced by dividing a graph by both a horizontal and a vertical axis. The word comes from the Latin *quadrans* meaning a quarter.

This last graph depicts data drawn *below* the *x*-axis. In fact, we may need to plot data in any of the four **quadrants** of the axes. Figure 9.3 names the four quadrants. We will not employ this notation, but the names are so common that we will certainly encounter them in other books. A straight-line graph will always occupy at least two quadrants.

Occasionally, we find that a line has been expressed in the form $y = c + mx$ rather than the more familiar $y = mx + c$. Having learnt the algebra in Chapter 2, we should now recognize how this form is identical to the more familiar version, $y = mx + c$ except its order has been changed.

■ Worked Example 9.4

The ionic conductivity Λ of a solute in solution depends on its concentration c according to the **Onsager equation** for strong electrolytes:

$$\Lambda = \Lambda^\circ - b\sqrt{c}$$

where b is a constant, Λ is the ionic conductivity and Λ° is the conductivity of the same compound at zero concentration. How would we obtain Λ° graphically?

Care: Do not confuse concentration c with the intercept, which is also symbolized as c.

Strategy:

(i) We first decide how the equation may be represented as a straight line.

(ii) Next, we analyse the nature of the general equation $y = mx + c$, and allocate the terms m and c to the component parts of the Onsager equation.

Solution:

(i) The variables in this example are concentration c and conductivity Λ. (Strictly the Onsager equation tells us the controlled variable is the *square-root* of concentration, $\sqrt{c}$.) Concentration c is the *controlled* variable here because we can readily prepare a series of solutions in the laboratory. Conductivity is the *observed* variable here because we do not measure a value of Λ until the solution of concentration c is available.

(ii) The term b is a constant; and Λ° is also a constant, because there can only be one value of conductivity at zero concentration. One of these constants will represent the intercept of a graph while the other will represent the gradient. To decide which is which, we note that if the Onsager equation is thought to describe a straight line, then the gradient will be the constant written adjacent to the controlled variable, and the intercept will be the constant that is not multiplied by any other term.

Accordingly, $-b$ is the gradient and Λ° is the intercept.

This choice of intercept makes physical sense, because it represents the conductivity at zero concentration, and the intercept on the y-axis by definition also occurs at zero concentration.

In summary:

$$\Lambda = \Lambda^\circ - b\sqrt{c}$$

$$y = c + mx$$

We sometimes say Λ° is the conductivity at *infinite* dilution.

Although we reversed the order of 'c' and 'mx' here, we have not altered the actual equation. With equal validity, we could have written, $\Lambda = -b\sqrt{c} + \Lambda^\circ$, but writing a minus sign in front of the constant might confuse.

Figure 9.4 shows an Onsager plot of Λ (as 'y') against $\sqrt{c}$ (as 'x') for KNO_3 in water at 298 K. The intercept of Λ° is labelled. ■

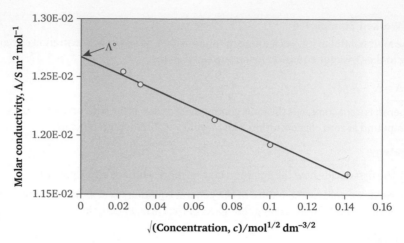

Figure 9.4 Onsager plot of molar conductivity of aqueous sodium nitrate (as '*y*') against the square root of concentration (as '*x*'). The intercept of Λ° represents the molar conductivity at 'infinite dilution.' (The fractional powers on the units are explained in Chapter 11.)

Self-test 9.3

Determine the intercept in each of the following graphs:

9.3.1   The velocity of an object decelerating: graph of velocity (as '*y*') against time (as '*x*')

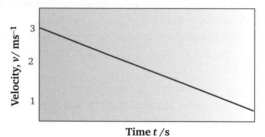

9.3.2   The amount of money earned: graph of salary (as '*y*') against duration of the work (as '*x*')

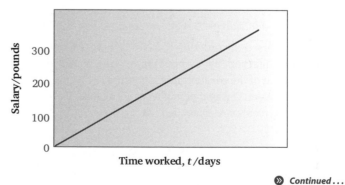

» *Continued . . .*

>> *Self-test 9.3 continued...*

9.3.3   The current drawn through an electrochemical cell during electroplating: graph of current (as '*y*') against the concentration of analyte (as '*x*').

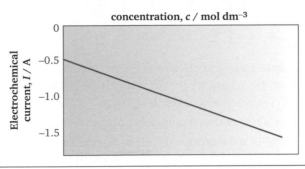

Quite often, we find we cannot just read off the intercept, because the *x*-axis does not extend as far as $x = 0$. In which case, we will need to redraw the graph, or work out the equation of the graph and *calculate* the value of *c*.

■ **Worked Example 9.5**

We give the symbol $E_{O,R}$ to an electrode potential of a redox couple. If $E_{O,R}$ is determined under standard conditions, we call it the *standard* electrode potential, $E_{O,R}^{\ominus}$. The two electrode potentials $E_{O,R}$ and $E_{O,R}^{\ominus}$ are related by the **Nernst equation**. For the Cu^0–Cu^{2+} couple, the Nernst equation is:

$$E_{Cu^{2+},Cu} = E_{Cu^{2+},Cu}^{\ominus} + \frac{RT}{2F}\ln\,[Cu^{2+}]$$

Determine the value of $E_{Cu^{2+},Cu}^{\ominus}$ from the data below:

Concentration $[Cu^{2+}]$ / mol dm^{-3}	10^{-2}	10^{-3}	10^{-4}	10^{-5}	10^{-6}
$E_{Cu^{2+},Cu}$ / V	0.281	0.251	0.222	0.192	0.162

We sometimes call the symbol $\ominus$ a **plimsol sign**. For this reason, we sometimes say that $E_{O,R}^{\ominus}$ is '*E* plimsoll.'

Strategy:

   (i) As with Worked Example 9.4 above, we must first decide which variable is the observed and which the controlled.

   (ii) With these variables, we decide which of $E_{Cu^{2+},Cu}^{\ominus}$ and $(RT \div 2F)$ is the gradient and which the intercept.

   (iii) We plot a graph with these variables, and determine the intercept.

Solution:

   (i) In this example, the electrode potential $E_{Cu^{2+},Cu}$ is the observed variable (which we plot as '*y*') and the natural logarithm of concentration will be the

We cover logarithms in Chapter 12.

controlled variable (and will be plotted as 'x'). In other words, we 'cut up' the equation as follows:

$$E_{Cu^{2+},Cu} \quad\Big|\quad = \quad E^{\ominus}_{Cu^{2+},Cu} \quad + \quad \Big|\quad \dfrac{RT}{2F} \quad\Big|\quad \ln[Cu^{2+}]$$

$$y \qquad\quad = \qquad c \qquad + \qquad m \qquad\quad x$$

(ii) From the arrangement of the equation, and with reasoning similar to that in Worked Example 9.4, we see that $(RT \div 2F)$ is the gradient m and $E^{\ominus}_{Cu^{2+},Cu}$ is the intercept c. Figure 9.5 shows a Nernst plot of $E_{Cu^{2+},Cu}$ (as 'y') against the logarithm of $[Cu^{2+}]$ (as 'x'), constructed using the data from the table.

(iii) The graph in Fig. 9.5 was drawn in a naïve way. In it, the line strikes the y-axis at 0.281 V. We would be wrong to say that 'the intercept c is 0.281 V.' To prove this statement, we only need to substitute 0.281 V into the Nernst equation above as '$E^{\ominus}_{Cu^{2+},Cu}$' and see how the data in the table do not fit the equation.

The reason why the intercept on this figure is not *the* intercept, i.e. $E^{\ominus}_{Cu^{2+},Cu}$ follows from the way we drew the x-axis: a careful look at the scale beneath the x-axis shows that it extends from -14.5 up to a value of -4.5, i.e. does not extend as far as 0. In fact, to determine the *correct* value of $E^{\ominus}_{Cu^{2+},Cu}$, we need to replot the graph, but ensure the x-axis extends as far as $x = 0$. Figure 9.6 shows such a graph. The intercept (that is, the value of y when $x = 0$) is seen to be 0.34 V. This is the correct value of $E^{\ominus}_{Cu^{2+},Cu}$. ■

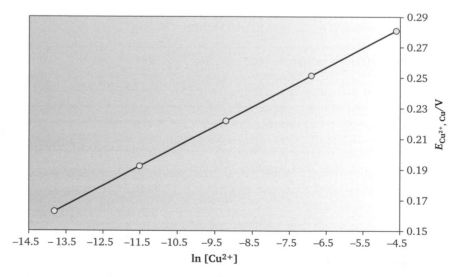

Figure 9.5 Nernst plot of $E_{Cu^{2+},Cu}$ (as 'y') against the logarithm of $[Cu^{2+}]$ (as 'x'). Its linearity shows the validity of the Nernst equation. The intercept is not the same as $E^{\ominus}_{Cu^{2+},Cu}$ because the y-axis does not join the x-axis at a value of zero.

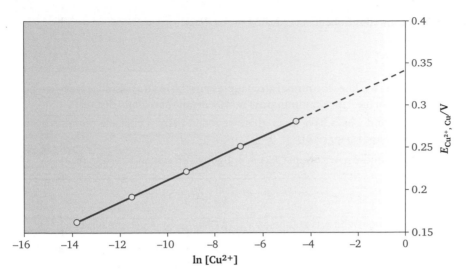

Figure 9.6 Nernst plot of $E_{Cu^{2+},Cu}$ (as 'y') against the logarithm of $[Cu^{2+}]$ (as 'x'), constructed with the same data as in Fig. 9.5. In this Figure, the intercept *is* the same as $E^{\ominus}_{Cu^{2+},Cu}$ because the y-axis joins the x-axis at a value of zero.

The gradient *m*

By the end of this section, you will know:

- How to determine the gradient of a graph by looking at it.
- How to determine a gradient if we know two points that satisfy the line.
- That a gradient will often have units.

We typically denote the gradient with the letter *m*. The gradient *m* represents a **rate of change**: in this case, it quantifies how fast *y* changes when we change *x*. The steepness of a hill is a simple example of such a gradient: when we see a sign describing the hill's gradient as '1 in 4', we should understand that the hill is steep, because it rises vertically by one unit for every four units we travel horizontally. A slope of '1 in 10' is gentle, and '1 in 100' is barely a hill at all.

Strategy to determine the gradient of a straight-line graph:

(i) We first draw the best straight line through our data.

(ii) At one end of the line, we choose a point on the line, and call it (x_1, y_1). We then choose a second point at the other end of the line, and call it (x_2, y_2). It does not matter where we choose the points (x_1, y_1) and (x_2, y_2), but in practice we tend to get a more accurate value if the two points are widely separated.

The points we use when determining the gradient do not have to be points for experimental data. Particularly when drawing a graph on graph paper, it is best to choose the points in any way that makes the calculation easier.

(iii) We define the gradient m according to eqn (9.2):

$$m = \frac{y_2 - y_1}{x_2 - x_1} \qquad (9.2)$$

where the top line (the **numerator**) represents the *vertical* distance travelled and the bottom line (the **denominator**) represents the *horizontal* distance.

(iv) If the x- and y-axes have units, then we retain them within the numerator and denominator in eqn (9.2).

Some people choose to write eqn (9.2) in a slightly different way, as eqn (9.3):

$$m = \frac{\Delta y}{\Delta x} \qquad (9.3)$$

We often meet the operator Δ in thermodynamics, for example in 'ΔH', which we define as the difference between the final enthalpy and the initial enthalpy.

where the symbol Delta Δ is an operator (see Chapter 2) which means 'change in'. We define Δ in eqn (2.1):

$$\Delta = \text{final value} - \text{initial value} \qquad (2.1)$$

Equations (9.2) and (9.3) are identical, and only differ in terms of the way we write them.

■ **Worked Example 9.6**

A smoker wishes to give up smoking. As part of his regime, he sticks a slow-release patch to his arm. The patch releases the addictive alkaloid nicotine (**III**) at a precise and controlled rate. Before the observation period, the patch had released 0.25 millimoles of (**III**) and after a further three days, it had released a total of 0.4 millimoles. Determine the daily rate of nicotine release, i.e. calculate the gradient of a graph of 'amount of nicotine released (as 'y') against time elapsing in days (as 'x').

III

Remember: The gradient of such a graph is a *rate*.

Before we start, we define terms. At the beginning, we say time $t = 0$, and at the end of the observation period $t = 3$ days. Similarly, at the beginning, $n_{(III)} = 0.25$ mol and $n_{(III)} = 0.4$ mol after 3 days.

Inserting values into eqn (9.2):

$$\text{rate} = \left(\frac{\text{change in observed variable}}{\text{change in controlled variable}} \right) = \frac{0.4 - 0.25}{3 - 0} \frac{\text{mmol}}{\text{day}}$$

so $$\text{rate} = \frac{0.15 \,\text{mmol}}{3 \,\text{days}} = 0.05 \,\text{mmol day}^{-1} \quad ■$$

Sometimes the gradient m is negative. Such a gradient tells us that the value of y DEcreases as the value of x INcreases.

■ **Worked Example 9.7**

During the course of a chemical reaction, the concentration of the reactant decreases from 0.1 mol dm^{-3} at the start of the reaction to 0.03 mol dm^{-3} after half an hour (30 min = 1800 s). What is the rate at which the reactant is consumed?

Inserting values into eqn (9.2):

$$\text{rate} = \frac{0.03 - 0.1 \text{ mol dm}^{-3}}{1800 - 0 \text{ s}}$$

so $$\text{rate} = \frac{-0.07 \text{ mol dm}^{-3}}{1800 \text{ s}}$$

and we calculate the rate as -3.9×10^{-5} mol dm^{-3} s^{-1}. ■

The minus sign of this rate (gradient) tells us the concentration *decreases* with time.

Self-test 9.4

Calculate the gradient *m* between the following points:

9.4.1	(1, 2)	and	(2, 4)
9.4.2	(10, 17)	and	(5, 22)
9.4.3	(3, 5)	and	(6, 14)
9.4.4	(−2, 3)	and	(0, 7)
9.4.5	(−3, −3)	and	(−5, −5)
9.4.6	(2, −3)	and	(4, 5)
9.4.7	(−2, −8)	and	(−4, −4)
9.4.8	(−5, 4)	and	(7, 12)

We often obtain a rate of change by measuring the magnitude of a gradient, rather than looking at two data points. In fact, a graph generally yields a superior gradient, because the line represents the best fit for many data points. We therefore minimize our errors. For example, it is usually clear which data are poor after plotting a graph. We can ignore the bad data provided we are confident they are **outliers**, by drawing a straight line though only the good data (Chapter 14 describes in detail how and when to do so.) Determining a gradient from two data is more risky: what if one or both were duff?

The gradient of a graph directly yields a rate of change. For example, the gradient of a graph of velocity against time tells us how fast an object speeds up ('accelerates') with time.

We describe a point as an **outlier** if statistical considerations suggest the data is incorrect.

■ **Worked Example 9.8**

The powerful magnets inside the chamber of a mass spectrometer cause the molecular ion derived from vitamin C (**IV**) to accelerate. Its velocity *v* increases steadily with time *t* as depicted in Fig. 9.7.

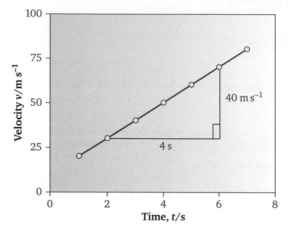

IV

Figure 9.7 A straight-line graph to demonstrate how we determine the velocity of a particle travelling inside a mass spectrometer. The gradient yields the acceleration, a.

The gradient of the graph directly yields the acceleration a. Determine from the graph
(i) The ion's acceleration and (ii) The units of acceleration a.

(i) We first obtain a *numerical magnitude* for a using the *numbers* within eqn (9.3). To determine a value of the gradient, we first construct a right-angled triangle against the straight-line trace. By measuring along the horizontal and vertical limbs of the triangle, and using the corresponding axis scales, we quantify how the velocity v increases by 40 m s⁻¹ during an interval of 4 s.

Inserting values from Fig. 9.7 into eqn (9.3), we calculate the acceleration:

$$a = \text{gradient} = \frac{40 \,\text{m s}^{-1}}{4\text{s}}$$

so **(IV)** accelerates at $a = 10 \text{ m s}^{-2}$

(ii) Secondly, we next derive the *units* of acceleration a from the arrangement of the *units* within the fraction. In this case, a has the units of the numerator divided by the units of the denominator:

units of $a = \text{m s}^{-1} \div \text{s}$

In words, we say the velocity increases by ten metres per second, for every second the molecule resides within the spectrometer. We manipulate these units algebraically to obtain m s⁻².

By looking carefully at the units of a gradient, we often possess a powerful method of working out a parameter. For example, a graph of distance covered (as y) against time (as x): the gradient has the units of 'distance ÷ the units of time', i.e. m s⁻¹, which helps us realize that the gradient of the graph represents a velocity. ■

We discuss the manipulation of units in Chapter 30.

Determining the full equation of a straight line

By the end of this section, you will know how to determine the equation of a straight line:

- By looking carefully at a graph.
- From the gradient and one point that satisfies the line.
- From two points that satisfies the line.
- Using computer software, such as *Excel*™.

Obtaining the equation of a straight line by looking at a graph:

As we start this section, we assume that all the lines are straight, so each follows the form $y = mx + c$. In practice, it is generally easier to start with the intercept since we merely read it from the y-axis. We need to be careful that we are reading the *real* intercept, though, because the intercept is the value of y when $x = 0$: if the x-axis does not go as far as zero, then we are looking at *an* intercept rather than *the* intercept, which is the value of c in the equation. If the graph is drawn in such a way, we may have to redraw it with a longer x-axis that does go as far as $x = 0$.

■ **Worked Example 9.9**

The graph in Fig. 9.8 below describes the temperature inside a calorimeter (as 'y') as a compound is consumed (as 'x'). Deduce the equation of the straight line.

First, the solid line in Fig. 9.8 strikes the y-axis at a value of $y = 10$, so the constant c has a value of '10'. An incomplete equation for the line is therefore $y = mx + 10$.

Next, we obtain the gradient m: we note how the line passes through the two points $(0, 10)$ and $(15, 40)$. We call these points respectively (x_1, y_1) and (x_2, y_2), and insert the respective values into eqn (9.2), which yields:

$$m = \frac{40 - 10}{15 - 0}$$

so $\quad m = \frac{30}{15} = 2$

We deduce the equation of the line to be, $y = 2x + 10$. ■

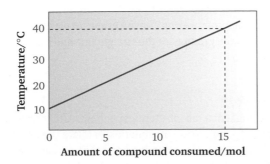

Figure 9.8 Energy is released when a compound is burnt within a bomb calorimeter. This energy causes the temperature inside the calorimeter to rise: graph of calorimeter temperature (as 'y') against amount of compound burnt (as 'x').

Self-test 9.5

Determine the equation of the straight line in each figure. Take care with the signs.

9.5.1

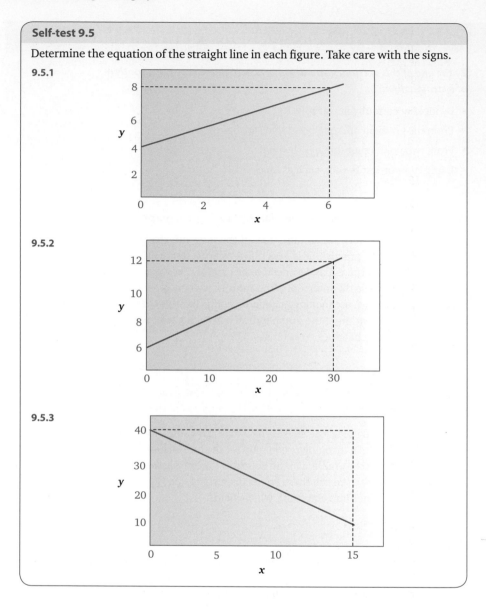

9.5.2

9.5.3

Obtaining the equation of a straight line from the gradient and one point that satisfies the line: Sometimes we don't have a graph, so we use the data directly. Most commonly, we possess the gradient of the straight line and know the coordinates of a single point that lies on it.

■ **Worked Example 9.10**

A factory produces sugar at a rate of 4 tonnes per hour. After one hour, the warehouse contains 12 tonnes of sugar. Write an equation to relate the amount of sugar in the warehouse and the process time, assuming none of the sugar is removed.

The amount of sugar is measured in tonnes, so any value of y has the units 'tonnes'.

We start by defining terms. We note there are two variables to think about: the amount of sugar produced, and the time length of the process. Let's say the amount of sugar in the warehouse is y. We'll call this quantity the observed variable.

Next, let's say that the time length of the process is x. Finally, we know the **rate** of sugar production is 4 tonnes per hour, so we equate the rate with the gradient m. While the numerical value of m is 4, it is a rate with units of 'tonnes per hour.'

The equation of a straight line is $y = mx + c$. Having equated the rate of production with the gradient m, we say $y = 4x + c$, which is an 'incomplete equation.'

We know one point that satisfies the equation of the straight line: the question says, 'After one hour, the warehouse contains 12 tonnes of sugar,' so we know the point $x = 1, y = 12$. Therefore, the second stage when deducing the equation of the line is to substitute values into the incomplete equation:

$$12 = 4 \times 1 + c$$

so $\quad 12 = 4 + c$

and $\quad c = 8$

The equation of the line is therefore $y = 4x + 8$. ∎

> The length of the process x is measured in hours, so any value of x has the units 'hours'.

Self-test 9.6

Deduce the equations of the following straight lines:

9.6.1 The gradient is 3 and the line passes through the point (1, 2)

9.6.2 The gradient is 10 and the line passes through the point (4, 4)

9.6.3 The gradient is –4 and the line passes through the point (3, 0)

9.6.4 The gradient is 2.5 and the line passes through the point (4, 5)

9.6.5 The gradient is –3.5 and the line passes through the point (–4, 0)

Determining the equation of a straight line from two points that satisfies the line: At other times, we do not have a gradient or a graph but just two of the data points. An obvious example is a chemist who has two masses from a tap-pan balance and subsequently two optical absorbances, having dissolved each sample in solvent and taken the spectrum.

■ **Worked Example 9.11**

An analytical chemist wants to prepare a calibration graph, relating the amount of the natural pigment β-Carotene (**V**) with its optical absorbance when in solution. The analyst dissolves 0.01 g of (**V**) and obtains an optical absorbance of 0.8, then weighs a mass of 0.03 g and obtains a higher absorbance of 2.0. What is the relationship between the absorbance (the observed variable, y) and the mass of (**V**) (the controlled variable x)?

V

It's usually a good idea to restate the data, as (0.01, 0.8) and (0.03, 2.0). We first determine the gradient m by inserting data into eqn (9.2)

$$m = \frac{2.0 - 0.8}{0.03 - 0.01\,\text{g}}$$

$$\text{so} \quad m = \frac{1.2}{0.02\ \text{g}}$$

and $m = 60\ \text{g}^{-1}$.

The incomplete equation is therefore $y = 60\,x + c$.

We then insert one of these data pairs into this equation to obtain the value of c. It does not matter which point we insert. Indeed, it is often a good idea to perform the calculation twice, in order to verify our answer. We will use the first point, (0.01, 0.8):

$$0.8 = 60 \times 0.01 + c$$

so $0.8 = 0.6 + c$,

and $c = 0.2$

We deduce the equation of the line is $y = 60\,x + 0.2$.

Chemists often find it useful to rephrase such a relationship in words, saying something like, 'absorbance $= (60 \times \text{mass}) + 0.2$'. ∎

Self-test 9.7

Deduce the equations of the straight lines connecting the following pairs of points:

9.7.1   (1, 2)   and   (2, 4)

9.7.2   (0, −2)   and   (3, −11)

9.7.3   (9, 12)   and   (28, 50)

9.7.4   (7, 9)   and   (6, 10)

9.7.5   (1, 2)   and   (−3, −4)

9.7.6   (−3, −3)   and   (−2, −6)

The trend line

The TRENDLINE represents the equation of the line that best fits the data. A computer can determine the best gradient and intercept using calculations similar to those in Chapter 15.

Obtaining the equation of a straight line using *Excel*™:

Sometimes, we have a full series of data. In such cases, we can determine the equation of the resulting straight line using computer software such as *Excel*™.

1. First plot the data, using the procedure on pp. 107 and 108 in the previous chapter.
2. Using the mouse, click on one of the data points on the graph. It will look like this:

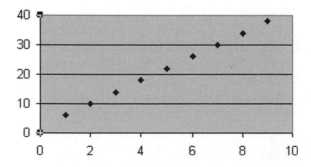

3. Choose the chart menu, and scroll to the 'Add Trendline' command:

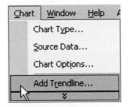

4. Choose a linear graph:

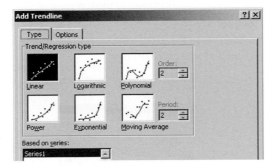

5. Before pressing OK, click on the 'options' sheet on this same screen, It will look like this:

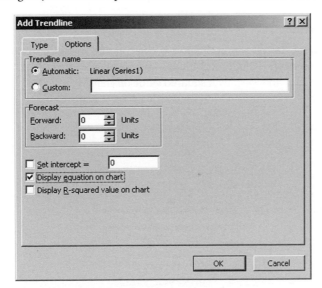

6. At the bottom of the page, click where it says, 'Display equation on the chart.' Also click where it says, 'display R-squared value on chart.' (Chapter 15 explains the meaning of the r^2 correlation coefficient.)

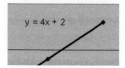

7. Press OK. The equation will be displayed at the top right of the graph. To move the location of the equation, click on it using the mouse and move it.

8. To change the size and font of the small text box containing the equation, click on it, and use the usual font and size commands.

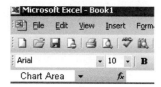

Excel 2007 version:

From the 'Layout' Tab shown in Chapter 8, click on the 'Trendline' button. It is actually best to click on 'More Trendline Options'—which allows us to select 'Linear', 'Display Equation on Chart' and 'Display R squared value on chart'.

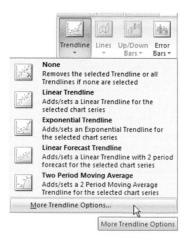

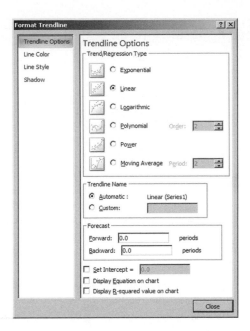

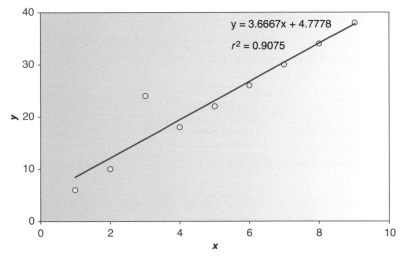

Figure 9.9 Graph drawn using *Excel*™. The point (3,14) was mistyped as (3,24) and now deviates from the best straight line. The computer programme could not distinguish between the eight good points and the one mistyped one, and treated each as equally valid. Accordingly, the computer-generated best-fit equation is wrong.

The human eye is actually very astute in its choice of the best line, so always check the computer has placed the line where it seems best.

Having analysed the data statistically (see Chapter 14, which discusses outliers), we delete those points that deviate from the best straight line, because they will ruin the equation algorithm. **(We should *never* delete points until we have performed these statistical analyses.)** The graph in Fig. 9.9 was obtained using exactly the same data as that above, except that (3,14) was mistyped as (3,24). This point clearly deviates from the line. The best straight line now misses almost half the points, and the computer thinks the equation of the straight line should be $y = 3.6667x + 4.7778$. Ignoring the rogue point, the straight line passes through *all* other points, and has the equation, $y = 4x + 2$.

An algorithm is a set of commands followed by the computer in order to calculate the best straight line.

Summary of key equations in the text

Equation of a straight line: $y = mx + c$ (9.1)

Definitions of the gradient, m: $m = \dfrac{y_2 - y_1}{x_2 - x_1}$ (9.2)

$$m = \frac{\Delta y}{\Delta x}$$ (9.3)

Additional problems

9.1 The amount of chemical converted per hour is 400 kg. The amount of the chemical in a container at the start of the day is 312 kg. Assuming none of the chemical is removed during the course of the day, derive an equation to describe the amount of chemical remaining in the container as the day progresses.

9.2 A lecture course is so easy that the exam marks M depend only on attendance A rather than ability. For every lecture attended, the exam mark increases by 1.5%. Prior knowledge means that a student attending nothing still gains a mark of 18%. Deduce an equation relating exam mark M and attendance A.

9.3 'Reduce' the following equation: $12y = 4x - 6$.

9.4 What is the equation of a straight line of gradient 4.2, which passes through the point (3,1)?

9.5 The **temperature voltage coefficient** of an electrode or cell is a rate of change:

Electrochemists usually call the right-hand side of this equation d*emf*/d*T*, using notation we introduce in Chapter 17.

$$\text{temperature voltage coefficient} = \frac{E_{(cell)2} - E_{(cell)1}}{T_2 - T_1}$$

Calculate the value of this rate of change (gradient) for the Clarke cell, if the value of $E_{(cell)}$ increases from 1.433 V to 1.456 V as the temperature is lowered from 56 °C to 25 °C.

9.6 Deduce the equation of the straight line that has a gradient of -4.9, and that passes through the point (−5,2).

9.7 A hill is steep. For every 20 m forward, it goes up 3 m. Calculate its gradient, m.

9.8 In a modified form of the **Beer–Lambert equation**, the optical absorbance Abs and concentration c of chromophore in solution as follows:

$$Abs = \varepsilon c l + k$$

where ε is the extinction coefficient, l is the optical path length, and k is the absorbance of the glass cell.

Deduce the value of k given that the absorbance Abs was 0.8 when the concentration c was 0.23 mol dm^{-3}, the pathlength l was 1 cm, and the extinction coefficient ε was 30 mol^{-1} dm^3 cm^{-1}.

9.9 A chemist using an ion-selective electrode notes the relationship between solution pH and the reading on a voltmeter, E. When the pH is 4.2, the reading is 400 mV, and when the pH is 5.5, the reading is 300 mV. Deduce the form of the values of m and c in the equation of a line of the form $E = \text{pH} \times m + c$.

9.10 The ideal-gas equation $pV = nRT$ suggests a correlation between the volume V of an ideal gas and its pressure p. The following data refer to carbon dioxide:

V/m^3	0.01	0.02	0.025	0.03	0.04	0.05	0.06	0.07	0.08	0.1	0.2
p/pa	50	25	20	16.5	12.5	10	8.3	7.2	6.2	5	2.5

Plot a graph of 1/pressure ($1/p$) (as 'y') against volume V (as 'x') and ascertain the equation of the straight line.

10

Algebra VII

Solving simultaneous linear equations

Introducing terms

By the end of this section, you will know:

- Solving simultaneous linear equations involves elimination and substitution.
- Elimination is achieved by *subtracting* like terms between the two equations.

If we wish to determine the value of a variable from an equation, we should manipulate and rearrange the equation until the variable is the subject. For example, we can say from the equation $x + 2 = 6$ that $x = 4$. We can solve an equation if it has a single variable.

But if we have an equation such as $x + y = 7$, it is not possible to solve it. While a wide variety of answers do exist ($x = 0$ and $y = 7$; $x = 1$ and $y = 6$; $x = 2$ and $y = 5$, $x = (10^{10} + 7)$ and $y = -10^{10}$, etc.), it is not possible to determine a *unique* pair of values for the two variables. We are stuck if we have two variables and only one equation: we cannot discover the value of either variable without additional information.

In order to find the unique solution for two variables, we require two separate, distinct equations. In fact, to find the unique solutions, we need one equation per variable: n variables require n equations.

The previous paragraph used the word 'distinct' for a reason. Consider the two equations $y = x + 2$ and $2y = 2x + 4$. While they initially look different, in fact they are the same equation. We can prove that they are identical by reducing (see p. 124) the second equation, which will show it is the same as the first. In other words, although written as two equations, we have only one unique equation and cannot solve for x and y.

So, in this chapter we will look at the situation in which we have a pair of equations and two unknowns. Because we need to solve both equations to obtain the two variables, the method we use is called solving **simultaneous linear equations**.

Solving simultaneous linear equations graphically

By the end of this section, you will know:

- Simultaneous linear equations can be solved using graphs.
- Two straight lines are parallel if they have the same gradient but different intercepts.
- The two straight lines will meet and cross over, unless they are parallel.
- The coordinates of the point where the two lines intersect gives the values of x and y that solve both equations.

In many cases, the simplest way to solve simultaneous linear equations is by drawing a graph. Think of two straight lines. The lines will be **parallel** if their intercepts c differ but their gradients m are identical. For example, the two lines (i) $y = 4x + 5$ and (ii) $y = 4x + 10$ in Fig. 10.1 are parallel because they have the same gradient of $m = 4$, but the lines are separated vertically on the page by five units because their intercepts are 5 and 10, respectively. Many calibration curves take the form of parallel lines.

When we first looked at straight lines in Chapters 8 and 9, we considered the equation $y = 4x + 2$, and saw how such an equation is satisfied by an infinite number of points, because the values of x can vary from $-\infty$ through to $+\infty$. For this reason, the line $y = 4x + 2$ is very long. In fact, a moment's thought should persuade us that all straight lines are infinitely long, again because there are an infinite number of values that x can take.

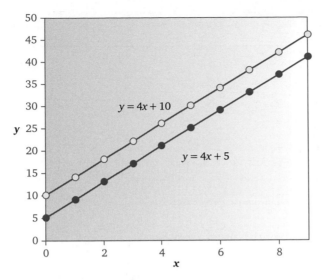

Figure 10.1 Parallel lines never meet: the two lines $y = 4x + 5$ and $y = 4x + 10$ are parallel because they have the same gradients m but different intercepts c.

But now think about two lines that are not parallel, by which we mean that their gradients differ. For example, (i) $y = 4x + 2$ and (ii) $y = 6x + 6$ are not parallel because line (i) has a gradient of 4 and line (ii) has a gradient of 6. The simplest definition of parallel lines says they are, 'lines that never meet.' By contrast, lines that are not parallel do meet: any pair of straight, non-parallel lines will overlap at a single value of x and y. We say the lines **cross over**, or **intersect**.

While each line is satisfied by an infinite number of points, only a single, unique point can simultaneously satisfy the equations of two straight lines that are not parallel. It should be obvious that the intersection occurs at a single point: two straight lines can cross only when the value of x is the same for both of them, and when the value of y is also the same for both. At no other values of x and y will a single point satisfy both equations. This is why we say the point is unique.

It is often crucial to know at what values of x and y two straight lines meet, and the point of intersection often represents crucial physicochemical parameters.

> The simplest definition of **parallel lines** says, 'Lines that never meet.'

■ **Worked Example 10.1**

Consider the molecule p-aminobenzoic acid, (**I**). The molecule is always ionized: in acidic solution with the amine portion being protonated. (**I**) is also ionized in alkaline solution with the acid moiety dissociating to form a carboxylate ion.

$$H_2N \!-\!\!\langle\bigcirc\rangle\!-\! COOH$$

I

Draw a schematic graph of the overall charge on (**I**) as a function of pH.

There are many simple ways to determine the isoelectric point. One of the simplest is to obtain data about the ionic charges at high pH and at low pH, i.e. at the extremities of the graph. We then draw straight lines through the data, and extend them both toward the centre of the graph: we say we **extrapolate** the lines. The isoelectric point represents the pH where the two **extrapolants** intersect on Fig. 10.2.

> **Schematic** means 'stylized' or 'general.' In the context of graphs and drawing, it can be taken to mean a qualitative 'sketch' rather than a more quantitative approach.

> The **isoelectric** point represents the pH when the molecule (**I**) exists as a zwitterion, bearing both a positive and a negative charge.

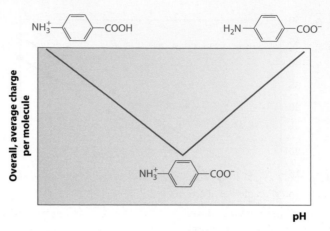

Figure 10.2 Two separate trends appear when we consider the overall, average charge per molecule: anions form at high pH and cations form at low pH. The two straight lines *intersect* at the **isoelectric point**, when a proportion of (**I**) exists as a **zwitterion**, that is, the molecule bears both a positive and a negative charge.

A simple way to obtain the charge data is from experimental methods that determine the number of ions in solution, e.g. ionic conductivity Λ. ∎

The simple example of an isoelectric point in Worked Example 10.1 suggests the simplest way to determine the point of intersection between two equations is to draw two straight lines on a single graph, and determine where they cross.

■ **Worked Example 10.2**

Consider the two lines (i) $y = x + 2$ and (ii) $y = 2x + 2$. By drawing both lines on a single graph, determine the single point of intersection.

Figure 10.3 shows both lines drawn on the same graph. The graph clearly shows the unique point of intersection is $(0, 2)$. ∎

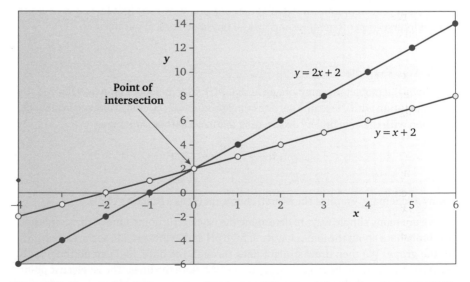

Figure 10.3 The two straight lines $y = 2x + 2$ and $y = x + 2$ intersect at the single, unique point $(0,2)$.

Self-test 10.1

By plotting a graph in each case, determine the point of intersection between the following pairs of straight lines:

(1)	(2)

10.1.1 $y = x + 2$ and $y = 4x + 5$

10.1.2 $y = 6x - 12$ and $y = -6x + 12$

Unfortunately, there are many good reasons why we generally choose not to solve two equations using graphical methods:

- Graphs take time to draw and we may be too busy. Drawing graphs when busy is a guaranteed way of making errors.

- Graphical methods are inherently imprecise. While it may be easy to read from a graph an intersection point like (1, 2), it will be far more difficult if the point does not represent integers, e.g. (1.23, 2.04). A chemist would need very good graph-plotting skills to obtain these values correctly to two decimal places.

Chemists prefer to solve simultaneous linear equations, i.e. using algebra.

> An **integer** is a whole number.

Solving simultaneous linear equations using algebra

By the end of this section, you will know algebraic solutions of simultaneous solutions:

- Is usually quicker than drawing a graph.
- Is considerably more accurate than drawing a graph.
- Involves elimination and substitution.

To solve two equations, we usually need a superior method to drawing a graph. The preferred method below involves algebra.

■ **Worked Example 10.3**

Consider the two lines $y = x + 2$ and $y = 2x + 4$. Determine their point of intersection using algebra.

Strategy:

  (i) We write the two equations to be solved with one above the other, much like a simple sum.

 (ii) By elimination, we manipulate one or both of the equations until there is a either no x or no y term.

Solution:

(i)

$$y = 2x + 4 \quad\quad\quad (1)$$
$$\underline{-y = x\ \ + 2} \quad\quad\quad (2)$$

(ii) In this example, the two source equations both have one lot of y. We will **eliminate** the y terms by simply subtracting the lower equation from the upper equation. To do this, we subtract one term at a time, so we say '$y - y = 0$', '$2x - x = x$' and '$4 - 2 = 2$'. Therefore, by elimination we obtain:

$$y = 2x + 4 \qquad (1)$$
$$\underline{-y = x \;\; + 2} \qquad (2)$$
$$0 = x \;\; + 2$$

So, since $x + 2 = 0$, the value of x that satisfies the two equations is $x = -2$. We say the value of $x = -2$ **satisfies** both equations. While there are an infinite number of x values that satisfy the first equation, and a infinite number of values that satisfies the second equation, there is only a single, unique value that satisfies both.

Now we know the value of x that satisfies both equations, we obtain the unique value of y by **substituting** $x = -2$ into either of the two equations that intersect:

Taking the equation $y = 2x + 4$ and inserting $x = -2$, we obtain $y = -4 + 4$, so the value of y at the point of intersection is 0. The point of intersection is $(-2, 0)$. We obtain the same result when substituting into the other equation, $y = x + 2$. ∎

In Worked Example 10.3 above, we chose to eliminate the y terms because the two values of y were the same. We can employ an identical approach if the two values of x are the same:

> **■ Worked Example 10.4**
>
> Use algebra to determine the point of intersection between the two lines (1) $y = x + 2$ and (2) $3y = x + 6$.

As before, we start by writing the two equations one above the other, and eliminate by subtracting. This time it is the x terms we will seek to remove:

$$3y = x + 6 \qquad (1)$$
$$\underline{-y = x + 2} \qquad (2)$$
$$2y = 0 + 4$$

By substituting into both equations and obtaining the same value of x, we corroborate that the value of y is common to both, i.e. is indeed y at the point of intersection.

$2y = 4$ and so $y = 2$. We determine the value of x at the point of intersection by substituting $y = 2$ into either of the two equations:

$$3 \times 2 = x + 6 \text{ so } x = 0$$
or
$$2 = x + 2 \text{ so } x = 0$$

so the two straight lines intersect at $(0, 2)$. ∎

> **Self-test 10.2**
>
> Using simple algebraic elimination, determine the values of x and y that satisfy the following pairs of straight lines:
>
> | | (1) | | (2) |
> |---|---|---|---|
> | **10.2.1** | $y = 2x + 5$ | and | $y = 11 - 4x$ |
> | **10.2.2** | $y = x + 2$ | and | $y = -3x - 14$ |
> | **10.2.3** | $y = 4x + 7$ | and | $y = 2x + 5$ |
> | **10.2.4** | $2y = x + 10$ | and | $2y = 6x + 10$ |
> | **10.2.5** | $y = 2x + 3$ | and | $y = 3x + 7$ |
> | **10.2.6** | $5y = 4x + 2$ | and | $3y = 4x + 6$ |

Solving simultaneous linear equations with negative coefficients

By the end of this section, you will know:

• When the coefficients are negative, an alternative scheme is preferable: elimination proceeds by the *addition* of like terms, rather than subtraction.

Sometimes we employ a variation of this method, because the coefficients of x or y are negative. The methodology is essentially identical except that the elimination procedure requires the *addition* of like terms, rather than subtraction.

■ **Worked Example 10.5**

Determine the point of intersection between the two lines $y = -x + 2$ and $2y = x + 4$ using algebra.

As before, we start by writing the two equations one above the other, but because one x term is positive and the other negative, this time we *add* the two equations:

$$2y = x + 4$$
$$+y = -x + 2$$
$$\overline{3y = 0 + 6}$$

If the factors on x are negative in one equation and positive in the other, we *add* the two equations to ensure the two x terms cancel.

$3y = 6$ and so $y = 2$. Substituting this value of y into either of the two starting equations yields $x = 0$. The point of intersection is $(0,2)$. ■

Self-test 10.3

Determine the point of intersection between the following pairs of straight lines algebraically:

	(1)		(2)
10.3.1	$y = 2x + 5$	and	$-y = -4x + 11$
10.3.2	$y = x + 2$	and	$y = -3x - 14$
10.3.3	$2x + y = 4$	and	$2x + 2y = 6$
10.3.4	$2x + 2y = 32$	and	$3x - 2y = 3$
10.3.5	$x + y = 16$	and	$5x + y = 60$
10.3.6	$2x - y = 2$	and	$x + y = 5.5$

Sometimes we can formulate a pair of equations from a physical problem:

■ **Worked Example 10.6**

Find two numbers x and y such that their sum is 12 and their difference is 4.

We first formulate a pair of simultaneous linear equations from the data in the question. First, the sum of the numbers is 12, so $x + y = 12$. Second, their difference is 4, so $x - y = 4$.

We can add the two equations in order to eliminate y:

$$x + y = 12$$
$$+x - y = 4$$
$$\overline{ 2x = 16}$$

and so $x = 8$. Clearly, substitution reveals that $y = 4$.

The equations we've just looked at are simple because we can eliminate either the value of x or the value of y by writing one equation above the other. This level of simplicity is rare. We usually need to manipulate the data a little before we can eliminate. ■

Solving simultaneous linear equations using factors

By the end of this section, you will know that:

- Harmonizing the coefficients can sometimes necessitate multiplying each equation by a suitable factor.
- The factors must be complementary.

It is sometimes quite difficult to decide which factors to use. We cannot harmonize the coefficients by simply multiplying one equation by a factor. We need to multiply both equations by complementary factors.

■ **Worked Example 10.7**

A mass spectroscopy experiment suggests that two molecular fragments contain only carbon and hydrogen. The empirical formula of the first is C_2H_6 and has a mass of 30 Da and the second has a formula of C_6H_{13} and a mass of 85 Da. What are the masses of carbon and hydrogen atoms?

> The SI unit of atomic mass is the **dalton**, which has the symbol **Da**.

We first re-express the data, saying $2C + 6H = 30$ and $6C + 13H = 85$. These equations only *look* more complicated than most of the equations above because we wrote H and C rather than x and y.

We could write the two sets of data in the form of algebraic equations, as follows:

$$2C + 6H = 30 \qquad (1)$$

$$6C + 13H = 85 \qquad (2)$$

Clearly, no simple algebraic solution is possible yet, but if we multiply (1) by a factor of 3, we obtain the pair:

$$6C + 18H = 90 \qquad 3 \times (1)$$

$$6C + 13H = 85 \qquad (2)$$

Subtracting (2) from '$3 \times (1)$' yields:

$$6C + 18H = 90 \qquad 3 \times (1)$$

$$\underline{-6C + 13H = 85} \qquad (2)$$

$$0 + 5H = 5$$

so $\qquad\qquad 5H = 5$ Da, so $H = 1$ Da

One hydrogen atom has a mass of 1 Da. Knowing the mass of one hydrogen atom, we then calculate the mass of carbon by substituting $H = 1$ into either equation to yield the result, $C = 12$. ∎

■ Worked Example 10.8

According to the Beer–Lambert Law, the absorbance of a single **chromophore** in solution is A and relates to the concentration c, the optical pathlength l and the extinction coefficient ε:

$$A = c\,\varepsilon\,l$$

A solution contains two chromophores, A and B. The extinction coefficient of dye A is ε_A, and ε_B relates to dye B. The pathlength l remains constant at 0.1 m, but the concentration of A and B vary. The absorbance is 1.0 when $c_A = 0.2$ mol dm^{-3} and $c_B = 0.4$ mol dm^{-3}; the absorbance drops to 0.7 when c_A is 0.1 mol dm^{-3} and $c_B = 0.3$ mol dm^{-3}. Use simultaneous linear equations to determine values of ε_A and ε_B.

Before we start, we note how a spectrometer measures the sum of the absorbance of A and B: $A_{(measured)} = A_A + A_B$. From the Beer–Lambert Law, we can substitute for the two absorbances, yielding:

$$A_{(measured)} = c_A\,\varepsilon_A\,l + c_B\,\varepsilon_B\,l$$

Next, we rewrite the data from the question concerning A, c and l terms, to obtain a pair of simultaneous linear equations:

$$1.0 = 0.2 \times 0.1 \times \varepsilon_A + 0.4 \times 0.1 \times \varepsilon_B \qquad (1)$$

$$0.7 = 0.1 \times 0.1 \times \varepsilon_A + 0.3 \times 0.1 \times \varepsilon_B \qquad (2)$$

These simultaneous linear equations are similar in form to those above. The first difference is cosmetic: we wrote ε_A and ε_B rather than x and y. Otherwise they are similar; in fact, such a difference is unimportant. A second and more serious difference is an inability to eliminate either ε_A or ε_B because their coefficients differ. Any algebraic solution would be easier if the coefficients on either ε_A or ε_B were the same.

But the coefficients *can* be made the same by multiplying eqn (2) by a factor of 2:

$$1.4 = 0.02 \times \varepsilon_A + 0.06\,\varepsilon_B \qquad 2 \times (2)$$

$$\underline{-1.0 = 0.02 \times \varepsilon_A + 0.04\,\varepsilon_B} \qquad (1)$$

$$0.4 = 0 \quad \times \varepsilon_A + 0.02\,\varepsilon_B$$

(The multiplication by 0.1 has been incorporated into the concentration terms here.)

Therefore, $0.4 = 0.02 \times \varepsilon_B$ and $\varepsilon_B = \dfrac{0.4}{0.02} = 20$.

We deduce by straightforward substitution into either (1) or (2) that $\varepsilon_A = 10$. Remember always to check the answers by inserting them into the original source equations. ∎

This use of simultaneous linear equations is quite common. For example, over the last half-century, the most frequently used assay for chlorophylls in higher plants and green algae, the Arnon assay, applies simultaneous linear equations to determine the concentrations of chlorophylls A and B in aqueous 80% acetone extracts of chlorophyllous plant and algal materials.

This example might yield data we knew already, but the calculation in this worked example demonstrates the power of the method.

A **spectrometer** is a device for measuring optical absorbances, A.

A **chromophore** is a molecule of species that imparts a colour, such as a dye.

■ Worked Example 10.9

The liquids benzene (**II**) and bromobenzene (**III**) are mixed. The vapour above this mixture contains both (**II**) and (**III**). The pressure above pure benzene is $p_{(\mathrm{II})}^{\ominus}$ and the pressure above pure bromobenzene is $p_{(\mathrm{III})}^{\ominus}$. According to **Raoult's Law**, the pressure of the gases above this mixture $p_{(\mathrm{total})}$ is given by the equation:

$$p_{(\mathrm{total})} = p_{(\mathrm{II})}^{\ominus} \times x_{(\mathrm{II})} + p_{(\mathrm{III})}^{\ominus} \times x_{(\mathrm{III})}$$

where x_i is the proportion of molecule i in the liquid, expressed as a fraction. The value of $p_{(\mathrm{total})}$ is 72 kPa when $x_{(\mathrm{II})} = 0.3$ and $x_{(\mathrm{III})} = 0.7$, and 80 kPa when $x_{(\mathrm{II})}$ and $x_{(\mathrm{III})}$ are both 0.5. Use simultaneous linear equations to determine values for $p_{(\mathrm{II})}^{\ominus}$ and $p_{(\mathrm{III})}^{\ominus}$.

The value of $p_i^{\ominus}$ is called the **saturated vapour pressure** of i. The higher value of $p^{\ominus}$ for benzene means it is more volatile than bromobenzene.

II   III

We first rewrite the data from the question in the form of two equations:

$$72 = 0.3\,p_{(\mathrm{II})}^{\ominus} + 0.7\,p_{(\mathrm{III})}^{\ominus} \qquad (1)$$

$$80 = 0.5\,p_{(\mathrm{II})}^{\ominus} + 0.5\,p_{(\mathrm{III})}^{\ominus} \qquad (2)$$

It is not possible to solve these two equations without additional factors. A moment's study suggests we cannot simply multiply one of the equations by a single factor: we will need to multiply both by additional factors; and the two factors will need to be different.

In a case like this:

(i) We first decide which variable we wish to eliminate. In this example, we will eliminate the two $p_{(\mathrm{II})}^{\ominus}$ terms. Each value of $p_{(\mathrm{II})}^{\ominus}$ already has a coefficient: in this case, $p_{(\mathrm{II})}^{\ominus}$ has been multiplied by 0.3 in eqn (1) and by 0.5 in eqn (2). We need to modify the two coefficients such that they become the same so the $p_{(\mathrm{II})}^{\ominus}$ term in one equation can subtract completely from the other.

(ii) To make them the same, we multiply the coefficient in eqn (1) by the factor in eqn (2), and multiply the coefficient in eqn (2) by the factor in eqn (1). (In fact, we often employ multiples of the factors.)

In this example, we multiply eqn (1) by 5, and multiply eqn (2) by 3, rather than 0.5 and 0.3, respectively, because the final numbers are easier to manipulate.

To ensure the two factors for $p_{(\mathrm{II})}^{\ominus}$ are the same, we multiply eqn (1) by 5 and eqn (2) by 3:

$$360 = 1.5\,p_{(\mathrm{II})}^{\ominus} + 3.5\,p_{(\mathrm{III})}^{\ominus} \qquad 5 \times (1)$$

$$240 = 1.5\,p_{(\mathrm{II})}^{\ominus} + 1.5\,p_{(\mathrm{III})}^{\ominus} \qquad 3 \times (2)$$

We can now subtract the lower equation from the upper:

$$360 = 1.5 p_{(II)}^{\ominus} + 3.5 p_{(III)}^{\ominus} \qquad\qquad 5 \times (1)$$

$$\underline{-240 = 1.5 p_{(II)}^{\ominus} + 1.5 p_{(III)}^{\ominus}} \qquad\qquad 3 \times (2)$$

$$120 = 0 \qquad + 2.0 p_{(III)}^{\ominus}$$

$$2.0 p_{(III)}^{\ominus} = 120, \text{ so } p_{(III)}^{\ominus} = 60 \text{kPa}.$$

Substitution into eqn (1) or eqn (2) yields $p_{(II)}^{\ominus} = 100 \text{kPa}$ and $p_{(III)}^{\ominus} = 60 \text{kPa}$. ∎

Self-test 10.4

Algebraically determine the point of intersection between the following pairs of straight lines:

	(1)		(2)
10.4.1	$y = -2x + 10$	and	$y = x + 1$
10.4.2	$y = 2x + 1$	and	$y = -3x + 11$
10.4.3	$2x = -3y + 48$	and	$3x = -2y + 37$
10.4.4	$y = x - 3$	and	$-4y = 5x - 6$
10.4.5	$5x + 2y = 4.5$	and	$3x + 4y = 5.5$
10.4.6	$0.5y = 2.5x + 7$	and	$0.25y = -5x + 1$

Solving simultaneous linear equations using generalized equations

By the end of this section, you will know:

- Occasionally, solving simultaneous linear equations is almost impossible by using elimination and substitution, particularly if the coefficients are decimal or otherwise irregular.
- Simultaneous linear equations can be solved using the two equations eqn (10.1) and eqn (10.2).

Occasionally, solution of the simultaneous linear equations *is* possible using the methods above but involves too great an expenditure of effort. They are simply too irregular. There always *will* be an answer—unless reduction reveals either that the two lines are parallel, or that they are in fact the same line (see p. 124 above).

Consider the two equations

$$a_1 x + b_1 y = c_1 \qquad\qquad (1)$$

$$a_2 x + b_2 y = c_2 \qquad\qquad (2)$$

The values of x and y are given, respectively, as:

$$x = \frac{c_1 b_2 - c_2 b_1}{a_1 b_2 - a_2 b_1} \qquad\qquad (10.1)$$

$$y = \frac{a_1 c_2 - a_2 c_1}{a_1 b_2 - a_2 b_1}$$ (10.2)

■ **Worked Example 10.10**

A pair of samples are burnt in excess oxygen and the heat evolved determined. Both samples contain naphthalene (**IV**) and anthracene (**V**). The first sample contains 0.15 mol of (**IV**) and 0.06 mol of (**V**) and liberates 1197.3 kJmol^{-1} when burnt. The second sample contains 0.12 mol of (**IV**) and 0.03 mol of (**V**) and liberates 830.7 kJmol^{-1}. How much of (**IV**) and (**V**) does each sample contain?

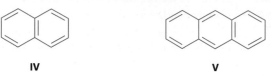

IV   V

We first rewrite the data in the form of linear equations. We will say $x_{(IV)}$ is the value of $\Delta H_c^{\ominus}$ for (**IV**) and $x_{(V)}$ is the value of $\Delta H_c^{\ominus}$ for (**V**):

For clarity, we will omit the units. We write:

$$0.15\, x_{(IV)} + 0.06\, x_{(V)} = -1197.3$$

$$0.12\, x_{(IV)} + 0.03\, x_{(V)} = -830.7$$

$x_{(IV)}$ is obtained by inserting values into eqn (10.1):

$$\frac{0.03 \times (-1197.3) - 0.06 \times (-830.7)}{0.15 \times 0.03 - 0.12 \times 0.06} = \frac{(-35.919) - (-49.842)}{0.0045 - 0.0072} = \frac{13.923}{-0.0027} = -5157 \,\text{kJ mol}^{-1}$$

And we obtain $x_{(V)}$ is by inserting values into eqn (10.2):

$$\frac{0.15 \times (-830.7) - 0.12 \times (-1197.3)}{0.15 \times 0.03 - 0.12 \times 0.06} = \frac{(-124.605) - (-143.676)}{0.0045 - 0.0072} = \frac{19.071}{-0.0027}$$

$$= -7063 \,\text{kJ mol}^{-1}$$

So values of $\Delta H_c^{\ominus}$ are −5157 kJmol^{-1} and −7063 kJmol^{-1} for (**IV**) and (**V**), respectively. ■

If a problem involved more than two equations, we *can* use simultaneous linear equations in this way, but it is far easier to employ matrices (see Chapter 27). Matrices can also be used to solve straightforward pairs of simultaneous linear equations.

Summary of key equations in the text

Generalized equations to solve simultaneous equations:

$$x = \frac{c_1 b_2 - c_2 b_1}{a_1 b_2 - a_2 b_1}$$ (10.1)

$$y = \frac{a_1 c_2 - a_2 c_1}{a_1 b_2 - a_2 b_1}$$ (10.2)

Additional problems

10.1–10.4 Which of the following pairs of lines are parallel?

10.1 (i) $y = 4x - 2$ and (ii) $y = 3x - 2$

10.2 (i) $7.1y = 9.2x + 5$ and (ii) $7.1y = 9.2x - 6.23$

10.3 (i) $y = 5x + 1$ and (ii) $2y = 10x$

10.4 (i) $0.2y = x + 11$ and (ii) $4y = 20x + 2$

10.5 Using a suitable graphical method, determine the point of intersection between the two straight lines (i) $y = 7x + 2$ and (ii) $y = 4x - 3$

10.6 The mass of a sample in a weighing boat is 12 g. The mass of a second batch comprising twice the amount of sample, but in the same weighing boat, is 23 g. Derive an equation relating the amount of sample to mass and hence determine the mass of the boat.

10.7 The **Onsager equation** relates ionic conductivity Λ and concentration c:

$$\Lambda = \Lambda° - b\sqrt{c}$$

where b is a constant and $\Lambda°$ is the conductivity at 'infinite dilution.' Two solutions are prepared: the first has a concentration c_1 of 1.100×10^{-4} mol dm^{-3} and a conductivity Λ of 0.0100 S m^{-1}, and the second has a concentration c_2 of 1.580×10^{-3} mol dm^{-3} and a conductivity of 0.0200 S m^{-1}.

Determine values for $\Lambda°$ and b.

10.8 The Gibbs function G is defined by the expression $G = H - TS$, and the enthalpy is given by $H = U + pV$. Eliminate H from the first expression. [Hint: first rearrange the second expression to make U the subject.]

10.9 A **pH meter** is in reality a pre-calibrated voltmeter. The relationship between pH and the *emf* of the voltmeter is given by the expression:

$$emf = K - 0.059\,pH$$

where the constant of 0.059 requires the units of volt.

A pH electrode is immersed in a solution of buffer at pH 7.0, and the *emf* measured as 276 mV. What is the pH (call it x) when the pH electrode is subsequently immersed in a solution and the *emf* is 502 mV?

10.10 The energy liberated when burning ethane C_2H_6 is 1560 kJmol^{-1}. The energy liberated when burning propane C_3H_8 is 2220 kJmol^{-1}. Use these data to calculate the mean enthalpies of C–C and C–H bonds.

11

Powers I

Introducing indices and powers

Introducing powers: the *base* and *index* of a number

By the end of this section, you will know:

- Basic nomenclature relating to base, power, index and exponent.
- The phrase a^b means the number 'a' has been multiplied by itself 'b' times.

Power terms dominate the algebra we encounter in chemistry. They appear in a large number of our equations, so we need to know how to manipulate them in order to use the equations effectively.

At heart, power terms can be reduced to units of the type a^b.

Concerning relationships of the type a^b, we need to know:

- We call a the **base**
- We call b the **index** of the base, or its **power**
- We sometimes see b called the **exponent**
- In words, we would say, 'a raised to the power of b', 'a to the power of b' or just 'a to the b'. Each means the same.
- We always write the exponent as a superscript, and place it to the *right* of the base, so a^b is correct but a_b or $^b a$ are not.
- Anything expressed with a power of '1' is the thing itself, i.e. $a^1 = a$.

■ **Worked Example 11.1**

Consider the molecules cyclobutadiene (**I**) and cubane, (**II**). The length of each C–C bond is l. Write an expression for the area of (**I**) and the volume of (**II**) in terms of l. For each expression, identify the base and the index.

I   II

Assuming compound (**I**) has a straightforward shape, we write its area A as, $l \times l = l^2$. Here, l is the base and the index is 2.

The volume of (**II**) V is $l \times l \times l = l^3$. This time, the base is l but the index is 3.

These examples illustrate the simplicity of indices. A few more examples:

- The expression 4^5 means 4 multiplied by itself five times: $4 \times 4 \times 4 \times 4 \times 4$.
- b^3 means b multiplied by itself three times, i.e. $b \times b \times b$.
- In general, we can write b^c, which means b multiplied by itself c times.

For simplicity, we sometimes omit the multiplication signs. We can write $c\,c\,c\,c$ instead of c^4; and XXXXX means the same as X^5. We can write the algebraic expression for the volume of a sphere as either $\left(\dfrac{4 \times \pi \times r \times r \times r}{3} \right)$ or as $\dfrac{4}{3}\pi r^3$, which most people

thinks looks tidier, but the two versions of the equation are identical. ■

Sometimes, manipulating exponents can greatly simplify an expression:

■ Worked Example 11.2

A chemical plant processes 10 000 litres of solution per hour. We wish to increase the rate a hundred-fold. Calculate the new processing rate.

We can answer this Worked Example using two different approaches:

- First, using simple arithmetic; or,
- Secondly, by manipulating the indices.

Simple algebra:

The new flow rate is $10\ 000 \times 100\ \text{l h}^{-1} = 1\ 000\ 000\ \text{l h}^{-1}$ so simple arithmetic says we process one million litres of solution per hour.

Indices:

We first write each number in terms of factors of 10, so '10 000' becomes '10 × 10 × 10 × 10', and '100' becomes '10 × 10'. We then simply sum the number of instances where '10' appears. We obtain:

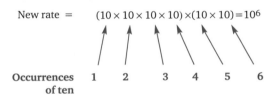

$$\text{New rate} = (10 \times 10 \times 10 \times 10) \times (10 \times 10) = 10^6$$

Occurrences of ten   1   2   3   4   5   6

The sum involves six instances of 'ten,' so the answer is 10^6 (i.e. one million).

A special case—an index of one:

Usually, we do not write the index against the base if its value is '1', because anything with an index of '1' is the thing itself. For example, we would write length $= l$ rather than l^1, although the latter is perhaps more useful when trying to understand the mathematics of indices. ■

Examples: seven to the power of 1 (i.e. 7^1) has a value of seven.
Further examples: $24^1 = 24$ $a^1 = 1 \times a$ $b^1 = 1 \times b$ $x^1 = 1 \times x$

Self-test 11.1

Write the following as a base and an index:

11.1.1 $T \times T \times T \times T \times T \times T \times T$

11.1.2 $c \times c \times c \times c \times c$

11.1.3 $a \times a \times a$

11.1.4 $q \times q \times q \times q \times q \times q \times q \times q \times q$

11.1.5 f

11.1.6 $g\,g\,g\,g\,g$

Multiple terms: We can join together whole series of mathematical phrases that incorporate powers.

■ **Worked Example 11.3**

Express $c \times c \times b \times b \times b$ in terms of bases and indices.

Strategy:

(i)   We first collect the like terms together.

(ii) We next determine the magnitude of the indices.

Solution:

(i)   Collecting together the like terms: $c \times c \times b \times b \times b = (c \times c) \times (b \times b \times b)$

(ii) Rewriting in terms of indices: $(c \times c) \times (b \times b \times b) = c^2 b^3$. ■

Self-test 11.2

Rewrite each of the following as a base and a single index:

11.2.1 $h\,h\,h\,g\,g\,h\,h$

11.2.2 $c\,c\,d\,d\,g\,g\,h\,h\,h\,h\,h$

11.2.3 $c\,d\,e\,f\,f$

11.2.4 $x\,x\,x\,x\,h\,h\,I\,I\,I$

11.2.5 $a^2 \times b \times a \times b \times a$

11.2.6 $x^3 \times x^2 \times x \times y \times y$

11.2.7 $d \times d \times d^3 \times e^4 \times e \times e^2$

11.2.8 $x^2 \times x^2 \times x \times y^6 \times y$

Negative indices

By the end of this section, you will know:

- A negative index indicates a fraction.
- The larger the number after the index sign, the smaller the number.

■ **Worked Example 11.4**

A solution of concentrated acid has a concentration of 1.0 mol dm⁻³. We dilute it ten-fold. Calculate the new concentration of the acid, expressing the answer in the form of ten to the power of an index.

Simple inspection alone tells us the new concentration is 0.1 mol dm⁻³, which we obtain as a simple fraction: $\dfrac{1\,\text{mol dm}^{-3}}{10} = 0.1\,\text{mol dm}^{-3}$.

In index form, we call this answer 10^{-1}. ■

This concentration is expressed incorrectly until we write the units after the number.

We need to know:

- The '10' merely tells us that we are working in base ten

- The minus sign tells us that a fraction is involved

- The index of '1' in this example tells us that the bottom line of the fraction contains one instance of 'ten.'

■ **Worked Example 11.5**

An analyst prepares a series of standard solutions of sodium, starting with a stock solution of concentration 1 mol dm^{-3}. He progressively dilutes the stock solution: ten-fold, ten-fold again, and then a further ten-fold.

What are the concentrations of sodium in the three solutions?

The concentration after the **first** dilution is $\dfrac{1}{10}$ mol dm^{-3} = 10^{-1} mol dm^{-3}

The concentration after the **second** dilution is $\dfrac{1}{10\times10}$ mol dm^{-3} = 10^{-2} mol dm^{-3}

The concentration after the **third** dilution is $\dfrac{1}{10\times10\times10}$ mol dm^{-3} = 10^{-3} mol dm^{-3}

We see how larger negative indices point toward smaller concentrations. ■

Self-test 11.3

Perform the following divisions, expressing the answer as a base and index:

11.3.1 $1 \div a^3$

11.3.2 $1 \div c^2$

11.3.3 $1 \div (b \times b \times b)$

11.3.4 $1 \div (d\,d\,d\,d)$

11.3.5 $1 \div z^{3.3}$

11.3.6 $1 \div (p^{4.1} \times p^{9.2})$

11.3.7 $1 \div (d \times d^4)$

11.3.8 $1 \div (j\,j\,j\,j\,k\,k\,k\,k^2\,h\,h\,h)$

Self-test 11.4

Write an expression for each of the following. Leave the answer as a fraction.

11.4.1 2^{-1}

11.4.2 10^{-2}

11.4.3 100^{-2}

11.4.4 6^{-3}

11.4.5 5^{-5}

11.4.6 13^{-3}

A special case: the index of zero

By the end of this section, you will know:

- Anything expressed with a power of zero has a value of 1.

The volume $V = l^3$ means $V = l \times l \times l$. Similarly, when we write an area A as l^2, we mean $A = l \times l$. When we write the length l, the (unwritten) index against the l is 1. In these simple examples, the indices are 3, 2 and 1, respectively. So what does the next member of this series—an index of 0—represent?

■ **Worked Example 11.6**

A solution of sulphuric acid has a concentration of 10.0 mol dm^{-3}. We progressively dilute the solution three times, first ten-fold, then ten-fold again, and then another ten-fold. Express the concentrations of the four acid solutions in index form.

(i)   Initially, the concentration is 1×10 mol dm^{-3}.

(ii)   The concentration after a ten-fold dilution is $(1 \times 10 \div 10) = 1$ mol dm^{-3}.

(iii) A further ten-fold dilution yields a concentration of $(1 \div 10) = 0.1$ mol dm^{-3}.

(iv)   Yet another ten-fold dilution yields a concentration of $(0.1 \div 10) = 0.01$ mol dm^{-3}.

In index form, the concentration of solution (i) is 10^1 mol dm^{-3}, the concentration of solution (iii) is 10^{-1} mol dm^{-3}, and solution (iv) has a concentration of 10^{-2} mol dm^{-3}.

From this trend, we see how the concentration of solution (ii) must be 10^0 mol dm^{-3} when written in index form. This is the only logical result. In fact, we derive the enormously important result that anything written to the power of zero has a value of 1. ■

Anything expressed with an index of **zero** has a value of 1.

Examples: $2^0 = 1$ $3^0 = 1$ $4000^0 = 1$ $(a\,b\,c)^0 = 1$ anything$^0 = 1$

Table 11.1 illustrates the trend for powers of ten in the range million to a millionth.

Table 11.1 Powers of ten from 10^6 to 10^{-6}, written in words, value and in index form.

Name	Value	10^n
million	1 000 000	10^6
hundred thousand	100 000	10^5
ten thousand	10 000	10^4
thousand	1000	10^3
hundred	100	10^2
ten	10	10^1
one	**1**	$\mathbf{10^0}$
tenth	0.1	10^{-1}
hundredth	0.01	10^{-2}
thousandth	0.001	10^{-3}
ten thousandth	0.0001	10^{-4}
hundred thousandth	0.000 01	10^{-5}
millionth	0.000 001	10^{-6}

Exploring the algebra of indices

By the end of this section, you will know:

- When **multiplying** terms expressed in terms of bases and indices, $a^x \times a^y = a^{(x+y)}$.
- When **dividing** terms expressed in terms of bases and indices $a^x \div a^y = a^{(x-y)}$.
- When we have anything a raised to a power x, that is itself raised to a further power y, then two powers are multiplied: $(a^x)^y = a^{(x \times y)}$.

■ **Worked Example 11.7**

A litre flask contains 1000 cm^3 of solution. What is the volume (in cm^3) contained within one hundred of the litre flasks?

As before, we can perform the calculation in two ways: by intuition and using bases and indices.

Intuition:

Simple arithmetic tells us the total volume is, 'volume per flask × number of flasks,' so the volume is 1000 cm^3 × 100 = 100 000 cm^3.

Bases and indices:

We first convert each number to its appropriate index form:

- The volume of a single flask is 1000 cm^3, i.e. 10^3 cm^3
- There are 100 flasks, i.e. 10^2
- Because 100 000 = $10 \times 10 \times 10 \times 10 \times 10$, the final volume is 10^5 cm^3

We should note the pattern emerging:

$$10^3 \times 10^2 = 10^5$$

Looking closely at the indices, we see how, $3 + 2 = 5$, so

$$10^3 \times 10^2 = 10^5 \quad \text{is the same as} \quad 10^{(3+2)} = 10^5$$

If we multiply together two numbers, and each has been expressed in terms of 10 to the power of an index, then we can express the answer as ten raised to a *new* index. We obtain this new index as the *sum* of the constituent indices. Note how we have, in effect, turned a multiplication problem into an addition problem. We can write the generalization for numbers expressed in base ten:

$$10^x \times 10^y = 10^{(x+y)} \qquad\qquad (11.1)$$

In fact, we can write eqn (11.1) in a more general form still for any base:

$$a^x \times a^y = a^{(x+y)} \qquad\qquad (11.2)$$

where a can be any number—indeed, the value of a need not be a whole number (an **integer**) but could be fractional. ■

Self-test 11.5

Rewrite the following multiplication problems. Express each in the form a^b:

11.5.1 $10^2 \times 10^4$

11.5.2 $10^3 \times 10^5$

11.5.3 $10^0 \times 10^2$

11.5.4 $10^{20} \times 10^{40}$

11.5.5 $6^3 \times 6^{12}$

11.5.6 $7^2 \times 7^4$

11.5.7 $b^9 \times b^2$

11.5.8 $z^{15} \times z^2$

11.5.9 $b^{4.1} \times b^{7.2} \times b^{3.8}$

11.5.10 $k^{6.22} \times k^{8.12}$

Working with exponents this way represents a very powerful way of simplifying.

■ **Worked Example 11.8**

Consider the unimolecular dissociation of ethane (**III**), which forms two methyl radicals:

III

The reaction proceeds in the gas phase with the following rate law

$$\text{rate} = k \times p(C_2H_6)$$

*Pa here is **Pascal**, the SI unit of pressure.*

where p is the partial pressure. k is the rate constant, which has a value of 5.36×10^{-4} s^{-1} at 700°C. What is the rate when $p(C_2H_6) = 10^{-2}$ Pa?

To simplify this calculation, we have omitted all the units.

Inserting numbers into the equation, rate $= 5.36 \times 10^{-4} \times 10^{-2}$

So rate $= 5.36 \times 10^{((-4) + (-2))}$

i.e. rate $= 5.36 \times 10^{-6}$ ■

■ **Worked Example 11.9**

A bottle of reagent holds a mass of 1000 g. We want to weigh out one tenth of this amount. What mass do we want?

To solve the problem longhand, we write:

$$\frac{1000\text{g}}{10} = 100 \text{ g}$$

Again, we could have written this sum in terms of indices:

$$\frac{10^3\text{g}}{10^1} = 10^2 \text{ g}$$

Once more, we notice a pattern: here, the index of the answer comes from the *difference* between the indices within the fraction, because $3 - 1 = 2$. ■

We write the general expression:

$$\frac{10^x}{10^y} = 10^{(x-y)} \tag{11.3}$$

This expression shows once more the power of this method of working. We can also write the algebraic expression in a yet more general way:

$$\frac{a^x}{a^y} = a^{(x-y)} \tag{11.4}$$

We can manipulate numbers in just the same way as we manipulate variables: for example, we can write $1 \div 2$ as 2^{-1}. Again, 10^{-1} is the same as $1 \div 10$.

We could have obtained this same result by noting how the term on the bottom could have been written instead as '$\times a^{-y}$'. In effect, then, $a^x \div a^y$ is the same as $a^x \times a^{-y}$. According to eqn (11.2), this *multiplication* sum yields $a^{(x-y)}$.

Self-test 11.6

Use eqn (11.4) to express each of the following division problems in the form a^b:

11.6.1 $10^2 \div 10^4$

11.6.2 $10^3 \div 10^5$

11.6.3 $10^0 \div 10^{12}$

11.6.4 $10^{-2} \div 10^4$

11.6.5 $4^3 \div 4^{12}$

11.6.6 $1.5^2 \div 1.5^4$

11.6.7 $b^9 \div b^{-2}$

11.6.8 $z^5 \div z^{2.5}$

11.6.9 $z^{1.3} \div z^{2.7}$

11.6.10 $c^{3.15} \div c^{2.93}$

■ **Worked Example 11.10**

How many cubic centimetres are contained within a cubic metre?

As before, we can solve this problem in two ways: a simple, intuitive approach, and using the laws of powers.

Intuitive approach:

The length 1 m contains 100 cm, therefore a cubic metre contains $(1\text{ m})^3 = (100\text{ cm})^3$ so $1\text{ m}^3 = 1\,000\,000\text{ cm}^3$
A cubic metre contains a million cubic centimetres.

The phrase **cubic metre**, i.e. $(1\text{ m})^3$, means the volume enclosed by the space $(1\text{ m} \times 1\text{ m} \times 1\text{ m})$; and a **cubic centimetre** is the volume $(1\text{ cm} \times 1\text{ cm} \times 1\text{ cm})$.

Approach using the laws of powers:

$$1\text{ m} = 10^2\text{ cm so } (1\text{ m})^3 = (10^2\text{ cm})^3$$

so $(1\text{ m})^3 = 10^2\text{ cm} \times 10^2\text{ cm} \times 10^2\text{ cm} = 10^6\text{ cm} \times \text{cm} \times \text{cm}$

and $1\text{ m}^3 = 10^6\text{ cm}^3$

If we look closely, we see a new pattern emerging: when we have a number raised to a power x, that is itself raised to a new power y, then we multiply the two powers:

$$(10^2)^3 = 10^6$$

We can manipulate units in much the same way as we manipulate numbers and letters

Or, in general

$$\left(a^x\right)^y = a^{(x \times y)} \tag{11.5}$$

■

Self-test 11.7

Use eqn (11.5) to express each of the following problems in the form a^b:

11.7.1 $(10^2)^5$

11.7.2 $(10^3)^{10}$

11.7.3 $(a^7)^3$

11.7.4 $(a^3)^7$

11.7.5 $(p^{4.4})^{1.2}$

11.7.6 $(7^{3.3})^{7.8}$

Fractional indices

By the end of this section, you will know:

- A fractional index of $\frac{1}{2}$ implies a square root.
- A fractional index of $\frac{1}{3}$ implies a cube root.
- In general, a fractional index of $1/n$ implies the nth root.
- The fraction of the index can be expressed as a decimal.

We have already encountered a few *fractional* indices above. It is now time to understand what they actually *mean*, physically. We start by looking at square roots.

■ Worked Example 11.11

Rewrite $\sqrt{16}$ in terms of a base and index.

In effect we are asking the question, 'what number do we need to multiply by itself to yield 16?' Clearly, the answer is '4'.

But we could have rephrased the question somewhat. In a square-root problem, the 'something to be square-rooted' is sixteen raised to some as yet unknown power, which we will call it x: we write 16^x instead of $\sqrt{16}$. Our problem here is to discover the appropriate value of x.

If 16^x is the square root of 16, then 16^x multiplied by itself will equal 16. And in terms of bases and indices, the number 16 can be written as 16^1. In summary $(16^x)^2 = 16^1$.

Using the relationship in eqn,

$$16^{2x} = 16^1$$

If the two halves of this equation are indeed the same, then the powers on either side must be the same, so

$$2x = 1$$

which can only be true if the value of x is $\frac{1}{2}$. Therefore, we indicate a square root by writing a power of $\frac{1}{2}$. ■

A few simple examples: $9^{1/2} = 3$ $144^{1/2} = 12$ $100^{1/2} = 10$

With similar reasoning, a power of a third $\frac{1}{3}$ indicates a **cube root**. The two expressions $\sqrt[3]{27}$ or $27^{1/3}$ are equally valid.

In general, we indicate the nth root of a number of phrase by writing the power $1/n$.

$$a^{1/b} = \sqrt[b]{a} \qquad (11.6)$$

A few examples to illustrate this last point:

- The fourth root of $p = p^{1/4}$
- The third root of $8 = 8^{1/3}$
- The yth root of $x = x^{1/y}$
- The 3.78th root of $q = q^{1/3.78}$

Notice that in each case, eqn (11.5) is still obeyed. For example:

$$16 = 4^2$$

so $16^{1/2} = (4^2)^{1/2}$

so $4 = 4^{2 \times 1/2} = 4^1$

and $4 = 4$

Self-test 11.8

Express each of the following word problems in the form $a^{1/n}$:

11.8.1   The third root of 27

11.8.2   The square root of 36

11.8.3   The fourth root of p

11.8.4   The ninth root of $(7b)$

11.8.5   The jth root of 12

11.8.6   The ith root of k

Summary of key equations in the text

Negative index:	$a^{-n} = \dfrac{1}{a^n}$	
Zero index:	$a^0 = 1$	
First law of power (in base 10):	multiplication: $10^x \times 10^y = 10^{(x+y)}$	(11.1)
First law of powers (general form):	$a^x \times a^y = a^{(x+y)}$	(11.2)
Second law of powers (in base 10):	division: $\dfrac{10^x}{10^y} = 10^{(x-y)}$	(11.3)
Second law of powers (general form):	$\dfrac{a^x}{a^y} = a^{(x-y)}$	(11.4)
Third law of powers: powers:	$\left(a^x\right)^y = a^{(x \times y)}$	(11.5)
Fractional powers:	$a^{1/b} = \sqrt[b]{a}$	(11.6)

Additional problems

11.1   The volume of a sphere is given by the equation:

$$V = \frac{4}{3}\pi r^3$$

Rearrange this expression to make r the subject, expressing the answer in terms of indices.

11.2   The current I induced in a conductor when a voltage V is applied across a resistor R is given by **Ohm's Law**:

$$V = IR$$

Rearrange the expression to make I the subject, and calculate the magnitude of the current I when a voltage V of 100 V is applied across a resistance R of $10^{12}\ \Omega$.

11.3   An electrochemist immerses a **rotated-disc electrode** RDE in a solution of analyte at a concentration of $c_{analyte}$. The kinematic viscosity of the solution is v. The RDE rotates at a frequency of ω. The current at the RDE is given by the **Levich equation**:

$$I = 0.62\ nFA\ c_{analyte}\ \omega^{\frac{1}{2}} D^{\frac{2}{3}} v^{\frac{-1}{6}}$$

where A is the area of the disc, D is the speed at which the analyte diffuses through solution, n is the number of electrons transferred and F the Faraday constant.

Rearrange this equation to make the diffusion coefficient, D, the subject of the equation.

11.4   Rewrite the Levich equation from Additional problem 11.3: change each index for the appropriate root sign.

11.5   The relationship between the energy of a photon, the speed of light c, the Planck constant h and the wavelength of light λ is:

$$E = h^1\ c^1\ \lambda^{-1}$$

Rewrite this expression using a more conventional notation.

11.6   The definition of concentration $c_{analyte}$ is

$$c_{analyte} = \frac{amount}{volume}$$

Starting from this definition, deduce the IUPAC units of concentration.

11.7   Two particles experience a force when they approach. The magnitude of the resultant potential energy varies with their separation r according to the **Lennard-Jones equation**

$$energy = 4\,\varepsilon\left\{\left(\frac{1}{r}\right)^{12} - \left(\frac{1}{r}\right)^{6}\right\}$$

where ε is the depth of the potential well. Multiply out the bracket, and then rewrite in terms of indices.

11.8   Chemical bonds can vibrate, much like two solid balls separated by a spring. The wavenumber of the vibration $\bar{v}$ is a function of force constant of the bond k and the reduced mass μ:

$$\bar{v} = \frac{1}{2\pi c}\sqrt{\frac{k}{\mu}}$$

c is the speed of light.

Rewrite the right-hand side as a single line rather than a fraction (i.e. use the appropriate indices).

11.9 Dynamite is an effective explosive because, during combustion, it produces a colossal volume of gas in a short period of time.

The volume of gas produced during a dynamite explosion is 100 dm³. Re-express the volume in cm³. (1 dm³ contains 1000 cm³.)

11.10 Charged species experience an electrostatic interaction ϕ when they approach, the magnitude of which is given by **Coloumb's law**:

$$\phi = \frac{z^+ z^-}{4 \pi \varepsilon_0 \varepsilon r^2}$$

where r is the interparticle separation, z^+ and z^- are the respective charges, and ε is the permittivity. Rewrite the right-hand side as a single line rather than a fraction, i.e. use the appropriate indices.

Powers II

Exponentials and logarithms

Introducing exponential functions

By the end of this section, you will know:

- An exponential function is generally written as e^x, but is occasionally written as $\exp(x)$.
- The rate of change of an exponential function is extremely rapid.

In everyday language, we often talk about 'exponential increases' like 'the prices are increasing exponentially.' In using this word, we imply an extremely fast rate of increase. Examples in everyday life of exponential function include:

- 'Price inflation' (that is, the rate at which shop prices rise).
- The mass of a bacterium as it grows in a pond or flesh wound.

   Examples in chemistry include:

- The increasing rate of a chemical reaction as the temperature increases.
- The decreasing rate of a first-order reaction with time since the reaction started.
- The decreasing amount of radio-nucleotide with time.
- The increasing number of neutrons formed during a nuclear chain reaction, e.g. within an atomic bomb based on uranium-235.

Mathematically, one of the simplest **exponential** equations is $y = a^x$, where a is a constant, x is the controlled variable and y is the observed variable.

■ **Worked Example 12.1**

During nuclear decay, a neutron hitting an atom of uranium-235 causes it to split. On splitting, each atom of U^{235} emits three more neutrons; see Fig. 12.1. Show that the number of neutrons increases exponentially.

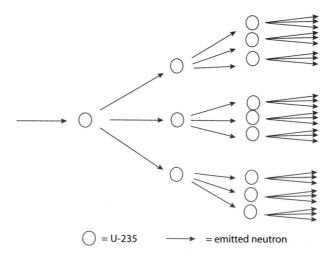

$\bigcirc$ = U-235 $\longrightarrow$ = emitted neutron

Figure 12.1 Schematic diagram to show how the number of U-235 atoms decaying during nuclear fission increases with the proportions, $1 \rightarrow 3 \rightarrow 9 \rightarrow 27 \rightarrow 81 \rightarrow 243 \rightarrow \ldots$

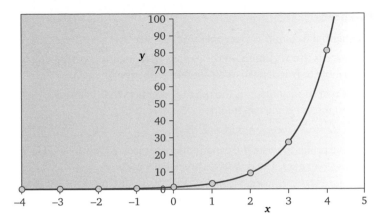

Figure 12.2 Graph of the exponential function $y = 3^x$.

Each atom of U^{235} that splits ejects three neutrons, each of which causes a further atom of U^{235} to split. Therefore, a further three atoms split during the 'second phase' of nuclear disintegration. Once again, each of these atoms emits three neutrons, so a total of nine neutrons are formed, and so on.

The number of atoms splitting therefore increase in the series, $1 \rightarrow 3 \rightarrow 9 \rightarrow 27 \rightarrow 81 \rightarrow 243 \rightarrow \dots$ In fact, the numbers of splitting atoms obey the simple function $y = 3^x$. In the sequence above, $x = 0, 1, 2, 3$ then 4. When plotting a graph of this function, $y = 3^x$, we need to include fractional and negative values of x. Figure 12.2 shows a graph of this function, which clearly increases rapidly. In fact, it illustrates a central feature of exponentials: the rate at which the gradient of the graph changes continually increases in proportion to the index x on 3. In other words, the graph becomes steeper at a rapidly increasing rate. ■

We need to notice on this graph:

- The graph passes through the point $(0,1)$, because any variable expressed to the power of 0 has a value of 1.

- The values of y are small and positive when x is negative.

- The values of y are large and positive when x is positive.

- The graph eventually increases at a very fast rate.

While we have used the word 'exponential' to describe this graph, in fact the word is generally used for a special type of function. In nature, we encounter many fundamental constants. The most common is pi (π), the ratio of a circle's circumference to its diameter. In Chapter 7, we met tau (τ), a constant that occurs often in nature. The third natural constant is e, and this one is, in many respects, the most commonly encountered.

Like the natural constants pi and tau, the value of e is **irrational**, which means we cannot express its value completely by writing it as a decimal number. To four significant figures, its value is 2.718. It might be more meaningful to write this as '2.718 …'

The simplest exponential function is $y = (2.718 \dots)^x$, although we generally write it as e^x. We write 'e' here merely as a shorthand notation. This notation style is intended to remove the need to write '2.718 …' each time.

We need to know:

- The exponential function 'e' is an operator.

- We say that 'x' is the **argument** of the exponential operator, where 'x' could be a number (whole or fractional), or an algebraic function of any sort.

- The **domain** of the argument is the range of numbers that x can take. For an exponential, x can be any real number.

- When we refer to e^x, we say aloud 'e to the x'.

- Some people prefer to write the exponential operator as 'exp' instead of 'e'. $y = e^x$ is the same as $y = \exp(x)$. The two styles are equally correct.

- We generally write 'exp' rather than 'e' to mean exponential when the argument is quite long, since the expression is less confusing and easier to read.

- Writing 'exp' instead of 'e' is often a good idea in chemistry, because 'e' can mean other things to a chemist:
  - An electron in electrochemistry.
  - (Occasionally) an electronic charge.
  - In organic chemistry, an electrophile.

Notice how we here write 'exp' rather than the shorter 'e', because the argument (the part of the equation enclosed within the bracket) is large and rather cumbersome.

■ **Worked Example 12.2**

The rate of a chemical reaction k depends exponentially on the temperature T, according to the **Arrhenius** equation:

$$k = A \exp\left(-\frac{E_a}{RT}\right)$$

where R, E_a and A are all constants. Give the mathematical name of each part of the equation.

- The operator is an exponential (written here as 'exp').
- The rate constant k on the left-hand side is the **observed variable**.
- The temperature T is the **controlled variable**.
- The portion of the equation within the bracket is the **argument**.
- The constant A is a **factor**. ■

We often call the term A the **pre-exponential factor**: 'pre' because it is written *before* the exponential operator, and 'factor' because the exponential has been multiplied by it.

■ **Worked Example 12.3**

What is the exponential of 2.5?

In the past, determining the value of the exponential $e^{2.5}$ entailed a time-consuming calculation, requiring a slide rule or a book of tables—we call such a book 'a logarithm table.' We can now employ a pocket calculator or personal computer.

Obtaining an exponential with a pocket calculator:

On most pocket calculators, the exponential key has the symbol $\boxed{e^x}$ (it is sometimes $\boxed{e^{\square}}$).

On most calculators, the $\boxed{e^x}$ key is often positioned directly above the $\boxed{\ln}$ key. We may have to access the $\boxed{e^x}$ function by pressing the $\boxed{\text{second}}$ $\boxed{\text{function}}$ or $\boxed{\text{inverse}}$ $\boxed{\text{function}}$ key.

(i)   Press the $\boxed{e^x}$ key

(ii)   Type in 2.5

The display will say 12.18 (ignore all the other, superfluous, significant figures.)
     Some calculators reverse steps (i) and (ii).

Obtaining an exponential with a PC:

This example assumes we wish to use the *Excel*™ spreadsheet programme.

(i)   Place the cursor in the cell in which the calculation is wanted.

(ii) Type '= EXP(2.5)' (Note: The letters 'EXP' here are not case sensitive).

(iii) Press the RETURN key. The display should say, 12.18249 ∎

■ **Worked Example 12.4**

One form of the **van't Hoff isotherm** is

$$K = \exp\left(-\frac{\Delta G^{\ominus}}{RT}\right)$$

where K is an equilibrium constant, $\Delta G^{\ominus}$ the corresponding value of the standard change in the Gibbs function, R the gas constant and T the absolute temperature. Using this equation, determine the value of K when $\Delta G^{\ominus} = 4\,\mathrm{kJmol^{-1}}$, $R = 8.314\,\mathrm{J\,K^{-1}}$ $\mathrm{mol^{-1}}$ and $T = 298\,\mathrm{K}$.

Before inserting values, we must remember to convert $\mathrm{kJ\,mol^{-1}}$ into $\mathrm{J\,mol^{-1}}$, by multiplying the energy by 1000. Inserting values into this expression yields:

$$K = \exp\left(-\frac{4000\,\mathrm{J\,mol^{-1}}}{8.314\,\mathrm{J\,K^{-1}\,mol^{-1}} \times 298\,\mathrm{K}}\right)$$

Incidentally, notice how all the units here have cancelled. This observation illustrates an important truth: we cannot obtain an exponential or a logarithm of anything except a number without units.

so $K = \exp\left(-\dfrac{4000}{2478}\right) = \exp(-1.614)$

i.e. $K = 0.199$. ∎

We inserted '4000' rather than just '4' on the top line because the question tells us that $\Delta \boldsymbol{G}^{\ominus}$ has a value of 4 kilo joules per mole; and the standard factor 'kilo' means '× 1000'.

Self-test 12.1

Determine a numerical value for each of the following four exponentials. Express each to four s.f.:

12.1.1 e^3

12.1.2 $e^{3.2}$

12.1.3 e^{14}

12.1.4 $e^{-3.2}$

12.1.5   Repeat the calculation in Worked Example 12.4, but use the different value, $\Delta G^{\ominus} = -12\,\mathrm{kJmol^{-1}}$

12.1.6   Repeat the calculation in Worked Example 12.4, but use the different value, $\Delta G^{\ominus} = -40.2\,\mathrm{kJmol^{-1}}$

Exponential graphs

By the end of this section, you will know:

- An exponential graph *in*creases rapidly when the argument is positive—we call it growth.
- An exponential graph *de*creases rapidly when the argument is negative—we call it decay.

Exponential graphs follow two general shapes, called 'growth' or 'decay,' which are shown schematically in Fig. 12.3. The cruial difference between the two graphs is the sign of the argument.

These two graphs illustrate how, when a graph is exponential, the gradient of the line will change rapidly:

- If the argument is *positive*, the graph gets *steeper* on going from left to right.
- If the argument is *negative*, the graph gets *shallower* on going from left to right.

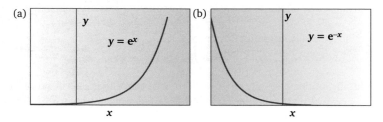

Figure 12.3 Schematic representations of exponential graphs (a) 'growth,' with a positive argument and (b) 'decay,' with a negative argument.

The inverse of exponentials: logarithm operators, and the algebra of their interconversion with exponentials

By the end of this section, you will know:

- The inverse function to exponential (e^x) is **natural logarithm** ($\ln x$).
- 10^x and logarithms expressed in base 10 are also algebraic inverses.

The letters **ln** stands for 'logarithm, natural.' The word 'natural' implies base 'e'. 'Unnatural' logarithms are expressed in a different base than 'e', and are introduced on p. 182.

It is always necessary to be able to reverse an operation in order to obtain the argument x on its own. To reverse an exponential function e^x, we need to consider its inverse, which is a **logarithm**.

We give the symbol **ln** to a logarithm, where the 'l' stands for 'logarithm' and the 'n' stands for 'natural.' The logarithm of an argument x is written as **ln** x.

It is easy to interconvert between e^x and its inverse $\ln x$ since they represent inverse functions:

$$e^x = y \quad \text{so} \quad x = \ln y \tag{12.1}$$

Since the exponential and the logarithm are inverse functions, the allowed values of x are all real numbers and those for y are all positive real numbers (zero is thus excluded) for both expressions.

■ **Worked Example 12.5**

Rearrange the van't Hoff isotherm from Worked Example 12.4, to make $\Delta G^\ominus$ the subject.

If we had MULTIPLIED a number by (for example) '4' and then performed the inverse function and DIVIDED by '4', we would have reobtained the original number. In the same way, by taking the inverse of a function, we obtain the argument on its own.

The inverse function to exponential is natural logarithm 'ln,' so to rearrange the equation, we must first take the logarithm of both sides:

$$\ln\left[\exp\left(-\frac{\Delta G^\ominus}{RT} \right) \right] = \ln K$$

The inverse of an exponential is a logarithm, so $\ln(e^x)$ becomes just 'x'. In this way, we simplify the left-hand side:

$$-\frac{\Delta G^\ominus}{RT} = \ln K$$

Finally, because the $\Delta G^\ominus$ term on the left-hand side has been divided by '$-RT$', we multiply both sides by this same factor:

$$\Delta G^\ominus = -RT \ln K \quad ■$$

■ **Worked Example 12.6**

At the start of a reaction, the concentration of reactant is $[R]_0$. During the course of the reaction, which occurs via first-order kinetics, the concentration of R decreases with time according to a so-called **rate equation**:

$$[R]_t = [R]_0\, e^{-kt}$$

Here, $[R]_t$ is the concentration of reactant R at time t after the reaction started, and k is a first-order rate constant. Rearrange this equation to make 't' the subject.

We will follow the BODMAS rules learnt previously from Chapters 3–5. First, we divide both sides of the equation by the concentration $[R]_0$:

$$\frac{[R]_t}{[R]_0} = e^{-kt}$$

Next, because '$-kt$' is the argument of an exponential, we perform the inverse operation to exponential. In words, we say, we 'take the natural logarithm':

$$\ln\left(\frac{[R]_t}{[R]_0} \right) = \ln\left(e^{-kt} \right)$$

But a function of its own inverse represents the original function. In other words, the right-hand side of the equation simplifies, and we obtain:

$$-kt = \ln\left(\frac{[R]_t}{[R]_0}\right)$$

Finally, we divide both sides by '$-k$':

$$t = -\frac{1}{k}\ln\left(\frac{[R]_t}{[R]_0}\right) \blacksquare$$

Care: ln (ell en) is often (incorrectly) written with the first letter as I (eye) or even as a numeral 1 (one).

We occasionally see the operator 'ln' written as '$\log_e$' to emphasize how it depends on e. We will avoid this notation style here: the two letters 'ln' automatically means logarithm in base e.

Logarithms in bases other than e

By the end of this section, you will know:

- We can express numbers in bases other than e, writing a^x, where a is a constant.
- The most common base other than e is base 10.
- We express each logarithm in terms of a **base** and an **argument**.
- The inverse of the function 10^x is a logarithm in base 10.
- Logarithms expressed in base 10 are written as '$\log x$'.

So far, we have talked about the natural logarithm as the inverse function of the exponential 'e', where e = '2.178 …' In fact, we could take many other numbers, and write a similar equation, e.g. $y = a^x$, where a is a constant.

We need to learn the terminology when using logarithms:

- a is the **base**.
- y is its **argument**.
- The **domain** of the argument is any positive number.

The only common example of equations of this sort is, $y = 10^x$, i.e. where the base is 10. Just as a (natural) logarithm to the base e, 'ln', is the inverse to the function $y = e^x$, so we find that a (different) logarithm is the inverse function to $y = 10^x$. We say, the inverse to '10^x' is **logarithm to the base 10**, which we symbolize as **log x**.

When speaking aloud, we call log x either 'log ex' or 'log to the base 10 of ex'

It is common to see '$\log x$' written as '$\log_{10} x$' to emphasize how the base differs from e. In fact the subscripted '10' here is superfluous since writing 'log' *means* a logarithm expressed in base 10. We will not employ the notation '$\log_{10} x$' here. In the same way that we can interconvert between e^x and $\ln x$, so we can readily interconvert between 10^x and $\log x$ since they are also inverse functions:

$$10^x = y \quad \text{so} \quad x = \log y \tag{12.2}$$

■ Worked Example 12.7

The pH of trifluoroethanoic acid (**I**) is 4.5. From the definition of pH

$$pH = -\log[H^+]$$

what is the concentration of the proton?

The definition of pH is

$$pH = -\log_{10}[H^+]$$

The p in 'pH' is short for *potenz*, the German word for 'power.' This definition of pH was introduced by Sørenson in 1909.

where the 'H' derives merely from the symbol for hydrogen, so we always give it a big letter. The 'p' is the **compound operator**, '$-1 \times \log_{10}$' of something. We always write it with a small letter. The phrase 'pH' means that we apply the operator 'p' to the argument, $[H^+]$. And the argument is abbreviated to just 'H'.

To obtain $[H^+]$ from the definition of pH, we first multiply both sides by '-1':

$$-pH = \log[H^+]$$

We then remember how 10^x and $\log x$ are inverse functions. To remove the logarithm, we therefore perform the reverse operation, which is 10^x. We write:

$$10^{-pH} = 10^{(\log[H^+])}$$

But, once again, a function of its own inverse yields the original function. The right-hand side becomes merely $[H^+]$, so:

$$[H^+] = 10^{-pH}$$

Then, inserting numbers,

$$[H^+] = 10^{-4.5}$$

so $[H^+] = 0.000\,031\,6$ mol dm^{-3}

Because this notation is quite difficult to read, we usually write the answer in terms of scientific notation (cf. Chapter 1). We say, $[H^+] = 3.16 \times 10^{-5}$ mol dm^{-3}. ∎

Some of the simpler values of $\log x$ can be calculated without a pocket calculator or computer, meaning we can interconvert from $\log x$ to 10^x without trouble. For example, when we are dealing with powers of 10 greater (or equal) to 1, the logarithm is the same as the number of zeros: the logarithm of 10 000 is 4, and the log of 10^9 is '9', and so on. Alternatively, look at the values in Table 12.1.

Table 12.1 The relationship between factors of 10 and their logarithms expressed in base 10.

Number, x	$\log x$	Rationale
1	0	because $10^0 = 1$
10	1	because $10^1 = 10$
100	2	because $10^2 = 100$
1000	3	because $10^3 = 1000$
$0.1 = 1/10$	-1	because $10^{-1} = 1/10^1 = 0.1$
$0.01 = 1/100$	-2	because $10^{-2} = 1/10^2 = 0.01$
$0.001 = 1/1000$	-3	because $10^{-3} = 1/10^3 = 0.001$

Self-test 12.2

In each case, rearrange to make x the subject:

12.2.1 $\ln x = 7$ **12.2.2** $\ln x^2 = y$

12.2.3 $\ln xt = y$ **12.2.4** $e^x = 3$

12.2.5 $e^{(-6x)} = h$ **12.2.6** $e^{(x^2)} = y$

How logarithmic functions work

By the end of this section, you will know:

- Logarithms are a powerful way to shrink a set of numbers that range over many orders of magnitude onto a much smaller scale, which is easier to understand (e.g. the concentration of H^+ ions can be stated using the pH scale).

- Logarithms can be used to simplify multiplication problems.

Logarithmic functions have a number of distinctive features:

- The logarithm of a number greater than 1 is positive.

- A logarithm of 1 has a value of zero.

- The logarithm of a positive number less than 1 is negative.

- We cannot take the logarithm of a negative number.

Logarithms are a powerful way to shrink a set of numbers which range over many orders of magnitude onto a much smaller scale, which is easier for us to understand. If we consider the pH scale then:

$$\text{pH} = -\log[H^+] \rightarrow [H^+] = 10^{-\text{pH}}$$

- When pH = 1 (acid), then $[H^+] = 10^{-1} = 0.1$ mol dm^{-3}.

- When pH = 14 (alkali), then $[H^+] = 10^{-14} = 0.000\,000\,000\,000\,01$ mol dm^{-3}.

Therefore, a change in concentration of the H^+ ions of 13 orders of magnitude (10^{-1} to 10^{-14}) can be represented by a set of numbers between 1 and 14.

If we look at Table 12.1, we can see that:

- $\log 1 = \log_{10} 10^0 = 0$

- $\log 10 = \log_{10} 10^1 = 1$

- $\log 100 = \log_{10} 10^2 = 2$

Thus, all of the numbers between 1 and 10 can be represented by an index between 0 and 1. We can calculate these indices by writing $\log_{10} a = x$. For example,

$$\log_{10} 2 = 0.301 \quad \text{so} \quad 2 = 10^{0.301}$$
$$\log_{10} 5 = 0.699 \quad \text{so} \quad 5 = 10^{0.699}$$

Both of these indices lie between 0 and 1. This amazing shrinking of scale is even more impressive when we look at how all of the numbers between 10 and 100 can be represented by an index between 1 and 2.

- $\log_{10} 20 = 1.301$ so $20 = 10^{1.301}$
- $\log_{10} 50 = 1.699$ so $50 = 10^{1.699}$

There are two key things we should note about these numbers:

(i)   Although $50 = \tfrac{1}{2} \times 100$, $\log 50 \neq \tfrac{1}{2} \times \log 100$

(ii) $\log 50 = 1 + 0.699 = \log 10 + \log 5 = \log (10 \times 5)$

(We will see more of this multiplication rule in the next section.)

Given that $\log_{10} 5 = 0.699$, we can predict, even without a calculator, that:

- $\log_{10} 500 = \log_{10} (100 \times 5) = \log_{10} 100 + \log_{10} 5 = 2 + 0.699 = 2.699$
- $\log_{10} 5000 = \log_{10} (1000 \times 5) = \log_{10} 1000 + \log_{10} 5 = 3 + 0.699 = 3.699$

To get back the original numbers, we just need to remember that we took logarithms in base 10:

$$10^{0.301} = 2$$
$$10^{0.699} = 5$$

In general, $\log_a a^x = x$.

■ **Worked Example 12.8**

Consider the following simple esterification reaction to make ethyl ethanoate (**II**):

for which equilibrium constant K has a value of 4 at 298 K. The value of K relates to the standard change in Gibbs function $\Delta G^{\ominus}$ according to the van't Hoff isotherm, $\Delta G^{\ominus} = -RT \ln K$ (see Worked Example 12.5). To what is the value of '$\ln K$' and what does this number actually related?

From a pocket calculator, the value of $\ln 4 = 1.386$.

  This number means that e, when raised to the power of $-\dfrac{\Delta G^{\ominus}}{RT} = 1.386$ (i.e. $e^{1.386}$), has a value of 4. ■

■ **Worked Example 12.9**

A solution of ethanoic acid (**III**) has a concentration of 10^{-4} mol dm^{-3}. What is the pH of the acid?

III

To obtain the pH, we insert numbers into the equation:

$$pH = -\log_{10} [10^{-4}]$$

Clearly, the number 10 is raised to the power of '−4' to obtain 10^{-4}, so the logarithm of 10^{-4} is '−4'. The pH of the aqueous solution of (**III**) is therefore

$$pH = -1 \times -4 = +4 \quad ■$$

It is usual to omit the positive sign at the end of this sum, and only cite a sign in the unusual cases when the pH is negative.

Logarithms provide a powerful way of simplifying multiplication problems. We have already seen in Chapter 11 that when multiplying together two algebraic power terms, we can predict the answer as a consequence of the mathematics:

$$10^x \times 10^y = 10^{(x+y)} = 10^z.$$

By writing the numbers in this way, we can convert a multiplication problem into an addition problem. Therefore we will greatly simplify the problem.

Imagine we wish to multiply together 8 and 5 to obtain 40. Obviously, we would usually just say, $8 \times 5 = 40$. We can look at the problem and write the answer straightaway. We cannot do so, however, if the numbers are more complicated.

The number '8' can be expressed as 10^x, where $x = \log 8 = 0.903$, and the number 5 can be expressed as 10^y, where $y = \log 5 = 0.699$. Following from the laws of powers in the previous chapter, we can write:

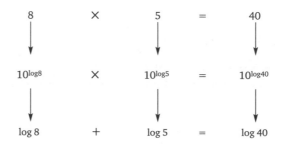

We can check that the last line of this equation is true:

$$\log 8 + \log 5 = 0.903 + 0.699 = 1.602 = \log 40 \quad \text{(to 2 s.f.)}.$$

From this example, we can see that:

$$\log 8 + \log 5 = \log(8 \times 5)$$

This is an example of one of the logarithm laws that will be discussed in more detail in the next section.

We need to note:

- We could have written these logarithmic terms in any other base we choose.
- The expression is only correct if we express each term in the same base.
- If no subscript is indicated, we assume base 10.
- If we use ln rather than log, we are working with natural logarithms, expressed in base e.
- The correct citation style always uses lower case letters: 'log'. We never write 'Log' with an upper-case L.
- Problems such as '$\log_6 4 + \log_5 3$' cannot be simplified because the two terms are expressed in different bases.

Care: Most modern pocket calculators offer a variety of logarithmic functions. Be sure to press the right key!

Self-test 12.3

Use a calculator to determine the values of the following logarithms:

12.3.1	ln 50	12.3.2	log 34
12.3.3	ln (−4)	12.3.4	log 340

 Continued . . .

Self-test 12.3 continued...

Without using a calculator, determine the values of the following logarithms;

12.3.5   ln 1

12.3.6   log 1000

12.3.7   ln 0

12.3.8   ln (e^1)

The laws of logarithms

By the end of this section, you will know:

- The sum of two logarithms ($\log a + \log b$) can be expressed as $\log (a \times b)$.
- The difference between two logarithms ($\log a - \log b$) can be expressed as $\log (a \div b)$.
- A logarithm of a power can be re-expressed such that $\log a^b = (b \times \log a)$.

Following directly from the way we defined logarithms above, we now formulate three laws of logarithms.

Law one: $\qquad \log a + \log b = \log (ab)$ (12.3)

Law two: $\qquad \log a - \log b = \log \left(\dfrac{a}{b} \right)$ (12.4)

Law three: $\qquad \log (a)^b = b \times \log a$ (12.5)

> While we have written these three laws in terms of log, they are equally valid for natural logarithms (ln), or indeed in terms of logarithms expressed in other bases.

■ Worked Example 12.10

Cis-2-butene (**IV**) readily isomerizes to form *trans*-2-butene (**V**), reaction 2:

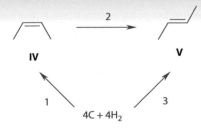

To determine the change in Gibbs function for the formation of (**IV**), we construct a simple Hess-law cycle and say $\Delta G_3^{\ominus}$ is the sum of the respective changes in Gibbs function $\Delta G_1^{\ominus}$ and the $\Delta G_2^{\ominus}$:

$$\Delta G_3^{\ominus} = \Delta G_1^{\ominus} + \Delta G_2^{\ominus}$$

Use the van't Hoff isotherm to derive an expression for the equilibrium constant of reaction 3.

The van't Hoff isotherm says, $\Delta G^{\ominus} = -RT \ln K$

so $\Delta G_1^{\ominus} = -RT \ln K_1$

$\Delta G_2^{\ominus} = -RT \ln K_2$

$\Delta G_3^{\ominus} = -RT \ln K_3$

Substituting for the $\Delta G^{\ominus}$ terms in the equation above, we obtain:

$$-RT \ln K_3 = -RT \ln K_1 + -RT \ln K_2$$

The factor of '$-RT$' is common to each term, so we cancel throughout by $-RT$, leaving:

$$\ln K_3 = \ln K_1 + \ln K_2$$

Then, because we have the sum of two logarithms, we rewrite the right-hand side using eqn (12.3):

$$\ln K_3 = \ln (K_1 \times K_2)$$

We can take the inverse of these logarithms, and discern the relationship:

$$K_3 = K_1 \times K_2 \quad \blacksquare$$

Self-test 12.4

Use eqn (12.3) to simplify each of the following:

12.4.1 $\ln 5 + \ln 3$

12.4.2 $\ln 5 + \ln 8 + \ln 2$

12.4.3 $\log 20 + \log 7$

12.4.4 $\log_p a + \log_p 7b$

12.4.5 $\log gh + \log j$

12.4.6 $\log n + \log m + \log p$

12.4.7 $\log_6 q + \log_6 r$

12.4.8 $\log_n 6 + \log_n 4t + \log_n 2f$

There is no limit to the number of logarithmic scales that can be formulated. These are merely a small variety.

■ **Worked Example 12.11**

When deriving the Clausius–Clapeyron equation, the last-but-one step is

$$\ln p_2 - \ln p_1 = -\frac{\Delta H^{\ominus}_{(vaporization)}}{R} \left(\frac{1}{T_2} - \frac{1}{T_1} \right)$$

where the two p terms are pressures and the two T terms are temperatures. R is the gas constant and $\Delta H^{\ominus}_{(vaporization)}$ is the mean enthalpy change on boiling. Simplify this equation using the laws of logarithms.

We first notice two logarithm terms, one of which is subtracted from the other. We therefore employ eqn (12.4). We rewrite the left-hand side of the equation as:

$$\ln\left(\frac{p_2}{p_1}\right) = -\frac{\Delta H^{\ominus}_{(\text{vaporization})}}{R}\left(\frac{1}{T_2} - \frac{1}{T_1}\right)$$

which is the more usual form of the Clausius–Clapeyron equation. ∎

Notice how we place the first logarithm on the numerator (*top*) of the fraction and the second resides on the denominator (*bottom*).

Self-test 12.5

Use eqn (12.4) to simplify each of the following:

12.5.1 $\ln 6 - \ln 3$

12.5.2 $\ln 5f - \ln 5$

12.5.3 $\log 12 - \log 4$

12.5.4 $\log y^2 - \log y$

12.5.5 $\log 6g - \log 3g$

12.5.6 $\log h - \log p$

■ **Worked Example 12.12**

The so-called **quinhydrone electrode** relies on the redox reaction between 'quinone' Q (1,4-benzoquinone, **VI**) and hydroquinone, H_2Q (**VII**):

Because the two species (**VI**) and (**VII**) always remain in the solid state, the Nernst equation for this couple is simply:

$$E_{Q,H_2Q} = E^{\ominus}_{Q,H_2Q} + \frac{RT}{2F}\ln\left(\left[H^+\right]^2\right)$$

where R is the gas constant, T is the thermodynamic temperature and F is the Faraday constant. Simplify this expression using the laws of logarithms.

The concentration term within the logarithm is a square, so we are taking the logarithm of a power. Accordingly, we look at the third law of logarithms in eqn (12.5). We therefore rewrite the logarithm term, saying:

$$E_{Q,H_2Q} = E^{\ominus}_{Q,H_2Q} + \frac{RT}{2F} \times 2 \times \ln\left(\left[H^+\right]\right)$$

We see how the factor against the proton concentration has moved, and is now positioned *outside* the logarithm, and written *before* it.

Now look more closely at the factors positioned *before* the logarithm: the factor of '2' on the bottom of the fraction will cancel with the new factor of '2'. Cancelling yields:

$$E_{Q,H_2Q} = E^{\ominus}_{Q,H_2Q} + \frac{RT}{2\!\!\!/F} \times 2\!\!\!/ \times \ln\left(\left[H^+\right]\right)$$

so $\quad E_{Q,H_2Q} = E^{\ominus}_{Q,H_2Q} + \frac{RT}{F}\ln\left(\left[H^+\right]\right)$ ∎

■ **Worked Example 12.13**

Probably the most all-embracing definition of an equilibrium constant K is that defined in terms of a Pi product (see Chapter 2):

$$K = \Pi \, [\text{concentration}]^{\upsilon}$$

where the Greek letter nu, υ, here is the stoichiometric number. We define υ as positive for products and negative for reactants. Write the equilibrium constant K for the reaction

$$a\text{A} + b\text{B} \rightarrow c\text{C} + d\text{D}$$

then write an expression to describe the logarithm of K.

In this example, the small italic letters are the stoichiometric numbers; and the upright capitals are the participating chemical species.

Strategy:

(i)   Write an expression for the equilibrium constant, and take its logarithm.

(ii)   Simplify using the first law of logarithms.

(iii)   Simplify further using the third law of logarithms.

Solution:

(i)   The equilibrium constant K is: $K = [\text{A}]^{-a} \times [\text{B}]^{-b} \times [\text{C}]^{c} \times [\text{D}]^{d}$

The two minus signs (before a and b) remind us that A and B are reactants. From Chapter 11, when we first looked at powers, we could have written K in its more familiar form as:

$$K = \frac{[C]^{c}[D]^{d}}{[A]^{a}[B]^{b}} \quad \text{so} \quad \ln K = \ln\left(\frac{[C]^{c}[D]^{d}}{[A]^{a}[B]^{b}}\right)$$

which illustrates the way that the bottom line of a fraction can be written as a number expressed to a negative power. To take apart the bracketed term, we generally find it easier to work in two stages, both of which illustrate the laws of logarithms:

(ii)   The first law of logarithms in eqn (12.3) yields:

$$\ln K = \ln\left([\text{A}]^{-a}\right) + \ln\left([\text{B}]^{-b}\right) + \ln\left([\text{C}]^{c}\right) + \ln\left([\text{D}]^{d}\right)$$

(iii)   The third law of logarithms, eqn (12.5), yields:

$$\ln K = -a \ln [\text{A}] - b \ln [\text{B}] + c \ln [\text{C}] + d \ln [\text{D}] \quad ■$$

Self-test 12.6

Use eqn (12.5) to simplify each of the following:

12.6.1 $\log a + \log a + 3 \log a$

12.6.2 $\log b + 2 \log b^{2}$

12.6.3 $\frac{1}{2} \log c^{3} + \log c$

12.6.4 $3 \ln c^{2}$

12.6.5 $6 \ln y + 2 \ln 4y$

The relationship between the related logarithmic functions ln and log

By the end of this section, you will know:

- To convert from $\log x$ to $\ln x$, multiply the value of $\log x$ by $\ln 10 = \text{‘}2.303\text{’}$.

It is easy to obtain values of either $\log x$ or $\ln x$ using a pocket calculator or PC. However, it is sometime necessary when we perform algebra to interconvert between the two forms of logarithm.

■ **Worked Example 12.14**

Determine the values of both ‘ln 45’ and ‘log 45’, then ascertain the value of their ratio.

From an electronic calculator, we obtain the value of ln 45 as 3.80666, and the value of log 45 as 1.60532. Taking their ratio, we find:

$$\frac{\ln 45}{\log 45} = \frac{3.80666}{1.65321} = 2.303$$

The ratio of these two logarithms is 2.303. In fact, the ratio of *any* number expressed in both ln and log has the same value of 2.303. We note that $e^{2.303} = 10$, and so $2.303 = \ln 10$. This generates an important result: to convert from $\log x$ to $\ln x$, just multiply the value of $\log x$ by $\ln 10 = 2.303$. ■

In fact, the ratio of *any* number expressed in both ln and $\log_{10}$ has a value of 2.303.

In natural **base e**, the ln 10 = 2.303. In other words, $e^{2.303} = 10$. This result has an important implication: to convert from a logarithm expressed in base 10 ($\log x$) to a logarithm expressed in base e (i.e. a natural logarithm, $\ln x$), we merely multiply the value of $\log x$ by $\ln 10$ (= 2.303).

$$\ln 10 \times \log x = \ln x \qquad (12.6)$$

Justification:

The conversion in eqn (12.6) works because $x = 10^{\log x}$

So $\ln x = \ln 10^{\log x}$

Using the third law of logarithms,

$\ln x = \log x \times \ln 10$

Therefore, $\ln x = \ln 10 \times \log x$

The discussions above have discussed logarithms in base 10 (‘log’) and in base e (‘ln’). The mathematics of chemistry frequently employs these two logarithms, but it is important to note that an infinite number of other logarithmic scales can be formulated, e.g. $\log_9$, $\log_{17.6}$, $\log_p$, etc.

Numerical calculations with logarithms

By the end of this section, you will know:

- Multiplication and division problems are simplified when using the first and second laws of logarithms, respectively.
- The third law of logarithms allow for a ready means of calculating roots.

Until the introduction of electronic calculators in the 1970s, and personal computers in the 1990s, the laws of logarithms were the only simple and swift means of calculating roots. They also simplified many other calculations. (True, slide rules also helped, but they work using the laws of logarithms.)

■ **Worked Example 12.15**

What is the value of 12.5×43.6?

At first sight, this problem is so simple it's trivial. We just pick up our pocket calculator and obtain the answer of 545. But what if the calculator did not work? Or we were uncertain if the answer on the calculator screen was correct? We would have to check it using a 'long hand' calculation.

Strategy:

(i)   We convert both numbers into logarithms, e.g. using a book of logarithm tables.

(ii)   We *add* together the two logarithms rather than multiply the original two numbers. This illustrates the first law of logarithms, eqn (12.3).

(iii) We take the answer, and obtain the inverse function.

       If we use natural logarithms ln in part 1, then the inverse function is exponential; if we used base-10 logarithms log, the inverse is 10^x. Either way, we often talk of taking the **antilog**.

Solution:

(i) $\ln 12.5 = 2.5257$

       $\ln 43.6 = 3.7751$

(ii)   Their sum is $\ln 12.5 + \ln 43.6 = 2.5257 + 3.7751 = 6.3008$

(iii) $\exp (6.3008) = 545.0077$, or 545 to 3 s.f.

       Overall the calculation that we have done is:

$$12.5 \times 43.6 = \exp(\ln(12.5 \times 43.6)) = \exp(\ln 12.5 + \ln 43.6)$$
$$= \exp (6.3008) = 545$$

We obtain the same result, but without the drudgery of long multiplication. ■

Notice how we usually work with far more significant figures than that required in the answer. If possible, work with two additional s.f.s.

■ **Worked Example 12.16**

What is the value of $\sqrt[4.2]{356}$?

We cannot just look at this problem and see its answer, although the value will certainly lie somewhere between 4 and 5, because $4^4 = 256$ and $5^4 = 625$.

Strategy:

(i)   We write the equation $x = \sqrt[4.2]{356} = 356^{\frac{1}{4.2}}$, where x is the value we seek.

(ii) We take the logarithm of both sides: $\ln x = \ln 356^{\frac{1}{4.2}}$. And, using the third law of logarithms, eqn (12.5), we can rewrite to simplify, saying $\ln x = \frac{1}{4.2} \ln 356$.

(iii) We convert 356 into a logarithm, again using a book of logarithm tables. The value is 5.8749.

(iv) We perform the simple mathematical calculation, $\ln x = \left(\frac{1}{4.2} \times 5.8749\right) = 1.39879$.

(v)   We take the antilog of both sides: $x = \exp(1.39879) = 4.05$.

A pocket calculator soon demonstrates that $4.05^{4.2} = 355.886$. To 3 s.f., we obtain the value of 356 we sought originally. ∎

Self-test 12.7

Using logarithms alone, obtain the value of the following:

12.7.1 65×41

12.7.2 $12^{3.5}$

12.7.3 $\sqrt[5.8]{1265}$

Summary of key equations in the text

In general:	$\log_a a^x = x$	
Interconversions:	$e^x = y$ so $x = \ln y$	(12.1)
	$10^x = y$ so $x = \log y$	(12.2)
The laws of logarithms:		
Law one:	$\log a + \log b = \log (ab)$	(12.3)
Law two:	$\log a - \log b = \log\left(\dfrac{a}{b}\right)$	(12.4)
Law three:	$\log (a)^b = b \times \log a$	(12.5)
Interconverting log and ln:	$\ln 10 \times \log x = \ln x$	(12.6)

Additional problems

12.1 The definition of pH is

$$pH = -\log[H^+]$$

Without using a calculator, what is the pH of a solution of HCl of concentration 10^{-5} mol dm^{-3}?

12.2 The decay of a radioactive atomic nucleus follows an exponential relationship of the form

$$\text{fraction remaining} = \left(\frac{1}{2}\right)^n$$

where n is the number of half-lives which have elapsed. Take the logarithm of both sides of the expression, and simplify it using the third law of logarithms.

12.3 Methyl ethanoate (**VIII**) is hydrolysed when dissolved in excess hydrochloric acid at 298 K. Its concentration [(**VIII**)] decreases with time during a chemical reaction, according to the **first-order integrated rate equation**:

$$\ln\left(\frac{[(\text{VIII})]_0}{[(\text{VIII})]_t}\right) = kt$$

where k is the rate constant and t is the time elapsing after the reaction has started. The initial concentration of (**VIII**) was 0.01 mol dm^{-3}, but 8.09×10^{-3} after 1260 s. Calculate the value of the rate constant, k.

VIII

12.4 The **Tafel equation** relates the current density I that flows when an electrode is immersed in a solution of analyte and the voltage is shifted by an amount η from its equilibrium potential:

$$\ln I = a + b\eta$$

where a and b are constants. Rearrange this equation to make I the subject.

12.5 The **Arrhenius equation** relates the rate constants of chemical reactions k and temperature T:

$$\ln\left(\frac{k_2}{k_1}\right) = -\frac{E_a}{R}\left(\frac{1}{T_2} - \frac{1}{T_1}\right)$$

where R is the gas constant and E_a is the activation energy. Using the laws of logarithms, expand the logarithm term on the left-hand side.

12.6 The **Debye–Hückel limiting law** relates the activity coefficient γ and a form of concentration known as ionic strength I:

$$\log\gamma = -Az^+z^-\sqrt{I}$$

where A, z^+ and z^- are constants. Rearrange the equation to make γ the subject.

12.7 The **Eyring equation** is a superior form of the Arrhenius equation, and relates the rate constants of chemical reactions k and temperature T:

$$\ln\left(\frac{k}{T}\right) = -\frac{\Delta H^{\#}}{RT} + \frac{\Delta S^{\#}}{R} + \ln\left(\frac{k_B}{h}\right)$$

where R is the gas constant, $\Delta H^{\#}$ is the enthalpy and $\Delta S^{\#}$ the entropy of activation, h is the Planck constant and k_B is the Boltzmann constant. Using the laws of logarithms, expand the equation and then group together all the log terms.

12.8 The **Nernst equation** relates the electrode potential $E_{\text{Cu}^{2+},\text{Cu}}$ for the copper couple and its standard electrode potential $E^{\ominus}_{\text{Cu}^{2+},\text{Cu}}$:

$$E_{\text{Cu}^{2+},\text{Cu}} = E^{\ominus}_{\text{Cu}^{2+},\text{Cu}} + \frac{RT}{2F} \times \ln[\text{Cu}^{2+}]$$

where T is the temperature, R the gas constant, and F the Faraday constant. Rearrange the Nernst equation to make the [Cu^{2+}] the subject.

12.9 The **Clausius–Clapeyron equation** concerns gases and liquids at equilibrium, and relates the pressure p and temperature T:

$$\ln p_2 - \ln p_1 = -\frac{\Delta H^{\ominus}_{(\text{vaporization})}}{R}\left(\frac{1}{T_2} - \frac{1}{T_1}\right)$$

Make p_2 the subject of the equation.

12.10 Molecules sometimes leave solution to adhere to a solid surface. We say they 'adsorb.' If the adsorption interaction is weak, the fraction m (amount of adsorbate ÷ mass of adsorbent) depends on the solute concentration c, according to the **Freundlich adsorption isotherm**:

$$m = kc^{\frac{1}{n}}$$

where k and n are constants. Take the natural logarithm of this equation to derive an expression having the form $y = mx + c$.

13

Powers III

Obtaining linear graphs from non-linear functions

Introducing the concept of linearization

By the end of this section, you will know:

- The rationale for linearizing curved graphs.
- The advantages of linearizing a curved graph.

Obtaining the equation of a *curved* line can be difficult. One of the simplest and most powerful ways is to replot the data of the line in such a way that we form a *straight* line. We call this process **linearization**, because we make the line straight.

Once we have replotted the data, we can then analyse the straight line and determine its gradient and intercept in the normal way (e.g. see Chapter 9). If we linearize a graph in a sensible way, the values of the gradient and intercept will tell us all we need to know in order to describe the original curved line.

To this end, we will test how to linearize a data set in cases for which the exact nature of the function is already known.

■ **Worked Example 13.1**

The so-called 'linear form' of the Gibbs–Helmholtz equation has the following form

$$\frac{\Delta G^{\ominus}}{T} = \Delta H^{\ominus} \times \frac{1}{T} + c$$

where T is the thermodynamic temperature, $\Delta H^{\ominus}$ is the change in enthalpy, $\Delta G^{\ominus}$ is the standard change in Gibbs function, and c is a constant. Obtain a linear graph from the data below, which relate to the reaction, $NH_3 + \frac{5}{4} O_2 \rightarrow NO + \frac{3}{2} H_2O$.

Temperature, T/K	250	300	350	400	450	500
$\Delta G^{\ominus}$/kJ mol^{-1}	−268	−240	−212	−184	−156	−129

A graph of $\Delta G^{\ominus}$ (as y) against T (as x) is not linear. This lack of linearity does not mean no relationship exists between $\Delta G^{\ominus}$ against T: the fact the resultant curve is regular and smooth suggests there *is* a relationship. Rather, the curvature indicates the relationship is not simply $\Delta G^{\ominus} \propto T$.

The symbol $\propto$ here means 'is proportional to'. It implies a *linear* proportionality.

If we want to linearize this data, we must learn how to 'read' the Gibbs–Helmholtz equation cited in the question. The equation of a straight line is:

$$y = mx + c \tag{13.1}$$

If we compare eqn (13.1) with the Gibbs–Helmholtz equation, we see a potential correlation:

General case	y	$= m$	x	$+c$
Specific case	$\dfrac{\Delta G^{\ominus}}{T}$	$= \Delta H^{\ominus}$	$\dfrac{1}{T}$	$+c$

Therefore, the equation tells us that a graph of $(\Delta G^{\ominus} \times T)$ (as 'y') against $1/T$ (as 'x') should be linear. Figure 13.1 shows such a graph, which is indeed linear. ■

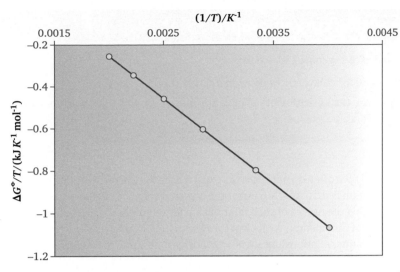

Figure 13.1 A linear graph drawn with data concerning $\Delta G^{\ominus}$ and T for the reaction $NH_3 + \frac{5}{4} O_2 \rightarrow$ $NO + \frac{3}{2} H_2O$: a Gibbs–Helmholtz graph of ($\Delta G^{\ominus} \div T$) (as '$y$') against $1/T$ (as 'x').

Self-test 13.1

Consider each of the following equations: how should we linearize each, i.e. how would we split them up to form an equation like $y = m x + c$?

13.1.1 The Nernst equation for the Cu^{2+}, Cu couple:

$$E_{Cu^{2+},Cu} = E^{\ominus}_{Cu^{2+},Cu} + \frac{RT}{2F}\ln[Cu^{2+}]$$

where $\dfrac{RT}{2F}$ and $E^{\ominus}_{Cu^{2+},Cu}$ are both constants.

13.1.2 The linear form of the Clausius–Clapeyron equation:

$$\ln p = -\frac{\Delta H^{\ominus}_{(vap)}}{R} \times \frac{1}{T} + c$$

where $\Delta H^{\ominus}_{(vap)}$, R and c are constants.

13.1.3 The linear form of the second-order kinetic rate equation:

$$\frac{1}{c_t} = \frac{1}{c_{t=start}} + k t$$

where k and $c_{t=start}$ is the start concentration, and is a constant.

13.1.4 Boyle's law relates the pressure and volume of a gas:

$$p = \frac{1}{V} \times constant$$

13.1.5 The linear form of the van't Hoff isochore:

$$\ln K = -\frac{\Delta H^{\ominus}_{(vap)}}{R} \times \frac{1}{T} + c$$

where $\Delta H^{\ominus}_{(vap)}$, R and c are constants.

Linearizing curved graphs

By the end of this section, you will know:

- That power functions follow two general types: $y = b\,a^x$ and $y = b\,x^a$.
- How to take the logarithm of the equation, and relate the reduced form to the equation of a straight line, $y = mx + c$.
- The best way to determine the values of the two constants a and b.

Unfortunately, we often do not know the exact mathematical form the graph should take. This observation is particularly true for the research chemist. In such a case, researchers must determine the nature of such relationship for themselves.

In the rest of this chapter, we look at the case of functions having indices. In their most elementary forms, relationships involving indices follow one of two forms:

$$y = a\,x^b \tag{13.2}$$

$$y = a\,b^x \tag{13.3}$$

where x is the controlled variable, y the observed variable. a and b are constants.

In the laboratory, it is quite common to obtain data that clearly follow a relationship involving indices, yet we do not know whether it follows eqn (13.3) or eqn (13.2). Even if we know which equation the data follow, we do not know the values of a and b. The purpose of this chapter is to develop strategies that allow us to answer these questions.

Our method will be to manipulate the data in such a way that the graphs become *linear*. We either say we **linearize** the graph, or that we **reduce** them to linear form. We then relate this linear line with the equation of a straight line, $y = mx + c$. Generally, we draw a series of graphs until a choice of axes yields a linear graph. We then determine the values of the gradient m and intercept c, and relate them to the values of a and b.

Care: In a chemical context, the word **reduce** also means acquisition of an electron.

Linearizing power equations of the type $y = a\,x^b$

By the end of this section, you will know:

- We first plot a graph of $\ln y$ (as 'y') against $\ln x$.
- The gradient of the graph is b.
- The intercept is '$\ln a$', so $a = e^{(\text{intercept})}$.

When seeking to linearize a graph of the type

$$y = a\,x^b \tag{13.2}$$

we first take its logarithm, as $\ln y = \ln a + \ln x^b$. We can rewrite the last term using the third law of logarithms, to yield,

$$\ln y = \ln a + b \ln x \tag{13.4}$$

This equation resembles the equation of a straight line in eqn (13.1) above. Indeed, we can compare the two equations:

We used natural logarithms **ln** in this example, but logarithms in base 10 (i.e. **log**10), would have worked equally well.

$$
\begin{array}{llll}
\ln y = & \ln a & +b & \ln x \\
y = & c & +m & x
\end{array}
$$

In summary:

- We plot a graph of $\ln y$ (as the ordinate) against $\ln x$ (as the abscissa)
- We analyse this linear graph, saying the **gradient** $= b$
- The intercept is '$\ln a$', so the value of $a = e^{(\text{intercept})}$

■ **Worked Example 13.2**

In an adsorption experiment, ethanoic acid **(I)** in water adsorbs weakly to the surface of alumina, Al_2O_3.

I

The mass of **(I)** adsorbed is m relates to its concentration at equilibrium, c:

$c/10^{-4}\,\text{mol dm}^{-3}$	6.8	13.9	22.7	37.5
$m/10^{-5}\,\text{mol}$	6.08	8.97	10.8	13.7

These data follow the **Freundlich adsorption isotherm**:

$$m = kc^{\frac{1}{n}}$$

where m is the fraction (mass of adsorbate ÷ mass of alumina adsorbent), c is the equilibrium concentration of substance adsorbed, and k and n are empirical constants. By means of a suitable graph, determine values for k and n.

We start by noting how the Freundlich equation has the form of eqn (13.2), i.e. $y = ax^b$. (In the Freundlich equation here, x is 'c' and y is 'm', a is called 'k', and b is called $\frac{1}{n}$.)

To obtain the values of a and b, we next take the logarithms of both m and c, and plot them. Figure 13.2 shows such a graph, with $\ln m$ (as y) plotted against $\ln c$ (as x).

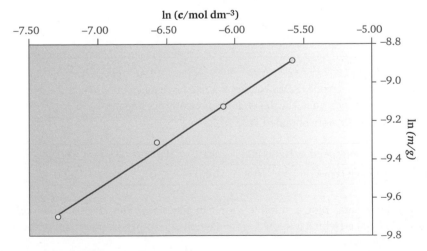

Figure 13.2 Freundlich isotherm plot of $\ln x$ (as y) $\ln c$ (as x) constructed with data for ethanoic acid **(I)** adsorbed on alumina.

While it shows a small amount of scatter, the graph is clearly linear. Its gradient is 0.47, and the intercept is –6.26.

When we compare the equation of a straight line eqn (13.1) with the logarithm of the exponential equation $y = a x^b$ (i.e. with eqn (13.4)), we see how:

- The value of the gradient $= \frac{1}{n}$, so $\frac{1}{n} = 0.47$ and $n = 2.13$

- The value of the intercept $= \ln k$, so $= \exp^{(intercept)}$, $k = e^{-6.26}$ and $k = 1.91 \times 10^{-3}$

The equation of the line describing the mass of (I) adsorbing on alumina as a function of concentration is therefore, $m = 1.91 \times 10^{-3} c^{0.47}$. ∎

To ensure dimensional correctness, k here will have units of $mol^{0.53}$ $dm^{1.41}$.

Self-test 13.2

The viscosity η of a linear-chain polymer in solution increases as the polymer's molar mass M increases, i.e. as the chain gets longer. The viscosity η and molar mass M are related according to the **Mark–Houwink equation**:

$$\eta = K M^\alpha$$

where K and α are constants that are characteristic of the solvent and the polymer.
   The following data relate to polystyrene (**II**) dissolved in benzene:

M/g mol^{-1}	500	1000	2000	3000	3500	4000
1000 η	1.91	2.95	4.57	5.89	6.50	7.06

II

Using a suitable graph, determine values for K and α for polymer (**II**) in benzene.

Self-test 13.3

A rotated disc electrode (RDE) rotates at a fixed frequency of ω. While the current I at the RDE depends on ω, the dependence is clearly not linear.
   The data below were obtained at an RDE:

Rotation speed ω/Hz	20	40	60	80	100
Current I/mA	9.8	14.0	17.0	19.7	22.0

Using these data, plot a suitable graph to determine the correct relationship between I and ω (which is known as the **Levich equation**), i.e. what is n in the following equation:

$$I = k \, \omega^n$$

Linearizing exponential equations of the type $y = a\,b^x$

By the end of this section, you will know:

- To linearize such equations, we first plot a graph of $\ln y$ (as 'y') against x (and *not* $\ln x$).
- The gradient of the graph is $\ln b$, so a is $e^{\text{(gradient)}}$.
- The intercept is $\ln a$, so $a = e^{\text{(intercept)}}$.

When seeking to linearize a graph of the type

$$y = a\,b^x \tag{13.3}$$

We first take the logarithm of this equation, yielding $\ln y = \ln a + \ln b^x$. The last term can be rewritten using the laws of logarithms to yield:

$$\ln y = \ln a + x \ln b \tag{13.5}$$

Again, we remind ourselves of the equation of a straight line in eqn (13.1), $y = mx + c$:
By comparing the two equations (13.1) and (13.5), we see:

$$\ln y = \boxed{\ln a} + x \boxed{\ln b}$$
$$y = \boxed{c} + x \boxed{m}$$

In summary:

- We plot a graph of $\ln y$ (as the ordinate) against x (as the abscissa), which will be linear.
- The gradient of this graph will be $\ln b$, so the value of $b = e^{\text{(gradient)}}$.
- The intercept is $\ln a$, so the value of $a = e^{\text{(intercept)}}$.

■ **Worked Example 13.3**

The pH of a solution of acid varies according to the concentration of solvated protons, $[H^+]$; see the table below:

$$[H^+] = 10^{-pH}$$

$[H^+]$/mol dm^{-3}	10^{-1}	10^{-2}	10^{-3}	10^{-4}	10^{-5}	10^{-6}	10^{-7}
pH	1	2	3	4	5	6	7

Use the data in this table, show we can linearize this equation by taking logarithms.

We first take the logarithm of both sides of the equation:

$$\ln[H^+] = \ln(10^{-pH})$$

so (from p. 191),

$$\ln[H^+] = -\ln 10 \times pH$$

Accordingly, we should be able to linearize the equation with an expression of the form $y = a\,b^x$, where $a = 1$, and $b^x = 10^{-pH}$. A graph of $\ln[H^+]$ as y against pH (as x)

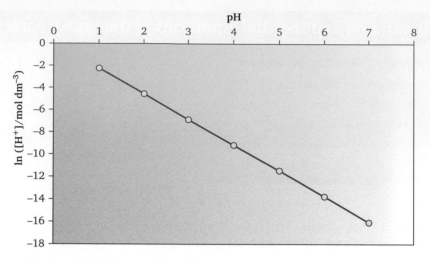

Figure 13.3 Graph of ln [H⁺] as y against pH (as x).

should be linear. Figure 13.3 shows such a graph, which is clearly linear. Accordingly, we confirm the relationship has been linearized.

Furthermore, we see how the gradient of this graph is $-2.303 = -\ln 10$ (see p. 191). This indicates that the relationship is:

$$\ln [H^+] = -pH \times \ln 10$$

$$\ln [H^+] = \ln 10^{-pH}$$

so $[H^+] = 10^{-pH}$ ∎

■ **Worked Example 13.4**

A radioactive material decomposes. We define the half-life $t_{1/2}$ as the time necessary for exactly half of the material to decompose. If the original amount of material is n_0, the amount of material remaining after x half-lives is n_t, according to the equation

$$n_t = n_0 \left(\frac{1}{2} \right)^{\left(\frac{t}{t_{1/2}} \right)}$$

where t is time. If we start with 1 mol of ^{60}Co, use the date below to determine the half-life $t_{1/2}$ for this radioisotope.

Time t/min	10	20	30	40	50
n_t/mol	0.574	0.330	0.189	0.109	0.063

We first take the logarithm of this expression, to yield:

$$\ln n_t = \ln n_0 + \left(\frac{t}{t_{1/2}} \right) \times \ln \left(\frac{1}{2} \right)$$

We will rewrite this expression slightly:

$$\ln n_t = \ln n_0 + t \times \left(\frac{1}{t_{1/2}} \right) \times \ln \left(\frac{1}{2} \right)$$

so a graph of $\ln n_t$ (as 'y') against time t (as 'x') will be linear.

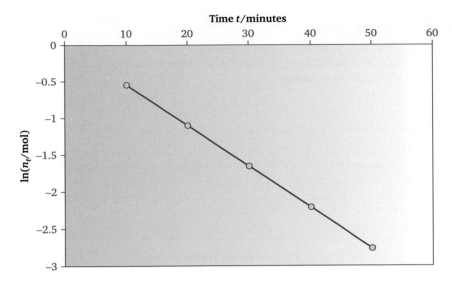

Figure 13.4 A radionucleotide such as ^{60}Co is unstable, and decomposes spontaneously, so the amount of ^{60}Co n_t decreases with time, t: graph of $\ln n_t$ (as y) against time t (as x).

- The intercept will be $\ln n_0$
- The gradient will be $\left(\dfrac{1}{t_{\frac{1}{2}}}\right) \times \ln\left(\dfrac{1}{2}\right)$.

Figure 13.4 shows such a plot, which is clearly linear.

- **The intercept:** From Fig. 13.4, intercept $=-0.0041$. The exponential of the intercept tells us the initial amount of material; we note how $\exp(-0.0041) = 1$, which confirms that we started with 1 mol of ^{60}Co.

- **The gradient:** From Fig. 13.5, gradient $=-0.0553$. Manipulation of the equation above says this gradient equals: $\left(\dfrac{1}{t_{\frac{1}{2}}}\right) \times \ln\left(\dfrac{1}{2}\right)$.

Therefore, $\qquad -0.0553 = \left(\dfrac{1}{t_{\frac{1}{2}}}\right) \times \ln\left(\dfrac{1}{2}\right)$

Accordingly, $\qquad t_{\frac{1}{2}} = -\dfrac{1}{0.0553} \times -0.693 \text{ min}$

so the half life $t_{\frac{1}{2}}$ of ^{60}Co is 12.5 min. ∎

Summary of key equations in the text

The equation of a straight line:	$y = mx + c$	(13.1)
If a line has the form,	$y = a x^b$	(13.2)
then	$\ln y = \ln a + b \ln x$	(13.4)
If a line has the form,	$y = a b^x$	(13.3)
then	$\ln y = \ln a + x \ln b$	(13.5)

Additional problems

13.1 In a kinetics experiment, a rate constant k varies with temperature T according to the linear form of the **Arrhenius equation**:

$$\ln k = -\frac{E_a}{R} \times \frac{1}{T} + c$$

where E_a and c are constants. How should we linearize the data of k and T?

13.2 (Following on from Additional problem 13.1): consider the data below, which relate to the rate of removing a naturally occurring protein with bleach on a kitchen surface. What is the activation energy of reaction?

These units for rate constant k assume a first-order reaction.

Temperature T/°C	20	30	40	50	60
Rate constant k/s^{-1}	2.20	2.89	3.72	4.72	5.91

Use the Arrhenius equation to show these data can be linearized.

13.3 When an electrode is immersed in a solution of analyte and polarized, the time-dependent current I_t decreases with time t according to the **Cottrell equation**:

$$I_t = nFAc\sqrt{\frac{D}{\pi t}}$$

where n is the number of electrons, F is the Faraday constant, A is the area of the electrode, c is the concentration of analyte, and D is the diffusion coefficient (a kind of velocity). How should data of I_t and t be linearized?

13.4 (Following on from Additional problem 13.3): the data in the following table relate to the one-electron reduction of an organic analyte at an electrode of area 0.21 cm^2, $c = 6.4 \times 10^{-7}$ mol cm^{-3}, and with a diffusion coefficient of 4.0×10^{-6} cm^2 s^{-1}.

Time t/s	1.0	2.0	3.0	4.0	5.0	6.0	7.0	8.0
Current I_t/μA	46.0	32.5	26.5	23.0	20.6	18.8	17.4	16.3

Use the Cottrell equation to show these data can be linearized.

13.5 During a first-order chemical reaction, the concentration of reactant $[A]_t$ decreases with time t according to the **first-order integrated rate equation**:

$$\ln [A]_t = -kt + c$$

where k is the rate constant and c is a constant. How should data of $[A]_t$ and t be linearized?

13.6 (Following on from Additional problem 13.5): excess cerium(III) ion promotes the decomposition of hydrogen peroxide H_2O_2, according to a first-order rate law. The following data were obtained at 298 K.

Time, t/s	2	4	6	8	10	12	14	16	18	20
$[H_2O_2]_t$/mol dm^{-3}	6.23	4.84	3.76	3.20	2.60	2.16	1.85	1.49	1.27	1.01

Show the data can be linearized using the first-order integrated rate equation.

13.7 The electrode potential of the cadmium(II)–cadmium couple $E_{Cd^{2+},Cd}$ is related to the concentration of cadmium $[Cd^{2+}]$ according to the **Nernst equation**:

$$E_{Cd^{2+},Cd} = E^{\ominus}_{Cd^{2+},Cd} + \frac{RT}{2F} \times \ln[Cd^{2+}]$$

where $E^{\ominus}_{Cd^{2+},Cd}$ is the standard electrode potential, T is the temperature, R the gas constant, and F the Faraday constant. How should data of $[Cd^{2+}]$ and $E_{Cd^{2+},Cd}$ be linearized?

13.8 (Following on from Additional problem 13.7): linearize the following data set relating the concentration $[Cd^{2+}]$ and the electrode potential $E_{Cd^{2+},Cd}$.

$[Cd^{2+}]$/mol dm^{-3}	0.1	0.05	0.02	0.01	0.005	0.002
$E_{Cd^{2+},Cd}$/V		−0.430	−0.438	−0.450	−0.459	−0.468	−0.480

Show the data can be linearized using the Nernst equation.

13.9 The pressure of gaseous iodine $p(I_2)$ above solid iodine is a function of the temperature T according to the **Clausius–Clapeyron equation**:

$$\ln p = -\frac{\Delta H^{\ominus}_{(vap)}}{R} \times \frac{1}{T} + c$$

where all other terms are constants. The following thermodynamic data refer to the sublimation of iodine.

T/K	270	280	290	300	310	320	330	340
$p(I_2)$/Pa	50	133	334	787	1755	3722	7542	14 659

Linearize this data using the Clausius–Clapeyron equation.

13.10 It is rare for a reaction to follow third-order kinetics. The only known examples involve nitrogen dioxide, NO_2. In such cases, the concentration of NO_2 varies with time t according to the **integrated third-order rate equation**:

$$\frac{1}{\left([NO_2]_t\right)^2} = 2kt + c$$

where k is the rate constant and c is a different constant. How should we linearize these $[NO_2]$ data?

Statistics I

Averages and simple data analysis

14

The concept of a data spread

By the end of this section, you will:

- Know that real experimental data are not exact.
- Understand that data will follow a Gaussian ('normal') distribution.
- Know the difference between 'accuracy' and 'precision'.

Reminder: The word *data* is a plural; the singular noun of *data* is *datum* and never *'datas'*

When we measure things in the laboratory, the reading we obtain will never be exact. Some pieces of equipment give better results than others—modern top-pan balances, for example, are likely to give trustworthy results; but undergraduate experiments involving calorimetry are notoriously poor at yielding the results similar to those in authoritative books of data.

Most chemistry text books contain many tables of data. Chemists generally call such data 'literature values.' It is easy to think of such values as being '*the* value,' by which we imply the result is entirely correct and hugely authoritative. Such a view also implies that any experimental chemist would obtain exactly the same result if only they could be sufficiently careful.

Such assumptions are often unsafe. The reasons for differences between our own values and those in the literature could include:

- Someone *computed* the data; they were not measured *experimentally*.
  - For example, many values of solubility product cannot be measured experimentally.
- The equipment used to obtain the data was more sophisticated, or completely different from that in an undergraduate laboratory.
  - For example, undergraduate calorimetry experiments often involve Dewar flasks, which can allow large amounts of heat to dissipate; sophisticated measurements require a modern bomb calorimeter, which loses considerably less heat to the environment.
- The scientists obtaining the data took additional precautions.
  - For example, when taking measurements with silver chloride, it is always advisable to shield the apparatus from light, because solid AgCl dissociates photolytically.
- The data in the book are in fact wrong.
  - There may have been errors in printing the book; errors during the computation of data; or even faithful transcriptions of data when the original sources were wrong.

While it is always wise to compare our experimental data with those in the literature, it is also wise to critique the literature data, asking how they were obtained.

When we look closely at the data in such books, the author has generally made a proviso for us. For example, a well-known data compendium cites the enthalpy of solution for potassium nitrate as 34.89 ± 0.02 kJ mol^{-1}. The additional term '± 0.02' tells us that legitimate values of ΔH in fact lie in a **range**: ΔH can extend from '$34.89 - 0.02$' kJ mol^{-1} through to '$34.89 + 0.02$' kJ mol^{-1}. Only if our value lies outside this narrow range should we question our own result.

In words, we say the symbol '$\pm$' means 'plus or minus.'

Competent chemists never perform a single analysis, but seek to *replicate* the data: imagine measuring a physicochemical parameter again and again and again—maybe dozens of times. In this way, we obtain a *series* of results, which we call a **data set**. Once this set is available, we then assess the data **statistically**. Only in this way will the reliability (or otherwise) of the data become clear.

The central idea behind the use of statistics lies in the way a series of data will differ. It is inconceivable that each value will be the same. *We should expect that even good-quality data shows a spread of results.* There are several reasons for the spread. For example:

- Different batches contains different impurities, and in different amounts.

- The conditions in the laboratory change between readings. For example:
  - Even in good-quality water baths, the temperature fluctuates very slightly.
  - The external air pressure can change, e.g. just before a storm.
  - If rain is imminent, the amount of air-borne water vapour will increase.
  - The amounts (and type) of tarnish on an electrode can vary between readings, if the electrode is not cleaned each and every time—which is tedious.

- Random human errors (e.g. the reading from a burette depends on the angle at which we look) can vary. Or think about the response time when measuring times with a stop-clock.

So we should *expect* a data spread when obtaining experimental data. Typically, the spread will describe a **Gaussian distribution**, like that depicted in Fig. 14.1.

The concept behind the Gaussian distribution curve is 'data spread.' The x-axis covers the full range of data values, while the notation on the y-axis '$n(x)$' means the number of times ('n') that each value of x occurs. Low values of x are not at all common, hence the graph is asymptotic at the far left-hand side. Similarly, large values of x are also unlikely, which explains the shape on the far right-hand side of the graph.

The centre of the peak in Fig. 14.1 corresponds to the *average* value of the data set (we will explore the full meaning of 'average' in subsequent sections). The 'bell' shape of the figure shows how most data will have values at (or near) this average value. In other words, the majority of the results will be similar. But some of the data have values that differ considerably—they lie at the two extremities of the x-axis. We give the name **outlier** to these points.

We often describe a Gaussian distribution as a **normal distribution** curve, where the qualifying term 'normal' warns us that we should *expect* our data to follow a similar pattern to that in Fig. 14.1. Data that do not follow such a distribution may be highly suspect.

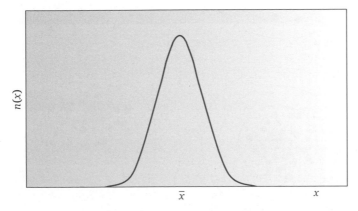

Figure 14.1 Schematic representation showing a Gaussian distribution of data. The graph shows a plot of number of data having a value of x (as 'y') against x. The peak occurs at a value of $\bar{x}$.

Because our data will follow a Gaussian distribution, and because we want our final results to be reasonable, we may wish to follow the example of many analysts and 'pre-select' data before we start our statistical analyses. We must therefore look at methods of **pre-selection**. This word means, in practice, that we seek to discard the data we think are unacceptably poor.

We now look further at the Gaussian distribution curve to see what it can tell the analyst. First, the shape of the distribution curve in Fig. 14.1 can be shallow and wide or narrow and sharp. We describe this aspect as the **precision** of the data. The value $\Delta H = 300 \pm 30$ kJ mol^{-1} is less precise than $\Delta H = 300 \pm 5$ kJ mol^{-1}: although the main value of ΔH is the same, 30 kJ mol^{-1} is a larger 'doubt' than 5 kJ mol^{-1}.

It is very common to see the terms **precision** and **accuracy** used interchangeably. Such a practice is wholly incorrect.

Secondly, the position of the central peak may coincide with the 'correct' value of the variable we wish to quantify, or may be completely different, for example are there any significant factors we failed to accommodate during our preparation such as an offset on our top-pan balance? The difference between the position of the peak and the 'true' value is termed the **accuracy**. A value can be very precise but inaccurate: $\Delta H = 300 \pm 0.5$ kJ mol^{-1} is extremely precise, but not worthwhile if the value of ΔH is actually $\Delta H = 200$ kJ mol^{-1}.

We will employ a different analogy to explore the similarities and differences between 'precision' and 'accuracy.' Think of a dart board: it will be circular, and marked with concentric circles. The exact centre of the board corresponds to the true value of the analysis. The position of each dart hole (as marked by a black dot) indicates a value obtained by an analyst. Figure 14.2 looks at the four common outcomes.

The four 'snapshots' in Fig. 14.2 clearly depict different situations. Analysts want measurements of high accuracy, that is, measurements in which the data are centred

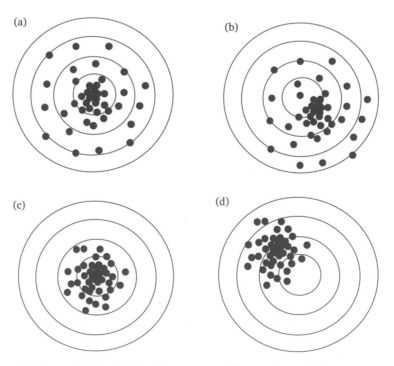

Figure 14.2 'Hits on a dart board': (a) High accuracy and low precision; (b) low accuracy and low precision; (c) high accuracy and high precision; and (d) low accuracy and high precision.

round the correct value. The dartboards in traces (a) and (c) both depict analyses of high accuracy because the results largely correspond with the centre of the board. In trace (a), the points are widely scattered, so they represent an imprecise measurement method. The precision is low. Conversely, in trace (c), the points are clustered more tightly around the centre, implying a higher precision. We will quantify the precision in later sections of this chapter.

Traces (b) and (d) depict analyses of poor accuracy. Even in trace (d) where all the points are clustered closely together, on the average they are positioned some distance from the centre of the board, which represents the correct value. Trace (b) is even worse: the accuracy is low and the precision is also poor. The traces in Fig. 14 (a)–(d) are extremes. In reality, the precision and accuracy may be intermediate.

Poor precision:

There are usually many causes of poor precision, such as human error (rushing, poor technique or an inappropriate choice of experimental method), but also the innate accuracy of the apparatus we use. We call such errors **random** or **indeterminate**. The data set will follow a Gaussian distribution because they are truly random.

- By 'innate accuracy of the apparatus,' we mean that some pieces of apparatus *cannot* determine a result to a high degree of precision. A good example concerns the apparatus used to weigh a sample. The old-fashioned kitchen scales in Fig. 4.1 cannot weigh a fraction of a gram, so very small weights are almost certain to be inaccurate: such a balance is *innately* inaccurate.

- By 'poor technique,' we mean a sloppiness and/or failure to take due care.

- By 'inappropriate choice of experimental method,' we mean that some methods are excellent at one experimental extreme but poor at the other. For example, titration is usually a good method when determining concentrations in the range $2 > c > 10^{-5}$ mol dm^{-3}. But titration is not usually a good method when quantifying a concentration of, say, 4 parts per billion (ppb); and we will need a different technique.

Poor accuracy:

The cause of poor accuracy is more likely to be the equipment we employ. For example, we incorrectly calibrated a pipette, or we incorrectly balanced the chemical equation. We call such an error a **systematic** or **determinate**, because (at least in principle) we can locate and/or compensate for the causes of the error. In such cases, the data set will *not* follow a Gaussian distribution because they are not truly random.

- Some pieces of apparatus *cannot* determine a result to a high degree of accuracy. A good example concerns the apparatus we chose to weigh a sample. A balance may cite a weight to 3 or 4 s.f., but the result will be wrong if we do not tare the weighing boat.

- Even a reliable method, if performed without following the instructions, leads to inaccuracy. A good example concerns a straightforward pipette: the volume is likely to be accurate if we touch the surface of the expelled liquid with the tip of the pipette (surface tension withdraws the right amount of solution). Failing to touch the surface or blowing out the entire volume, will cause the expelled volume to be too small or too large, respectively.

We now look at quantitative methods of analysing our data set.

Averages

By the end of this section, you will know:

- The word 'average' is a statistical construct.
- The word 'average' is ambiguous, because there are three common types of average: the mean, the mode, and the median.

Because analysts perform multiple measurements, and obtain a *series* of results, the simplest way of obtaining 'the answer' is to take the average of the data (or, strictly, of the data *series*).

The word 'average' is ambiguous, so we will deliberately avoid using it on its own in this chapter. Where it does appear, its use will always be qualified. The ambiguity arises because 'average' can mean one of three distinct types of calculation: the **mean**, the **mode**, and the **median**. In everyday language, 'average' generally implies the mean, which explains why, if we type 'average' as the function in an *Excel*™ spreadsheet, the program computes a mean.

The mean

By the end of this section, you will know:

- The mean is also called the *arithmetic* mean.
- The mean is the weighted average of a set of data.

The word 'average' generally means the mathematical function known as a **mean**. Such a 'mean' is a statistical way of looking at a series of data, and allows us to represent a *series* of data with a single value.

The mean is calculated using eqn (14.1):

$$\bar{x} = \frac{\sum_{i=1}^{N} x_i}{N} \tag{14.1}$$

In words, we add up all the data, and divide by the number of data in the set. Notice the small 'overbar' positioned above the x, as $\bar{x}$. This overbar tells us at a glance that we have a mean. The value we compute will always lie between the extremes of the data we employed when computing the mean, so $\bar{x}$ will always lie between the largest and smallest data points. We have miscalculated $\bar{x}$ if its value does not lie within this range.

■ **Worked Example 14.1**

An analytical chemist wishes to determine the accuracy of a 1.0 cm³ pipette. The method chosen is to measure out '1.0 cm³' of water ten times, and weigh the amount of water in each case. In grammes, the mass of water is 1.003, 0.983, 1.022, 1.031, 1.011, 1.001, 0.939, 0.993, 0.992, and 0.988 g.

If the density of water is 0.998 cm³ g⁻¹, what is the mean mass of water $\bar{m}$ and hence what is the mean volume $\bar{V}$?

Inserting data into eqn (14.1) yields:

$$\bar{m} = \frac{(1.003 + 0.983 + 1.022 + 1.031 + 1.011 + 1.001 + 0.939 + 0.993 + 0.992 + 0.988)\text{g}}{10}$$

so $\bar{m} = \dfrac{9.963\text{g}}{10} = 0.9963$ g

To three significant figures, the volume of the pipette is therefore

$\bar{V} = \bar{m} \times$ density, so $\bar{V} = 0.9963$ g $\times 0.998$ cm^3 g$^{-1} = 0.994$ cm^3.

In other words, the pipette is 0.6% smaller than it claims to be. ∎

The mean is also called the **arithmetic mean**, because we calculate it as a *sum* of the component data. A different mean is the **geometric mean**. Chemists only rarely employ the *geometric* mean, which is defined in terms of a Pi product Π (rather than the Sigma sum in eqn (14.1), so we will not consider it further.

Note it is common and acceptable to cite a mean to more **significant figures** than the original data within the set. In general, the mean is cited to $(n + 1)$ s.f. if the data are cited to n s.f.

Self-test 14.1

Calculate the mean of each series of data using eqn (14.1):

14.1.1   Mass m (in g) = 12.1, 12.2, 12.1, 11.8, 11.4, 11.6, 12.3, 12.0 and 12.2.

14.1.2   Concentration c (in ppm) = 712, 654, 733, 656, and 704.

14.1.3   Enthalpy ΔH (in kJ mol^{-1}) = 32.0, 32.1, 32.0, 32.4, and 32.3.

The mode

By the end of this section, you will know:

- The *mode* is the value that occurs most frequently.
- Statistically, the mode is rarely as useful as the mean.

For the analyst, the **mode** is not as generally as useful as the mean, and is not employed so often. As before, the analyst measures a variable several times to obtain a series. The mode is, then, the value that occurs most frequently. In other words, if we draw a bar graph, the mode corresponds to the highest bar.

■ **Worked Example 14.2**

An industrial effluent is found to contain the banned insecticide dichlorodiphenyl trichloroethane, DDT (**I**). The amount of [DDT] (in parts per million, ppm) was found to be, 21.2, 21.3, 20.9, 21.4, 21.3, 21.7, 20.6, 20.3, 22.3, 21.4, 21.5, 21.3, 21.0, 21.5, 21.3, 21.2, and 21.1. Display the data as a bar graph, and hence determined the mode.

The highest peak in the bar chart in Fig. 14.3 is that for [DDT] = 21.3 ppm, so the mode is 21.3 ppm. As expected, the mode lies at the centre of the spread of data. Note how the shape of the graph in Fig. 14.3 broadly follows the Gaussian distribution in Fig. 14.1. ∎

I

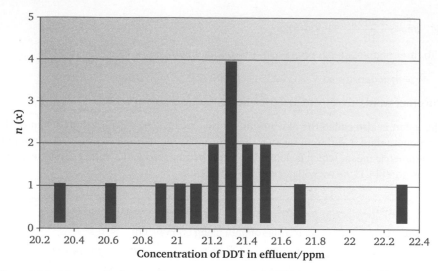

Figure 14.3 A bar graph showing the concentration of DDT (I) in an industrial effluent (as 'x') against number of times this concentration was obtained in an analysis (as 'y').

Self-test 14.2

Determine the mode of each series of data:

14.2.1   Mass (in mg): 9.21, 9.19, 9.20, 9.18, 9.21, 9.17, 9.20, 9.18, 9.22, 9.24, 9.22, 9.21.

14.2.2   Concentration (in ppm): 215, 213, 212, 215, 219, 221, 211, 227, 220, 215, 212, 215, 214, and 213.

14.2.3   Concentration in $\mu g\ dm^{-3}$: 32, 33, 32, 34, 35, 33, 34, 33, 30, 29, 31, 32, 33, and 30.

The median

By the end of this section, you will know:

- The median of a range of data lies in the middle of the spread of values.
- We can only properly take the median of an odd number of data.
- If we want the mode of an even number of data, we take the mean of the *two* central data.

The **median** is another form of average. Analysts rarely use the median because its value is often of questionable worth.

Any series of data will show a spread of values. The median is defined as the middle value. Accordingly, if there are five data, then the median is the third; if there are 99 data, the median is the fiftieth, and so on. Clearly, we must be strict and employ only an odd number of data if a median is to be determined.

The median is a reliable average if there are many data that follow a Gaussian distribution. Its usefulness decreases as the number of data decreases, and as the data deviate from a normal distribution.

■ **Worked Example 14.3**

Several students sit an exam in analytical chemistry. In order of increasing percentage, their exam marks are 12, 22, 30, 40, 45, 50, 53, 55, 56, 58, 60, 70, and 90%. What is the median mark?

There are thirteen marks, so the median corresponds to the middle one, the seventh. Accordingly, the median mark is 53%.

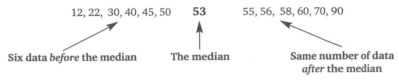

12, 22, 30, 40, 45, 50   **53**   55, 56, 58, 60, 70, 90

Six data *before* the median   **The median**   **Same number of data *after* the median**

Roughly half of our multiple analyses will yield a data set comprising an even number. To obtain the mode of such a data set, we again arrange the data in ascending order, and then choose the *two* central points. The median is then the mean of these two. ■

Self-test 14.3

Determine the median of each data set.

[Hint: always place the numbers in increasing order before starting].

14.3.1   The length of a sample, in cm: 1.90, 1.93, 1.90, 1.98, 1.99, 1.93, 1.93, 1.90, and 1.93.

14.3.2   The energy of a $N=N$ bond, for example in azine dyes, in kJ mol^{-1}: 473, 474, 472, 476, 475, and 475.

The standard deviation

By the end of this section, you will know:

• How the sample standard deviation relates to the relative precision of a result.

• A high value of s implies poor precision.

• How to calculate the sample standard deviation s.

Indeterminate (random) errors can normally be analysed by simple statistical methods. We now understand that chemical analysts measure a variable several times, and that the results will vary. Having looked at averages, that is, at ways of abbreviating a series to form a single number, we now look at the validity of that single number, asking 'how accurate is the mean: how wide is the spread of results within the data set?'

Most of the simpler statistical methods assume the data set follow a Gaussian distribution profile. The first statistical tool is the **sample standard deviation s**. By the very nature of the sample standard deviation, a large value of s means a wide spread of data and a smaller value of s means a narrow spread. In other words, large values of s imply imprecision, and a small value of s means a more precise data set. To reiterate, the precision tells us nothing about the accuracy: an answer may be precise but wrong.

In terms of a formal definition, the sample standard deviation s is:

$$s = \sqrt{\frac{\sum_{i=1}^{N}(x_i - \bar{x})^2}{N-1}}$$ (14.2)

The sample standard deviation s describes the spread of data around the mean datum within a data set.

The **root mean square** (RMS) is the square root of an arithmetic mean of a series of square terms. The RMS error is similar to eqn 14.2 but the '$N-1$' term in the denominator is replaced by 'N'.

where N, the number of data within the set is typically less than 10, and $\bar{x}$ is the mean.

This definition looks rather complicated, but mathematical rearrangement allows us to rewrite it in a more 'user friendly' way:

$$s = \sqrt{\frac{\sum_{i=1}^{N} x_i^2 - \left(\sum_{i=1}^{N} x_i\right)^2 / N}{N-1}} \qquad (14.3)$$

This version is considerably easier to use, because we calculate two sums, and then take the root of their difference.

Also, notice how s has the same units as the original variable, so if our data set relate to concentrations in parts per billion (ppb), then we also express s in ppb.

■ Worked Example 14.4

An electrochemist repeatedly measured the capacitive charging current i_{cap} (in μA) as: 19.4, 20.6, 18.7, 19.2, 21.6, 18.9, and 19.9. Calculate the sample standard deviation s of this data set

Strategy:

We calculate each term separately:

(i) $\sum_{i=1}^{N} x_i^2$,

(ii) $\left(\sum_{i=1}^{N} x_i\right)^2$,

(iii) Lastly, we assemble the terms within eqn (14.3).

Solution:

(i) To calculate $\sum_{i=1}^{N} x_i^2$:

$$\sum_{i=1}^{N} x_i^2 = (19.4)^2 + (20.6)^2 + (18.7)^2 + (19.2)^2 + (21.6)^2 + (18.9)^2 + (19.9)^2$$
$$= 376.36 + 424.36 + 349.69 + 368.64 + 466.56 + 357.21 + 396.01$$
$$= 2738.83$$

(ii) To calculate $\left(\sum_{i=1}^{N} x_i\right)^2$:

$$\left(\sum_{i=1}^{N} x_i\right)^2 = (19.4 + 20.6 + 18.7 + 19.2 + 21.6 + 18.9 + 19.9)^2$$
$$= (138.3)^2$$
$$= 19\,126.89$$

(iii) There are seven data. Inserting these numbers within eqn (14.3) yields:

$$s = \sqrt{\frac{2738.83 - (19\,126.89)/7}{7-1}}$$

$$s = \sqrt{\frac{2738.83 - 2732.41}{6}}$$

$$s = 1.1 \ \mu A$$

Notice how the standard deviation has the same units (here μA) as the data it refers to.

The magnitude if the sample standard deviation here is not large—about 5% of the mean (in this example, $\bar{x} = 19.76 \ \mu A$). ■

We are generally concerned when an experiment yields a value of s of 20% or more; but occasionally, we actually want a large value of s. For example, the standard deviation for an academic examination ought to have a high value of s to demonstrate a clear separation between the students on this course.

When a data set comprises a larger number of data points (typically greater than 10), we employ a slightly different version of s, known as the **population standard deviation** σ. Note how the two standard deviations have different symbols.

$$\sigma = \sqrt{\frac{\sum_{i=1}^{N} x_i^2 - \left(\sum_{i=1}^{N} x_i\right)^2 / N}{N}} \qquad (14.4)$$

The only mathematical difference between eqns (14.3) and (14.4) lies in the denominator within the root: because there are more data within the set, we divide by the total number N. We say the expression in eqn (14.4) has an **extra degree of freedom**.

We only rarely replicate a measurement more than ten times, so the sample standard deviation s is encountered and used far more frequently than the population standard deviation σ.

<aside>
The two expressions in eqns (14.3) and (14.4) describe essentially the same quality, i.e. the variations within a data set.
</aside>

Obtaining standard deviation values using a pocket calculator:

Some 'scientific' calculators will calculate the standard deviations automatically. It's always wise to read the calculator manual before use, because every calculator is different. The necessary keys will generally have the labels $\boxed{\sigma_n}$ and $\boxed{\sigma_{n-1}}$. Some calculators use a more advanced notation, with two sets of keys: labels $\boxed{x\sigma_n}$ and $\boxed{x\sigma_{n-1}}$, and $\boxed{y\sigma_n}$ and $\boxed{y\sigma_{n-1}}$.

On the *Casio fx-85ES* calculator, go to MODE → STAT (option 2) → 1-VAR (single variable entry - option 1) → ENTER DATA → Press AC at the end of the data list → STAT (SHIFT and '1') → VAR (option 5) → $x\sigma_{n-1}$ = STDEV.

<aside>
Aside

The calculation in Worked Example 14.4 is longwinded, time consuming, and prone to errors and slips. Luckily, we can speed up the process by using either a calculator or a computer spreadsheet.
</aside>

Obtaining standard deviation values using a computer spreadsheet:

(i)   Open *Excel*™.

(ii) Enter the data in a single column.

(iii) In a cell below the column, type '=STDEVA()'

(iv) Press RETURN twice, i.e. cancelling the '*error*' message.

(v)   Using the mouse, highlight all the cells to be considered.

(vi) Press RETURN once more.

The cell should display the standard deviation. (It may be necessary to manually change the number of significant figures.)

<aside>
Ensure no gap appears after the = sign here. The letters are not case sensitive. Also note how STDEVA() is the more general form of standard deviation, and allows cells to include logical values TRUE/FALSE.
</aside>

Self-test 14.4

Determine the standard deviations for each of the two data series in Self-test 14.3.

The Q-test

By the end of this section, you will know:

- That outlier points are statistically rare.
- That including outlier points in a calculation can ruin the result.
- How to use a simple Q-test to identify outlier points.

It is now clear how the data within a set have a range of values. The majority have a value close to the mean, a few could have values that are extreme and seriously distort a mean or standard deviation calculation. We need to identify such data and (possibly) omit them before they distort our final result. We need to stress the following crucial condition: *we must exercise great care when rejecting data because, each time we reject a datum, we effectively introduce a bias.*

We need to have confidence in our data: are they reliable? If not, then we are foolish if we continue using them. Unfortunately, there is no guaranteed method for rejecting or retaining data, but the simplest method, which assesses the reliability of a single datum at a time, is the so-called **Q-test**.

In the *Q*-test method, we first calculate the quotient *Q* of a single experimental point, according to the equation:

$$Q_{(\text{exp})} = \frac{\left| x - x_{(\text{next})} \right|}{x_{(\text{maximum})} - x_{(\text{minimum})}} \tag{14.5}$$

where the value of *x* without a subscript in the numerator relates to the point under scrutiny. The other three terms relate to other members of the data set, which have already been arranged in order of increasing value: $x_{(\text{next})}$ is the value of the datum next in the series, and (in the denominator), $x_{(\text{minimum})}$ and $x_{(\text{maximum})}$ are, respectively, the data of the two extreme ends of the set.

Having determined a value of $Q_{(\text{exp})}$, we compare its value with calculated data in a table. Table 14.1 contains such a compendium, as explained in Worked Example 14.5.

■ **Worked Example 14.5**

A chemist determines the absorbance of the dye Indigo Carmine (**II**) solutions in order to determine its concentration. The absorbances *A* within the data set are, 0.104, 0.113, 0.114, 0.114, 0.115, and 0.117. We suspect the first point is anomalous.

Using a *Q*-test, with what confidence can the first point be rejected?

II

Table 14.1 *Q*-test table, indicating confidences for replicate measurements, allowing for the confident assessment of outlying data.

No of replicate measurements	Reject with 90% confidence	Reject with 95% confidence	Reject with 99% confidence
3	0.941	0.970	0.994
4	0.765	0.829	0.926
5	0.642	0.710	0.821
6	0.560	0.625	0.740
7	0.507	0.568	0.680
8	0.468	0.526	0.634
9	0.437	0.493	0.598
10	0.412	0.466	0.568

Strategy:

(i)   Calculate $Q_{(exp)}$ using eqn (14.5)

(ii) Compare the value of $Q_{(exp)}$ with values in Table 14.1.

Solution:

(i)   The suspect datum is 0.104; in sequence, the neighbouring point $x_{(next)}$ has a value of 0.113; the maximum point $x_{(maximum)}$ is 0.117, and the minimum point $x_{(minimum)}$ is 0.104 again. Inserting values into eqn (14.5) yields:

$$Q_{(exp)} = \frac{|0.104 - 0.113|}{0.117 - 0.104} = \frac{0.009}{0.013}$$

Therefore, $Q_{(exp)} = 0.692$.

(ii) We now look at Table 14.1.

Our data set comprises six values, so we will concentrate our attention to the row starting '6'. If $Q_{(exp)}$ had a value of 0.56, we could have rejected it with 90% certainty; in other words, we are uncertain about the value 10% of the time. Conversely, if the value of $Q_{(exp)}$ was 0.625, we would have 95% certainly the value was erroneous, so we would have halved the uncertainty. And if $Q_{(exp)}$ had a value of 0.740, we would have a 99% confidence the datum was an outlier that should be rejected. The uncertainty is only 1%. The higher the value of $Q_{(exp)}$, the safer we are when rejecting a datum.

To return to Worked Example 14.5, the value of $Q_{(exp)}$ was 0.692. We can reject this datum with a confidence of more than 95%, but with a confidence of less than 99%. In other words, the point is almost certainly unreliable, so we can safely reject it before any statistical manipulation of the data. ∎

> More comprehensive tables are available if we want to know the confidence to other degrees of certainty.

In summary, before we manipulate the data, calculating mode, mean and median, we need to ascertain whether the data are reliable, and omit those that are likely to be suspect.

Self-test 14.5

Each of the following data sets contains a suspect point. Identify the suspect point, and then use a simple Q-test to ascertain the confidence with which we could reject the suspect point.

14.5.1   Using a Karl–Fischer titration apparatus, the amount of water in a sample of ethanol is: 0.65, 0.68, 0.71, 0.72 and 0.92%.

14.5.2   Five samples of an ore were analysed for their lead content. The content (in ppm) is: 214, 217, 219, and 226.

14.5.3   A chemist titrates a sample of the acid found in the gut of a mouse, and determines its concentration (in mmol dm^{-3}) as: 32.2, 34.3, 35.2, 35.2, 35.4, and 35.7.

Summary of key equations in the text

Definition of the mean:

$$\bar{x} = \frac{\sum_{i=1}^{N} x_i}{N} \tag{14.1}$$

Definitions of sample standard deviation:

- Definition:

$$s = \sqrt{\frac{\sum_{i=1}^{N}(x_i - \bar{x})^2}{N-1}} \tag{14.2}$$

- The form we actually use:

$$s = \sqrt{\frac{\sum_{i=1}^{N} x_i^2 - \left(\sum_{i=1}^{N} x_i\right)^2 / N}{N-1}} \tag{14.3}$$

Population standard deviation:

$$\sigma = \sqrt{\frac{\sum_{i=1}^{N} x_i^2 - \left(\sum_{i=1}^{N} x_i\right)^2 / N}{N}} \tag{14.4}$$

The Q-test:

$$Q_{(exp)} = \frac{\left| x - x_{(next)} \right|}{x_{(maximum)} - x_{(minimum)}} \tag{14.5}$$

Additional problems

14.1–14.5

Nine students each perform an acid–base titration. Their titres of alkali solution are: 17.0, 18.3, 18.4, 18.5, 18.9, 19.3, 19.3, 19.3 and 20.0 cm³.

14.1 Perform a simple Q-test analysis of the data.

14.2 Calculate the mean titre using those data that we are confident are reliable.

14.3 Using those data that we are confident are reliable, calculate the standard deviation associated with the titre.

14.4 Calculate the *median* titre using those data that we are confident are reliable.

14.5 Calculate the *mode* titre using those data that we are confident are reliable.

14.6–14.10

A thermodynamicist determined the enthalpy change associated with the reaction:

$$2H_2S(g) + SO_2(g) \rightarrow 2H_2O(g) + 3S(s)$$

and found the enthalpy change associated with reaction was: −230.0, −230.2, −230.2, −230.3, −230.6, −230.6, −230.6, −230.8, −230.9, and −232.7 kJ mol⁻¹.

14.6 Perform a simple Q-test analysis of the data.

14.7 Calculate the *mean* energy using those data that we are confident are reliable.

14.8 Calculate the standard deviation associated with the energy measurement; use those data that we are confident are reliable.

14.9 Calculate the *median* energy using those data that we are confident are reliable.

14.10 Calculate the *mode* energy using those data that we are confident are reliable.

Statistics II

Treatment and assessment of errors

Readings, errors and the concept of signal-to-noise

By the end of this section, you will learn:

- That all real experimental data have an associated error.
- Because of this error, all literature data actually comprise a *range* of values.
- The meaning of signal-to-noise ratio.
- All experimental apparatus has a recommended range of values.
- Analysts should ensure the experimental readings lie within the recommended range of the apparatus.

Any reading ever made by a chemist has an associated error. The error may be so small that we can effectively ignore it, but the error may so large that we must seriously question the meaningfulness of the reading.

We usually cite a reading in the form given in eqn (15.1):

$$\text{reading} = \text{value} \pm \text{error} \tag{15.1}$$

In words, we say $\pm$ means 'plus or minus.'

The symbol $\pm$ means the reading actually possesses a range of values from 'value − error' through to 'value + error.'

■ **Worked Example 15.1**

The enthalpy of the reaction

$$Fe^{3+}(aq) + SCN^-(aq) \rightarrow [FeSCN]^{2+}(aq)$$

is $\Delta H_r^{\ominus} = -35.5 \pm 0.7$ kJ mol^{-1}. What are the range of values for $\Delta H_r^{\ominus}$?

The minimum value of $\Delta H_r^{\ominus}$ is $-35.5 - 0.7$ kJ mol$^{-1} = -36.2$ kJ mol^{-1}.

The maximum value of $\Delta H_r^{\ominus}$ is $-35.5 + 0.7$ kJ mol$^{-1} = -34.8$ kJ mol^{-1}. ■

We sometimes refer to the relative magnitudes of the reading and the error in terms of the **signal-to-noise ratio**, where the reading is the signal. (This ratio is alternatively symbolized as SN, S/N or even SNR.)

The signal-to-noise ratio is determined simply as the quotient in eqn (15.2):

We usually indicate a **ratio** in the form 'number : 1'. The colon tells us the number is a ratio; the last number is always '1'.

$$\text{signal-to-noise ratio} = \frac{\text{reading}}{\text{error}} \tag{15.2}$$

The signal-to-noise ratio is often cited as a ratio, i.e. as 'signal-to-noise $= x : 1$.'

There are no absolute rules, but most scientists would say a signal-to-noise ratio of 3:1 or less means the reading is too unreliable to use.

■ **Worked Example 15.2**

A solution of the intensely blood-red coloured complex [FeSCN]$^{2+}$(aq) = 0.05 mol dm^{-3} is prepared (cf. Worked Example 15.1), and its optical absorbance is determined to be 3.6 ± 0.4; see Fig. 15.1. What is the signal-to-noise ratio?

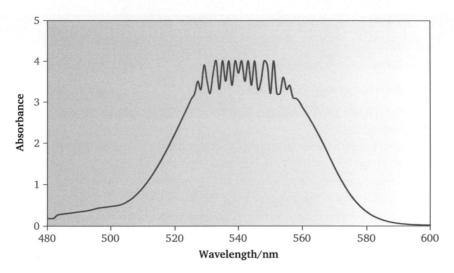

Figure 15.1 The UV-visible spectrum of $[FeSCN]^{2+}(aq) = 0.05 \text{ mol dm}^{-3}$. The jagged appearance at the top, near the spectrum centre is a manifestation of the large signal-to-noise ratio, and is caused by the detector 'not coping' with the extremely large absorbances.

Inserting numbers into eqn (15.2) yields:

$$\text{signal-to-noise ratio} = \frac{3.6}{0.4} = 9$$

So the signal-to-noise ratio is 9:1. This error is large for spectometry. ∎

Equation (15.2) suggests the best way to enhance the signal-to noise ratio is to ensure the reading is huge, even if the error is large. Occasionally, sensitive analytical equipment cannot cope with large readings, though, meaning that we actually worsen the signal-to noise ratio. Every piece of scientific equipment has an optimum range: below a certain threshold the reading is too small, and above a different threshold, the error is too large.

An experienced analyst will always look up the extremities of this range before making a measurement. If the reading lies outside this recommended range, the analyst should (i) employ a different method, or (ii) modify the analyte (dilute it, pre-concentrate it, etc.) to change the reading, so it enters the recommended range. For example, in Worked Example 15.2, we should modify the dye solution until its absorbance is considerably smaller by diluting it, since the Beer–Lambert law suggests that absorbance is directly proportional to concentration.

In optical spectroscopy, the best way to minimize the signal-to-noise ratio is to ensure the absorbance reading has a magnitude in the range 0.75 to 1.0.

Self-test 15.1

Calculate the signal-to-noise ratio, and decide whether the reading is useable:

15.1.1 $\Delta H_r^{\ominus} = 4.6 \text{ kJ mol}^{-1}$ error $= 1.9 \text{ kJ mol}^{-1}$

15.1.2 $\lambda_{max} = 556 \text{ nm}$ error $= 20 \text{ nm}$

15.1.3 absorbance $= 2.5$ error $= 0.02$

15.1.4 transmittance $= 89\%$ error $= 3\%$

15.1.5 $emf = 1.104 \text{ V}$ error $= 12 \text{ mV}$ (note the standard factor)

15.1.6 time $= 34 \text{ s}$ error $= 0.5 \text{ s}$

Estimating the magnitude of an error for linear functions

By the end of this section, you will learn:

- A reading's cited error represents the spread of legitimate values obtained experimentally.
- If the spread of data values is Gaussian, the error is half the separation between the extreme values.
- When a single parameter x is measured, the minimum error is the quotient of innate error σ_x and the actual reading.
- When several parameters are measured, the minimum error σ is the square root of the sum of the squares of the individual errors.

When asked, 'what is the error in a reading,' we must be truthful and say, 'I don't know.' Even after doing a full error analysis, we will not know the error in any particular experiment. Rather, we know a *statistical value* of the error. By this phrase, we mean that we have a good idea of the average error if we perform the same calculation n times; but we do not know individual errors.

The simplest way to find the average magnitude of an error is to read the manual written for a particular instrument. The manufacturer will have calculated the error and published it in the manual or dedicated website. But sometimes the data are unhelpful or not available in a meaningful way. Or they might be wrong. Accordingly, we must be able to estimate errors for ourselves.

■ **Worked Example 15.3**

An electrochemist measures a cell *emf* using a digital voltmeter (DVM). The *emf* reading fluctuates between 110 mV and 130 mV, as shown by a graph of *emf* (as '*y*') against time (as '*x*') in Fig. 15.2. What is the error?

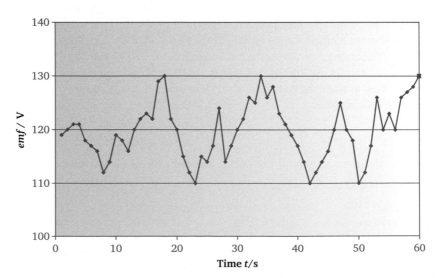

Figure 15.2 The *emf* read on a voltmeter can fluctuate with time in a more-or-less random manner.

In this simple example, the *emf* fluctuates with time. The reason for the fluctuation could be merely the random behaviour of the electronic components within the DVM, or the movement of air over the cell in the laboratory, stirring the solution. It does not indicate an electrochemical reaction, itself causing the composition of the cell to alter, because there is no overall *trend:* with time, the *emf* is neither increasing nor decreasing.

While the *emf* varies between 110 and 130 mV, the median value is 120 mV, so the *emf* fluctuates between '120 − 10 mV' and '120 + 10 mV'. The error in the *emf* is therefore 10 mV. In this example, we could have written *emf* = 120 ± 10 mV. ■

In this example, the spread of data is 20 mV, so we say the error is 10 mV. In fact, whenever the data follow a Gaussian distribution of values, we cite the error as half the separation between the extreme values, provided the extremes are statistically acceptable (for details, see Chapter 14).

Occasionally, we don't have a fluctuation. Even when the reading appears unchanging with time, it has an **innate** (or 'intrinsic') error, which we symbolize as σ. Generally, the existence of the error σ points to our inability to read an instrument with *perfect* accuracy. For example, any straightforward ruler made to measure lengths has centimetres and millimetres markers printed along its edge. These markers may be as much as $\frac{1}{3}$ mm thick, in which case the experimenter must decide how to measure with the ruler: from the left-hand side of the mark, from a subjective measure of its centre, or from its right-hand edge. For this reason, an *intrinsic* error σ exists when reading from this ruler.

> In the context, **innate** means 'natural,' 'inborn,' or 'pre-determined.'

The error when measuring with a ruler will decrease in proportion as the markers on the ruler edge get thinner. Unfortunately, while an infinitely thin mark will have no error in this respect, it will be too faint to actually see it. And even if we could see it, we then start the next discussion: is the marker printed in the right place: how accurate is the method for printing the markers on the ruler edge.

Most of these questions cannot be answers from first principles. Rather, we must look carefully at each type of measurement, deciding the extent to which we can trust the experimental values. The rules are very simple.

■ **Worked Example 15.4**

What is the innate error σ when measuring a length with a pocket ruler on which the separation between the printed marks is 1 mm?

Because the marks on the ruler are placed at 1 mm intervals, we cite a length to the nearest mm. Therefore, each time we cite a length l, we have made a decision, choosing which of the marks on the ruler is closest to the real length, l. If this length is closest to the 11 mm mark, we say $l = 11$ mm, but if the length is minutely longer and now closer to the 12 mm mark, we say $l = 12$ mm.

The shortest distance we can choose with confidence is 1 mm, so the innate error is half this distance i.e. $\sigma = 0.5$ mm. the measurement is 11 ± 0.5 mm. ■

Determining the innate error of simple functions: In this context, a *linear* function has the form $y = mx + c$, i.e. has no powers, exponential or logarithmic component.

The result we obtained in Worked Example 15.4 is general: when measuring a variable with a pre-calibrated instrument, we define the innate error as half the separation between the gradations on the printed scale.

But we need now to decide the magnitude of an error. For a function y that only depends on a variable x, eqn (15.3) describes the minimum error associated with y, which is equivalent to a relative error (σ_y/y):

$$\text{minimum error on } y = \left(\frac{\text{innate error on } x, \sigma_x}{\text{reading, } x} \right) \qquad (15.3)$$

Clearly, we wish to minimize the error, so we aim for a small innate error and employ an instrument with a finely divided scale, and we aim for a large reading.

■ Worked Example 15.5

A chemist is diluting a solution ten-fold. To this end, 1 cm³ of solution is discharged from a burette and made up to 10 cm³. What is the minimum error if the etched divisions on the burette are 0.1 cm³?

The gradations are 0.1 cm³ apart, so a competent chemist will work with an error of 0.05 cm³. To obtain the error, we insert figures into eqn (15.3):

$$\text{minimum error} = \left(\frac{0.05}{1.0} \right) = 0.05$$

minimum error = 0.05, or 5%

An additional error will arise with preparing the new solution, and is associated with making up to the mark in the graduated ('volumetric') flask.

The best way to minimize the error is to work with larger volumes than 1.0 cm³, so rather than running 1.0 cm³ to make 10 cm³ of solution, the chemist should run in 10.0 cm³ and make up to 100 cm³. The error will also probably increase considerably if the chemist is inexperienced or simply rushing. ■

In many instances, chemists need to determine several parameters, and then perform a calculation. Each measurement has its own associated errors. The overall minimum error depends on each.

If all the parameters are simple functions, then the overall minimum error σ will depend on the individual errors σ_n, according to eqn (15.4):

$$\sigma^2 = \sum_{i=1}^{n} \left(\frac{\text{innate error, } \sigma_n}{\text{reading, } x_n} \right)^2 \qquad (15.4)$$

Conceptually, this equation is the sum of the squares of the minimum errors for *each* piece of equipment, found using eqn (15.3).

■ Worked Example 15.6

In order to determine the solubility product of silver chloride, a chemist titrates $AgNO_3$ (0.1 mol dm⁻³) against aqueous KCl (0.1 mol dm⁻³), and measures the change in *emf* after the addition of each aliquot. Each aliquot has a volume of 1.0 cm³ and the burette scale has divisions of 0.1 cm³. The *emf* at the end point is 0.255 V; the minimum division on the digital voltmeter (DVM) is 1 mV. The mass of KCl was 0.745 g, as determined on a top-pan balance whose LCD display cites the mass to the nearest 1 mg. Calculate the innate error in the solubility product, σ_{Ksp}.

We first rewrite eqn (15.4):

$$(\sigma_{Ksp})^2 = \left(\frac{\text{burette error}}{\text{burette volume}}\right)^2 + \left(\frac{\text{DVM error}}{\text{DVM reading}}\right)^2 + \left(\frac{\text{mass error}}{\text{overall mass}}\right)^2$$

The overall innate error is therefore the square root of the right-hand side:

$$\sigma_{Ksp} = \sqrt{\left(\frac{\text{burette error}}{\text{burette volume}}\right)^2 + \left(\frac{\text{DVM error}}{\text{DVM reading}}\right)^2 + \left(\frac{\text{mass error}}{\text{overall mass}}\right)^2}$$

Burette: The error is half the scale on the burette, so $\sigma_{\text{burette}} = \pm 0.05$ cm^3

DVM: The error is half the scale on the DVM, so $\sigma_{\text{DVM}} = \pm 0.5$ mV

Top-pan balance: The error is half the scale on the balance, so $\sigma_{\text{balance}} = \pm 0.5$ mg

We then insert these values into the following table:

Instrument	Measured to the nearest	Innate error	Maximum reading	$\dfrac{\sigma_x}{x}$
Burette	0.1 cm^3	± 0.05 cm^3	1 cm^3	$\dfrac{0.05}{1} = 0.05$
DVM	1.0 mV	± 0.5 mV	0.255 V = 255 mV	$\dfrac{0.5}{255} = 1.96 \times 10^{-3}$
Balance	1 mg	± 0.5 mg	0.745 g = 745 mg	$\dfrac{0.5}{745} = 6.71 \times 10^{-4}$

$$\left(\sigma_{Ksp}\right)^2 = \left(\frac{0.05 \text{ cm}^3}{1.0 \text{ cm}^3}\right)^2 + \left(\frac{0.5 \text{ mV}}{255 \text{ mV}}\right)^2 + \left(\frac{0.5 \text{ mg}}{745 \text{ mg}}\right)^2$$

$$\left(\sigma_{Ksp}\right)^2 = (0.05)^2 + \left(1.96 \times 10^{-3}\right)^2 + \left(6.71 \times 10^{-4}\right)^2$$

$$\left(\sigma_{Ksp}\right)^2 = 2.5 \times 10^{-3} + 3.84 \times 10^{-6} + 4.5 \times 10^{-7}$$

i.e. $(\sigma_{Ksp})^2 = 2.50 \times 10^{-3}$

so $\sigma_{Ksp} = \sqrt{2.50 \times 10^{-3}}$

$\sigma_{Ksp} = 0.05$

Notice how the each of the units has cancelled.

This value means the final value of solubility product will have an error σ_{Ksp} of 5.0%, assuming each stage of the measurement is performed with care. The error will of course be much larger if the chemist fails to take proper precautions, and is careless.

The calculation shows that most of the error σ_{Ksp} (about 99.9%) is associated with the volume of solution delivered by the burette. For this reason, the burette should be used with especial care. The error associated with the *emf* is negligible compared with the other errors. ∎

Self-test 15.2

Calculate the minimum errors associated with the following measurements:

15.2.1 Measuring a length l of 11.2 cm with a ruler whose divisions are 1 mm apart.

15.2.2 Measuring a temperature T of a water bath with a mercury-in-glass thermometer: $T = 24.9\,°C$. T is measured to the nearest $0.1\,°C$.

15.2.3 Measuring an enthalpy of reaction ΔH: the temperature change was $0.7\,°C$ and measured with a thermometer with divisions of $0.1\,°C$; and weighing a mass of chemical of 3.503 g on a balance with divisions of 0.001 g.

15.2.4 Measuring the extinction coefficient ε via the Beer–Lambert law:

$$Abs = \varepsilon\, c\, l$$

where l is the path length and c is the concentration.

- The mass of analyte was 0.504 g, and was determined on a balance with divisions of 0.001 g.
- The volume of solvent was 100 cm³ in a volumetric flask with an innate error of 0.5 cm³.
- The path length l was 1.0 cm, as determined with a ruler with 0.5 mm divisions.
- The absorbance Abs was 0.74, as determined with an instrument with an innate error of 0.05.

Estimating the magnitude of an error for non-linear functions

By the end of this section, you will learn:

- Non-linear functions comprise powers, exponentials or logarithms.
- The mathematical function associated with a parameter affects the way we determine its innate error.

We now consider how to treat the errors of other, non-linear functions. The functions we will consider are powers, logarithms or exponentials. In context, by 'non-linear functions,' we mean here functions that do not have the form $y = mx + c$.

Propagation of errors:

If the measurement of an observed variable y is dependent on the quantities A and B, then something known as a *Taylor series* can be used to expand the function about the most probable errors:

The notation in this subsection depends on the concept of partial differentiation; see Chapter 22.

$$y = f(A,B) \approx f(\bar{A},\bar{B}) + \left(\frac{\partial f}{\partial A}\right)_B dA + \left(\frac{\partial f}{\partial B}\right)_A dB$$

The difference in the observed variable y—call it dy—is the same as the difference, $f(A,B) - f(\bar{A},\bar{B})$,

Table 15.1 Common non-linear functions and their uncertainty formulae.

Function form	Formula	Uncertainty formula for σ_y
Product	$y = a\,b$	$y\left(\dfrac{\sigma_a}{a} + \dfrac{\sigma_b}{b}\right)$
Simple powers	$y = a\,x^b$	$y\left(\lvert b \rvert \left(\dfrac{\sigma_x}{x}\right)\right)$
Product of powers	$y = a\,x^b\,z^c$	$y\left(\lvert b \rvert \left(\dfrac{\sigma_x}{x}\right) + \lvert c \rvert \left(\dfrac{\sigma_z}{z}\right)\right)$
Exponential	$y = a\,e^{bx}$	$y\,(b\,\sigma_x)$
Logarithm	$y = a \ln (bx)$	$a\left(\dfrac{\sigma_x}{x}\right)$
Log 10	$y = a \log (bx)$	$\left(\dfrac{a}{\ln 10}\right)\left(\dfrac{\sigma_x}{x}\right)$

so

$$dy = \left(\frac{\partial f}{\partial A}\right)_B dA + \left(\frac{\partial f}{\partial B}\right)_A dB$$

therefore,

$$\sigma_y = \left(\frac{\partial y}{\partial A}\right)_B \sigma_A + \left(\frac{\partial y}{\partial B}\right)_A \sigma_B$$

The deviation is then, $\sigma_y^2 = \left(\dfrac{\partial y}{\partial A}\right)_B^2 \sigma_A^2 + \left(\dfrac{\partial y}{\partial B}\right)_A^2 \sigma_B^2$

Methodology of this kind allows us to use calculus (Chapters 17–22) to determine the errors involved when dealing with non-linear functions. Table 15.1 shows the most common non-linear functions and their **uncertainty formulae**.

Functions of powers and exponentials:

If obtaining y requires that we take a simple power of an observed variable x, and the power function has the form $y = a\,x^b$, we obtain the error on y (call it σ_y) as,

$$\sigma_y = y\left(\lvert b \rvert \left(\frac{\sigma_x}{x}\right)\right) \tag{15.5}$$

Exponentials are discussed in great detail in Chapter 12.

where σ_x is the error associated with measuring the controlled variable, x.

We need to note:

- The factor before the power a does not feature in either uncertainty formula.
- If the power b is 1, eqn (15.5) straightforwardly collapses to yield eqn (15.3) because $b = 1$ means the function is now *linear*.

  If the function involves a product of two distinct power functions, e.g. $y = ax^b\,z^c$, we employ an amended uncertainty formula:

$$\sigma_y = y\left(\lvert b \rvert \left(\frac{\sigma_x}{x}\right) + \lvert c \rvert \left(\frac{\sigma_z}{z}\right)\right) \tag{15.6}$$

The form of eqns (15.5) and (15.6) means that errors can be considerably larger than with a linear function, cf. the error for a linear function in eqn (15.3).

The form of the uncertainty formula for an exponential function such as $y = a\, e^{bx}$ is:

$$\sigma_y = y\big(|b|\ \sigma_x\big) \tag{15.7}$$

■ **Worked Example 15.7**

In a neutron-scattering experiment, the radius r of a ^{13}C nucleus was determined as 61.9 fm. The uncertainty is 7.27×10^{-3}. Assuming the atom is spherical, determine the innate error σ_A in the atomic cross-sectional area A.

We start by assuming the area $A = \pi r^2$, so the power is '2'. The value of $(\sigma_r/r) = 7.27 \times 10^{-3}$.

We insert the data into eqn (15.5):

$$\sigma_A = A \times |b| \left(\frac{\sigma_r}{r}\right) = \underbrace{12037.4}_{\text{Area},\,A} \times \underbrace{|2|}_{\text{power on } r} \times \underbrace{7.27 \times 10^{-3}}_{\sigma_r/r} = 175.0\,\text{fm}^2$$

so the area of a ^{13}C atom is $A = \pi r^2 = (\pi \times 61.9^2) = 12037 \pm 175 \text{ fm}^2$.

The relative error is: $\dfrac{\sigma_A}{A} = \dfrac{175}{12037} = 0.0145$ or 1.5%. ■

Logarithms: If obtaining y requires that we take the logarithm of an observed variable x, and the logarithmic function has the form $y = a \ln (bx)$, we obtain the error on y (call it σ_y) as,

$$\sigma_y = a \left(\frac{\sigma_x}{x}\right) \tag{15.8}$$

where σ_x is the error associated with measuring x.

■ **Worked Example 15.8**

An electrochemist wants to determine an *emf* from a measured concentration using the Nernst equation.

$$E_{Cu^{2+},Cu} = E^{\ominus}_{Cu^{2+},Cu} + \frac{RT}{2F}\ln[Cu^{2+}]$$

Determine the innate error in $E_{Cu^{2+},Cu}$ (call it σ_E) if the concentration of copper sulphate is 0.05 mol dm^{-3}, and the error associated with $[Cu^{2+}]$ (call it σ_c) is 2×10^{-4} mol dm^{-3}? Take $T = 298$ K.

The numerical value of the factor $RT/2F$ is 1.284×10^{-2} V. In the template expression eqn (15.8), $a = 1.284 \times 10^{-2}$ V and $b = 1$.

Inserting terms into eqn (15.8), we obtain:

$$\sigma_E = \left(1.284 \times 10^{-2}\,\text{V}\right) \times \left(\frac{0.0002}{0.05}\right) = 5.14 \times 10^{-5}\,\text{V}$$

In other words, the innate error in this example is tiny. The electrode potential $E_{Cu^{2+},Cu}$ is quite insensitive to fluctuations in concentration. Because rearrangement to make concentration the subject generates an exponential, we see the converse situation is not true: even a small fluctuations in electrode potential can indicate a huge changes in concentration. We need to determine $E_{Cu^{2+},Cu}$ with care.

The analysis above assumed there were no errors except the concentration. If we assume there is also an error in the standard electrode potential $E^{\ominus}_{Cu^{2+},Cu}$, the right-hand side of the error expression would require an additional term, of $\sigma\left(E^{\ominus}_{Cu^{2+},Cu}\right)$. This term has no factor of $RT/2F$ because $E^{\ominus}_{Cu^{2+},Cu}$ is a single linear variable, and is not a logarithmic function. ∎

Graphical treatment of errors: error bars

By the end of this section, you will learn:

- When plotting a graph, we plot the reading as a point, and the errors are indicated by error bars either side and/or above and below this point.

- Error bars may be indicated on both the x- and y-axes.

- The length of the error bars either side of the point indicates the error of each reading, so the total length of the error bar is 2σ.

- The best straight line through the data should accommodate all the error bars.

We often generate graphs that show scatter. Figure 15.3 shows kinetic data relating to the uptake of *Balsalazide*™ (**I**), a drug used to treat ulcerative colitis. While sometimes tempting, *we should never join the points in a dot-to-dot fashion*. The gradients m for each linear segment clearly differ, which implies many different relationships between x and y. In extreme cases, if we draw a dot-to-dot graph, we generate a zigzag pattern in which the gradient even changes sign before and after each data point.

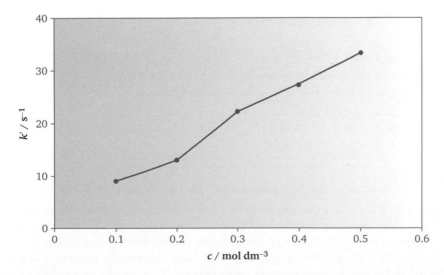

Figure 15.3 Plot of k' (as 'y') against c (as 'x') for the uptake of *Balsalazide*™: the experimental data are shown as bold points •. The use of dot-to-dot lines indicates a chemist who does not understand error analyses.

Furthermore, it makes no physicochemical sense to draw dot-to-dot graphs. If we do join adjacent dots with a straight line, as here, we are saying there is an exact linear relationship between the controlled and observed variables. Indeed, the different gradients imply that a different uptake regime operates between each pairs of points—a *very* unlikely situation!

I

We cannot cite an experimental value on its own: experimental data always have an associated error. In other words, each value experimental value represents a *range* of values. We indicate this range by writing, 'reading ± error', where the magnitude of the error σ was the subject of the previous two sections.

We now consider how we accommodate these errors when we draw graphs. In practice, we indicate this range using **error bars**.

How to draw error bars for the graph in Fig. 15.3:

(i) We plot each value of k' in the usual way, for example using a bold • point to indicate each value we obtained experimentally.

(ii) For each value of k', we plot two more points, $(k' + \sigma)$, and $(k' - \sigma)$, and draw a short horizontal line through each at the same value of c. These additional points correspond to the maximum and minimum values of k' for each value of c, having considered the innate errors associated with measurement.

(iii) We form the **error bars** by drawing a vertical line between the two horizontal lines – . This line will pass through the point, •.

■ **Worked Example 15.9**

A kineticist determines a pseudo-first-order rate constant k' as a function of concentration c. Each value of k' has an associated error, as follows:

c/mol dm^{-3}	0.1	0.2	0.3	0.4	0.5
k'/s^{-1}	9.0 ± 4.0	13.0 ± 2.0	22.2 ± 4.5	27.2 ± 2.0	33.3 ± 4.2

Plot these data with k' (as 'y') against c (as 'x'), and determine a value for k_2.

Figure 15.3 plots the data with the points joined together as a dot-to-dot picture—which is the wrong representation. Figure 15.4 shows the data with error bars. The gradient of this graph yields the second-order rate constant k_2. Unfortunately, the data • clearly do not show a straightforward linear relationship between k' and c, which makes any determination of k_2 more difficult.

The best straight line through this data set must accommodate the errors (as indicated by the error bars –) as well as the readings (the bold points, •). In practice, we draw

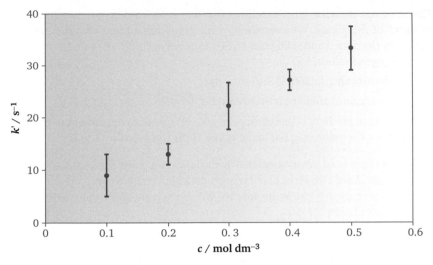

Figure 15.4 Plot of k' (as 'y') against c (as 'x'): the bold points • are the readings and the vertical lines joining the short horizontal lines—are the error bars.

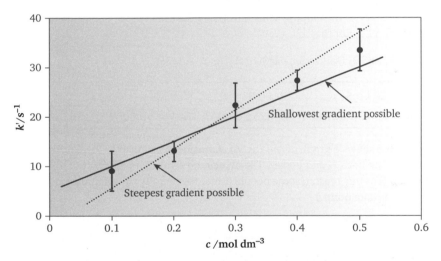

Figure 15.5 Plot of k' (as 'y') against c (as 'x') showing error bars on the y-axis: the dotted blue line has the maximum gradient possible with a line cutting all five error bars; the solid blue line has the minimum gradient possible.

two gradients on the graph, representing the steepest and the shallowest gradients possible. Each gradient must cut through all five of the error bars.

Figure 15.5 redraws Fig. 15.4, with the lines of steepest and shallowest gradients superimposed. These gradients need to pass through the error bars. The steepest gradient (the dotted line . . . . .) just touches the error bars for the first, fourth, and fifth points. If it was steeper, it would fail to cut across one of more of these error bars; if the line was less steep, its gradient would not be the *maximum*.

In the same way, the solid line (———) on Fig. 15.5 only just cuts the error bars of the second and fourth points. If the line was steeper it would not be the minimum gradient; and if it were shallower still, it would fail to cut across all five error bars.

Conventional chemical kinetics tells us that the second-order rate constant k_2 is the gradient of this figure. We now see how the graph has a *range* of values, as represented by the dotted and solid lines on Fig. 15.5. Accordingly, the value of k_2 will also have a range of values.

- The **minimum** gradient is 51.0 $(mol\ dm^{-3})^{-1}\ s^{-1}$,
- The **maximum** gradient is 81.3 $(mol\ dm^{-3})^{-1}\ s^{-1}$

The maximum gradient is therefore nearly 60% larger than the minimum gradient. In summary, the value of k_2 lies in the range 51–81 $(mol\ dm^{-3})^{-1}\ s^{-1}$. ∎

Sometimes there is an error associated with the data in both the *x*- and the *y*-axes, in which case we need two error bars for each data point, •. We construct the error bars for the *x*-axis in exactly the same way we constructed them for the *y*-axis in Worked Example 15.9, above.

■ **Worked Example 15.10**

An electrochemist wants to determine the change in entropy ΔS_{cell} associated with a cell discharging. To this end, we plot a graph of cell *emf* (as '*y*') against temperature (as '*x*'). The gradient is termed the **voltage coefficient**. We calculate a value of ΔS_{cell} as 'voltage coefficient × *nF*', where *n* is the number of electrons in the balanced cell reaction, and *F* is the Faraday constant. Using the data below, determine a range of values for $\Delta S_{(cell)}$ for this cell.

⊖ Zn | ZnSO₄ (sat'd soln.)HgSO₄(s) | Hg ⊕

The error in temperature is an innate error, i.e. using a relatively insensitive thermometer. The error in *emf* is probably due to fluctuations in the voltmeter reading.

$T/°C$	20 ± 0.5	30 ± 0.5	40 ± 0.5	50 ± 0.5
emf/V	1.4430 ± 0.0004	1.4426 ± 0.0003	1.4421 ± 0.0003	1.4417 ± 0.0003

Figure 15.6 shows a graph of *emf* (as '*y*') against temperature (as '*x*'), and containing error bars on both *x*- and *y*-axes. The associated error of a point is contained within the bars. (For this reason, some people prefer to mentally superimpose an oval, centred on the data point.)

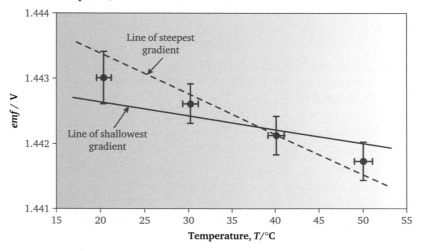

Figure 15.6 Plot of *emf* (as '*y*') against *T* (as '*x*') for a Clark cell, showing error bars on both *x* and *y*-axes: the dotted line (- - - -) has the *maximum* gradient possible with a line cutting all five error bars; the solid line (——) has the *minimum* gradient possible.

The line of maximum gradient is shown as a dotted line, and the minimum gradient is drawn as a solid line.

- The **maximum** gradient is 6.17×10^{-5} V K^{-1}
- The **minimum** gradient is 2.67×10^{-5} V K^{-1}

In this example, the maximum gradient is nearly 2.3 times larger than the minimum gradient. ■

In this last example, the spread of values is so large that we ought to stop and wonder, what then do we cite as 'the' gradient? We address this question in the next section.

Self-test 15.3

The heat capacity C_p of chloroform (**II**) varies with temperature, according to the data below:

T/K	20 ± 0.5	30 ± 0.5	40 ± 0.5	50 ± 0.5
C_p/J K^{-1} mol^{-1}	91.47 ± 0.08	92.25 ± 0.07	93.02 ± 0.08	93.86 ± 0.06

Plot a graph of C_p (as 'y') against temperature T (as 'x'). Show all error bars.

II

The correlation coefficient *r*

By the end of this section, you will learn:

- The correlation coefficient *r* relates to the scatter of data.
- The value of *r* decreases as the scatter gets more random.
- A correlation coefficient of 1 relates to a line with symmetrical scatter, with as many points on one side of the line as on the other.
- A correlation coefficient of 0 indicates a data set that is completely random.

Sometimes we need to obtain the error of our data for calculations like those above. At other times, though, having obtained our data set, we wish to plot them and obtain a physicochemical law. Unfortunately, we have no completely fail-safe method for deciding the best smooth line to draw though them—and we need the straight line to deduce the law that describe the data. We want a **correlation** between the controlled and observed variables, *x* and *y*, respectively, in order, or use the data in a predictive capacity, e.g. to draw calibration curves. In the discussion below, we talk only of *straight* lines.

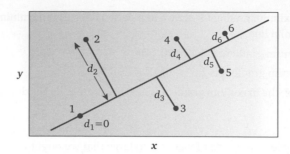

Figure 15.7 Schematic representation of a graph showing scatter. The solid line is the line of best fit if the sum of the distances $(d_2 + d_4 + d_6)$ equals the sum $(d_3 + d_5)$. In this example, $d_1 = 0$ in consequence of the point residing on the line.

Probably the simplest method of finding the best straight line concerns the **correlation coefficient, r**:

1. We start by plotting data, and draw a straight line through them. Next, we consider the distances between the points either side of the line. Figure 15.7 clearly shows some scatter, as do most graphs drawn using genuine data.

2. We draw the 'line of best fit' through the data, such that some points lie above the line and others are below it. The perpendicular distance between the point '1' and the line is d_1 (in this example, point 1 lines on the line) and the perpendicular distance between the point '2' and the line is d_2, etc. A graph with little scatter will have small distances, d_n, but graphs with more scatter will have larger distances.

3. We now consider two sets of points: those above the line and those below it. For each set, we can consider a sum of their distances from the line. We call them, respectively, $\Sigma d_{n(above)}$ and $\Sigma d_{n(below)}$. If the line we drew through the data set is the *best* line, these two sums will be more-or-less the same. The calculation of these sums is a straightforward process, although somewhat laborious.

Having drawn a line through the graph of a data set, the correlation coefficient r acts as a quantitative measure of how good is the choice of line. In effect, it informs our choice of line. The value of r tells us how close we are to the equation of the best line for our data set.

We define the correlation coefficient r as,

$$r = \frac{\sum_{i=1}^{i}\left[(x_i - \bar{x})(y_i - \bar{y})\right]}{\left\{\left[\sum_{i=1}^{i}(x_i - \bar{x})^2\right]\left[\sum_{i=1}^{i}(y_i - \bar{y})^2\right]\right\}^{1/2}} \qquad (15.9)$$

While this expression certainly looks unpleasant, it is readily broken down into manageable chunks. We have defined each of the terms previously.

The value of r can have any value between -1 and $+1$. A value of $r = +1$ indicates the data set show an excellent correlation, with all points lying on a straight line having a positive gradient. A value of $r = -1$ defines a perfect negative correlation, i.e. a straight line with a negative slope. A value of r close to 0 tells us that no correlation exists between respective values of x and y in the data set. They are random.

Commonly, we cite a value of r^2 rather than r alone. This practice has two advantages: first, it ensures values are always positive. Secondly, by squaring the value, we enhance the sensitivity of assessments based on the correlation coefficient; squaring accentuates the difference between the experimental value of r and the maximum value of 1.

We define the **best** straight line through a data set as that for which the sum of the perpendicular distances of the points *above* the line essentially equals the corresponding sum for the points *below* it.

The value of r obtained using eqn (15.9) is also called the **Pearson** correlation coefficient, and the **product-moment** correlation coefficient.

We usually start by drawing a graph. Three situations can be anticipated:

1. **The graph is linear, and shows little scatter:** the value of r^2 will probably lie in the range 0.95–1.00. A value of $r^2 \geq 0.98$ indicates that we are safe to assume a good correlation exists between x and y.

2. **The graph is linear, but shows extensive scatter:** in this case, the correlation is likely to be poorer. The value of r^2 can still be high, though. This time, we might wish to obtain a duplicate set of data, choose to recalibrate the apparatus, or find an alternative technique. We might even decide to discontinue the investigation.

3. **The graph shows a smooth *curve*:** In this case, we assume a correlation exists, but not a *linear* correlation. We must first linearize the data (with the techniques in Chapter 13), and only then compute a correlation coefficient for the resulting straight line.

■ **Worked Example 15.11**

We draw a graph of K (as 'y') against temperature T (as 'x'). The graph clearly shows a smooth *curve*. We suspect the data will follow the linear form of the van't Hoff isochore:

$$\ln K = -\frac{\Delta H^{\ominus}}{R}\frac{1}{T} + c$$

When calculating the correlation coefficient r, what parameter(s) should we include within eqn (15.9)?

The linear form of the van't Hoff isochore follows a line of the form, $y = mx + c$, where $y = \ln K$, $m = -\Delta H^{\ominus}/R$, and $x = 1/T$. Accordingly:

Controlled variable		Observed variable	
Symbol:	x	Symbol:	y
Choice of x:	$1/T$	Choice of y:	$\ln K$
Symbol for the mean of x	$\bar{x}$	Symbol for the mean of y:	$\bar{y}$
Choice of $\bar{x}$:	$\overline{\left(\dfrac{1}{T}\right)}$	Choice of $\bar{y}$:	$\overline{(\ln K)}$ ■

r^2 via a pocket calculator: the method we include here concerns the popular *Casio fx-85ES*. The method below will probably differ for different makes and/or models, although the logic implied will probably not differ greatly. Always check the instruction manual, looking up 'linear regression' in the index.

The calculator will probably follow a similar approach to that below.

1. **Mode → Stat** (option 2)
2. Select **A + Bx** (option 2)
3. Fill in the (x, y) data in the table shown:

	X	Y
1		
2		

   Use the '=' key to enter the data.

   Use the arrow keys to switch between the x and y columns.
4. Press '**AC**' when all of the data has been entered.
5. Press '**Shift**' and '**1**' to activate the '**STAT**' options.
6. Press '**7**' for the '**Reg**' = Regression options.

Aside:
Obtaining r^2 via a computer spreadsheet:
In recent times, most chemists obtain values of r^2 using computer software. For example, using the programme *Excel* ™. The procedure to obtain r with a graph is:

1. Draw the graph within *Excel* ™ in the usual way.

2. Place the mouse on one of the data points, and click once. This action will highlight the data.

3. With the mouse, click on the CHART menu (or the LAYOUT menu in *Excel* 2007), and choose ADD TREND-LINE (and MORE TRENDLINE OPTIONS in *Excel* 2007).

4. Click on the icon labelled 'LINEAR.'

5. At the top of the box, now click on the second page, labelled 'OPTIONS.' In *Excel* 2007, the options are listed at the bottom of the window.

6. Click the box labelled 'DISPLAY R-SQUARED VALUE ON CHART'

7. Click on OK.

A small text box will appear on the graph, citing the value of r^2.

7. Press '1' and '=' for 'A' = Intercept of the line of best fit.

8. Repeat options 5 and 6, then press '2' and '=' for 'B' = Gradient of the line of best fit.

9. Repeat options 5 and 6, then press '3' and '=' for 'r' = Correlation coefficient, which measures how good the fit is of the data to the line of best fit.

10. **Mode → Comp** (option 1) to return to the standard calculation environment. 'From linear regression, it can be shown that the gradient = ... , the intercept = ... and the correlation coefficient of the data, r, is ... and r^2 is .... '

■ **Worked Example 15.12**

An electrochemist determines values of the electrode potential $E_{Cd^{2+},Cd}$ as a function of cadmium-ion concentration, $[Cd^{2+}]$, as follows:

$[Cd^{2+}]/mol\ dm^{-3}$	0.1	0.05	0.02	0.01	0.005
$E_{Cd^{2+},Cd}/V$	−0.433	−0.437	−0.450	−0.461	−0.468

The electrochemist seeks to determine a value of the standard electrode potential $E^{\ominus}_{Cd^{2+},Cd}$, so plots a graph of $E_{Cd^{2+},Cd}$ (as 'y') against ln $[Cd^{2+}]$ (as 'x'). Calculate the correlation coefficient, r.

> We sometimes refer to this set of axes as a **Nernst plot**. We obtain $E^{\ominus}_{Cd^{2+},Cd}$ as the intercept on the y-axis.

Figure 15.8 shows a graph $E_{Cd^{2+},Cd}$ of (as 'y') against ln $[Cd^{2+}]$ (as 'x'). The graph is clearly linear, but contains some slight scatter. It already shows the line of best fit, as chosen electronically using *Excel*™.

Strategy:

(i) Calculate the mean of both x and y (in this case, the means of $E_{Cd^{2+},Cd}$ and ln $[Cd^{2+}]$, respective.

(ii) Calculate the terms $(x_i - \bar{x})$ and $(y_i - \bar{y})$, and hence determine $\sum_{i=1}^{i} \left[(x_i - \bar{x})(y_i - \bar{y}) \right]$.

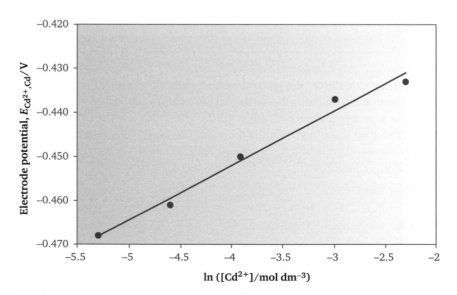

Figure 15.8 Nernst plot of $E_{Cd^{2+},Cd}$ (as 'y') against ln $[Cd^{2+}]$ (as 'x').

(iii) Calculate the terms $(x_i - \bar{x})^2$ and $(y_i - \bar{y})^2$, and hence determine values of

$$\sum_{i=1}^{i}(x_i - \bar{x})^2, \quad \sum_{i=1}^{i}(y_i - \bar{y})^2, \text{ and thence } \left\{\left[\sum_{i=1}^{i}(x_i - \bar{x})^2\right]\left[\sum_{i=1}^{i}(y_i - \bar{y})^2\right]\right\}^{1/2}$$

(iv) Insert the terms from parts (i) and (ii) into eqn (15.9).

Solution:

(i) The mean of ln $[Cd^{2+}] = \bar{x} = -3.823$, and the mean of $E_{Cd^{2+},Cd} = \bar{y} = -0.4498$ V.

(ii) Concerning the **controlled variable**:

$[Cd^{2+}]$ mol dm^{-3}	$x = (\ln [Cd^{2+}]/\text{mol dm}^{-3})$	$(x_i - \bar{x})$	$(x_i - \bar{x})^2$
0.1	-2.303	1.520	2.310
0.05	-2.996	0.827	0.684
0.02	-3.912	-0.089	0.008
0.01	-4.605	-0.782	0.612
0.005	-5.298	-1.475	2.176

Concerning the **observed variable**:

$y = E_{Cd^{2+},Cd}$	$(y_i - \bar{y})$	$(y_i - \bar{y})^2$
-0.433	0.017	2.822×10^{-4}
-0.437	0.013	1.638×10^{-4}
-0.450	0.000	4.000×10^{-8}
-0.461	-0.011	1.254×10^{-4}
-0.468	-0.018	3.312×10^{-4}

Working with data from the penultimate column in the two tables, we calculate:

$$\sum_{i=1}^{i}\left[(x_i - \bar{x})(y_i - \bar{y})\right]$$

So the numerator of eqn (15.9) = 0.0717.

(iii) Working with data in the far right-hand columns in the two tables, we calculate:

$$\sum_{i=1}^{i}(x_i - \bar{x})^2 = 5.790$$

$$\sum_{i=1}^{i}(y_i - \bar{y})^2 = 9.026 \times 10^{-4}$$

> **Remember:** The **numerator** is the top line of a quotient or fraction The **denominator** is the bottom line of a quotient or fraction.

Therefore, $\left\{\left[\sum_{i=1}^{i}(x_i - \bar{x})^2\right]\left[\sum_{i=1}^{i}(y_i - \bar{y})^2\right]\right\}^{1/2} = \sqrt{(5.792) \times (9.029 \times 10^{-4})}$

So the denominator of eqn (15.9) = $\sqrt{5.2296 \times 10^{-3}} = 0.0723$

(iv) Finally, we insert values into eqn (15.9):

$$r = \frac{0.0717}{0.0723} = 0.9917$$

In summary, $r = 0.9917$, so $r^2 = 0.9835$.

This value looks reasonable because it is close to unity; and the data show a clear correlation, though with some slight scatter. ∎

Self-test 15.4

The freezing temperature of water depends on its *purity*, as expressed by the concentration of sodium chloride it contains.

The following melting temperatures were measured:

$[NaCl]/mol\ dm^{-3}$	0	0.1	0.2	0.3	0.4	0.5
T_{freeze}/K	273.15	273.01	272.94	272.88	272.80	272.72

Determine a value for the correlation coefficient r for this data set.

Summary of key equations in the text

Simple error definitions:

$$\text{reading} = \text{value} \pm \text{error} \tag{15.1}$$

$$\text{signal-to-noise ratio} = \frac{\text{reading}}{\text{error}} \tag{15.2}$$

Error for a linear function:

$$\text{minimum error on } y = \left(\frac{\text{innate error on } x, \sigma_x}{\text{reading, } x} \right) \tag{15.3}$$

$$\sigma^2 = \sum_{i=1}^{n} \left(\frac{\text{innate error, } \sigma_n}{\text{reading, } x_n} \right)^2 \tag{15.4}$$

Errors for a non-linear function:

Functional form	Formula	Uncertainty function	
Product of powers	$y = a\,x^b\,z^c$	$\sigma_y = y\left(|b|\left(\frac{\sigma_x}{x}\right) + |c|\left(\frac{\sigma_z}{z}\right) \right)$	(15.6)
Exponential	$y = a e^{bx}$	$\sigma_y = y\left(|b|\ \sigma_x \right)$	(15.7)
ln	$y = a \ln(bx)$	$\sigma_y = a\left(\frac{\sigma_x}{x} \right)$	(15.8)
$\log_{10}$	$y = a \log(bx)$	$\sigma_y = \frac{a}{\ln 10}\left(\frac{\sigma_x}{x} \right)$	

If more than one variable contains errors, then we combine them using:

$$\sigma_y = \left(\frac{\partial z}{\partial n_1} \right)_{n_2,n_3,\dots} dn_1 + \left(\frac{\partial z}{\partial n_2} \right)_{n_1,n_3,\dots} dn_2 + \dots$$

Deviation: $\sigma_y^2 = \left(\dfrac{\partial z}{\partial n_1}\right)_{n_2, n_3, \ldots}^2 (dn_1)^2 + \left(\dfrac{\partial z}{\partial n_2}\right)_{n_1, n_3, \ldots}^2 (dn_2)^2 + \ldots$

Definition of correlation coefficient, r:

$$r = \dfrac{\displaystyle\sum_{i=1}^{i}\left[(x_i - \bar{x})(y_i - \bar{y})\right]}{\left\{\left[\displaystyle\sum_{i=1}^{i}(x_i - \bar{x})^2\right]\left[\displaystyle\sum_{i-1}^{i}(y_i - \bar{y})^2\right]\right\}^{1/2}} \tag{15.9}$$

Additional problems

15.1 A voltmeter fluctuates when the analyte solution is stirred. When the *emf* is 0.340 V, the maximum fluctuation is 10 mV. What is the signal-to-noise ratio?

15.2 The progress of a simple kinetics experiment is followed with a normal wristwatch. The watch has the three usual hands of hours, minutes and seconds, and the watch face has a printed mark for every second. Determine the innate error in the time measurement.

15.3 The optical absorbance of a dye in solution is 3.2, but the amplitude of the interference is 0.7. Calculate the signal-to-noise ratio.

15.4 Consider the data below, which relate to the second-order racemization of glucose (**III**) in aqueous hydrochloric acid at 17 °C. The concentration of glucose and hydrochloric acid are the same, '[A]'. The concentration [A]$_t$ decreases with time according to the linear form of the integrated second-order rate equation:

$$\frac{1}{[A]_t} = kt + c$$

Time, t/s	0	600	1200	1800	2400
[A]$_t$/mol dm^{-3}	0.400	0.360	0.301	0.289	0.244

Linearize the data and plot a graph. What is the correlation coefficient, r^2?

III

15.5 The innate error in the absorbance reading with a modern spectrophotometer can be as low as 0.001. If the absorbance is 0.670, calculate the minimum error.

15.6 A chemist determines a rate constant by removing an aliquot of solution every ten minutes, and titrating it with base. Calculate the minimum error if:

- The aliquot volume V is 10.0 cm^3, removed with a pipette having an innate error σ_V of 0.005 cm^3.

- The concentration of base is 0.104 ± 0.001 mol dm^{-3}.

- Because we did not remove the solution instantaneously, but had to fill the pipette, the innate error in the time reading is 15 s.

15.7   Consider the isomerization of 1-butene (**IV**) to form *trans* 2-butene (**V**). The equilibrium constants of reaction are given in the table below:

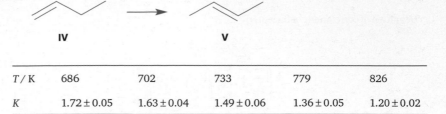

IV				V

$T\,/\,\mathrm{K}$	686	702	733	779	826
K	1.72 ± 0.05	1.63 ± 0.04	1.49 ± 0.06	1.36 ± 0.05	1.20 ± 0.02

Plot a graph of $\ln K$ (as 'y') against $1/T$ (as 'x'). Determine the maximum and minimum gradients possible.

15.8   A cell is constructed to help determine the standard electrode potential of the zinc(II)–zinc couple, and using a variant of the **Nernst equation**:

$$emf = \left(E^{\ominus}_{Zn^{2+},An} + \frac{RT}{2F}\ln\left[Zn^{2+}\right] \right) - E_{SCE}$$

The *emf* is 28.3 mV, with an innate error of 1 mV. The concentration $[Zn^{2+}]$ is 4.4×10^{-3} mol dm^{-3}, with an innate error of 1×10^{-5} mol dm^{-3}. What is the minimum error associated with this measurement?

15.9   Return to the graph in Self-test 15.3, and determine the maximum and minimum gradients.

15.10 A kineticist determines a reaction rate constant k via the integrated first-order rate equation.

$$\frac{c_0}{c_t} = \exp\left(kt\right)$$

where c_0 is the initial concentration, and c_t is the concentration at time t after the start of the reaction. The initial concentration was 0.1 mol dm^{-3}. If the concentration of reactant is 0.06 ± 0.005 mol dm^{-3} at time $t = 10$ s, what is the error in rate constant k, call it σ_k?

Trigonometry

16

Introduction: naming a right-angled triangle

By the end of this section, you will know:

- The elementary nomenclature defining each part of a right-angled triangle.
- How to determine the sine, cosine and tangent of a right-angled triangle.

Consider the right-angled triangle in Fig. 16.1. We need to learn a small amount of nomenclature. First, we indicate the angles with capital letters and the sides with small letters.

The angles:

- The internal angles are designated with capital letters.
- The sum of the three internal angles *A, B* and *C* will always equal 180°.
- We write the Greek letter **theta** θ (which, properly, is pronounced *thay-tah*) if a single angle is indicated.
- Angles *A* and *B* can have any value between 0 and 90°. Since angle *C* is 90°, the other angles *A* and *B* will be θ, and $(90° - \theta)$.
- We say that any angle θ less than 90° is **acute**.
- By convention, the angle *A* is positioned opposite the side *a*, angle *B* is opposite the side *b,* and angle *C* is positioned opposite the side *c*.
- Some texts indicate the angle *A* as $\widehat{BAC}$ while others employ this notation only if the vertices are labelled *A, B, C* rather than the angles. To understand this notation, we note that angle *A* lies between angles *B* and *C*, so angle B is therefore $\widehat{ABC}$; and angle *C* is $\widehat{ACB}$. We only really need this nomenclature when we go round a triangle in a clockwise or anticlockwise direction, so we will not use it further.

The sides:

- Properly, we call the angle of interest θ the **angle of focus**. We define it in terms of its position relative to the two sides that touch it.
- We call the side positioned next to the angle of focus the **adjacent** (from the Latin *adjacere* 'lie near to'). We usually draw it *horizontally*.
- Generally, we draw the side of the triangle that does not touch the angle of focus θ vertically. Its position explains why we call it the **opposite**.
- We call the longest side of the triangle the **hypotenuse**, and is generally drawn sloping at an angle, and touching θ at its lower end.

These names are shown schematically on Fig. 16.2.

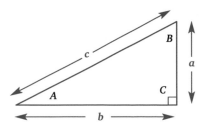

Figure 16.1 General depiction of a right-angled triangle.

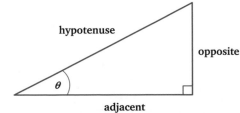

Figure 16.2 The names of the sides making up a right-angled triangle.

The simple trigonometric functions

By the end of this section, you will know:

- The definitions of sine, cosine and tangent, and their reciprocals.
- How to calculate a sine, cosine and tangent.
- How to calculate an angle of focus θ if the sine, cosine and tangent is already known.
- The notation for trigonometric functions raised to powers.
- How to manipulate and rearrange algebraic phrases involving trigonometric functions.

Our definitions of sine, cosine and tangent start with the three sides of a right-angled triangle in Fig. 16.2:

- We define a **sine** as the ratio of the adjacent to the hypotenuse:

$$\sin \theta = \frac{\text{opposite}}{\text{hypotenuse}} \qquad (16.1)$$

  Figure 16.3 shows a graph of sin θ (as 'y') as a function of θ (as 'x'). The reciprocal of sine is called a cosecant (or 'cosec').

- We define a **cosine** as the ratio of the adjacent to the hypotenuse:

$$\text{cosine } \theta = \frac{\text{adjacent}}{\text{hypotenuse}} \qquad (16.2)$$

  Figure 16.4 shows a graph of cosine θ (as 'y') as a function of θ (as 'x'). The reciprocal of cosine is called a secant (or 'sec').

- We define a **tangent** as the ratio of the adjacent to the hypotenuse:

$$\text{tangent } \theta = \frac{\text{opposite}}{\text{adjacent}} \qquad (16.3)$$

  Figure 16.5 shows tangent θ (as 'y') as a function of θ (as 'x'). The reciprocal of a tangent is called a cotangent (or 'cot').

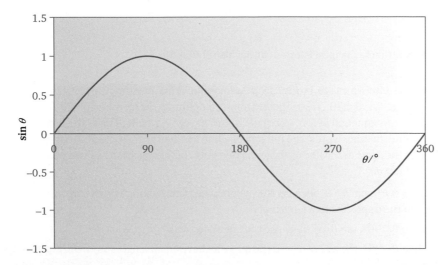

Figure 16.3 Graph of sin θ (as 'y') as a function of θ (as 'x').

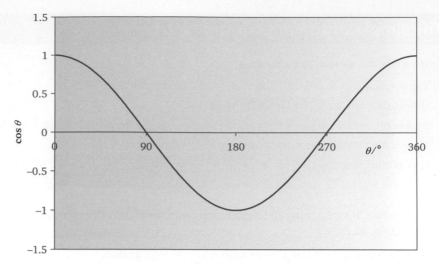

Figure 16.4 Graph of cosine θ (as 'y') as a function of θ (as 'x').

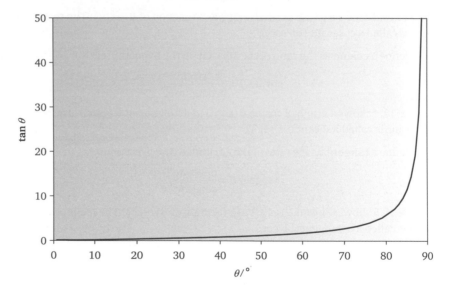

Figure 16.5 Graph of tangent θ (as 'y') as a function of θ (as 'x').

Notice the abscissa range in Fig. 16.5: we cannot think meaningfully of tan θ except in the range $0 \leq \theta < 90$. This follows from the definition of tangent in eqn (16.3). We have seen this equation before as eqn (9.2) in Chapter 9, which defines the gradient of a straight line. In other words, the tangent of a line inclined at an angle of θ to the adjacent is exactly the same as the line's *gradient*. Following this observation, we can predict three values of tan θ:

- When $\theta = 0$, tan $\theta = 0$, because the hypotenuse is horizontal, which can only happen when the opposite has a length of 0.

- When $\theta = 90°$, tan $\theta = \infty$, because the line is vertical.

- When $\theta = 45°$, the adjacent and the opposite have equal lengths so (from eqn (16.3)) tan $\theta = 1$.

Some people find it easier to remember the three definitions in eqns (16.1) – (16.3) using a simple rule: if sine = s, cosine = c and tangent = t, and if opposite = o, adjacent = a and hypotenuse = h, then the three definitions above are summarized in the word:

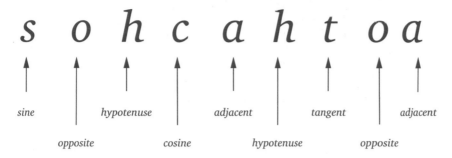

$$s \quad o \quad h \quad c \quad a \quad h \quad t \quad o \quad a$$

sine	hypotenuse	adjacent	tangent	adjacent
	opposite	cosine	hypotenuse	opposite

■ **Worked Example 16.1**

A right-angled triangle has a hypotenuse of 5 cm, an adjacent of 4 cm and an opposite of 3 cm. Calculate the values of sin θ, cos θ and tan θ.

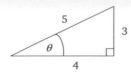

From eqn (16.1), the value of the sine θ is $\dfrac{3\ \text{cm}}{5\ \text{cm}} = 0.6$

From eqn (16.2), the value of the cosine θ is $\dfrac{4\ \text{cm}}{5\ \text{cm}} = 0.8$

From eqn (16.3), the value of the tangent θ is $\dfrac{3\ \text{cm}}{4\ \text{cm}} = 0.75$ ■

Generally, we don't write 'sine,' 'cosine' or 'tangent,' but abbreviate them, respectively, as 'sin,' 'cos' and 'tan.'

A right-angled triangle with sides of length 3, 4, 5 is known as a Pythagorean triple. Another example has sides with lengths, 5, 12, 13. Such triangles are fairly common and are useful to know.

Table 16.1 lists a few values of the trigonometric functions that are worth memorizing.

Table 16.1 A few values of the three trigonometric functions sin θ, cos θ and tan θ.

θ	0°	30°	45°	60°	90°
Sin θ	0	0.5	$\dfrac{1}{\sqrt{2}}$	$\dfrac{\sqrt{3}}{2}$	1
Cosine θ	1	$\dfrac{\sqrt{3}}{2}$	$\dfrac{1}{\sqrt{2}}$	0.5	0
Tangent θ	0	$\sqrt{\dfrac{1}{3}}$	1	$\sqrt{3}$	∞

■ **Worked Example 16.2**

The ions in a crystal lie in layered planes. The successive layers are separated by a distance d. We can determine the value of d with the technique of X-ray diffraction. Here, X-ray light of wavelength λ strikes the solid at an angle of θ, and is diffracted (see Fig. 16.6), according to the Bragg equation:

$$n\lambda = 2d \sin \theta$$

where n represents the number of diffractions.

Derive Bragg's Law using the definition of sin θ in eqn (16.1).

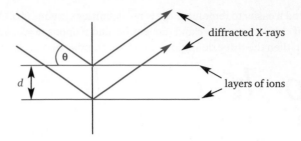

Figure 16.6 In the technique of **X-ray diffraction**, X-rays strike the sample at a glancing angle of θ degrees.

We first look closely at Fig. 16.7, which represents a magnification of Fig. 16.6, above.

From the definition of a sin in eqn (16.1), $\sin\theta =$ (opposite ÷ hypotenuse). The length of the hypotenuse is d and the length of the opposite is MN. Therefore,

$$\sin\theta = \frac{MN}{d}$$

The path length difference between the ray reflected at **O** and the ray reflected at N is $(MN + NP) = 2 \times MN$. Diffraction only occurs successfully when the length $2 \times MN$ comprises an integral number of wavelengths, call it $n \times \lambda$. We therefore substitute for the length MN, saying it is $n\lambda$:

$$2 \times d \sin\theta = n\lambda$$

This yields the Bragg equation, $2d \sin\theta = n\lambda$. ∎

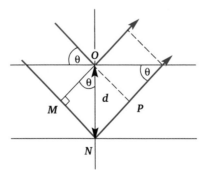

Figure 16.7 Inset of Fig. 16.6 highlighting the lengths involved during diffraction.

It is generally rare to calculate a sin, cos or tan using the respective values of adjacent, opposite, and hypotenuse. More usually, we know the angle of focus, and want to calculate a length from it.

If we know θ and the length of the hypotenuse (call it h), then, we can calculate the length of both the adjacent and the opposite:

$$\text{length of the opposite} = h \sin\theta \tag{16.4}$$

$$\text{length of the adjacent} = h \cos\theta \tag{16.5}$$

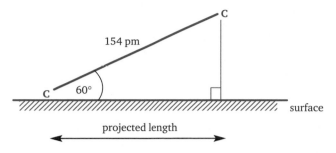

Figure 16.8 We can define the **projected length** as, 'The perceived bond length when the bond is positioned at an angle to a surface or to another part of a molecule.'

■ Worked Example 16.3

A carbon–carbon single bond has a length of 154 pm. If the bond is positioned at an angle of 60° to a surface (see Fig. 16.8), what is the **projected length** of the bond?

To answer this question, we first note that the overall shape in Fig. 16.5 is that of a right-angled triangle. Accordingly, the projected length of the bond is the same as the adjacent of the triangle. We will therefore use eqn (16.5).

Projected length = adjacent = 154 pm × cos 60°

Projected length = 154 pm × 0.5

Projected length = 77 pm

So, if a bond is positioned obliquely to a surface at an angle of 60°, its apparent length is half its real length. ■

Self-test 16.1

Use a calculator or book of tables to determine values of the following:

16.1.1 sin 54° **16.1.2** cos 33°

16.1.3 tan 12° **16.1.4** sin 30°

16.1.5 cos 45° **16.1.6** tan 80°

■ Worked Example 16.4

A molecule of methyl viologen (**I**) adheres strongly to a platinum electrode. If the surface area of a flat molecule is 8800 pm², but the *projected* area is only 3800 pm², calculate the angle between the molecule of methyl viologen and the platinum.

$$H_3C-\overset{+}{N}\underset{}{\bigcirc}-\bigcirc-\overset{+}{N}-CH_3$$

2Cl⁻

I

In this example, we need to perform a similar calculation to that in Worked Example 16.3, but in reverse. We work again with eqn (16.5). The 'real area' corresponds to the hypotenuse and 'projected area' corresponds to the adjacent. Therefore:

Projected area = (real area) × cos θ

A simple way of visualizing a **projected length** is to think, 'What would be the length of the shadow cast by the bond when illumined from a light positioned directly above it?'

Aside

To obtain the value of a sin, cos or tan of an angle using a pocket calculator.

1. Ensure the calculator is set for *degrees*: at the top of the screen it will say 'DEG', 'RAD' or 'GRAD'. Press the $\boxed{DRG}$ button until the display reads 'DEG'.

2. Enter the angle of focus as θ in degrees, °. For more advanced mathematics, it is normal for θ to be in radians.

3. Press the appropriate function button, usually labelled $\boxed{sin}$, $\boxed{cos}$ or $\boxed{tan}$.

Steps 2 and 3 are reversed on some calculators.
    On the *Casio fx-83ES* calculator, press SHIFT + SET UP + DEG. A box saying $\boxed{D}$ will appear at the top of the screen. Alternatively, press e.g. sin (30 + shift + DRG + °) = 0.5. The DRG button is located above the ANS key.

Inserting numbers yields:

$$3800 \text{ pm}^2 = 8800 \text{ pm}^2 \times \cos \theta$$

We first rearrange by dividing throughout by 8800 pm²:

$$\frac{3800 \text{ pm}^2}{8800 \text{ pm}^2} = \cos \theta$$

so $\cos \theta = 0.432$

To obtain the value of θ, we perform the inverse function to cosine, which we usually denote as $\cos^{-1}$. Therefore:

$$\theta = \cos^{-1}(0.432)$$

Finally, we find from a calculator or book of tables, the angle having a cos of 0.432 is 64°. The molecule of methyl viologen (**I**) is inclined to the platinum surface at an angle of 64°. ∎

> The **inverse function** to cosine θ is $\cos^{-1}\theta$,
> We occasionally write 'arc cos θ'; we never write 'inverse cos θ'

■ Worked Example 16.5

It is usually difficult to obtain the NMR spectrum of *solid* samples. In order to analyse samples in the solid state rather than in solution, we use **magic-angle NMR**. Here, the tube containing the sample is spun fast while tilted at an angle of θ to the spectrometer's magnetic field (rather than perpendicular to it); the value of θ is given by the equation:

$$3 \cos^2 \theta - 1 = 0$$

Calculate the value of θ that satisfies this expression.

Before we solve this algebraic expression, we must introduce a new way of reading trigonometric functions. In this example, we see $\cos^2\theta$, which is an alternative way of writing $(\cos \theta)^2$. $\sin^3 \theta$ is the same as $(\sin \theta)^3$, etc. Most chemists agree that '$\cos^2 \theta$' is not a logical way to write a square, but the usage is intended to help us distinguish for clarity between, say, $(\sin \theta)^2$ and $\sin (\theta^2)$. We will have to learn this notation style.

To solve this expression and make θ its subject, we first rearrange slightly:

$$3\cos^2 \theta = 1$$

$$\cos^2\theta = \tfrac{1}{3}$$

so $\cos\theta = \sqrt{\tfrac{1}{3}} = 0.577$

we then find the inverse cosine:

$$\theta = \cos^{-1}(0.577)$$

so $\theta = 54.7°$

We should spin the NMR sample at an angle of 54.7°. ∎

All these problems assume that θ lies in the first quadrant, i.e. $0 \le \theta < 90°$. In chemistry, we usually employ the most acute angle, but a second or even multiple solutions are possible. Particularly in quantum mechanics, we expect multiple answers; so, for example, $\sin kl = 0$, so kl is a multiple of π radians. We deduce the useful result, $kl = n\pi$, so $k = n\pi/l$.

Radians are explained on p. 253.

Self-test 16.2

Rearrange each of the following to obtain the value of θ in degrees:

16.2.1	$\sin \theta = 0.3$	**16.2.2**	$\cos \theta = 0.92$
16.2.3	$2 \cos \theta = 0.4$	**16.2.4**	$\tfrac{1}{2} \sin \theta = 0.45$
16.2.5	$\cos \theta + 2 = 2.1$	**16.2.6**	$\sin^2 \theta = 0.9$
16.2.7	$2 \tan^2 \theta = 9.0$	**16.2.8**	$4 \sin^3 \theta = 0.76$

Radians

By the end of this section, you will know:

- Degrees are not the only way of subdividing angles within a circle.
- There are 2π radians in a complete circle.

We are familiar with a circle comprising 360° (so a semicircle is 180°, a quadrant is 90°, and so on). Figure 16.9 shows these angles diagrammatically. In the alternative system of radians, we say a complete circle comprises 2π radians, where π is the familiar constant 3.142. One full circle therefore comprises 2×3.142 radians = 6.284 radians.

Help is at hand:

- Most calculators have the facility to work in radians rather than degrees. Look at your calculator's manual and see how to switch from degrees (probably labelled as 'DEG' on the LCD screen) to radians. Alternatively, the display could indicate D, R or G: D = degrees; R = radians; G = grads.

- *Excel*™ automatically works in units of radians within trigonometric functions. For example, it will assume a function such as '=SIN(3)' means we want the sine of the angle, 'three radians,' and will produce the result, '0.14'.

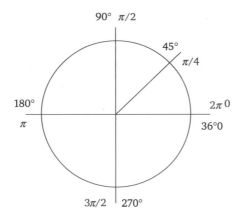

Figure 16.9 The interconversion between radians and degrees. The angles in radians are shown in **bold** print and the degrees in lighter print.

By the Greek letter π (pi), we mean the ratio of a circle's circumference to its diameter.

■ Worked Example 16.6

What is an angle of 22° in radians?

The easiest way to calculate radians from degrees is to employ ratios. We say, $360° = 2\pi$,

so $\dfrac{22°}{360°} = \dfrac{x}{2\pi}$

and therefore $x = \dfrac{22° \times 2 \times \pi}{360°}$ (Just press the π button on your calculator.)

i.e. $x = \dfrac{44 \times 3.142}{360}$

and $x = 0.384$ rad ■

We have omitted the degree signs ° at this stage because the unit of 'degree' on the top and bottom will cancel.

Self-test 16.3

Convert each of the following from, degrees to radians

16.3.1 60° **16.3.2** 300°

16.3.3 160° **16.3.4** 34°

Pythagoras' theorem

By the end of this section, you will:

- Appreciate that Pythagoras' theorem requires a right-angled triangle.
- Learn that the theorem can be applied to two or three dimensions.
- Appreciate how Pythagoras' theorem is most useful to a chemist when looking at molecules and molecular arrangements in which a right-angled triangle is involved.

Pythagoras' theorem relates to right-angled triangles like that in Fig. 16.1. If we know the lengths a and b, we can calculate the length of the hypotenuse c according to eqn (16.6):

$$a^2 + b^2 = c^2 \tag{16.6}$$

c here must be the longest side, but it does not matter which of a and b are the opposite and which the adjacent.

■ Worked Example 16.7

The original route to manufacturing nylon involved reacting adipic acid (**II**) with diaminohexame (**III**) to form a series of amide bonds. During the reaction, the carboxyl group of (**II**) comes level with the amine of (**III**) according to Fig. 16.10. In the transition-state structure, the distance between the carboxyl carbon and the amino nitrogen is 300 pm, and the length of the N–H bond is 103.8 pm long. Assuming the C–N–H bond is a right angle, what is the distance between the carboxyl carbon and the amino hydrogen—call it c?

We will use Pythagoras' theorem in eqn (16.6), taking $a = 300$ pm and $b = 103.8$ pm.

$$c^2 = (300\ \text{pm})^2 + (103.8\ \text{pm})^2$$

so $$c^2 = 90\,000\ \text{pm}^2 + 10\,774\ \text{pm}^2$$

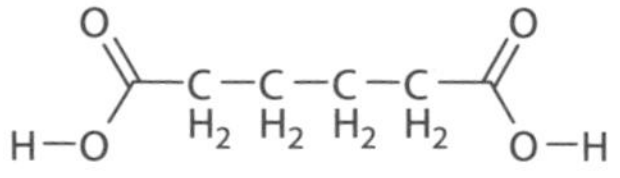

II

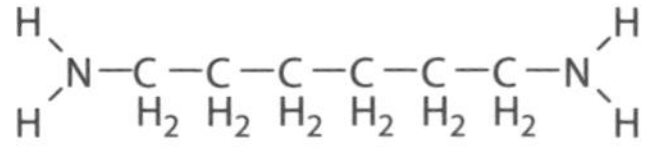

III

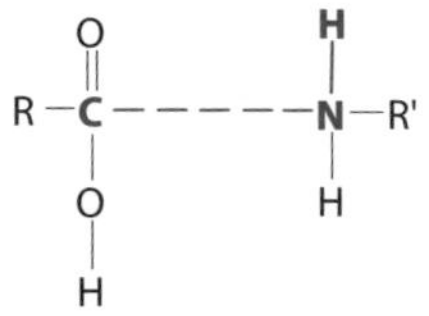

Figure 16.10 In the transition-state complex, the carbon of the carboxyl must approach close enough to the amine nitrogen for electrons to transfer along the path of the dotted line.

i.e. $c^2 = 100\ 774\ \text{pm}^2$

Therefore, $c = \sqrt{100\ 774\ \text{pm}^2}$

so $c = 317\ \text{pm}$ ∎

We can also apply Pythagoras' theorem to a *three*-dimensional (3D) system.

■ **Worked Example 16.8**

Solid sodium chloride adopts a simple cubic structure when it crystallizes (see Fig. 16.11). If the shortest distance between adjacent chloride and sodium ions is 282 pm, then what is the next-shortest distance, i.e. what is the length of the diagonal drawn on Fig. 16.9?

The angles between the sides a, b and c are all right angles, so we employ an advanced form of the Pythagoras theorem. The length of the diagonal d is given by eqn (16.7):

$$a^2 + b^2 + c^2 = d^2 \tag{16.7}$$

In this simple example, $a = b = c$, so we simplify slightly and say,

$3a^2 = d^2$

To make d the subject, we merely take the square root of both sides. We obtain:

$d = \sqrt{3a^2}$

Inserting numbers yields

$d = \sqrt{3 \times (282\ \text{pm})^2}$

so $d = \sqrt{3 \times (79524\ \text{pm}^2)}$

i.e. $d = \sqrt{238572\ \text{pm}^2} = 488\ \text{pm}$

We see how the diagonal is 73% longer than the sides a, b and c. ∎

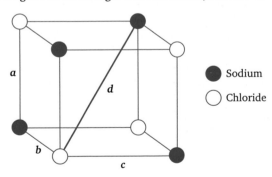

Figure 16.11 The ions in solid sodium chloride adopts a simple cubic structure, with $a = b = c = 282\ \text{pm}$.

Self-test 16.4

Consider the following triangles using Pythagoras' theorem. In each case, assume the triangle has a right angle. The length c always relates to the *hypotenuse*.

16.4.1 $a = 2\ \text{cm}$ $b = 5\ \text{cm}$.   What is c?

16.4.2 $a = 7\ \text{km}$ $b = 13\ \text{km}$.   What is c?

» *Continued...*

> **Self-test 16.4 continued...**

16.4.3 $c = 150$ pm $a = 130$ pm.   What is b?

16.4.4   Consider the strained molecule, cyclobutadiene (**IV**): assuming all the internal angles are 90°, calculate the length of the *diagonal* in (**IV**). Take the bonds lengths as C–C is 150 pm and C=C is 140 pm.

$$HC \!\!-\!\!\!-\!\!\!-\!\! CH$$
$$HC \!\!-\!\!\!-\!\!\!-\!\! CH$$

IV

Polar coordinates

By the end of this section, you will know:

- It is possible to represent a graph using polar coordinates.

- Polar coordinates are especially useful when considering spherical or cylindrical spaces, when we call them, respectively, spherical polar coordinates and cylindrical polar coordinates.

- In two dimensions, the coordinates are the distance r and the angle θ. And additional coordinate, ϕ is required in situations involving three dimensions.

Polar coordinates:

If we need to consider spherical or cylindrical spaces, Cartesian coordinates can be quite cumbersome; worse, they can actually obscure the relationship(s) involved. In such circumstances, polar coordinates are generally superior because they pre-suppose a circular or spherical geometry.

Polar coordinates employ trigonometry to describe two- or three-dimensional spaces. Figure 16.12 shows a schematic 2D graph drawn in terms of polar coordinates. The position of the point **P** has not been defined using Cartesian coordinates of x and y, but in terms of a distance from the origin r and the angle θ formed between the x-axis and the line that joins **P** to the origin. The coordinates (r, θ) define the position of the point **P**. When citing the coordinates of **P**, we always write the length r before the angle θ.

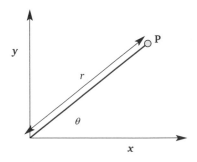

Figure 16.12 Polar coordinates are useful when considering functions that are circular, spherical or cylindrical. In 2D, the position of the polar coordinates of the point **P** is (r, θ).

■ Worked Example 16.9

What are the polar coordinates of the point **P**, whose Cartesian coordinates are (4,3)?

- **Length, r :** The length r relates to the line that joins **P** to the origin. We obtain the magnitude of r using Pythagoras' theorem, eqn (16.6): $r^2 = 3^2 + 4^2 = 9 + 16 = 25$, so $r = 5$.

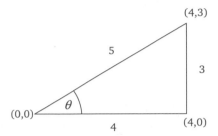

- **Angle θ:** The angle θ again relates to the line joining **P** to the origin. This time, θ is the angle between this line and the x-axis. Taking the origin as one vertex of a right-angled triangle, we can obtain the angle of focus θ by using the definition of tangent in eqn (16.3). The adjacent of this triangle has a length of 4, and the opposite has a length of 3.

From eqn (16.3), we obtain the angle θ:

so $\tan \theta = \frac{3}{4} = 0.75$.

and $\theta = \tan^{-1}(0.75) = 36.9°$.

The position of the point, expressed in polar coordinates, is (5, 36.9°). ■

Self-test 16.5

Determine the polar coordinates in 2D of the following points:

16.5.1 (2, 5)   **16.5.2** (−3, 6)   **16.5.3** (4.2, 1.9)

Figure 16.12 describes the position of a point **P** in *two* dimensions. We can adapt this coordinate system to describe spaces in *three* dimensions, in two different ways.

Spherical polar coordinates:

Spherical polar coordinates define the position of a point **P** relative to the centre of a sphere in 3D. We define the position of a point **P** as (r, θ, ϕ); see Fig. 16.13.

In this coordinate system, the point **P** describes a sphere if r is the radius, and θ and ϕ describe its position in the x–y plane, and vertically above the x–y plane, respectively. The angle ϕ gives the position of **P** in the third dimension, that is, above or below the plane in Fig. 16.12.

The following equations supply the relationships between spherical polar coordinates and Cartesian coordinates:

$$x = r\sin\theta\cos\phi \qquad (16.8)$$

$$y = r\sin\theta\sin\phi \qquad (16.9)$$

$$z = r\cos\theta \qquad (16.10)$$

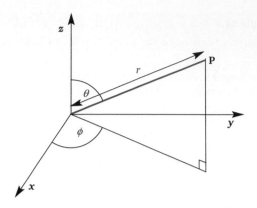

Figure 16.13 Spherical polar coordinates for a 3D function: the polar coordinates describing the position of the point P are (r, θ, ϕ).

■ Worked Example 16.10

Determine the Cartesian coordinates of the point whose spherical polar coordinates are $(r, \theta, \phi) = (2, 30°, 45°)$.

These three simple calculations require the values of sin and cos from Table 16.1.

From eqn (16.8), $x = 2 \times (\sin 30°) \times (\cos 45°) = 2 \times 0.5 \times \dfrac{1}{\sqrt{2}} = 0.707$.

From eqn (16.9), $x = 2 \times (\sin 30°) \times (\sin 45°) = 2 \times 0.5 \times \dfrac{1}{\sqrt{2}} = 0.707$.

From eqn (16.8), $x = 2 \times (\cos 30°) = 2 \times \dfrac{\sqrt{3}}{2} = 1.732$.

So the position of the point in Cartesian space is, $(0.707, 0.707, 1.732)$. ■

Cylindrical polar coordinates:

As the name suggests, the system of cylindrical polar coordinates requires a three-dimensional space based on a cylinder, and requires a third dimension of height z in addition to R and ϕ. Figure 16.14 demonstrates the 3D arrangement these three coordinates require (R, ϕ, z).

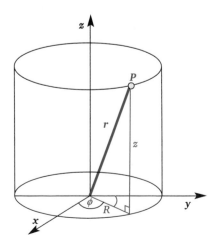

Figure 16.14 Cylindrical polar coordinates for a 3D function: the polar coordinates describing the position of the point P are (R, ϕ, z).

- The length R: We obtain R via Pythagoras' theorem, as $\sqrt{x^2 + y^2}$
- The height z: we simply read off the value of z from the z-axis.
- The angle ϕ: we obtain ϕ as $\tan^{-1}(y/x)$, according to the inverse of eqn (16.3).
- Cylindrical polar coordinates are readily converted into Cartesian coordinates:

Cylindrical polar coordinates	Cartesian coordinates
• R	• $x = R \cos \phi$
• ϕ	• $y = R \sin \phi$
• z	• $z = z$

In contrast to spherical polar coordinates, cylindrical polar coordinates are employed only very rarely in chemistry. Accordingly, we will not discuss them further.

The cosine rule

By the end of this section, you will know:

- The cosine rule.
- That the cosine rule does not require a right-angled triangle.

Sometimes we want to determine the properties of triangles that do not have a right angle. One of the more powerful equations to describe such triangles is the **cosine rule** (which we also call the 'cosine formula'). Again, we frame the definition in using the letters from Fig. 16.1. The rule says:

$$c^2 = a^2 + b^2 - 2ab \cos C \qquad (16.11)$$

When we use the cosine rule, we simply ensure that the angle of focus C is the angle opposite the side of unknown length, c.

Note how the cosine rule simplifies to yield the Pythagoras' theorem if the angle of focus is 90°, because $\cos 90° = 0$.

■ Worked Example 16.11

Consider the primary amine $R–NH_2$ in Fig. 16.15. The angle between the two N–H bonds is 109.3°. The length of each N–H bond is 103.8 pm. What is the interatom separation between the two hydrogen atoms?

In this triangle, the angle of focus is greater than 90°. We say it is **obtuse**. An angle less than 90° is called **acute**.

The sides a and b here clearly have the same length. Side c of the triangle is the interatom separation. Inserting values into eqn (16.11) yields:

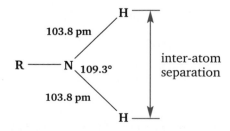

Figure 16.15 Interatom separations are easily calculated with the cosine rule.

$$\left(\frac{\text{interatom}}{\text{separation}}\right)^2 = (103.8\,\text{pm})^2 + (103.8\,\text{pm})^2 - 2 \times (103.8\,\text{pm}) \times (103.8\,\text{pm}) \times \cos 109.3°$$

$$\left(\frac{\text{interatom}}{\text{separation}}\right)^2 = 2 \times (103.8\,\text{pm})^2 - 2 \times \left((103.8\,\text{pm})^2 \times \cos 109.3°\right)$$

This sum simplifies by factorizing, to yield:

$$\left(\frac{\text{interatom}}{\text{separation}}\right)^2 = 2\,(103.8\,\text{pm})^2 \times [1 - \cos 109.3°]$$

so $$\left(\frac{\text{interatom}}{\text{separation}}\right)^2 = 21\,549\,\text{pm}^2 \times [1 - (-0.331)]$$

$$\left(\frac{\text{interatom}}{\text{separation}}\right)^2 = 21\,549\,\text{pm}^2 \times [1.331]$$

$$\left(\frac{\text{interatom}}{\text{separation}}\right)^2 = 28\,682\,\text{pm}^2$$

so interatom separation $= \sqrt{28\,682\,\text{pm}^2}$

i.e. interatom separation $= 169\,\text{pm}$

In fact, we could have avoided the cosine rule altogether if we had bisected the obtuse triangle, to form two right-angled triangles, each with an angle of focus of $109.3° \div 2 = 54.65°$. ∎

Self-test 16.6

Consider the following triangles. In each case, calculate the unknown variable using the cosine rule, eqn (16.11).

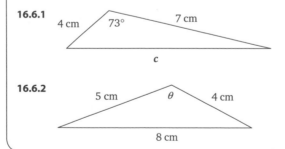

16.6.1

4 cm   73°   7 cm

c

16.6.2

5 cm   θ   4 cm

8 cm

Trigonometric identities

By the end of this section, you will know:

- The sum of the squares of $\cos \theta$ and $\sin \theta$ is equal to 1.

The so-called trigonometric identities are a series of simple mathematical relationships, which allow us to perform many simple trigonometric interconversions, and greatly simplify trigonometric derivations.

As a simple consequence of Pythagoras' theorem, we can manipulate eqn (16.6) when applied to a triangle like that in Fig. 16.2, and obtain the relation:

$$\sin^2 \theta + \cos^2 \theta = 1 \tag{16.12}$$

■ **Worked Example 16.12**

Demonstrate that eqn (16.12) is true for the angle 39°.

$\sin 39° = 0.6293$ $\sin^2 39° = 0.3960$ (to 4 s.f.)
$\cos 39° = 0.7771$ $\cos^2 39° = 0.6040$ (to 4 s.f.)
so $\sin^2 39° + \cos^2 39° = 1.$ ■

■ **Worked Example 16.13**

Derive eqn (16.12) using eqn (16.6) and Fig. 16.2.

If the length of the hypotenuse is h, the length of the adjacent is a and the length of the opposite is o, then the Pythagoras theorem says that $o^2 + a^2 = h^2$.

Dividing throughout by h^2 yields, $(o/h)^2 + (a/h)^2 = 1$. From eqn (16.1), $o/h = \sin \theta$ and eqn (16.2) says $a/h = \cos \theta$. Therefore, $\sin^2 \theta + \cos^2 \theta = 1$. ■

Other trigonometric identities exist. Although we employ them only rarely in chemistry, perhaps the most useful are the so-called **double-angle formulae**:

$$\sin (\theta \pm \varphi) = \sin \theta \cos \varphi \pm \sin \varphi \cos \theta \tag{16.13}$$

$$\cos (\theta \pm \varphi) = \cos \theta \cos \varphi \mp \sin \theta \sin \varphi \tag{16.14}$$

If we need them, the proofs of these two equations are available in many textbooks but, as chemists, we do not need to know to know them.

There are two special cases that are useful in quantum mechanics, as will be discussed in the integration chapters:

$$\cos 2\theta = \cos^2 \theta - \sin^2 \theta \tag{16.15}$$

This equation can be amended via eqn (16.12):

$$\cos 2\theta = 1 - 2\sin^2 \theta \tag{16.16}$$

Again, via eqn (16.12), we could change eqn (16.15) to:

$$\cos 2\theta = 2\cos^2 \theta - 1 \tag{16.17}$$

Also,

$$\sin 2\theta = 2 \sin \theta \cos \theta \tag{16.18}$$

We can derive these equations from eqns (16.13) and (16.14) above.

Summary of key equations in the text

The definitions of the simple trigonometric functions:

$$\sin \theta = \frac{\text{opposite}}{\text{hypotenuse}} \tag{16.1}$$

$$\cos \theta = \frac{\text{adjacent}}{\text{hypotenuse}} \tag{16.2}$$

$$\tan \theta = \frac{\text{opposite}}{\text{adjacent}} \tag{16.3}$$

Pythagoras' theorem:

in   2D: $a^2 + b^2 = c^2$ (16.6)

in   3D: $a^2 + b^2 + c^2 = d^2$ (16.7)

Relationships between spherical polar coordinates (x, y, z) and Cartesian coordinates (r, θ, ϕ):

$$x = r \sin \theta \cos \phi \qquad (16.8)$$

$$y = r \sin \theta \sin \phi \qquad (16.9)$$

$$z = r \cos \theta \qquad (16.10)$$

The cosine rule: $c^2 = a^2 + b^2 - 2ab \cos C$ (16.11)

$$\cos^2 \theta + \sin^2 \theta = 1 \qquad (16.12)$$

The double-angle formulae:

$$\sin (\theta \pm \varphi) = \sin \theta \cos \varphi \pm \sin \varphi \cos \theta \qquad (16.13)$$

$$\cos (\theta \pm \varphi) = \cos \theta \cos \varphi \mp \sin \theta \sin \varphi \qquad (16.14)$$

Special cases: $\cos 2\theta = \cos^2 \theta - \sin^2 \theta$ (16.15)

$$\cos 2\theta = 1 - 2\sin^2 \theta \qquad (16.16)$$

$$\cos 2\theta = 2\cos^2 \theta - 1 \qquad (16.17)$$

$$\sin 2\theta = 2 \sin \theta \cos \theta \qquad (16.18)$$

Additional problems

Questions 16.1–16.3 relate to the simplest of the hydrocarbons, methane (V), which is tetrahedral about the central carbon.

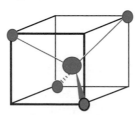

V

In a perfect **tetrahedral** arrangement such as that in methane (CH_4), the central atom is located at the centre of a cube, and the four hydrogen atoms sit on alternate vertices (see the figure, below)

16.1   If the length of each side of the cube is $2a$, calculate the length of a single C–H bond.

16.2   If the length of each side of the cube is $2a$, calculate the distance between the two hydrogen atoms at the top of the cube (i.e. determine the length of a diagonal across one of the cube faces).

16.3   Calculate the angle of focus θ, i.e. the bond angle $\widehat{H - C - H}$.

16.4 Consider the copper aquo ion (**VI**): calculate the distance between the oxygen atoms in the two adjacent water molecules. Assume all angles are 90°, and each Cu–OH$_2$ bond has the same length of 167 pm.

VI

16.5 Consider the cyclopentadienyl anion (**VII**), which is a regular pentagon: calculate the internal $\overset{\frown}{C-C-C}$ angles, expressing it both in degrees and in radians.

VII

16.6 Consider the transition-state structure (**VIII**) formed during the reaction of saponification of an ester with base:

VIII

The central carbon of (**VIII**) is nearly tetrahedral, with a bond angle of 112°. The length of the bond between the central carbon and the ionized oxygen is 139 pm, and the length of the bond between the central carbon and the methoxide is 125 pm. What is the distance between the two oxygen atoms?

16.7 Show how the cosine rule (eqn (16.11)) simplifies to form Pythagoras' theorem (eqn (16.6)) when the triangle is a right-angled triangle.

16.8 In a molecule of benzene (**IX**), what is the distance across the ring from carbon-1 to carbon 4? Assume the molecule is a perfect hexagon, each side having a length of 143 pm.

IX

16.9 (Following on from Additional problem 16.8): benzidine (**X**) is a powerful carcinogen. The bond between the two rings has a length of 153 pm. From Worked Example 16.11, the angle between the two hydrogen atoms is 109.3° and the length of each N–H bond is 103.8 pm and each C–N bond length is 139 pm. What is the length of a molecule of benzidine (**X**)?

X

16.10 Consider the anticancer drug, *cis*-platin (**XI**), which is planar: calculate the distance between the top Cl and the nitrogens of the NH$_3$ groups, assuming each angle is 90°, and that the Pt–Cl bond is 170 pm, and the Pt–N bond is 162 pm.

XI

17

Differentiation I

Rates of change, tangents, and differentiation

Introduction: tangents and the gradients of curves

By the end of this section, you will know:

- When we determine the gradient of a curve, we first draw a tangent.
- We determine the gradient of a tangent in the same way as for a straight line.

It is easy to determine the gradient of a *straight* line. Chapter 9 described the process whereby we take the ratio of the distances travelled vertically (call it Δy) and horizontally (Δx). The gradient is then $\Delta y \div \Delta x$; eqn (9.2). The gradient is merely a rate of change.

Look at Figure 17.1, which represents the amount of reactant remaining during the course of a reaction. The curve has the mathematical form $y = e^{kt}$, where t is the time elapsing after the reaction commences, and k is the rate constant. The heavy straight line drawn against the curve represents the curve's gradient at a time of 1 min. we call the line a **tangent**, and define it as, 'a straight line that meets a curve at a single point.'

A true tangent touches the curve only at one point—in this case at $x = 1$ min—and never *crosses* the line. It's now easy to determine the gradient of the curve at $x = 1$: we determine the tangent's gradient by measuring Δx then Δy for this straight line, and taking their ratio.

Self-test 17.1

Draw the following curves and then determine the gradient by first drawing a tangent:

17.1.1 $y = x^2$ Determine the gradient at $x = 3$

17.1.2 $y = x^3$ Determine the gradient at $x = 4$

17.1.3 $y = x^2 + 4$ Determine the gradient at $x = 2.5$

17.1.4 $y = x^3 + x^2 + 2$ Determine the gradient at $x = 2$

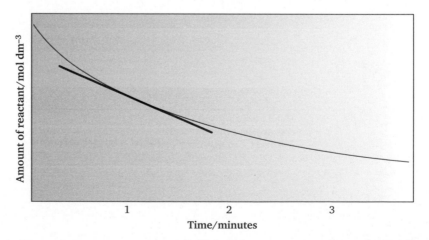

Figure 17.1 During a chemical reaction, the amount of reactant remaining (plotted as 'y') decreases as the reaction progresses, i.e. as a function of time (plotted as 'x'): to determine the rate of reaction, we determine the gradient by drawing a tangent at a point.

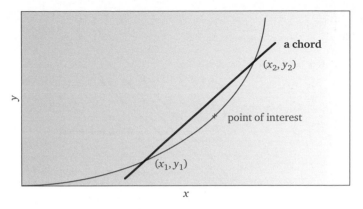

Figure 17.2 Determining the gradient of a chord cutting a smooth curve. When the chord just touches the line, we call it a tangent.

With a little practice, it becomes quite easy to determine the gradient of a curve at a point. But problems remain with this approach. First, we must note that a curve has an infinite number of gradients: the rate at which y changes will itself change as a function of x. In Fig 17.1, the gradient at $x = 2$ is about half as steep as the gradient at $x = 1$. The gradient at $x = 3$ is about a quarter as steep, and so on. And there are also different gradients at $x = 2.1, 2.2, 2.3$, etc., all of which differ.

The second problem is the way we actually measure the gradient of a tangent in practice. Most people find it quite hard to draw a tangent, so they are recommended to draw two points on the curve, placed at equal distances along the curve, either side of the point where we want the gradient; see Fig. 17.2. We then join together these two points with a straight line. This line is called a **chord**. The chord becomes the tangent when it no longer cuts the curve, but sits on the surface at a single point.

Having drawn a chord to the curve in Fig. 17.2, we determine a numerical value for its gradient (and an estimate for the gradient of the tangent), saying:

$$\text{gradient} = \frac{\Delta y}{\Delta x} = \frac{y_2 - y_1}{x_2 - x_1} \qquad (17.1)$$

The only difference between eqn (17.1) and the equation for a gradient, eqn (9.2) in Chapter 9, is the way the gradient here refers to a *portion* of a straight line, rather than a line that is simply linear.

In practice, it does not matter if we write $(y_2 - y_1)$ or $(y_1 - y_2)$: what does matter is that the subscripted numbers agree, so if we write $(y_1 - y_2)$ then the bottom line of the fraction must be $(x_1 - x_2)$.

This definition ensures the gradient has a *sign* as well as magnitude.

Introduction to differentiation: 'taking the limit'

By the end of this section, you will know:

- It is difficult to draw a tangent accurately enough.

- To counter this difficulty, a line is drawn between two equally spaced points placed either side of the point **x**.

- As the two points get closer together, the gradient of the chord becomes closer to the gradient of the tangent.

- When the two points coincide (one on top of the other), the chord is equivalent to the tangent, and we say we 'take the limit'.

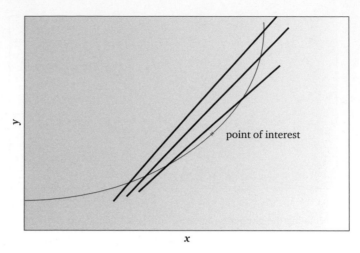

Figure 17.3 The magnitude of a chord's gradient depends on where we draw it.

The method of determining a gradient outlined above will work well after a little practice, but it does have one major drawback. Unfortunately, the way we draw the heavy line can affect the value of the gradient. The numerical value of the gradient changes according to where we draw the chord. Even if we position x_1 and x_2 at an equal distance either side of the point of interest, the actual distance chosen will itself change the value of the gradient. Look at Fig. 17.3.

Three possible gradients have been drawn in Fig. 17.3. In each case, the chord cuts the curve twice, with the two points of intersection being equidistant above and below the point of interest. The gradients of the three chords are clearly very different. In fact the steepest (on the left) is about 30% steeper than the shallowest (on the right).

This situation is very unsatisfactory. We want to know the rate of change of the curve, and even drawing a chord close to the point of interest appears not to yield a good-quality gradient. After a moment's thought, we realize that the best gradient (by which we mean the most accurate) will be the one where the two points of intersection are closest to the point of interest. If we could make the distance between the intersection points **infinitesimally** small, then the gradient would be *exact*, and we would not need to worry any further.

Infinitesimal means '$1 \div \infty$' i.e. so small that its value is effectively zero.

Differentiation from first principles

By the end of this section, you will know:

- When differentiating, we employ algebra rather than drawing a graph.
- We start by taking two coincident points (x, y) and $(x + \delta x, y + \delta y)$, and write an expression for the gradient between them.
- The separation between the two points is in fact zero ($\delta x = \delta y = 0$).

As trainee mathematicians, we learn the gradient of the tangent becomes exact *as we take the limit*, i.e. as the separation between x_1 and x_2 becomes zero. When the separation between x_1 and x_2 is indeed zero, the tangent becomes a perfect tangent. And we can determine the gradient this way using algebra, which saves us the time and trouble of actually drawing a graph.

■ Worked Example 17.1

We accelerate a molecule inside the vacuum of a mass spectrometer. A graph of its velocity (as 'y') against time (as 'x') follows a curve of the type $y = x^2$. What is the gradient of the curve $y = x^2$ when $x = 4$?

We first draw the curve as Fig. 17.4. We draw a tangent that crosses the line at two points. We will call these two points (x, y) and $(x + \delta x, y + \delta y)$, where the Greek letter δ means an infinitesimal increment. In other words, we use the following mathematical procedure, which allows us to consider two points that are in fact the same point, one on top of the other.

We obtain the gradient of the tangent with eqn (17.1):

$$\text{gradient} = \frac{\Delta y}{\Delta x} = \frac{y_2 - y_1}{x_2 - x_1}$$

Inserting the two points $(x + \delta x, y + \delta y)$ and (x, y) into this equation as (x_2, y_2) and (x_1, y_1) respectively, we obtain:

$$\text{gradient} = \frac{(y + \delta y) - y}{(x + \delta x) - x}$$

We then remember that the function of y is a square, $y = x^2$ so we can therefore substitute into the top line (the **numerator**) of the equation: y becomes x^2 and $(y + \delta y)$ becomes $(x + \delta x)^2$:

$$\text{gradient} = \frac{(x + \delta x)^2 - x^2}{(x + \delta x) - x}$$

We next multiply out the bracket on the top line of the equation:

$$\text{gradient} = \frac{(x^2 + 2x\,\delta x + \delta x^2) - x^2}{(x + \delta x) - x}$$

It is now time to simplify: on the bottom line, $(x + \delta x - x)$ becomes just δx, and on the top line the two x^2 terms also cancel. We are left with:

$$\text{gradient} = \frac{2x\,\delta x + \delta x^2}{\delta x}$$

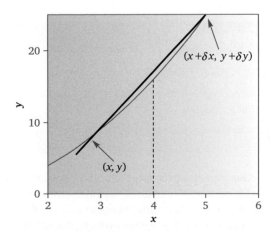

Figure 17.4 Determining the gradient to the curve $y = x^2$.

We can then simplify further, since there are δx terms on top and bottom. Perhaps it's easier if we first rewrite the fraction in two parts:

$$\text{gradient} = \frac{2x\,\delta x}{\delta x} + \frac{\delta x^2}{\delta x}$$

The two δx terms in the left-hand fraction cancel, and the right-hand fraction becomes just δx. We are left with:

$$\text{gradient} = 2x + \delta x$$

And finally, we investigate how the value of the gradient simplifies further as the value of δx becomes *infinitesimal*, i.e. zero. In fact, the gradient of the curve $y = x^2$ becomes simply $2x$, because $\delta x = 0$.

   By this procedure, we have **differentiated** the function $y = x^2$ to obtain a rather surprising result: whatever the value of x, the gradient of the curve is always $2x$. For example, the gradient when $x = 4$ is $2 \times 4 = 8$; the gradient at $x = 6$ is 12, and so on. ■

We give the name **differential coefficient** to the numerical value of the gradient obtained in this way.

Self-test 17.2

Take the following functions and differentiate from first principles:

17.2.1 $y = x^2 + 3$ **17.2.2** $y = x^3$

17.2.3 $y = 4x^3$ **17.2.4** $y = x^3 + 4$

Important terminology:

- We call the resultant gradient the **differential**, so the differential of $y = x^2$ is $2x$.

- We call the methodology employed in Worked Example 17.1 **differentiation from first principles**.

- Instead of writing the word 'gradient' each time, we write $\dfrac{dy}{dx}$.

- The y on top and the x on the bottom will remind us of eqn 17.1. Instead of a capital Greek delta Δ on top and bottom, we write a small 'd', which stands for 'difference.' The difference d implies an *infinitesimal* difference.

> Concerning this fraction, when reading aloud we would say, 'dee why by dee ex.'

Differentiation by rule

By the end of this section, you will know:

- It is possible to differentiate without the time-consuming necessity for determining an answer from first principles.

- Learn the simple rules for differentiation by rule.

Differentiating a function from first principles is usually a hard slog. We therefore want a simpler method of obtaining a gradient. Luckily we rarely ever need to differentiate from first principles. It's too time consuming and, being so tricky, we run the risk of making mistakes. From now on, we will differentiate—that is, obtaining a gradient (a rate of change)—by a simpler route: we will differentiate by rule.

Look again at Worked Example 17.1: we started with $y = x^2$ and obtained the differential as $2x$. Similarly, the differential of x^3 was $3x^2$. If we had differentiated x^4, we would have obtained a differential of $4x^3$. Hopefully we can now discern a pattern: when we differentiate a polynomial of the type x^n, we obtain $n \times x^{(n-1)}$. In mathematical notation we write:

$$\text{If} \quad y = x^n \quad \text{then} \quad \frac{dy}{dx} = n \times x^{(n-1)} \tag{17.2}$$

This equation does not work when $x = 0$.

■ **Worked Example 17.2**

What is the differential of $y = x^5$?

Comparing this example with the template answer in eqn (17.2) shows the value of 'n' is 5. Accordingly, the power will be $(5-1) = 4$. The differential is therefore $5 \times x^4$. In fact, we normally omit the multiplication sign, writing the answer as '$5x^4$'. ■

In many cases, we start with the x (or function of x) positioned within the *denominator*. In such cases, we first rewrite the function using the notation in Chapter 11. For example, we would write $1/x^2$ as x^{-2}, $1/x^4$ as x^{-4}, etc. We then differentiate in the normal way via eqn (17.2).

■ **Worked Example 17.3**

According to Coulomb's Law, the magnitude of the attractive force ϕ between two charges decays according to the inverse square of the intervening distance r. Mathematically, we say:

$$\phi = \frac{1}{r^2}$$

If we drew a graph of ϕ (as 'y') against r (as 'x'), what would be the gradient, i.e. what is the differential of $\phi = \dfrac{1}{r^2}$?

The version of **Coulomb's Law** here is greatly simplified. In reality, the law includes several additional constant terms.

Strategy:

(i) To date, we have only employed the algebraic characters x and y. Since algebra is itself a form of shorthand, we are free to choose the shorthand notation of our choice. When we write a differential, we always write the variable of the vertical axis on the top as 'd variable' (here, as $d\phi$) and the variable of the horizontal axis on the bottom, so here we write dr.

(ii) It should not worry us that we've written $d\phi/dr$ rather than dy/dx: the only difference is that the observed variable is ϕ instead of y, and the controlled variable is r instead of x.

(iii) Next, we recognize that we must rewrite this function without the fraction, i.e. in terms of scientific notation. Instead of $\phi = 1/r^2$, we write $\phi = r^{-2}$.

Solution:

Inserting values into the template in eqn (17.2), we say

$$\phi = r^{-2} \quad \text{so} \quad \frac{d\phi}{dr} = -2 \times r^{(-2-1)}$$

The '-2' at the beginning of the answer comes from the original power on r. The new power is -3 from $(-2-1)$. The answer is therefore

$$\frac{d\phi}{dr} = -2r^{-3}$$

We could alternatively write this result as, $-\dfrac{2}{r^3}$ ■

We need to note:

If we differentiate a function of the type 'ax', the 'x' vanishes to leave just 'a'.

(i) When we differentiate x, we need first to write it as x^1. When we decrease the power by '1' (as before in accordance with eqn (17.2)), we obtain x^0. Since anything raised to the power of '0' has a value of 1, the differential of $y = x$ has the value of 1, so the line's gradient is 1.

(ii) We need not always differentiate y with respect to x. For example, in Self-test questions 17.3.6–17.3.10, below, the algebraic letters differ from x and y, so the resultant differentials will *look* different. In fact, we employ exactly the same pattern in each: the differential has the observed variable on the top (the numerator) and the controlled variable (the denominator) on the bottom.

In effect, then, the variable on the bottom line of the differential will be the variable from the right-hand side of the source equation, and the variable on the top line of the differential will be the variable from the left-hand side of the source equation.

Self-test 17.3

Differentiate the following by rule using eqn (17.2):

17.3.1	$y = x^6$	**17.3.2**	$y = x^{12}$
17.3.3	$y = x^5$	**17.3.4**	$y = x^3$
17.3.5	$y = x$	**17.3.6**	$h = 5T$
17.3.7	$w = p^{-11}$	**17.3.8**	$c = \dfrac{1}{q^7}$
17.3.9	$h = \dfrac{1}{k^3}$	**17.3.10**	$z = p^{2.73}$

■ Worked Example 17.4

The potential energy of 2 bonded atoms, when shifted a distance x away from their equilibrium positions, is given by the following equation

$$U = \tfrac{1}{2}kx^2$$

where k is a constant. If we were to plot a graph of U (as 'y') against x (as 'x'), what would be its gradient?

Strategy:

(i) Because $U \propto x^2$, the graph of U against x will clearly be curved rather than linear. We need to differentiate if we want the gradient.

(ii) The word 'gradient' in the question alerts us that we want a rate of change.

(iii) We cannot use eqn (17.2) because we have a more complicated situation—the function of x has been multiplied by a constant (in this case, by '$\tfrac{1}{2}k$'). Accordingly, we employ a more realistic variant of eqn (17.2):

If $y = a x^n$, then

$$\frac{dy}{dx} = a \times n \times x^{(n-1)} \tag{17.3}$$

where 'a' represents any constant.

Solution:

Inserting values from the question into eqn (17.3), we obtain:

$$U = \left(\tfrac{1}{2}k\right)x^2$$

so $\dfrac{dU}{dx} = \left(\tfrac{1}{2}k\right) \times 2 \times x^{(2-1)}$

Omitting the multiplication signs, we write

$\dfrac{dU}{dx} = kx$ ∎

> The differential dU/dx is force, $-F$.

Self-test 17.4

Differentiate the following by rule, using eqn (17.3):

17.4.1 $y = 3x^3$ **17.4.2** $y = 5x^{14}$

17.4.3 $y = 4x^5$ **17.4.4** $y = 1.2x^{-7}$

17.4.5 $y = 4.3x^6$ **17.4.6** $y = 10^6 x^{-4}$

■ **Worked Example 17.5**

The rotated-disc electrode (RDE) is rotated at a constant speed of ω and the resultant current is I. The Levich equation defines the relationship between I and ω:

$$I = k \times \omega^{\frac{1}{2}}$$

where k represents a jumble of constants. If we were to plot a graph of I (as 'y') against ω (as 'x'), what would be its gradient?

This example looks more complicated than those we saw earlier because the power is a fraction. In fact, we can still use eqn (17.3).

Strategy:

(i) The word 'gradient' in the question again alerts us to expect a rate of change.

(ii) Because $I \propto \omega^{\frac{1}{2}}$, the graph will clearly be curved rather than linear, so we need to differentiate.

Solution:

To differentiate, we insert values into eqn (17.3), so

$$\dfrac{dI}{d\omega} = \tfrac{1}{2}k\omega^{-\frac{1}{2}}$$

We obtain the power of $-\tfrac{1}{2}$ here because, $\tfrac{1}{2} - 1 = -\tfrac{1}{2}$.

So the equation is, $\dfrac{dI}{d\omega} = \dfrac{k}{2\sqrt{\omega}}$. ∎

> The value of k in the **Levich equation** accommodates many variables such as temperature T, number of electrons n, analyte concentration c, electrode area A, etc.

> In this example we have written I instead of y and ω instead of x, but will otherwise employ an identical methodology.

Self-test 17.5

Differentiate the following by rule, applying eqn (17.3) in each case:

17.5.1 $y = 3x^3$ **17.5.2** $y = 4x$

17.5.3 $y = 3.2\sqrt{x}$ **17.5.4** $y = 6 \times 10^5 x^2$

17.5.5 $y = 4.65\sqrt[3]{x}$ **17.5.6** $y = 10x^{9.2}$

Whatever we do to the variables in an equation, constants do not change. That's why we call them constants. Consider the three curves $y = x^3$, $y = x^3 + 20$ and $y = x^3 + 40$, which are plotted in Fig. 17.5. The figure clearly shows the three lines in the graph are *parallel*.

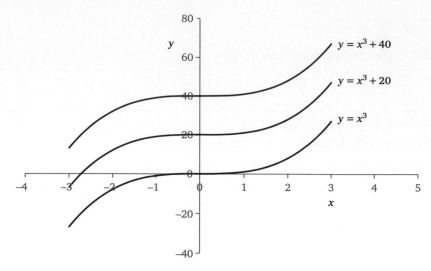

Figure 17.5 Graph showing the three parallel curves, $y = x^3$, $y = x^3 + 20$ and $y = x^3 + 40$.

After a moment, we realize that parallel lines have the same gradients, precisely *because* they are parallel. In fact, they are merely shifted up or down the page. Close inspection of Fig. 17.5 should persuade us that the gradient at (say) $x = 2$ is the same on each of the three curves. A constant in an equation therefore merely shifts the line up or down a graph, and in no way alters the gradient. *For this reason, we obtain a zero whenever we differentiate a constant.*

$$\text{If} \quad y = a, \quad \text{then} \quad y = a \times x^0, \quad \text{then} \quad \frac{dy}{dx} = a \times 0 = 0 \tag{17.4}$$

■ Worked Example 17.6

The conductivity Λ of a simple ionic salt in solution follows the so-called **Onsager equation**:

$$\Lambda = \Lambda^\circ - b\sqrt{c}$$

> Λ° is the conductivity at a single concentration (infinite dilution) and can have only one value. Accordingly, Λ° is a constant.

where b is a constant, c is the concentration and Λ° is the limiting conductivity, i.e. it is also a constant. If we were to plot a graph of Λ (as 'y') against c (as 'x'), what would be its gradient?

Strategy:

(i)   We will differentiate each term, one at a time.

(ii) We will treat the Λ° term as a constant, so its differential is zero.

Solution:

We first rewrite the equation slightly, as $\Lambda = \Lambda^\circ + (-b) c^{\frac{1}{2}}$. The Λ° term does not vary because it is a constant, so it vanishes when we differentiate. The term, '$(-b) c^{\frac{1}{2}}$' is a simple function, and can be approached in the same way as other examples. Overall

$$\frac{d\Lambda}{dc} = 0 + \tfrac{1}{2} \times (-b) c^{-\frac{1}{2}}$$

so $\dfrac{\mathrm{d}\Lambda}{\mathrm{d}c} = -\dfrac{1}{2} bc^{-\frac{1}{2}}$

We could have written this answer in a variety of other ways:

so $\dfrac{\mathrm{d}\Lambda}{\mathrm{d}c} = -\dfrac{\frac{1}{2}b}{c^{\frac{1}{2}}}$ or $= -\dfrac{1}{2}bc^{-\frac{1}{2}}$ or $= -\dfrac{b}{2c^{0.5}}$ or $= -\dfrac{b}{2\sqrt{c}}$

Each of these four alternatives is equally valid. ■

Self-test 17.6

Differentiate the following by rule:

17.6.1 $y = x^3 + 12$

17.6.2 $y = 2x^2 - 3$

17.6.3 $y = 6 - x^{\frac{1}{2}}$

17.6.4 $y = 8 + \dfrac{1}{x^3}$

17.6.5 $2y = x^2 - 2$

17.6.6 $y = 8x^4 + \dfrac{3}{x^2}$

(Remember we may first need to 'reduce the equation,' to yield an equation of the form '$y = \ldots$')

■ **Worked Example 17.7**

The potential of a saturated calomel electrode (SCE) varies with temperature T. We can calculate its potential E_{SCE} with a power series:

$$E_{SCE} = 0.242 - (7 \times 10^{-3})\, T + (4 \times 10^{-7})\, T^2$$

where the bracketed terms are merely constants. What is the gradient of a graph of E_{SCE} (as 'y') against temperature T (as 'x')?

E_{SCE}	=	0.242	−	$(7 \times 10^{-3})\, T$	+	$(4 \times 10^{-7})\, T^2$
		↓		↓		↓
$\dfrac{\mathrm{d}E_{SCE}}{\mathrm{d}T}$	=	0		$-(7 \times 10^{-3})$		$+(4 \times 10^{-7}) \times 2T$

so $\dfrac{\mathrm{d}E_{SCE}}{\mathrm{d}T} = -(7 \times 10^{-3}) + (8 \times 10^{-7})\, T$ ■

Self-test 17.7

Differentiate the following by rule, eqn (17.3):

17.7.1 $y = x^4 + x^2$

17.7.2 $y = x^3 - x - 4$

17.7.3 $y = x^7 - 6x^2 + 2$

17.7.4 $y = \sqrt{x} + 12x - 3$

Summary of key equations in the text

Definitions of differentiation by rule:

If $y = ax^n$ then $\dfrac{dy}{dx} = a \times n \times x^{(n-1)}$ (17.3)

If $y = ax$, then $\dfrac{dy}{dx} = a$ because $x^0 = 1$

If $y = a$, then $\dfrac{dy}{dx} = 0$ (17.4)

Additional problems

17.1 A molecule is moving through the high vacuum of a mass spectrometer at a velocity of v, defined as:

$$v = \frac{l}{t}$$

where l is the distance travelled in a time t. Electromagnets accelerate the particle. Write an expression for the acceleration a, $\dfrac{dv}{dt}$.

17.2 The pressures at the solid–liquid **phase boundary** for propane are given by the empirical expression

$$p = -718 + 2.386 \times T^{1.283}$$

Empirical means 'obtained from experimental observation.'

Write an expression for the gradient of this phase boundary, i.e. of $\dfrac{dp}{dT}$.

17.3 In the electrochemist's technique of cyclic voltammetry, the peak current I is a simple function of the scan rate v, according to the **Randles–Sevčik equation**:

$$I = 0.4463nFAc\left(\frac{nF}{RT}\right)^{\frac{1}{2}} D^{\frac{1}{2}} v^{\frac{1}{2}}$$

where n is the number of electrons transferred, F is the Faraday constant, D the diffusion coefficient, and c the concentration of analyte. Write an expression to describe the gradient of a graph of I (as 'y') against v (as 'x'), i.e. $\dfrac{dI}{dv}$. [Hint: treat the equation as $I = k\, v^{\frac{1}{2}}$, where k is a constant.]

17.4 Molecules sometimes leave solution to adhere to a solid surface. We say they '*adsorb*'. When the bond causing the adsorption is weak, the amount adsorbed M depends on the concentration c according to the **Freundlich adsorption isotherm**:

$$M = kc^{\frac{1}{n}}$$

where M is the fraction (mass of adsorbate ÷ mass of substrate), and k and n are constants obtained experimentally. Write an expression for $\dfrac{dM}{dc}$.

17.5 If two electric dipoles are aligned in a parallel and adjacent arrangement, the potential energy V of their interaction is given by the expression:

$$V = \frac{\mu_1 \mu_2}{4\pi\varepsilon_0 r^3}$$

where r is the interdipole separation, and all the other terms are constants. Write an expression for $\dfrac{dV}{dr}$.

17.6 The speed at which electrons move through a semiconductor relates to their mobility μ. The temperature dependence T of μ is complicated but for many semiconductors, μ shows the following relationship:

$$\mu = kT^{3/2}$$

where k represents a collection of constants. Write an expression for $\dfrac{d\mu}{dT}$.

17.7 In a hydrogenic atom, the single electron circulates around a central nucleus. The energy of this electron can be subdivided into the radial and angular components of the relative motion of an electron about the nucleus, and the Coulombic potential. We can combine the latter two terms to generate an **effective potential energy** V_{eff}:

$$V_{\text{eff}} = -\frac{Ze^2}{4\pi\varepsilon_0 r} + \frac{l(l+1)\hbar^2}{2\mu r^2}$$

where the first term describes the Coloumbic electron–nucleus interaction and the second term describes the energy of the electron due to angular component of motion, and r is the distance of the electron from the nucleus. All other terms are constants. Write an expression for $\dfrac{dV_{\text{eff}}}{dr}$.

17.8 Light is **scattered** as it passes through a solution containing microscopically small particles. For example, milk looks white because it contains a suspension of tiny spheres of insoluble oils and fats. The intensity of the scattered light I relates to the wavelength of the light scattered λ:

$$I = I_0 \frac{\pi\alpha^2}{\varepsilon_r^2 \lambda^4 r^2}\sin^2\phi$$

where all the other terms may be considered to be constant. Write an expression for $\dfrac{dI}{d\lambda}$.

17.9 The ideal gas equation ($pV = nRT$) often fails to describe the behaviour of real gases. One of the better alternatives is the so-called **virial equation**

$$pV_m = RT\left(1 + \frac{B}{V_m} + \frac{C}{V_m^2}\right)$$

where V_m is merely $V \div n$. Multiply out the brackets, then write an expression for $\dfrac{dp}{dV_m}$.

17.10 Worked Example 17.6 described the **Onsager equation**:

$$\Lambda = \Lambda^\circ - b\sqrt{c}$$

This equation is empirical, but it can be derived from first principles using the Debye–Hückel laws. In this way, the constant b has the form

$$b = \frac{qz^3\varepsilon F}{24\pi\varepsilon_0 RT}\left(\frac{2}{\varepsilon RT}\right)^{\frac{1}{2}}$$

where z is the charge on the ions, T is the temperature, and other terms are constants.

Combine the T terms together, and then write an expression describing the temperature dependence of b, i.e. write an expression for $\dfrac{db}{dT}$.

18

Differentiation II

Differentiating other functions

Differentiating more functions

By the end of this section, you will know:

- Functions other than straightforward polynomials can be differentiated. such as exponentials, logarithms, sines and cosines.
- How to 'read' these functions, that is, see which part is the operator, which the argument, and which the factor(s).

There are a many functions we cannot differentiate by the rule defined in the previous chapter. The simplest are exponential functions of the form e^{ax}.

Differentiating exponential functions

When we differentiate an exponential of the type $y = e^{ax}$ (where a is a constant) we employ the relationship in eqn (18.1):

$$\text{If } y = e^{ax} \quad \text{then} \quad \frac{dy}{dx} = a e^{ax} \qquad (18.1)$$

Nomenclature:

- The exponential is an **operator**.
- The operator operates on an **argument**: the argument is the variables, constants and maybe secondary operators on which the exponential operates. In the example above, the argument is 'ax', but it can be more complicated.
- The values of $(a \times x)$ that we are allowed to insert into the equation are the **domain**. For exponentials, the domain is essentially unrestricted.
- In this example, the final differential has been multiplied by a **factor** of 'a'.

We need to note:

- For clarity, we often need to write the argument within brackets.
- The differential of the exponential *always* contains the original exponential, whatever its argument. Neither the operator nor the argument changes: in fact, only the factor that precedes the exponential will change.

■ **Worked Example 18.1**

What is the gradient of the exponential function $y = e^{4x}$ when $x = 2$?

Strategy:

(i) The word 'gradient' in the question alerts us that we want a rate of change.

(ii) We insert respective terms into eqn (18.1).

Solution:

Because $\quad y = e^{4x}$,

Therefore, $\quad \dfrac{dy}{dx} = 4 e^{4x}$

Inserting $x = 2$ into this expression gives a gradient of $4e^8 = 1.192 \times 10^4$ (to 4 s.f.). ■

■ **Worked Example 18.2**

1,2Dichloroethane (**I**) decomposes in the gas phase at 780 °C with a rate constant k of 4.4×10^{-3} s^{-1}.

The amount of (**I**) at any time during the course of the reaction is $[(\mathbf{I})]_t$, and equals

$$[(\mathbf{I})]_t = [(\mathbf{I})]_0\, e^{-kt}$$

where t is the time since the reaction commenced. The term $[(\mathbf{I})]_0$ is the concentration of reactant when the reaction starts, and is therefore a constant. What is the rate of reaction as a function of time?

By asking for the *rate* of reaction, we are really asking for the differential. For simplicity, we rewrite the equation slightly, as $c_t = c_0 e^{-kt}$ to make it look less frightening.

$$\frac{dc_t}{dt} = (-k) \times c_0\, e^{-kt}$$

We then reinsert the concentration c_0 and the numerical value of the rate constant k, saying

$$\frac{dc_t}{dt} = \left(-4.4 \times 10^{-3}\right) \times \left[(\mathbf{I})\right]_0\, e^{(-4.4 \times 10^{-3})t} \quad ■$$

This example is marginally more complicated than that Worked Example 18.1 because:

- The function was itself multiplied by a constant (in this case, a concentration).
- It illustrates the way we often simplify an equation by first substituting.

In general:

If $y = be^{ax}$ then

$$\frac{dy}{dx} = (a \times b)\, e^{ax} \tag{18.2}$$

Self-test 18.1

Differentiate each of the following exponential functions using eqn (18.2):

18.1.1 $y = e^{5x}$ **18.1.2** $y = e^{3.4x}$

18.1.3 $y = 7e^{fx}$ **18.1.4** $y = 9.3e^{4.2x}$

18.1.5 $y = d\, e^{dx}$ **18.1.6** $y = 8.73\, e^{(4 \times 10^{-7})x}$

18.1.7 $y = e^x - e^{-x}$ **18.1.8** $y = 3x^4 - 7x^2 + e^{5x}$

Differentiating logarithmic functions

The differential of a logarithmic function is not intuitively obvious: if $y = \ln ax$, then

$$\frac{dy}{dx} = \frac{1}{x} \tag{18.3}$$

We need to note:

- The constant of 'a' disappears completely, whatever its value.

- Because the argument of a logarithm must be positive, the **domain** of the logarithm must be positive, i.e. $ax > 0$. This criterion does not mean we cannot use negative values of x; but it *does* mean that if x is negative, a must be negative also, to ensure a positive domain.

Rationale
The rationale for the disappearing constant in eqn (18.3) comes from the laws of logarithms, which says we can split the expression '$\ln 7x$' into two: we say

$$\ln 7x = \ln 7 + \ln x$$

Because 7 is merely a number, '$\ln 7$' is also just a number—in other words, '$\ln 7$' is a constant, and the differential of a constant is always zero.

   An alternative explanation is given in terms of the chain rule on p. 294

■ **Worked Example 18.3**

What is the differential of $y = \ln 7x$?

Inserting values into the template in eqn (18.3),

$$\frac{dy}{dx} = \frac{1}{x}$$

Notice that the factor of '7' has disappeared because the function in the question was a natural logarithm. ■

■ **Worked Example 18.4**

The relationship between pH and hydrogen ion concentration is given by the equation:

$$pH = -\log_{10}[H^+]$$

If we made a series of solutions of the strong acid trifluoroethanoic acid (**II**), and measured the gradient of a graph of pH (as 'y') against concentration of the proton [H^+] (as 'x'), what would be its gradient?

II

Strategy:

(i) Before we can differentiate this example, we must convert from a logarithm expressed in base 10 (that is, **log**) to a natural logarithm (that is, **ln**).

(ii) We then differentiate as normal, using an extended version of eqn (18.3):

If $y = b \times \ln ax$ then

The differential in eqn (18.4) is effectively, $b \times (1/x)$

$$\frac{dy}{dx} = \frac{b}{x} \tag{18.4}$$

Again, the value of the argument $ax > 0$.

Solution:

(i) From Chapter 12, we say, $\ln x = \ln 10 \times \log x$, so

$$pH = -\frac{\ln[H^+]}{\ln 10}$$

(ii) Inserting terms into the template differential in eqn (18.4), we obtain:

$$\frac{d\,pH}{d[H^+]} = -\frac{1}{\ln 10 \times [H^+]} = \frac{-0.434}{[H^+]} . \quad \blacksquare$$

■ **Worked Example 18.5**

In 1905, the electrochemist Tafel devised a simple model to describe the rate of an electron transferring between an analyte and an electrode. The equation he derived still bears his name:

$$\eta = c + d \ln I$$

where I is the current density we measure as a consequence of the electrons transferring, and η is the deviation of the potential away from its equilibrium value (the so-called overpotential). c and d are simply constants. What is the rate of change of a graph of η (as 'y') against I (as 'x')?

Strategy:

(i) We will differentiate the Tafel equation one term at a time in just the same way as we differentiated an equation incorporating a number of simple polynomial terms in Chapter 17 (see p. 275).

(ii) The second part of the Tafel equation is a logarithm, so following eqn (18.4), we differentiate to obtain the relation:

Solution:

$$\frac{d\eta}{dI} = 0 + d \times \frac{1}{I}$$

which we would normally write as: $= \dfrac{d\eta}{dI} = \dfrac{d}{I} \quad \blacksquare$

Self-test 18.2

Differentiate each of the following logarithmic functions using eqn (18.3):

18.2.1 $y = \ln x$	**18.2.2** $y = \ln 7x$
18.2.3 $y = 5.4 \ln x$	**18.2.4** $y = 45 - \ln x$
18.2.5 $y = \ln [(4 \times 10^{-6})\, x]$	**18.2.6** $y = ab \ln gx$
18.2.7 $y = 4x - 3 \ln x$	**18.2.8** $y = x^4 - \dfrac{1}{x^3} + \ln 4x$
18.2.9 $y = 1000x + \ln 3x$	**18.2.10** $y = \ln ax - c \ln bx$

We need to note:

- We call the function written to the right of the symbol 'ln' its **argument**. For clarity, we often need to write the argument within brackets.

- The argument of the logarithm remains unaltered following differentiation.

- The argument of the logarithm must be positive, so we restrict the domain to all values of ax that are positive, since logarithmic functions are not defined for negative numbers.

Differentiating trigonometric functions

By the end of this section, you will:

- Recall that angles may be expressed in radians rather than degrees.
- Know that we cannot (readily) obtain a differential coefficient of a sine or cosine unless we work in radians.

The general name **trigonometry** (see Chapter 16) describes the mathematics of sines, cosines, etc. Before we can learn the correct way to differentiate trigonometric functions, we need to consider a technicality. The differentials of a sine or a cosine of an angle *must* be expressed not in straightforward degrees but in **radians**.

By definition, there are 2π radians in a circle: an angle of $360°$ is equivalent to 2π radians, $180°$ is equivalent to π radians, $90°$ equates to $\pi/2$, and so on. Because we *must* use radians when differentiating trigonometric functions, we need to convert to radians from degrees:

$$\text{Angle in radians} = \text{Angle in degrees} \times \frac{\pi}{180} \qquad (18.5)$$

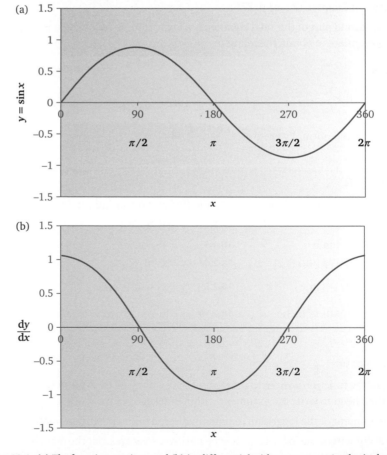

Figure 18.1 (a) The function $y = \sin x$, and (b) its differential with respect to x. Angles in degrees are included for reference only: the value of x must be expressed in radians.

We can now consider the differentiation of trigonometric functions. Figure 18.1(a) shows the shape of the simple function $y = \sin x$, and Fig. 18.1(b) shows the differential of $y = \sin x$ (which is just the rate of change or *slope* of the line shown in trace (a) of Fig. 18.1).

We can deduce by inspection some of the gradients along this sin curve:

- At $x = 0$, $y = \sin x$ looks very close to $y = x$, so the slope of the graph (its rate of change $dy/dx = 1$).

- At $x = \pi/2$, the function $y = \sin x$ is a tangent to the horizontal line $y = 1$, so the slope $dy/dx = 0$.

- At $x = \pi$, the function $y = \sin x$ looks parallel to $y = -x$, so the slope $dy/dx = -1$.

Figure 18.2(a) shows the simple function, $y = \cos x$, and Fig. 18.2(b) shows the differential of $y = \cos x$. We now look critically at the four graphs on pages 284 and 285. The differential of $y = \sin x$ looks identical to $\cos x$ (that is, Figs. 18.1(b) and 18.2(a), respectively, look the same). Again, the differential of $y = \cos x$ is almost identical to $y = \sin x$. The only difference between the two graphs is their sign, because the shape of the differential in Fig. 18.2(b) is the *mirror image* in the horizontal direction of the function $y = \cos x$ in Fig. 18.1(a).

We can express these relationships mathematically by saying the differential of a sine generates a function of a cosine:

We should always check that our calculator assumes the angle x is being expressed in radians. Usually, the display will show a small **R** in one corner.

If $y = \sin ax$ then

$$\frac{dy}{dx} = a \cos ax \qquad (18.6)$$

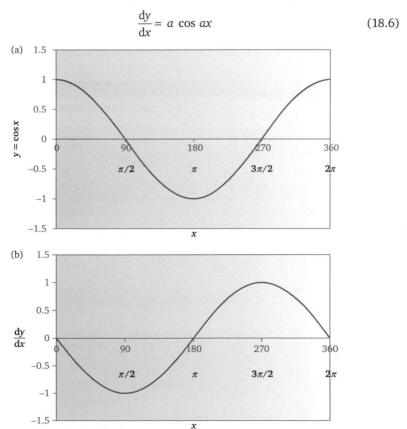

Figure 18.2 (a) The function $y = \cos x$, and (b) its differential with respect to x. Angles in degrees are included for reference only: the value of x must be expressed in **radians**.

If $y = \cos ax$ then

$$\frac{dy}{dx} = -a \sin ax \tag{18.7}$$

We need to note:

- The sine or cosine operators each operate on an argument. For clarity, we often need to write the argument within brackets.
- We always write the argument to the right of the operators 'sin' or 'cos'.
- The domain of the operators is the set of real numbers ($\sin 4\pi = \sin 2\pi = 0$).
- The arguments of the sine or cosine functions *always* remain unaltered. This is a universal rule: we *always* retain the argument, whatever its complexity.
- The differentiation of either a sine or of a cosine will follow the same rules in just about every way. The only real exception is the way the differential of a sine is a cosine, but the differential of a cosine is a sine multiplied by '−1'.

■ Worked Example 18.6

In X-ray crystallography, the Bragg equation allows us a simple means of determining the distance d between successive layers in a crystal (see Fig. 18.3). In its simplest form, it relates the X-rays' wavelength, λ, the number of reflections n and the angle through which the X-rays are scattered, θ, as:

$$\lambda = \frac{2d}{n} \sin \theta$$

What is the rate of change of λ with θ?

Inserting terms into the template in eqn (18.6), we obtain:

$$\frac{d\lambda}{d\theta} = \frac{2d}{n} \cos \theta$$

We must remember that θ is expressed in *radians*. ■

■ Worked Example 18.7

This Worked Example assumes the differentiation was performed with θ in **radians**.

(Following Worked Example 18.6): What is the rate of change of λ with θ when $d = 190$ pm, $n = 1$, and $\theta = \pi/4$ radians?

From Worked Example 18.6, $\dfrac{d\lambda}{d\theta} = \dfrac{2d}{n} \cos \theta$.

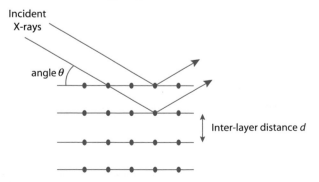

Figure 18.3 A beam of X-rays diffract when striking a crystal with a repeat lattice.

Inserting numbers, $\dfrac{d\lambda}{d\theta} = \left[\dfrac{2\times(190\times10^{-12}m)}{1}\right]\cos\left(\dfrac{\pi}{4}\right)$

$\pi/4$ is 45°, so the value of $\cos\left(\dfrac{\pi}{4}\right)$ is 0.707.

$\dfrac{d\lambda}{d\theta} = 3.8\times10^{-10}$ m $\times 0.707 = 2.69\times10^{-10}$m. ∎

At first sight, this value appears very small. However, since a typical interplane distance is 150–300 nm, the value of d λ/d θ could be as much as 2% of d. For this reason, X-ray crystallographers use constant values of λ with **monochromated** X-rays.

■ Worked Example 18.8

What is the differential of the equation $y = 4\cos(-5x)$?

Inserting terms into the template in eqn (18.7), we obtain:

$\dfrac{dy}{dx} = -4\times(-5)\sin(-5x)$

so $\dfrac{dy}{dx} = +20\sin(-5x)$ ∎

Because we obtained this result via eqn (18.7), the angle x is expressed in radians.

Self-test 18.3

Differentiate each of the following trigonometric functions with respect to x:

18.3.1 $y = \sin 4x$ 18.3.2 $y = 4\sin 3x$

18.3.3 $y = -12\sin(8.1x)$ 18.3.4 $y = \cos(44x)$

18.3.5 $y = 6.3\cos(-7.8x)$ 18.3.6 $y = d\cos(dx)$

18.3.7 $y = \sin x + \cos x$ 18.3.8 $y = \dfrac{\cos x}{2} - \dfrac{\sin 3x}{4}$

Table 18.1 lists many of the simple standard differentials we need to remember.

■ Worked Example 18.9

What is the rate of change of the function $y = \sin 3x$ at an angle of 45°?

Strategy:

(i) Before we start, we must convert the angle into radians, otherwise, we cannot employ eqns (18.6) or (18.7).

(ii) By 'rate of change' we mean the gradient of the graph of $\sin 3x$ (as 'y') against angle (as 'x').

(iii) We differentiate the function, and insert $\pi/4$ for 'x', and calculate its value. The argument is $3x$, which has a value of $3\times\pi/4$.

Solution:

(i) From Fig. 16.9 in Chapter 16, we can quickly see that 45° is equivalent to an angle of $\pi/4$.

(iii) The function is, $y = \sin 3x$

Therefore we obtain from eqn (18.6),

$\dfrac{dy}{dx} = 3\cos 3x$

We next insert values:

$$\frac{dy}{dx} = 3 \times \cos\left(\frac{3\pi}{4}\right)$$

The value of $\cos\left(\frac{3\pi}{4}\right)$ is -0.7076.

Accordingly, $\frac{dy}{dx} = 3 \times (-0.707) = -2.12$

so the gradient of the graph, $y = \sin 3x$, is -2.12 at $\pi/4$ (45°). ∎

Self-test 18.4

Evaluate the differential coefficient for each of the following:

18.4.1 $y = 4 \cos 5x$,   at an angle of $\pi/6$ rad

18.4.2 $y = 2 \sin 7x$,   at an angle of $\pi/8$ rad

Remember, in each case we must express the value of x in radians, not degrees.

Summary of key equations in the text

Table 18.1 Summary of standard differentials.

y	$\dfrac{dy}{dx}$	Equation number
constant	0	
$a x^n$	$a n x^{(n-1)}$ (where $n \neq 0$)	(17.1)
e^x	e^x	(18.1)
e^{ax}	$a\, e^{ax}$	(18.1)
be^{ax}	$(a \times b)\, e^{ax}$	(18.2)
$\ln ax$	$\dfrac{1}{x}$ (where $ax > 0$)	(18.3)
$b \times \ln ax$	$\dfrac{b}{x}$ (where $ax > 0$)	(18.4)
$\sin ax$	$a \cos ax$	(18.6)
$b\sin ax$	$ab \cos ax$	
$\cos ax$	$-a \sin ax$	(18.7)
$b\cos ax$	$-ab \sin ax$	

Additional problems

18.1 The exponential function e^x can be represented by a **power series** in x. One of the most useful concerns exponential functions, which we write as the infinite series:

$$y = e^x = 1 + \frac{x^1}{1!} + \frac{x^2}{2!} + \dots + \frac{x^\infty}{\infty!} = 1 + \sum_{i=1}^{\infty} \frac{x^n}{n!}$$

Some people prefer to write the second term on the right-hand side as just x. We write it here as $x^1/1!$ merely to emphasize the logic of the series. We recognize the function in the denominator of each fraction as a *factorial* from p. 29. Differentiate the expression and thereby prove eqn (18.1).

18.2 Show that the two equations below are the same:

(i) $\dfrac{d\ln p}{dT} = \dfrac{\Delta H_{vap}}{RT^2}$

(ii) $\dfrac{dp}{dT} = \dfrac{p\Delta H_{vap}}{RT^2}$

18.3 The **Sackur–Tetrode equation** relates the molar entropy S_m of a gas to the temperature T

$$S_m = R\ln\left(\frac{e^{\frac{5}{2}}kT}{p^{\ominus}\Lambda^3}\right)$$

where, for a pure gas, all the others terms are constants. Write an expression describing the temperature dependence of S_m, i.e. $\dfrac{dS_m}{dT}$:

18.4 The wavefunction of particle in a 1-dimensional box can be approximated by the following:

$$\psi = \left(\frac{2}{L}\right)\sin\left(\frac{n\pi x}{L}\right)$$

where L is the length of the box's side, and x is distance of the particle from the wall of the box at $x = 0$. (The internal walls of the box correspond to $x = 0$ and $x = L$.)

What is the rate of change of ψ with the position in the box, x?

18.5 During a first-order chemical reaction, the concentration of reactant $[A]_t$ decreases with time t according to the **first-order integrated rate equation**:

$$[A]_t = [A]_0 \exp(-kt)$$

where $[A]_0$ and k are constants. Write an expression for $\dfrac{d[A]_t}{dt}$.

18.6 In quantum mechanics, the wavefunction ψ for an electron in a molecular orbital can often be written in the general form:

$$\psi = A \sin kx + B \cos kx$$

where x is a distance and k is a constant. Write an expression for $\dfrac{d\psi}{dx}$.

18.7 When an ideal gas changes volume V, its entropy increases by an amount ΔS:

$$\Delta S = nR \ln V + c$$

where n, R and c are constant. Write an expression for the rate of change of entropy with volume, i.e. write an expression for $\dfrac{d\Delta S}{dV}$.

18.8 The change in entropy of an ideal gas can be written in the form

$$\Delta S = C_V \ln T + R\ln V$$

Write an expression for $\dfrac{d\Delta S}{dT}$.

[Hint: C_V is the heat capacity *at constant volume* so the second term is a constant.]

18.9 The **van't Hoff isochore** defines the temperature T dependence of equilibrium constants K:

$$K = c \exp\left(-\frac{\Delta H^{\ominus}}{RT}\right)$$

where c and all other terms are constants. What is the rate at which K changes with temperature, (i.e. write an expression for $\frac{dK}{dT}$)?

18.10 Using an orbital constructed using Slater-type functions, the wavefunction ψ of a 1s orbital can be simplified to:

$$\psi = \left(\frac{\zeta^3}{\pi}\right)^{\frac{1}{2}} \exp(-\zeta r)$$

where r is the distance between an electron and the nucleus, and all other terms are constants.

Write an expression for $\frac{d\psi}{dr}$.

19

Differentiation III

*Differentiating functions of functions:
the chain rule*

Functions of functions

By the end of this section, you will know:

- How to recognize a function of a function.
- How to discern which is the first function and which the second.

Sometimes, an equation does not contain a straightforward function, but displays a function of another function, with the first function acting as the **argument** of the second. For example, if we say $y = \sin(x^2)$, then the *first* function is x^2. And, in this example, x^2 also acts as the argument of the *second* function, sine.

Before learning how to differentiate such composites, we must learn how to identify the two functions involved, and which is the first and which the second:

- Generally, if brackets are involved, the first function always lies *inside* the brackets while the second operates on the bracket, treating it as an argument. The symbol identifying the second argument will be written outside the bracket, so a power will be printed to the top right of the bracket, and a sine will have the letters 'sin' written immediately before the bracket.

- If there is no explicit bracket written, we should look for a different way of writing the expression—one that does contain a bracket. For example:
  - $\ln x^2$ is the same as $\ln(x^2)$
  - $e^{x^2} = \exp(x^2)$.

- It is rare that we cannot easily rewrite an expression in this way.

- We need to learn a convention for trigonometric functions. When we square, cube, etc. such functions, we don't write $(\cos x)^2$ but $\cos^2 x$; and we write $(\cos x)^3$ as $\cos^3 x$. The reasons for this notation style lie buried in the history books. So each time we meet either $\sin^n$ (argument) or $\cos^n$ (argument), we rewrite them with brackets as $(\sin(\text{argument}))^n$ and $(\cos(\text{argument}))^n$. In this way, it's obvious which is the first function, and which the second.

■ **Worked Example 19.1**

Identify the first and second functions in the expression, $y = (x^3 + x^2 + 1)^4$.

The function within the bracket, $x^3 + x^2 + 1$, is the first function. The second function uses this function as its argument. The second function is to take the fourth power of $x^3 + x^2 + 1$. ■

Self-test 19.1

In each case, identify the functions, stating which is the first function, and which the second.

19.1.1 $y = (x^2 + 2)^2$	**19.1.2** $y = (\ln x)^2$
19.1.3 $y = \sin(e^{3x})$	**19.1.4** $y = \exp(x^4)$
19.1.5 $y = \ln(x^3)$	**19.1.6** $y = \sin^5 x$
19.1.7 $y = \cos(x^4 - x^2)$	**19.1.8** $y = 3\ln(\sqrt{x})$
19.1.9 $y = \ln(3x^{3/2})$	**19.1.10** $y = \exp\left(\dfrac{3}{x^2} - \dfrac{2}{x^3}\right)$

The chain rule

By the end of this section, you will know:

- A function of a function cannot be differentiated straightforwardly.
- The chain rule allows the differentiation of a function of a function.
- When using the chain rule, we identify each of the functions within the overall equation.

The **chain rule** tells us to:

1. Identify the functions.

2. Differentiate each in turn.

Having identified the two functions, we separate them in order to use the chain rule. We substitute, saying the argument of the second function is also a form of variable—call it 'u'. u here is itself the function in the first expression. We will obtain two equations, $u = f_1(x)$ and $y = f_2(u)$, where we included the subscripts 1 and 2 merely to emphasize that two separate functions are involved.

> The notation of '$f(\ )$' in these two equations simply means 'is a function of …'

Next comes the clever part: we write 'the magic line,' which lets us differentiate each function in turn:

$$\left(\frac{dy}{dx}\right) = \left(\frac{dy}{du}\right) \times \left(\frac{du}{dx}\right)$$

(19.1)

It is usual to omit the brackets here since they serve no useful purpose except to clarify the expression. Having differentiated each in turn, we reassemble the equations.

■ **Worked Example 19.2**

Differentiate $y = (x^3 + x^2 + 1)^4$.

We could differentiate this function by first multiplying out the bracket: $(x^3 + x^2 + 1) \times (x^3 + x^2 + 1) \times (x^3 + x^2 + 1) \times (x^3 + x^2 + 1)$, then differentiating each term by rule. And if we were careful, we would obtain the correct answer. Such an approach would, however, be time consuming and fiddly. We want a simpler method, so we will use the chain rule.

Strategy:

We must start by identifying the two functions, one of which operates on the other, hence the phrase 'function of a function.' Therefore:

(i) We identify the first function, generally looking to see if a bracket is involved.

(ii) We identify the second function, looking to see what is happening to the bracket.

(iii) We separate the two functions, calling the first u.

(iv) We differentiate each in turn.

(v) We insert terms into the magic line in eqn (19.1) and, if necessary, tidy up by rearrangement, and substitute back in for the first function.

Solution:

(i) The first function of x lies within the bracket, so '$x^3 + x^2 + 1$' is the first function.

(ii) This function acts as the argument of the second function. The second function is a fourth power.

(iii) We now start the answer. We merely separate the two functions in order to use the chain rule: we substitute, saying the argument is 'u':

We say: $y = u^4$ where $u = x^3 + x^2 + 1$

(iv) We then differentiate each:

If $y = u^4$ then $\left(\dfrac{dy}{du}\right) = 4u^3$

and $u = x^3 + x^2 + 1$ then $\left(\dfrac{du}{dx}\right) = 3x^2 + 2x$

(v) We use 'the magic line' from eqn (19.1) and combine these two differentials:

$$\left(\frac{dy}{dx}\right) = 4u^3 \times (3x^2 + 2x)$$

This expression is correct but not useful until we back-substitute, replacing the term u with its original argument of $(x^3 + x^2 + 1)$.

> We usually omit the central multiplication sign '×', and write '$4\,(x^3 + x^2 + 1)^3\,(3x^2 + 2x)$'.

Therefore, $\left(\dfrac{dy}{dx}\right) = 4(x^3 + x^2 + 1)^3\,(3x^2 + 2x)$

Simplifying this expression is not straightforward. There is no need to multiply the two brackets. ∎

When differentiating a bracket, this procedure can be summarized as:

$$\text{power} \times \text{differential of the terms in the bracket} \times (\quad)^{\text{power}-1}$$

■ Worked Example 19.3

What is the differential of the equation, $y = \ln(x^3 + 2)$?

We first identify the functions. It is clear after inspecting the equation that we have a logarithm for which the argument is $(x^3 + 2)$. We next note how this argument contains a simple polynomial function.

Usually, we substitute for the argument, so we rewrite the equation, saying:

$y = \ln u$ and $u = x^3 + 2$

We can differentiate each of these expressions in turn without any great difficulty:

If $y = \ln u$ then $\left(\dfrac{dy}{du}\right) = \dfrac{1}{u}$

and $u = x^3 + 2$ then $\left(\dfrac{du}{dx}\right) = 3x^2$

Then, using 'the magic line' from eqn (19.1):

$$\left(\frac{dy}{dx}\right) = \frac{1}{u} \times 3x^2$$

We then back-substitute for u:

$$\left(\frac{dy}{dx}\right) = \frac{1}{x^3 + 2} \times (3x^2)$$

This equation is completely correct, but we would normally rewrite it slightly by placing the term '$3x^2$' on top of the fraction:

$$\left(\frac{dy}{dx}\right) = \frac{3x^2}{x^3 + 2} \quad ∎$$

We need to note:

- When we employ the chain rule on a logarithmic function, the contents of the argument u remain absolutely unchanged in every particular.

- We always find the unchanged argument positioned as the denominator (bottom line) of a reciprocal (i.e. a fraction of the form '1 ÷ argument').

- It is sometimes possible to simplify the resultant differential further by cancelling and/or factorizing, so the expression may change in appearance in subsequent lines of the problem.

■ **Worked Example 19.4**

Differentiate the equation, $y = \sin(3x^4 + 5x)$?

We first identify the functions. After inspecting the equation:

- The first function within the bracket is $(3x^4 + 5x)$.

- This function then acts as the argument of the sine function.

We substitute for the argument of the sine, thereby rewriting the equation and say:

$$y = \sin u \qquad \text{and} \qquad u = 3x^4 + 5x$$

We then differentiate each of expression:

$$\text{If} \quad y = \sin u \qquad \text{then} \qquad \left(\frac{dy}{du}\right) = \cos u$$

$$\text{and} \quad u = 3x^4 + 5x \quad \text{then} \qquad \left(\frac{du}{dx}\right) = 12x^3 + 5$$

Then, using 'the magic line' from eqn (19.1), we combine the two expression to obtain:

$$\left(\frac{dy}{dx}\right) = \cos u \times (12x^3 + 5)$$

This expression is incomplete since it still contains u. We must back-substitute for u:

$$\left(\frac{dy}{dx}\right) = (12x^3 + 5)\cos(3x^4 + 5x) \quad ■$$

Notice how we have swapped the order here: we placed the bracket *before* the cosine. We did this to avoid any ambiguity. If we had written $y = \cos(3x^4 + 5x)(12x^3 + 5)$, some people might have thought the equation meant the cosine of all of '$(3x^4 + 5x) \times (12x^3 + 5)$', and obtain a different, *wrong* result.

We need to note:

- Differentiating a sine or cosine using the chain rule *always* results in the argument remaining absolutely unchanged.

- By convention, if we have a trigonometric function and a polynomial function, we write the trigonometric function last. This convention is designed to avoid ambiguity.

■ **Worked Example 19.5**

What is the differential of the equation, $y = \cos^3 x$?

Following the trigonometric notation we learnt on pp. 252 and 292, this Worked Example is asking us to differentiate, $y = (\cos x)^3$.

We start by identifying the functions:

- The first function (within the bracket) is cos x.
- This function is itself cubed.

In mathematical notation: $y = u^3$ and $u = \cos x$

We can differentiate each of these expressions without difficulty:

If $\quad y = u^3 \quad$ then $\quad \left(\dfrac{dy}{du} \right) = 3u^2$

and $\quad u = \cos x \quad$ then $\quad \left(\dfrac{du}{dx} \right) = -\sin x$

Then, using 'the magic line' from eqn (19.1):

$$\left(\frac{dy}{dx} \right) = 3u^2 \times (-\sin x)$$

We then back-substitute for u:

$$\left(\frac{dy}{dx} \right) = 3(\cos x)^2 \times (-\sin x)$$

It is neater to write the final answer as

$$\left(\frac{dy}{dx} \right) = -3\sin x (\cos x)^2$$

We again rearrange slightly, placing the sin x term before the bracket, as a buttress to avoid ambiguity. We could have rewritten this answer in terms of the old-fashioned notation, saying, '$-3 \sin x \cos^2 x$'. ∎

■ **Worked Example 19.6**

In electrochemistry, the electrode potential of the proton–hydrogen redox couple E_{H^+,H_2} is a function of the concentration of the proton, according to the **Nernst equation**,

$$E_{H^+,H_2} = E^{\ominus}_{H^+,H_2} + \frac{RT}{2F} \ln([H^+]^2)$$

The equation also assumes a standard pressure of hydrogen gas, i.e. $p^{\ominus}$.

where R, T, F are all constants. The standard electrode potential, $E^{\ominus}_{H^+,H_2}$, is also a constant. Determine the gradient of a graph of E_{H^+,H_2} (as 'x') against $[H^+]$ (as 'y').

We will start by simplifying the appearance of the equation, saying:

$$E = E^{\ominus} + k \ln(H^2)$$

where $\quad k = RT \div 2F$.

We can now analyse the equation, and see the functions are a logarithm and a square. Using the chain rule, we will say $E = E^{\ominus} + k \ln u$ and $u = H^2$.

We can differentiate each of these expressions without difficulty:

if $\quad E = E^{\ominus} + k \ln u \quad$ then $\quad \left(\dfrac{dE}{du} \right) = k \times \dfrac{1}{u}$

and $\quad u = H^2 \quad\quad$ then $\quad \left(\dfrac{du}{dH} \right) = 2H$

Then, using 'the magic line' from eqn (19.1):

$$\left(\frac{dE}{dH} \right) = k \times \frac{1}{u} \times 2H$$

After back-substituting for u, we obtain

$$\left(\frac{dE}{dH}\right) = k \times \frac{1}{H^2} \times 2H$$

which simplifies by cancelling the H on top with one of the two H terms within the square on the bottom, to yield:

$$\left(\frac{dE}{dH}\right) = \frac{2k}{H}$$

Finally, we back-substitute for $k = RT \div 2F$ and reintroduce the standard nomenclature (i.e. 'H' is really [H$^+$]). We say:

$$\left(\frac{dE_{H^+,H_2}}{d[H^+]}\right) = \frac{\cancel{2}RT}{\cancel{2}F[H^+]}$$

We see how the factor of '2' on top and bottom will cancel, so

$$\left(\frac{dE_{H^+,H_2}}{d[H^+]}\right) = \frac{RT}{F[H^+]}$$

A graph of electrode potential E_{H^+,H_2} (as 'y') against [H$^+$] (as 'x') therefore has a simple gradient of RT/F [H$^+$]. ∎

This example could have been performed without the chain rule if we had used the laws of logarithms (see Chapter 11) after recognizing how the Nernst equation $E = E^{\ominus} + k \ln(\text{H}^2)$ could have been written more simply as $E = E^{\ominus} + 2k \ln(\text{H})$.

■ **Worked Example 19.7**

At the heart of quantum mechanics lies a paradox: an object such as a photon can exist both as a wave of wavelength λ, and as a particle. One (non-standard) form of the **de Broglie equation** relates respective values of λ and its mass m:

$$\lambda = \sqrt{\frac{h^2}{2\pi m T k_B}}$$

where T is the absolute temperature, h is the Planck constant and k_B is the Boltzmann constant. What is the differential of λ with respect to mass m?

λ is also called the thermal de Broglie wavelength.

The equation looks very intimidating, so we first simplify it by rewriting:

$$\lambda = \left(\frac{k}{m}\right)^{1/2} \quad \text{where} \quad k = \frac{h^2}{2\pi T k_B}$$

We can now analyse the equation, and see the functions are a square root and a reciprocal. We can now use the chain rule: we will say $\lambda = u^{1/2}$ and $u = k\,m^{-1}$.

We can differentiate each of these expressions without difficulty:

If $\quad \lambda = u^{1/2} \quad$ then $\quad \left(\frac{d\lambda}{du}\right) = \frac{1}{2}u^{-1/2}$

and $\quad u = k\,m^{-1} \quad$ then $\quad \left(\frac{du}{dm}\right) = -k\,m^{-2}$

Then, using 'the magic line' from eqn (19.1):

$$\left(\frac{d\lambda}{dm}\right) = \frac{1}{2}u^{-1/2} \times (-k\,m^{-2})$$

After back-substituting for u, we obtain:

$$\left(\frac{d\lambda}{dm}\right) = \frac{1}{2}\left(\frac{k}{m}\right)^{-\frac{1}{2}} \times \left(-k\,m^{-2}\right)$$

We can simplify the expression further: $\left(-k\,m^{-2}\right) = \left(-\frac{k}{m^2}\right)$, and $\left(\frac{k}{m}\right)^{-\frac{1}{2}} = \left(\frac{m}{k}\right)^{\frac{1}{2}}$,

so $\left(\frac{d\lambda}{dm}\right) = \frac{1}{2}\left(\frac{m}{k}\right)^{\frac{1}{2}} \times \left(-\frac{k}{m^2}\right) = -k^{\frac{1}{2}} \times \frac{1}{2\,m^{\frac{3}{2}}}$

Finally, we substitute for k:

so $\left(\frac{d\lambda}{dm}\right) = \left(\frac{h^2}{2\pi T k_B}\right)^{\frac{1}{2}} \frac{1}{2\,m^{\frac{3}{2}}} = \frac{1}{2}\left(\frac{h^2}{2\pi T k_B}\right)^{\frac{1}{2}} \frac{1}{m^{\frac{3}{2}}}$ ∎

Self-test 19.2

Using the chain rule, differentiate each of the equations in Self-test 19.1.

Summary of key equations in the text

The chain rule: $\left(\dfrac{dy}{dx}\right) = \left(\dfrac{dy}{du}\right) \times \left(\dfrac{du}{dx}\right)$ (19.1)

Additional problems

19.1 Electrochemists often employ the **ac impedance** technique, applying a time-dependent voltage V across a sample and measuring the ensuing current. The voltage magnitude alters periodically in a sinusoidal way, so

$V = \sin \omega t$

where t is time and ω is the frequency of the ac signal. Use the chain rule to obtain an expression for $\dfrac{dV}{dt}$.

19.2 Light is *scattered* as it passes through a solution containing microscopically small particles. Milk is a good example, for it contains minute spheres of insoluble oils and fats. The intensity of the scattered light I relates to the angle θ between the plane of polarisation and the incident beam:

$I = I_o \dfrac{\pi \alpha^2}{\varepsilon_r^2 \lambda^4 r^2} \sin^2 \theta$

Write an expression for $\dfrac{dI}{d\theta}$, assuming all the other terms remain constant.

19.3 The acronym **laser** stands for 'light amplification by stimulated emission of radiation.' The emitted light should be monochromatic, i.e. has a single wavelength λ. Radiation of a particular frequency is emitted in consequence of the dipoles in the host solid oscillating at a constant frequency of ω. But the emission behaviour can be complicated by so-called **non-linear effects**, in which case the electric field $\mathcal{E}$ of the incident light follows the expression:

$$\mathcal{E} = \sqrt{\tfrac{1}{2}\,\mathcal{E}_0^2\,(1 + \cos 2\omega t)}$$

where t is the time during an oscillation cycle. All other terms are constant. Write an expression for $\dfrac{d\mathcal{E}}{dt}$.

19.4 **Magic-angle NMR** was described in Worked Example 16.5, in which the tube containing a sample is spun fast while tilted at an angle of θ to the spectrometer's magnetic field (rather than perpendicular to it). The extent of the distortion in the NMR spectrum d relates to θ according to the equation:

$$d = 3\cos^2\theta - 1$$

Write an expression for $\dfrac{dd}{d\theta}$.

19.5 The **Bragg equation** relates the angle of diffraction θ with the interplane distance d in a regular crystal:

$$n\lambda = 2d\sin\theta$$

where n is an integer and λ the wavelength of the diffracted light. The usual practice is to vary θ and look at d, so assume that λ is constant. Rearrange the expression to make d the subject, then differentiate to obtain $\dfrac{dd}{d\theta}$.

19.6 The **van't Hoff isochore** defines the temperature T dependence of equilibrium constants K:

$$K = c\exp\!\left(-\frac{\Delta H^{\ominus}}{RT}\right)$$

where c and all other terms are constants. Rather than using eqn (18.1) in Chapter 18, use the chain rule to obtain an expression for $\dfrac{dK}{dT}$.

19.7 Using an orbital constructed using Gaussian functions, the wavefunction ψ of a 1s orbital can be simplified to:

$$\psi = \left(\frac{2\alpha}{\pi}\right)^{\!3/4}\exp(-\alpha r^2)$$

where r is the distance between an electron and the nucleus, and other terms are constants. Write an expression for $\dfrac{d\psi}{dr}$.

19.8 The general form of a wavefunction ψ for a free particle moving along the x-axis is:

$$\psi = K\cos\!\left(\frac{x\sqrt{2mE}}{\hbar}\right)$$

where E is the energy, and all other terms are constant. Write an expression for $\dfrac{d\psi}{dE}$.

19.9 The potential U between two atomic nuclei is given by a form of the **Morse curve**:

$$U = D_e\,[1 - \exp(-\beta r)]^2$$

where D_e is the bond dissociation energy, r is the deviation from the equilibrium bond separation, and β is a constant. Write an expression for $\dfrac{dU}{dr}$.

19.10 When X-rays strike a regular crystal, they diffract by an angle θ. If the X-rays are of wavelength λ, and the interplane distance is a, then

$$\lambda = \sqrt{\frac{4a^2}{n^2\left(h^2 + k^2 + l^2\right)}\sin^2\theta}$$

where h, k and l are the so-called Miller indices are constants, and n is also a constant. Write an expression for $\dfrac{d\lambda}{d\theta}$.

20

Differentiation IV

The product rule and the quotient rule

Reintroducing products and quotients

By the end of this section, you will:

- Know the meanings of the mathematical terms 'product' and 'quotient'.
- Learn how to distinguish between simple functions, and products or quotients.

We have met the mathematical term 'product' in several chapters. For example, we met the Π product operator in Chapter 2. The word **product** means simply 'multiplied together.' For example, the product of 5 and 2 is 10, and the product of a and b is ab.

The word **quotient** also has a precise mathematical usage: it means a fraction. The quotient of 15 and 5 is '$15 \div 5$' $= 3$; the quotient of x and x^2 is $x \div x^2 = 1/x$. Incidentally, this last example also illustrates the way we can often simplify a quotient by cancelling.

In this chapter, we explore the ways of differentiating products and quotients, that is, differentiating two functions that have either been multiplied together, or one has been divided by the other. We will use, respectively, the 'product rule' and the 'quotient rule'.

It is easy to tell when we have a quotient, because we see a fraction with two different functions, one on its top and another on its bottom. But it is sometimes not so easy to spot a product. The problem lies in the way we generally write maths with the multiplication signs omitted. Even if we mentally superimpose a multiplication sign, we often need to look critically at the expression and ask ourselves, 'which functions are involved.' It is easy with a little practice, but can take time.

> We probably first met the word quotient in the initials **IQ**, which stands for **intelligence quotient**. The basis of an IQ test is therefore a mathematical fraction.

Rules to help distinguish between simple functions and products:

1. If a variable only occurs once in an expression, then we have a single function. For example, in this context $\sin x^2$ and $\ln x^3$ are both single function.

2. If we see a variable occurring more than once, but each time it appears within the argument of the same function, then again we have a single function. For example, $\exp(x^2 + x^3)$ is a *single* function, albeit with a complicated argument.

3. If there are two functions and both are simple polynomials, we should combine them using the laws of powers (see Chapter 11) to form a single function. For example, if $y = x^2 \times 4x^3$, then we rewrite saying, $y = 4 \times x^{(2+3)} = 4x^5$.

4. If one term is a polynomial, and the other is enclosed within a bracket, and if we can easily multiply the terms, we should do so, and then differentiate term by term. For example, if the function was $y = 4x^3(3x - 13)$, we would first multiply out the bracket to obtain $y = 12x^4 - 52x^3$, which we can easily differentiate term by term.

5. If one term is a polynomial and the other is best treated within the BODMAS scheme as a function OF (for example sine, cosine, exponential, or logarithm), then we have a genuine product.

6. If we see more than one of the functions above, we will certainly have a product.

■ Worked Example 20.1

How should we split up the product, $y = x^5 \times \sin 4x$?

The question tells us the overall expression is a product, so we are looking for two separate functions. We have written the sine function in the usual way, with its argument to the right of the operator, so we could have written, $y = x^5(\sin 4x)$.

By enclosing the entire sine term within brackets, we see how the sine is one function, which means the polynomial x^5 is the other. In fact, bracketing a function in this way is a good way of identifying the two functions.

Note how we wrote the sine function *after* the polynomial according to standard usage. ∎

Self-test 20.1

Identify whether the following are single functions, products or quotients.

20.1.1 $y = x^4 \ln x$

20.1.2 $y = \dfrac{x^3}{x^5}$

20.1.3 $y = \dfrac{\sin 4x}{\ln x}$

20.1.4 $y = x^4 \, 4x^5$

20.1.5 $y = \sin x \cos x$

20.1.6 $y = x^3 \sin 3x^2$

20.1.7 $y = \ln x \cos 4x$

20.1.8 $y = \dfrac{4x^5}{\cos 3x}$

The product rule

By the end of this section, you will know:

- The product rule.
- How to differentiate a product with the product rule.
- That the final answer can often be simplified further, particularly when one function involves an exponential.

When we differentiate a product, it is usual to say the two component functions are u and v, i.e. the product is '$u \times v$'. We differentiate the product $(u\,v)$ via the template differential in eqn (20.1):

$$\frac{d(u \times v)}{dx} = u \times \left(\frac{dv}{dx}\right) + v \times \left(\frac{du}{dx}\right) \tag{20.1}$$

Notice the pattern on the right-hand side here:

'1st function × the differential of the 2nd + 2nd function × the differential of the 1st'.

We need to note:

- We often see the product rule written without either multiplication signs or brackets.
- This expression is strongly symmetrical, which may make it easier to remember.

■ Worked Example 20.2

Differentiate the simple product, $y = x^2 \ln x$.

Strategy:

(i) Because (usually) no multiplication sign separates the product's two functions, we must first identify the two component parts. We usually call them u and v.

(ii) We differentiate u and v separately, away from the overall equation. We shall call these differentials $\dfrac{du}{dx}$ and $\dfrac{dv}{dx}$, respectively.

(iii) We enter the four terms, u, v, $\dfrac{du}{dx}$ and $\dfrac{dv}{dx}$, into the product rule in eqn (20.1).

Solution:

(i) In this example, we have a polynomial function together with a second function. In such cases, we follow the convention that says we should write the operator of the second function *after* the polynomial. This practice decreases the scope for misunderstanding. For example, if we had written the function as '$\ln x \, x^2$', it might be mistaken for the different function $\ln (x \times x^2)$ which not only equals $\ln (x^3)$, but is not a product. We avoid this source of confusion by writing the polynomial *before* the operator of the second function. So, $u = x^2$ and $v = \ln x$.

(ii) We differentiate each of the two component functions. To aid comprehension, we write the two derivatives *within* square brackets, and the original functions u and v *without* brackets:

$$u = x^2 \quad \text{so} \quad \frac{du}{dx} = [2x]$$

$$v = \ln x \quad \text{so} \quad \frac{dv}{dx} = \left[\frac{1}{x}\right]$$

(iii) We insert terms into the product rule, eqn (20.1) generates:

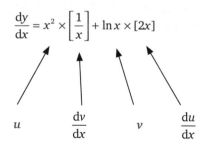

$$\frac{dy}{dx} = x^2 \times \left[\frac{1}{x}\right] + \ln x \times [2x]$$

$$u \qquad \frac{dv}{dx} \qquad v \qquad \frac{du}{dx}$$

If both functions are polynomials, we can usually simplify the answer further. The first half of the answer, $x^2 \times \frac{1}{x}$, can be simplified to yield just 'x'.

It is wise to subsequently rewrite the second half of an answer like this in a different order, as '$2x \ln x$', to avoid ambiguity (because '$\ln x \, 2x$' could be mistaken for '$\ln (x \times 2x)$', which simplifies to the *different* (wrong) result: $\ln (2x^2)$.

So, having used the product rule the result is $dy/dx = x + 2x \ln x$, which simplifies readily to $x(1 + 2 \ln x)$. ∎

The question sometimes arises, 'What if I mistake a simple function for a product, and use the product rule?'

■ **Worked Example 20.3**

Differentiate $y = x^5 \times 4x^2$ using the product rule.

The example here should properly be differentiated as a single function, because $x^5 \times 4x^2$ is the same as $4x^7$; and the differential of $y = 4x^7$ is $28x^6$.

Nevertheless, we shall pretend this example is a product, with $u = x^5$ and $v = 4x^2$.

$$u = x^5 \quad \text{so} \quad \frac{du}{dx} = 5x^4$$

$$v = 4x^2 \quad \text{so} \quad \frac{dv}{dx} = 8x$$

Inserting terms into the product rule expression in eqn (20.1) yields:

$$\frac{dy}{dx} = x^5 \times [8x] + 4x^2 \times [5x^4]$$

so

$$\frac{dy}{dx} = 8x^6 + 20x^6$$

i.e.

$$\frac{dy}{dx} = 28x^6$$

In summary, if we mistook this simple pair of functions for a product, and differentiated using the product rule, we obtain the same differential of $28x^6$ that we would have obtained by differentiating normally. ∎

■ **Worked Example 20.4**

The probability of finding an electron in a 1s orbital of a hydrogen atom is ψ^2_{1s}, and is obtained via the expression:

$$\psi^2_{1s} = \frac{4}{a_0^3} r^2 \exp\left(-\frac{2r}{a_0}\right)$$

where a_0 is the atomic diameter and is a constant; and r, being the distance of an electron from the nucleus, is a variable. Differentiate this expression with respect to r.

Before we differentiate this product, we must identify which are the two functions. Perhaps it would help if we rewrite the expression as:

$$\psi^2_{1s} = kr^2 \exp(k'r)$$

where $k = (4 \div a_0^3)$, and $k' = (-2 \div a_0)$. Clearly, then, kr^2 is the first term (which we will call u) and $\exp(k'r)$ is the second (which we will call v).

To differentiate:

$$u = kr^2 \quad \text{so} \quad \frac{du}{dr} = 2kr$$

$$v = \exp(k'r) \quad \text{so} \quad \frac{dv}{dr} = k' \exp(k'r)$$

Inserting terms into eqn (20.1) yields:

$$\frac{d\psi^2_{1s}}{dr} = kr^2 \times [k'\exp(k'r)] + \exp(k'x) \times [2\,kr]$$

$$\begin{array}{cccc} \uparrow & \uparrow & \uparrow & \uparrow \\ u & \dfrac{dv}{dr} & v & \dfrac{du}{dr} \end{array}$$

To aid comprehension, as before we write the two derivatives within square brackets. This answer is correct, but it would not be wise to leave it in this form. We can factorize, which will result in a considerably shorter and simpler-looking expression.

When differentiating any exponential expression of the form 'exp (cr)' (where c is a constant), the exponential will itself remain intact and unchanged, although there is usually an additional term obtained in consequence of the chain rule. Accordingly, when we differentiate a mathematical product in which one term is an exponential function, this same exponential will appear in the two halves of the answer, e.g. both within the v term and within dv/dr. In this example, the term 'exp $(k'\,r)$' does indeed appear twice.

We can therefore factorize the expression:

$$\frac{d\psi^2_{1s}}{dr} = \{kr^2 \times k' + 2kr\}\,\exp(k'r)$$

where the curly brackets { } indicate that we have factorized. We may even wish to simplify further by factorizing out the two k terms within the square bracket. We can write:

$$\frac{d\psi^2_{1s}}{dr} = k\{k'\,r^2 + 2r\}\,\exp(k'r)$$

Finally, we back-substitute with the values of k and k':

$$\frac{d\psi^2_{1s}}{dr} = \left(\frac{4}{a_0^3}\right)\left\{\left(-\frac{2}{a_0}\right)r^2 + 2r\right\}\exp\left(-\frac{2r}{a_0}\right)$$

While each of these answers is correct, the last version is simpler on the eye and therefore easier to 'read.' It would also make it easier to insert numbers when determining a value for the differential coefficient. ■

Sometimes we choose to leave one of the differentials as a differential.

> **■ Worked Example 20.5**
>
> Use the product rule to differentiate the product, G/T, where G is the Gibbs function, and T is temperature.

Let $u = 1/T = T^{-1}$, and let $v = G$. Therefore, $du/dT = -1/T^2$ and $dv/dT = (dG/dT)$. We will not alter this differential term further.

Inserting terms into eqn (20.1) yields:

$$\frac{d(G/T)}{dT} = \left(\frac{1}{T}\left(\frac{dG}{dT}\right) - \frac{1}{T^2}G\right)$$

so $$\frac{d(G/T)}{dT} = \left(\frac{1}{T}\left(\frac{dG}{dT}\right) - \frac{G}{T^2}\right)$$

The right-hand side forms the basis of the Gibbs–Helmholtz equation. ■

> **Self-test 20.2**
>
> Differentiate the following products. The answers on p. 492 are given in two forms: first, in the form generated immediately after using eqn (20.1), and secondly following factorizing.
>
> 20.2.1 $y = 2x^4 \sin 3x$ 20.2.2 $y = \ln(2x)\exp(2x)$
>
> 20.2.3 $y = x^5 \ln x$ 20.2.4 $y = 2x^{-4} \cos 4x$
>
> 20.2.5 $y = \sin^2 x \cos^2 x$ 20.2.6 $y = \ln x^2 \sin^3 2x$

If one term within the product is an exponential function, we will be able to factorize the immediate answer obtained with the product-rule expression in eqn (20.1).

This example illustrates the way that we can use the product rule with simple quotients.

The quotient rule

By the end of this section, you will know:

- The quotient rule.
- How to differentiate a quotient with the quotient rule.
- That the final answer can usually be simplified further.

A **quotient** always takes the form of a fraction, with one function divided by another. In a similar way to the 'splitting up' of a product, we will call the function on the top 'u' and the bottom will be called 'v'. We differentiate a quotient using the following template formula:

$$\frac{dy}{dx} = \frac{v\left(\dfrac{du}{dx}\right) - u\left(\dfrac{dv}{dx}\right)}{v^2} \tag{20.2}$$

Notice the pattern on the right-hand side here:

$$\frac{\text{'bottom'} \times \text{differential of top} - \text{'top'} \times \text{differential of bottom}}{\text{'bottom' squared}}$$

■ **Worked Example 20.6**

Differentiate the quotient, $y = \dfrac{x^3}{\exp(2x)}$.

Strategy:

(i) Identify the component functions u and v.

(ii) Differentiate each function in turn.

(iii) Insert terms into the quotient rule expression, eqn (20.2).

(iv) We simplify the expression by cancelling and factorizing.

Solution:

(i) The component functions are: $u = x^3$ and $v = \exp(2x)$.

(ii) $u = x^3$ $\dfrac{du}{dx} = 3x^2$

     $v = \exp(2x)$ $\dfrac{dv}{dx} = 2\exp(2x)$

(iii) Insert terms into the template expression in eqn (20.2) yields,

$$\frac{dy}{dx} = \frac{\exp(2x) \times \left[3x^2\right] - x^3 \times \left[2\exp(2x)\right]}{\left(\exp(2x)\right)^2}$$

Again, we write the two derivatives within square brackets.

(iv) While this expression is wholly correct, it is long and looks daunting. We can generally simplify such expressions by factorizing. In this example, we first take out a factor of $\exp(2x)$:

$$\frac{dy}{dx} = \frac{\exp(2x)\left[3x^2 - 2x^3\right]}{\left(\exp(2x)\right)^2}$$

We then cancel the '$\exp(2x)$' on the top with one of the two '$\exp(2x)$' terms represented by the square on the bottom:

$$\frac{dy}{dx} = \frac{3x^2 - 2x^3}{\exp(2x)}$$

The polynomials on the top row can also be factorized, to read '$x^2 (3 - 2x)$', although this last factorization is not essential. ∎

The question sometimes arises, 'What if I mistake a simple function for a quotient, and use the quotient rule?'

■ Worked Example 20.7

Differentiate the function $y = \dfrac{x^3}{4x^2}$ via the quotient rule.

While this pair of functions are certainly written in the form of a quotient, it is easy to cancel one with the other using the laws of powers (see Chapter 11), to yield $y = \frac{1}{4} x$. The differential of this expression is merely $\frac{1}{4}$. We will, however, pretend it *is* a quotient and differentiate using the quotient rule. We start by saying $u = x^3$ and $v = 4x^2$.

$$u = x^3 \quad \frac{du}{dx} = 3x^2$$

$$v = 4x^2 \quad \frac{dv}{dx} = 8x$$

We then insert respective terms into the quotient rule, eqn (20.2):

$$\frac{dy}{dx} = \frac{4x^2 \times \left[3x^2\right] - x^3 \times \left[8x\right]}{\left(4x^2\right)^2}$$

Again, the derivatives here are written within square brackets as an aid to comprehension. We multiply out the terms on the top row, and square the term on the bottom:

$$\frac{dy}{dx} = \frac{12x^4 - 8x^4}{16x^4}$$

The terms on the top row can be collected together:

$$\frac{dy}{dx} = \frac{4x^4}{16x^4}$$

The x^4 terms on top and bottom cancel, leaving '$4 \div 16$', which has a value of $\frac{1}{4}$.

In summary, if we mistook this simple pair of functions for a quotient, and differentiated using the quotient rule, we obtain the same differential of $\frac{1}{4}$ that we would have obtained by differentiating normally. ∎

■ Worked Example 20.8

During the reaction $N_2O_4 \rightleftharpoons 2NO_2$, the partial pressure of N_2O_4 is given by the expression

$$p_{(N_2O_4)} = \frac{1 - \zeta}{1 + \zeta}$$

where ζ is the extent of reaction. Write an expression for $\dfrac{dp}{d\zeta}$.

The function, p, is a quotient:

$$u = 1 - \zeta \qquad \frac{dp}{d\zeta} = -1$$

$$v = 1 + \zeta \qquad \frac{dp}{d\zeta} = 1$$

Inserting terms: $\qquad \dfrac{dp}{d\zeta} = \dfrac{(1+\zeta)[-1] - (1-\zeta)[1]}{(1+\zeta)^2}$

Multiplying out the brackets: $\qquad \dfrac{dp}{dz} = \dfrac{-1 - \zeta - 1 + \zeta}{(1+\zeta)^2}$

So $\qquad \dfrac{dp}{d\zeta} = \dfrac{-2}{(1+\zeta)^2}$ ∎

Self-test 20.3

Differentiate the following products using eqn (20.2). (The answers on p. 492 are given in two forms: first, in the form generated immediately after using eqn (20.2), and secondly following factorizing.)

20.3.1 $\quad y = \dfrac{x^2}{\sin x}$

20.3.2 $\quad y = \dfrac{\ln x}{x^4}$

20.3.3 $\quad y = \dfrac{\exp 2x}{3x^3}$

Summary of key equations used in the text

The product rule: $\qquad \dfrac{d(u \times v)}{dx} = u \times \left(\dfrac{dv}{dx}\right) + v \times \left(\dfrac{du}{dx}\right)$ $\qquad$ (20.1)

The quotient rule: $\qquad \dfrac{dy}{dx} = \dfrac{v\left(\dfrac{du}{dx}\right) - u\left(\dfrac{dv}{dx}\right)}{v^2}$ $\qquad$ (20.2)

Additional problems

Questions 20.1–20.3: Which of the following are products and which are quotients? Explain the answer.

20.1 The Planck function, $\dfrac{G}{T}$

20.2 During a first-order reaction, the rate of reaction follows an equation of the type:

$\qquad$ rate $= k\,[A]$

where k is the rate constant and t is the time.

20.3 The conductivity λ of an ion through a solution is a function of the mobility μ and the ion charge, z:

$$\lambda = zF\mu$$

20.4 All ions in solution are **solvated**, that is, they are covalently bound to molecules of solvent. And ions will also associate with each other. We say the ion has an 'atmosphere.' This atmosphere causes the potential of the ion to decrease. The modified potential is ϕ_{atm}:

$$\phi_{atm} = Z_i\left(\frac{\exp(-r/r_D)}{r} - \frac{1}{r}\right)$$

where r is the distance between a test charge and the centre of the ion. Write an expression for $\frac{d\phi_{atm}}{dr}$.

20.5 Consider the following two-part expression:

$$y = \frac{\sin 2x}{x^3}$$

(i) Differentiate this expression using the quotient rule with, $u = \sin 2x$ and $v = x^3$.

(ii) Differentiate this same expression but using expression using the product rule, with $u = \sin 2x$ and $v = x^{-3}$.

(iii) Demonstrate how the answers to parts (i) and (ii) are the same.

20.6 Using the first law of powers on p. 167, the exponential product $y = \exp(ax)\exp(bx)$ can be rewritten as $y = \exp((a+b)x)$. Differentiate both expressions and demonstrate they are the same.

20.7 The equilibrium constant K for a reaction is given by the expression

$$K = \frac{4\zeta_{(eq)}^2}{1 - \zeta_{(eq)}^2}$$

where ζ is the extent of reaction. Write an expression for $\frac{dp}{d\zeta}$.

20.8 A **polymer** comprises a chain of smaller 'monomer' units. Consider a single-strand polymer chain of N monomer units. The length of a single monomer is l. The structure can coil in three-dimensional space, so it will not have a length of $N \times l$. In practice, the probability P that the distance between the ends is nl is given by the expression:

$$P = \left(\frac{2}{\pi N}\right)^{1/2} \exp\left(-\frac{n^2}{2N}\right)$$

Write an expression for $\frac{dP}{dN}$.

20.9 Remembering that $\tan\theta$ is the quotient $\sin\theta \div \cos\theta$, obtain an expression for $\frac{d(\tan\theta)}{d\theta}$.

20.10 The Arrhenius equation represents an empirical way of relating rate constant k and temperature T. We obtain the **Eyring equation**

$$k = \frac{k_B T}{h} \exp\left(\frac{\Delta S^{\#}}{R}\right) \exp\left(-\frac{\Delta H^{\#}}{RT}\right)$$

theoretically. Only T on the right-hand side is a variable. $\Delta H^{\#}$ is the enthalpy of activation and $\Delta S^{\#}$ is the entropy of activation, and are both constants. Other terms are also constants. Write an expression for $\frac{dk}{dT}$.

21

Differentiation V

Maxima and minima on graphs: second differentials

Introducing turning points: maxima, minima and inflections

By the end of this section, you will know that:

- All graphs of polynomial functions $y = x^n$ have maxima and/or minima.
- The gradient at a maximum or a minimum has a value of zero.
- These maxima and minima represent the turning points of the graph: before a maximum, the gradient is positive and afterwards the gradient is negative; before a minimum, the gradient is negative and afterwards the gradient is positive.
- If a graph has an inflection (with horizontal tangent) the gradient at the turning point does not change sign, but merely decreases to zero then increases again.

We saw in Chapter 9 how a straight line follows an equation of the type $y = mx + c$, where m is the gradient and c the intercept on the y-axis. The value of m is constant because the line is linear. But consider the example function, $y = x^2$, a graph of which is shown in Fig. 21.1. The most obvious feature of the graph is its curvature. It is not linear, so the gradient continually changes. In fact, the gradient is not constant for any set of points.

Simple differentiation of this function tells us the gradient. The differential $dy/dx = 2x$, so the gradient at $x = 2$ is 4; the gradient at $x = -6$ is -12. And we notice an unusual feature: the gradient of $y = x^2$ has a value of 0 when the x is 0, i.e. when the line passes through the origin.

After a moment's thought, we may see a trend emerging. The graph has a bell shape. In technical terms, its shape is a **parabola**. The two arms are fairly steep, and become progressively steeper as x becomes either more positive or more negative. Between these two extremes, the curve shows a **minimum**. Before the minimum, the gradient is negative: the curve travels 'downhill' as we move from left to right. Conversely, the gradient is positive after the minimum, and starts its journey 'uphill.' We say the minimum represents a **turning point**, because the sign of the gradient turns from negative to positive.

By looking at this very simple example, we see how the gradient changes sign either side of the minimum. Logic dictates that if the sign changes smoothly as we move from

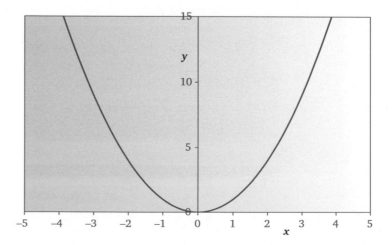

Figure 21.1 The function $y = x^2$ has a **minimum** at (0,0).

left to right (it was negative and becomes positive), then the gradient must transiently have a value of 0, i.e. when the graph is horizontal for a moment. As we saw above, this horizontal portion has a gradient of 0.

Now look at Fig. 21.2, which shows a graph of the related function $y = -x^2$. This graph also has a turning point, which this time is a **maximum**. And being a maximum rather than a minimum, the way the gradient changes its sign differs from that in Fig. 21.1. The gradient is positive before the turning point and negative afterwards i.e. follows the opposite trend to that seen before. Again, the gradient at the turning point is 0.

Figure 21.3 shows another graph, and depicts the function $y = x^3$. This graph also has a turning point, which again occurs at $x = 0$. Even a quick look at the gradients, though, shows how this third graph differs from those in Figs. 21.1 and 21.2. It shows neither a maximum nor a minimum: the gradient is positive on both the left-hand side of the graph and its right-hand side. At no time is the gradient negative. We say the graph shows a **point of inflection**; as described on p. 118. Points of inflection need not have a gradient of zero (but the second differential must equal zero).

Table 21.1 summarizes these changes in gradient.

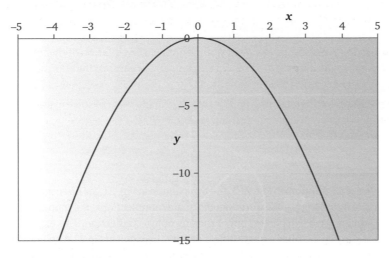

Figure 21.2 The function $y = -x^2$ has a **maximum** at $(0, 0)$.

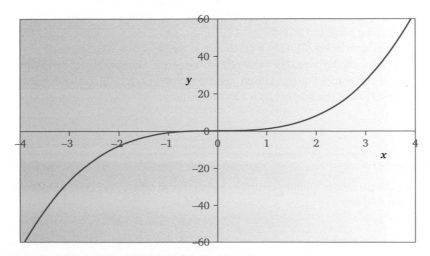

Figure 21.3 The function $y = x^3$ has an **inflection** at $(0, 0)$.

Table 21.1 Analysis of the gradients either side of a turning point.

Type of turning point	Gradient before turning point	Gradient after turning point	Examples*		
Maximum	positive	negative	$y=-x^2$	$y=-x^4$	$y=-x^6$
Minimum	negative	positive	$y=x^2$	$y=x^4$	$y=-x^6$
Point of inflection	(i) positive	positive	$y=x^3$	$y=x^5$	$y=x^7$
	(ii) negative	negative	$y=-x^3$	$y=-x^5$	$y=-x^7$

*Notice how graphs of even-numbered powers display a maximum or minimum, and graphs of odd-numbered powers yield an inflection.

Determining the coordinates of a turning point

By the end of this section, you will know that:

- To determine a turning point, we first differentiate the equation of the function and equate its differential to zero. Subsequent algebraic manipulation allows for determination of x at the turning point.

- When a graph has two turning points, we will usually have to solve a quadratic equation in x to determine the turning points.

The presence of a minimum in Fig. 21.1 and a maximum in Fig. 21.2 is no accident: all graphs of polynomial functions possess turning points and/or inflections. Polynomial expressions involving a single term will possess only 1 turning point or inflection. This explains why the graph of $y=x^3$ in Fig. 21.3 has a single inflection. But then look at Fig 21.4, which depicts the more complicated function $y=x^3+3x^2-9x+10$: this graph has *two* turning points. In fact, the general rule says graphs of polynomial functions of the type, $y=x^n+x^{(n-1)}$ ... will have $(n-1)$ turning points. So, for example, the quadratic equation, $y=ax^2+bx+c$ will always have one turning point, but no more.

While the graph in Fig. 21.4 has a maximum at the point $(-3, 37)$, the value of y is not the largest value on the page. The value of y increases considerably higher at the far right-hand side of the graph. For this reason, we say the point $(-3, 37)$ is a **local** maximum. For the same reason, the point $(1,5)$ is a **local** minimum.

Self-test 21.1

What is the maximum number of turning points we can expect in each of the following graphs?

21.1.1	$y=x^4$	21.1.2	$y=x^2+x+2$
21.1.3	$y=x^4+3x^3+12$	21.1.4	$y=x^5+x^4-3x^3+6x^2-8$

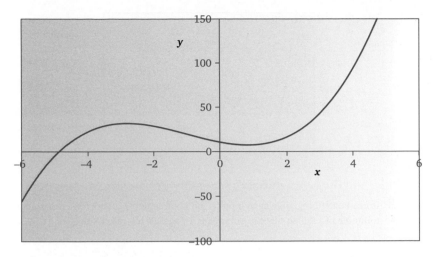

Figure 21.4 The function $y = x^3 + 3x^2 - 9x + 10$ has two turning points, both a local maximum at the point $(-3, 37)$ and a local minimum at $(1, 5)$.

■ **Worked Example 21.1**

This graph of the curve $y = x^2 + x - 4$ has a single turning point. What are its coordinates?

Strategy:

(i)   At the turning point, the gradient has a value of zero.

(ii)   We differentiate the function, and equate the differential coefficient to zero.

(iii)   We rearrange this equation to make x the subject. We say this value of x is the x-coordinate of the turning point.

(iv)   Knowing x at the turning point, we back-substitute into the original equation to determine the value of y at the turning point.

Solution:

(i)   The differential of the function $y = x^2 + x - 4$ is:

$$\frac{dy}{dx} = 2x + 1$$

(ii)   We note how the gradient at a turning point is zero, and say $\frac{dy}{dx} = 0$:

$$2x + 1 = 0$$

(iii)   If $2x + 1 = 0$, then $2x = -1$, so $x = -\frac{1}{2}$. We calculate the value of x at the turning point as $-\frac{1}{2}$.

(iv)   Inserting this value of x into the original function allows us to compute the value of y at the turning point:

$$y = x^2 + x - 4,$$

so $$y = \left(-\frac{1}{2}\right)^2 - \frac{1}{2} - 4$$

and $y = 0.25 - 0.5 - 4 = -4.25$

The turning point is at $(0.5, -4.25)$. ■

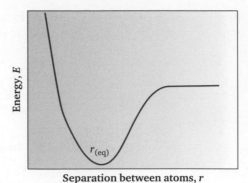

A more realistic equation to describe a Morse curve takes the form:

$$E = \varepsilon\left[\left\{1 - \exp\left(\beta\left(r - r_{(eq)}\right)\right)^2 - 1\right\}\right]$$

where β and ε are constants, and r_{eq} is the equilibrium bond length.

Figure 21.5 Atoms bonded together have a preferred bond length r_{eq}, which corresponds to the length at the minimum of a Morse curve of energy (as 'y') against interatom separation r (as 'x').

■ **Worked Example 21.2**

A single bond connects two atoms. The potential energy of the two atoms depends on their separation according to Fig. 21.5, which is known as a **Morse curve**. A greatly simplified equation for the graph is

$$E = 2r^2 - 4r + 3$$

where r is the normalized bond length. Determine the minimum on this graph.

We first differentiate the expression, saying

$$\frac{dE}{dr} = 4r - 4$$

At the minimum, the gradient of the curve is 0, so

$$0 = 4r - 4$$

Slight rearranging yields:

$$4r = 4$$

so $r = 1$

Knowing the value of r at the minimum, we calculate the corresponding value of E:

$$E = 2 \times (1)^2 - 4 \times (1) + 3$$

so $E = 1$

i.e. the minimum of the graph occurs at the point (1,1).

In summary, we determine the minimum of the Morse curve. This minimum corresponds to the preferred separation between the atoms. So the preferred bond length r_{eq} corresponds to the minimum energy in a Morse curve. ■

■ **Worked Example 21.3**

The radial distribution function is a probability density because, when multiplied by dr, it defines the probability of finding the electron anywhere in a shell of thickness dr at a radius r from the nucleus. For a 1s orbital,

$$P(r) = \left(\frac{4Z^3}{a_0^3}\right)r^2 \exp\left(-\frac{2Zr}{a_0}\right)$$

where P is probability, Z is the atomic number and a_0 is the Bohr radius = 53 pm. Calculate the most probable radius at which an electron will be found for the hydrogen atom, H$^{\bullet}$.

Strategy:

(i) We differentiate the expression for the radial distribution function with respect to r, i.e. obtain $\dfrac{dP}{dr}$.

(ii) To find the turning point, we equate the expression to zero.

(iii) We rearrange to make r the subject.

Solution:

(i) Using the product rule, we obtain, $\dfrac{dP}{dr} = \left(\dfrac{4Z^3}{a_0^3}\right)\left(2r - \dfrac{2Zr^2}{a_0}\right)\exp\left(-\dfrac{2Zr}{a_0}\right)$

(ii) At the turning point, this expression equates to zero, so

$$0 = \left(\dfrac{4Z^3}{a_0^3}\right)\left(2r - \dfrac{2Zr^2}{a_0}\right)\exp\left(-\dfrac{2Zr}{a_0}\right)$$

(iii) If we call the position of the electron, r^*, then we substitute for r with r^*:

$$0 = \left(\dfrac{4Z^3}{a_0^3}\right)\left(2r^* - \dfrac{2Zr^{*2}}{a_0}\right)\exp\left(-\dfrac{2Zr^*}{a_0}\right)$$

There are only two straightforward ways for the right-hand side to equal zero: for the exponential to equal zero, in which case r^* is infinite; and when the middle bracket is 0. The first situation is not meaningful.

We say, $\left(2r^* - \dfrac{2Zr^{*2}}{a_0}\right) = 0$, so $2r^* = \dfrac{2Zr^{*2}}{a_0}$. Cancelling yields, $r^* = a_0/Z$.

At first sight, this result is surprising, for it suggests that electrons are closer to the nucleus as the nuclear charge increases. A moment's thought says the result is likely, for coulombic attractions will pull the electron closer to the nucleus as Z increases.

Inserting $Z = 1$ yields, $r^* = 53$ pm. ■

Self-test 21.2

Determine the coordinates of the minimum or maximum for each of the following functions:

21.2.1 $\quad y = x^2 + x + 5$	**21.2.2** $\quad y = 3x^2 - 12x - 7$
21.2.3 $\quad y = -5x^2 + 4x + 2$	**21.2.4** $\quad y = 3.2x^2 - 9.1x$

Note: when a graph has more than one root, the differential has the form of a quadratic equation. Solution of this quadratic then yields two roots, which identify the two turning points.

Again, determine the coordinates of the minimum or maximum for each of the following functions:

21.2.5 $\quad y = \frac{1}{3}x^3 + 5x^2 + 24x + 7$	**21.2.6** $\quad y = x^3 + 9x^2 + 24x$

Determining the nature of the turning point using second differentials

By the end of this section, you will know:

- We can discern whether a turning point is a maximum or minimum by considering the gradients near a turning point.
- The double-differentiation method allows us to quantify the rate at which the gradient changes. We differentiate the differential: the differential coefficient of a maximum is negative and the differential coefficient of a minimum is positive.

Look again at the simple function $y = x^2$ depicted in Fig. 21.1. We have seen already how the turning point occurs at the origin, (0, 0). We saw briefly in the introduction how the gradient changes sign at a turning point. We now analyse ways in which this change of sign can be used critically, to allow us to determine whether a turning point is a maximum, minimum or inflection.

The simplest way to tell the nature of a turning point is **inspection** of a drawn curve. Such an approach is time consuming but otherwise reliable.

The method of second differentiation:

We have consistently emphasized how a differential represents a *rate of change*. We now extend the idea to look at the way the gradients of these graphs itself shows a trend—in effect, we analyse how the rate of change itself changes.

As we have seen, the best way to determine the rate at which a function changes is to differentiate. The resultant differential yields directly the gradient of the function, dy/dx. Therefore, if we want to know how fast a gradient changes, we differentiate the differential. We obtain the *second* differential.

The **first differential** of a function y yields the gradient: $\dfrac{dy}{dx}$

The **second differential** of y yields the rate of change of the gradient:

$$\frac{d\,(dy/dx)}{dx} = \frac{d^2 y}{dx^2}$$

The reason we can use this method follows from the trends in the gradients, as summarized in Table 21.2.

When we describe a double differential, we say aloud, 'dee two why by dee ex squared.'

Table 21.2 The trends in the signs of a gradient either side of a maximum, minimum of point of inflection.

Maxima:	The gradient shows the trend: positive → zero → negative, i.e. decreases
Minima:	The gradient shows the trend: negative → zero → positive, i.e. increases
Inflection:	(i) The gradient shows the trend: positive → zero → positive, i.e. no sign change
	(ii) The gradient shows the trend: negative → zero → negative, i.e. no sign change

Table 21.3 The trends in the signs of a second differential d^2y/dx^2 either side of a maximum, minimum, or point of inflection.

Maximum:	$\dfrac{d^2y}{dx^2}$	is negative
Minimum:	$\dfrac{d^2y}{dx^2}$	is positive
Inflection:	$\dfrac{d^2y}{dx^2}$	has a value of 0

- If an observed overall trend is a decrease (positive via 0 to negative), as seen for a maximum, then the second differential will be negative.

- If an observed overall trend is an increase (negative via 0 to positive), as seen for a minimum, then the second differential will be positive.

- If the observed overall trend shows no overall change in sign, as seen for an inflection, then the second differential has a value of zero.

These results are summarized in Table 21.3.

CARE: The method of double differentiation to test for inflections does not work for simple polynomial functions: for example, the value of d^2y/dx^2 for $y = x^4$, $y = x^6$, $y = x^8$ etc., will each be zero at the turning point, although the turning point in each case is not an inflection but a minimum. To determine whether a turning point is an inflection, or a maximum or minimum may require more advanced methods.

■ **Worked Example 21.4**

Consider the simple function, $y = 4x^2 + 1$. Determine its turning point. Is it a maximum, minimum or an inflection?

The *first* differential provides the gradient of this function: $\dfrac{dy}{dx} = 8x$.

The turning point occurs when this gradient is 0, so the turning point occurs when $x = 0$. Back-substitution yields the value of y is 1. Therefore, the coordinates of the turning point are (0,1).

The *second* differential provides the rate at which the gradient changes: $\dfrac{d^2y}{dx^2} = 8$.

In this case, the second differential is always positive regardless of the value of x, which means this turning point is a *minimum*. ■

■ **Worked Example 21.5**

Consider the simple function, $y = 4x^5$. First determine the coordinates of its turning point, then decide if the turning point is a maximum, minimum or an inflection.

The gradient of this function is given by the *first* differential: $\dfrac{dy}{dx} = 20x^4$

The turning point occurs when the gradient is 0, so the turning point occurs when $x = 0$. Back-substitution says the value of y is also 0. Therefore, the turning point is (0,0).

The rate of change of this gradient is given by the *second* differential: $\dfrac{d^2y}{dx^2} = 80x^3$

When we substitute the value $x = 0$ into this, the second differential, we obtain the result, $d^2y/dx^2 = 0$, so this turning point represents an inflection. ■

Self-test 21.3

Determine the turning point for each of the following functions. Differentiate a second time to discern whether the turning point(s) are maxima, minima or points of inflection.

21.3.1 $y = x^3 + 2$ **21.3.2** $y = 3x^2 + 6x$

21.3.3 $y = 4x^3 + 4x^2 + 12$

Additional problems

The majority of chemical examples used to illustrate the concepts of second differentials and turning points are so complicated that the mathematic concepts become lost. These examples are purely mathematical.

21.1–21.3 These questions relate to the equation $y = x^3 + 4x^2 - 5x$

21.1 Determine graphically the turning points of the function, and identify which are maxima and which are minima.

21.2 Factorize the equation in order to determine the points where the function crosses the x-axis.

21.3 Determine the coordinates of the turning points by obtaining the first and second differentials.

21.4 Write the double differential of the equation, $y = x^4 - x^3 + 12 \ln x + \dfrac{1}{x^2}$.

21.5 Write the double differential of the equation, $y = (x^3 + x)^5$.

21.6 Write the double differential of the equation, $y = \sin x \cos x$.

21.7 Consider the equation, $y = 2x^3 + 6x^2 + 6x$. Using differentiation, confirm the identity of the turning points.

21.8 Write the double differential of the equation, $y = x^3 + x^2 + 1$, and hence determine the coordinates of its turning point. From the double differential, is the turning point a maximum or a minimum?

21.9 Write the double differential of the equation, $y = \sin(12x)$, and hence determine the coordinates of its first turning point.

21.10 Write the double differential of the equation, $y = \ln(3x) - x^2$, and hence determine the coordinates of its turning point. Is the turning point a maximum or a minimum?

Differentiation VI

Partial differentiation and polar coordinates

Introducing partial differentiation

By the end of this section, you will know:

- A function represented in a three-dimensional graph forms a surface rather than a line.
- It is possible to take the gradient at any point on this 3D surface, but it is first necessary to fix one of the variables.

We saw in Chapter 9 how to determine the gradient of a straight line: we simply take the ratio of the vertical distance travelled and the horizontal distance, as defined by eqn (9.2) in Chapter 9.

We discussed the gradients of curved lines in Chapter 17. It is usually quite easy to determine such gradients: we either take the tangent as discussed on p. 266 or, if we know the actual function described by the line, we differentiate the equation of the line and thence calculate the differential (see p. 268 ff.).

It is *not* so easy to determine the gradient of a curve if there are three or more dimensions. In progressing from two dimensions to three, we need to notice several changes:

	Two dimensions	Three dimensions
What the function looks like	The function forms a *line* in a 2D graph	The function forms a *surface* on a 3D graph
About the space under the line	Below the line is an *area*	Below the line is a *volume*

From looking at the contours of a hill or any other curved surface, we should appreciate there are an infinite number of gradients possible. Look at the curved surface in Fig. 22.1, which shows the three-dimensional (3D) surface obtained by plotting pressure–volume–temperature data for one mole of an ideal gas. This 3D surface is clearly not flat.

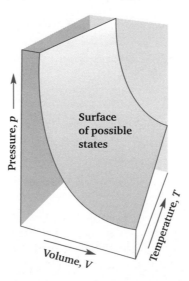

Figure 22.1 The volume, pressure and temperature of an ideal gas are interrelated according to the ideal gas equation, $PV = nRT$. This three-dimensional graph shows the surface obtained by plotting volume–pressure–temperature data for one mole of an ideal gas.

Therefore, if someone asks for 'the gradient' of this graph, the correct response will be to ask, '*which* gradient?'

We cannot determine any meaningful gradients from the surface of such a 3D graph, so need to simplify it. In practice, we simplify by turning a *three*-dimensional graph into a series of *two*-dimensional graphs. We do this because we already know how to take the gradient of a two-dimensional (2D) graph.

■ **Worked Example 22.1**

Look at the 3D graph in Fig. 22.1. In what ways can we simplify the graph?

The graph looks complicated because it shows three dimensions, which relate to the three variables p, V and T. We need to generate a 2D graph from this 3D graph. To do so, we ensure one of the variables stays constant. We say it is 'fixed' or 'pegged.'

By fixing, say, the variable T, we generate a single graph of p against V. In this way, we have in effect drawn a *cross-section* through the 3D graph in Fig. 22.1, cutting through the 3D surface at a fixed value of T.

This procedure is useful, because we *can* take the gradient of a 2D graph. There are three possible types of 2D graphs we can generate:

1. If we hold the pressure constant, then volume and temperature remain the only variables. A graph of T against V is a cross-section through the 3D graph in Fig. 22.1. The resultant cross-section in Fig. 22.2(a) illustrates Charles' Law.

2. If we hold the volume constant, then pressure and temperature remain the only variables. A graph of T against p is again a cross-section through the 3D graph in Fig. 22.1, but in a different dimension. Figure 22.2(b) illustrates the alternative version of Charles' Law.

3. If we hold the temperature constant, then volume and pressure remain the only variables. A graph of V against p illustrates the third way we can obtain a cross-section through the 3D graph in Fig. 22.1. Figure 22.2(c) illustrates Boyle's Law.

We obtained each of these three graphs by fixing one of the three variables of p, T or V. But the value of the fixed variable dictates the graph we obtain. If we changed the value of the fixed variable, we would obtain a different cross-section through the 3D graph. The shape might not alter much, but we would obtain a *quantitatively* different graph. ■

The gradients we determine from cross-sectional graphs such as those in Fig. 22.2 are genuine and (usually) measurable.

We denote the gradients by writing an expression of the sort $(\partial x/\partial y)_z$. We need to unpack this notation very carefully:

- Since this expression looks like a fraction with 'd something' on top and 'd something else' on the bottom, this tells us we have a differential. Therefore, it is a gradient or rate of change.

- The mathematics used by chemists requires five different d characters. Table 22.1 lists them all with a brief description of each. We call the strange symbol on top and bottom of this differential **curly d**.

- Curly d tells us that the system involves more variables than two, but that only two of the variables are allowed to change. We have fixed all the other variables.

- The small z nestling at the foot of the right-hand bracket tells us which of the variables we have fixed. In this case, we fixed the value of z. The value of $(\partial y/\partial x)$ will probably vary according to the actual value of z chosen. The variation with z could be dramatic, but we have no way of knowing from this expression alone.

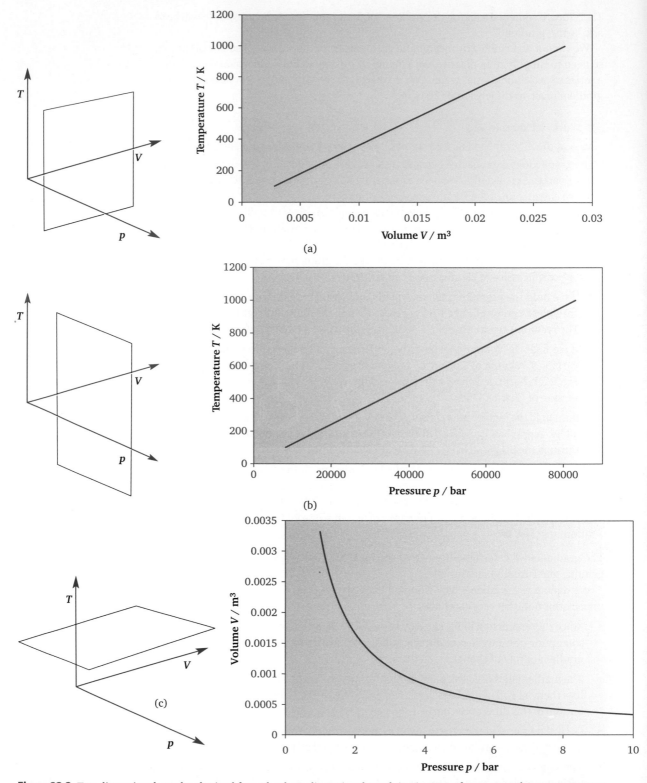

Figure 22.2 Two-dimensional graphs obtained from the three-dimensional graph in Fig. 22.1 of pressure–volume–temperature for one mole of an ideal gas. (a) Graph of temperature T against volume V at a fixed pressure of 3 bar. (b) Graph of temperature T against pressure p at a fixed volume of 1 m³. (c) Graph of volume V against pressure p at a fixed temperature of 400 K.

Table 22.1 The five d symbols employed in the mathematics of chemistry.

Symbol	Common name	Meaning	Example
δ	Small Greek d	A small change or increment.	The electronic charges in an induced dipole, such as a hydroxyl bond
Δ	Big Greek d	The delta operator described by eqn (2.1).	Enthalpy change during reaction $= \Delta H = H_{final} - H_{initial}$
d	Small Roman d	Symbol representing a straightforward differential.	The gradient of a simple 2D graph of y against $x = \dfrac{dy}{dx}$
D	Big Roman d	Deuterium or diffusion coefficient.	Fick's second law of diffusion: $\dfrac{\partial c}{\partial t} = D\dfrac{\partial^2 c}{\partial x^2}$, where $t =$ time, $x =$ distance and $c =$ concentration.
∂	Curly d	Symbol representing a partial differential.	In multidimensional graphs, when all the variables but y and x have been fixed: the gradient of y against $x = \dfrac{\partial y}{\partial x}$

We call expressions of the type $(\partial y/\partial x)$ a **partial differential** because we took the gradient from only one part of the overall graph. The variables defining the rest of the graph were fixed. Because the value of $(\partial y/\partial x)$ varies according to the choice of z, the differential is not straightforward.

Self-test 22.1

Write an expression for the partial differential gradients of the three graphs in Fig. 22.2:

22.1.1 Graph of T (as y) against p (as x).

22.1.2 Graph of V (as y) against p (as x).

22.1.3 Graph of T (as y) against V (as x).

In complicated situations, a function can depend on a very large number of variables. To continue with the simple example above, the behaviour of an ideal gas actually depends on four variables, so any graph depicting the behaviour of the gas requires *four* dimensions. (The discussion above simply ignored the amount of gas n.)

As an immediate result of such complexity, it is most unlikely that any pictorial representation could adequately describe such behaviour. But we could write four separate partial differentials. This time, we need to fix *two* of the variables, and then analyse how the remaining two interact. To indicate that we have fixed two variables, we merely insert two subscripted letters against the foot of the right-hand bracket, separating each with a comma.

■ Worked Example 22.2

The physicochemical properties of an ideal gas are described by four variables: p, T, V and n. Write an expression for the rate of change of volume as a function of the amount of gas. Assume that all other variables remain fixed.

Strategy:

(i) Write a bracket with a curly d on top and bottom.

(ii) Decide which is the controlled and which the observed variable. Let the change in the controlled variable be the denominator.

(iii) Against the foot of the right-hand bracket, write subscripted letters representing the variables that we fix.

Solution:

(i) $\left(\dfrac{\partial \ldots}{\partial \ldots} \right)$

(ii) from the wording of the question, amount of gas n is the controlled variable and volume V is the observed variable. ∂V therefore goes on the top of the differential and ∂n goes on the bottom. Therefore, $\left(\dfrac{\partial V}{\partial n} \right)_{\ldots}$.

(iii) If the variables are V and n, then p and T must be fixed. We write, $\left(\dfrac{\partial V}{\partial n} \right)_{p,T}$. Note the comma between the p and the T. ■

Self-test 22.2

22.2.1   The value of x if a function of a, b, c and d. Write an expression for the partial differential of x if all but c are fixed.

22.2.2   A particle accelerates through the chamber of a mass spectrometer. The acceleration a depends on the time t the initial speed u and the strength of the magnetic field B. Write an expression for the rate of change of acceleration with magnetic field strength.

Differentiating with partial differentials is essentially identical to normal differentiation.

■ Worked Example 22.3

Consider the expression $f(x,y,z) = 2x^2 + 4y^3 + 6z^2 + 2xy + 5xz - yz$. Obtain the three partial differentials of this equation.

$$f(x,y,z) = 2x^2 + 4y^3 + 6z^2 + 2xy + 5xz - yz$$

$$\begin{array}{ccccccc} 1 & 2 & 3 & 4 & 5 & 6 & 7 \end{array}$$

$$\left(\dfrac{\partial f}{\partial x} \right)_{y,z} = 4x \;+\; 0 \;+\; 0 \;+\; 2y \;+\; 5z \;-\; 0$$

Rationale:

1. $(\partial f / \partial x)$ is the partial differential of f with respect to x. Note the small subscripted y and z, which mean that y and z are constant. We no longer regard them as variables.
2. x is a straightforward variable. Therefore, the differential of $2x^2$ is $4x$.
3. y is a constant, so $4y^3$ is a constant. The differential is 0.
4. z is a constant, so $6z^2$ is a constant. The differential is 0.
5. Within $2xy$, only x is a variable; y is merely a factor. The differential is $2y$.
6. Within $5xz$, only x is a variable; z is merely a factor. The differential is $5z$.
7. y and z are constants. The differential is 0.

So $\left(\dfrac{\partial f}{\partial x}\right)_{y,z} = 4x + 2y + 5z$

With similar reasoning, $\left(\dfrac{\partial f}{\partial y}\right)_{x,z} = 12y^2 + 2x - z$

$\left(\dfrac{\partial f}{\partial z}\right)_{x,y} = 12z + 5x - y$ ∎

Self-test 22.3

Write the three differentials for each of the following:

22.3.1 $f(x,y,z) = 4x^2 + 5xy^2 - 6z$

22.3.2 $f(x,y,z) = 3.2x^{5.3} + \dfrac{5}{yz} - 2\ln z^2$

Partial derivative relationships:

There are a number of useful relationships that can be derived for partial derivatives. These include:

$$\left(\frac{\partial y}{\partial x}\right)_z = \frac{1}{(\partial x / \partial y)_z} \qquad (22.1)$$

$$\left(\frac{\partial y}{\partial x}\right)_z = -\left(\frac{\partial y}{\partial z}\right)_x \left(\frac{\partial z}{\partial x}\right)_y \qquad (22.2)$$

$$\left(\frac{\partial f}{\partial x}\right)_z = \left(\frac{\partial f}{\partial x}\right)_y + \left(\frac{\partial f}{\partial y}\right)_x \left(\frac{\partial y}{\partial x}\right)_z, \text{ for a function } f(x,y) \text{ and } y(x,z) \qquad (22.3)$$

■ **Worked Example 22.4**

Show that eqn (22.2) is valid for the ideal gas equation: $pV = nRT$ by setting $x = p$, $y = V$ and $z = T$.

First, we calculate the separate differentials:

$$\left(\frac{\partial V}{\partial p}\right)_T = -\frac{nRT}{p^2} \qquad \left(\frac{\partial V}{\partial T}\right)_p = \frac{nR}{p} \qquad \left(\frac{\partial T}{\partial p}\right)_V = \frac{V}{nR}$$

Substituting in for V:

$$\left(\frac{\partial T}{\partial p}\right)_V = \frac{1}{nR} \times \frac{nRT}{p} = \frac{T}{p}$$

Substituting into eqn (22.2):

$$\left(\frac{\partial y}{\partial x}\right)_z = -\left(\frac{\partial y}{\partial z}\right)_x \left(\frac{\partial z}{\partial x}\right)_y$$

Right-hand side of the equation:

$$\left(\frac{\partial V}{\partial p}\right)_T = -\left(\frac{\partial V}{\partial T}\right)_p \left(\frac{\partial T}{\partial p}\right)_V = -\frac{nR}{p} \times \frac{T}{p} = -\frac{nRT}{p^2}$$

This is the same as the expression derived directly for $\left(\frac{\partial V}{\partial p}\right)_T$, proving that eqn (22.2) is indeed valid for the ideal gas equation. ∎

Self-test 22.4

22.4.1   Given the following definitions:

$$\frac{1}{V}\left(\frac{\partial V}{\partial T}\right)_p = \alpha, \text{ the thermal expansivity,}$$

$$-\frac{1}{V}\left(\frac{\partial V}{\partial p}\right)_T = \kappa, \text{ the isothermal compressibility,}$$

derive an expression for the partial differential, $(\partial p/\partial T)_V$.

The total differential

By the end of this section, you will know that the total differential:

- Expresses the change in a physicochemical variable.
- Is expressed in terms of the changes in the variables that constituent parameter under scrutiny.

When purchasing goods at a shop, the total amount we must pay depends on both the *identity* of the items we buy and the *number* of each. For example, when buying apples and pears, the total price will depend on the price of each item, and the amounts of each that we purchase.

We could write this obvious finding in a general way:

$$\text{d(money)} = \left(\begin{array}{c} \text{price per apple} \\ \times \text{ number of apples} \end{array}\right) + \left(\begin{array}{c} \text{price per pear} \\ \times \text{ number of pears} \end{array}\right)$$

While more mathematical in form, we could have rewritten this equation in a more rigorously way, as:

$$\text{d(money)} = \left(\frac{\partial \text{ money}}{\partial \text{ item 1}}\right) \times N(1) + \left(\frac{\partial \text{ money}}{\partial \text{ item 2}}\right) \times N(2) + ...$$

where the notation $N(1)$ represents the number of items of type (1) that we choose to buy; each differential term represents the price of each item: it is the amount by which the total amount of money in our purse or wallet changes each time we purchase an item on the shopping list.

We give the name **total differential** to equations following this style. It will remind us of the equations we encountered in Chapter 15, when estimating the errors during a physicochemical measurement.

In more familiar mathematical notation, we write the total differential for a function f the value of which depends on the two variables x and y:

$$df = \left(\frac{\partial f}{\partial x}\right)dx + \left(\frac{\partial f}{\partial y}\right)dy \qquad (22.4)$$

This seems quite abstract, so we now look at a real example:

■ **Worked Example 22.5**

The Gibbs function G for one mole of a pure substance is a function of temperature and pressure. Write the total differential dG for this system.

Strategy:

(i) The change in G depends on *two* variables, so we need to generate *two* pairs of terms—one each for pressure and temperature.

(ii) Each of these pairs will follow the same simple template: $(\partial G/\partial(\text{variable}))$ multiplied by $d(\text{variable})$.

(iii) Against the foot of each right-hand bracket, write subscripted letter(s) representing the variables that we fix.

(iv) Add together the terms.

Solution:

(i) and (ii): The terms will be $(\partial G/\partial p) \times dp$ and $(\partial G/\partial T) \times dT$.

(iii) The terms become $(\partial G/\partial p)_T \times dp$ and $(\partial G/\partial T)_p \times dT$.

(iv) $dG = \left(\dfrac{\partial G}{\partial p}\right)_T dp + \left(\dfrac{\partial G}{\partial T}\right)_p dT$

This equation is the total differential of G. ■

Self-test 22.5

Write an expression for the total differential for each of the following.

22.5.1   The kinetic energy E of a gas molecule depends on the molecular velocity v and the molecular mass, m.

22.5.2   The entropy S of a pure material depends on its heat capacity at constant pressure C_p and the absolute temperature T.

22.5.3   The temperature T of an ideal gas depends on the amount of gas n, the volume V and the pressure p.

In fact, the total differential is rarely useful in its own right, but it becomes increasingly useful when we use it to derive other equations. For example, it is informative to compare the Gibbs–Duhem equation, $dG = Vdp - S\,dT$, and the total differential obtained during Worked Example 22.5:

The Gibbs–Duhem equation: $dG = \quad V \quad dp \quad -S \quad dT$

The total differential of G: $dG = \left(\dfrac{\partial G}{\partial p}\right)_T dp + \left(\dfrac{\partial G}{\partial T}\right)_p dT$

If these two equations are both correct (and they are), then the coefficient terms can be equated: $(\partial G/\partial p)_T = V$ and $(\partial G/\partial T)_p = -S$.

In fact, we call these two partial differential equations the **Maxwell relations**. We could have obtained them in other ways, but this is the simplest derivation. This sort of manipulation demonstrates that we can sometimes treat partial differentials in much the same way as more familiar fractions, although this is not mathematically rigorous and should be avoided except in examples similar to the one shown in Worked Example 22.6.

■ **Worked Example 22.6**

Write the total differential of enthalpy, which is a function of temperature and pressure, and derive an expression in terms of heat capacity C_p.

[Hint: $C_p = (\partial H/\partial T)_p$]

The total differential of H is : $dH = \left(\dfrac{\partial H}{\partial T}\right)_p dT + \left(\dfrac{\partial H}{\partial p}\right)_T dp$

If we divide throughout by dT, we obtain: $\left(\dfrac{\partial H}{\partial T}\right)_V = \left(\dfrac{\partial H}{\partial T}\right)_p + \left(\dfrac{\partial H}{\partial p}\right)_T \left(\dfrac{\partial p}{\partial T}\right)_V$.

If we apply the rules of mathematics strictly, division by an infinitesimally small quantity such as dT is not possible. However, we can easily see that the same result can be found by using the partial derivative relationship stated in eqn (22.3).

By substituting with the definition of C_p, we obtain: $\left(\dfrac{\partial H}{\partial T}\right)_V = C_p + \left(\dfrac{\partial H}{\partial p}\right)_T \left(\dfrac{\partial p}{\partial T}\right)_V$

In fact, using the answer to Self-test 22.4.1, we can simplify the equation even further,

$$\left(\dfrac{\partial p}{\partial T}\right)_V = \dfrac{\alpha}{\kappa}$$

Accordingly, the equation becomes, $\left(\dfrac{\partial H}{\partial T}\right)_V = C_p + \left(\dfrac{\partial H}{\partial p}\right)_T \dfrac{\alpha}{\kappa}$. ■

Derivations such as that immediately above demonstrate the power of partial differential equations when deriving equations. Table 22.2 lists many of the useful partial differential parameters in chemical thermodynamics.

Table 22.2 A few useful partial differential parameters from chemical thermodynamics.

Differential	Name	Notes
$\left(\dfrac{\partial G}{\partial n}\right)_{T,p}$	Chemical potential, μ	Often called the partial molar Gibbs function
$\left(\dfrac{\partial G}{\partial \zeta}\right)_{T,p}$	Reaction free energy, and 'reaction affinity' in older texts	ζ (zeta) is the extent of reaction. This differential only corresponds to the change in Gibbs function ΔG under certain, well defined, and precisely controlled experimental conditions.
$\left(\dfrac{\partial G}{\partial T}\right)_{p}$	$-1 \times$ entropy, S	
$\left(\dfrac{\partial G}{\partial p}\right)_{T}$	Volume, V	
$\left(\dfrac{\partial H}{\partial T}\right)_{p}$	Heat capacity at constant pressure, C_p	
$\left(\dfrac{\partial U}{\partial T}\right)_{V}$	Heat capacity at constant volume, C_v	
$\left(\dfrac{\partial V}{\partial n}\right)_{T,p}$	The molar volume, V_m The molar volume	

Higher-order partial differentials

By the end of this section, you will know:

- It is possible to differentiate a partial differential.

- The notation for such higher differentials follows exactly the same pattern as found for more straightforward differentials, except of course we write ∂ instead of d.

- The Euler reciprocity relation says that if we differentiate a function twice, with respect to two different variables, it does not matter in which order we differentiate.

In Chapter 21, we saw how it is often useful to differentiate a differential. As a simple example, if $y = 4x^4$, then $dy/dx = 16x^3$, $d^2y/dx^2 = 48x^2$, $d^3y/dx^3 = 96x$, and $d^4y/dx^4 = 96$. The fifth differential, d^5y/dx^5 is not useful, because it equates to zero. We call each of these results (except the last) a **higher differential**.

In exactly the same way, it is often useful to obtain a higher partial differential, and analyse the result.

The symbolism we use for higher differentials is very similar to that seen in Chapter 21: the first higher-order differential of $(\partial y/\partial x)$ is $(\partial^2 y/\partial x^2)$, and so on. Such differentials are immensely useful in chemistry. For example, by looking at a second-order differential such as $\partial^2 y/\partial x^2$, we can tell whether a differential is increasing or decreasing, or approaching a maximum or minimum.

There is an additional complication not encountered with simple higher differentials: a partial differential with respect to x, $(\partial f/\partial x)$, can itself be differentiated, but with respect to a different variable.

■ **Worked Example 22.7**

Consider the function $f(x, y, z) = 6x^2yz^2 + 4xy + 2yz^3$. Obtain the first differential with respect to x and then the second differential with respect to z.

The straightforward rules of calculus in Chapter 17 yields, $\left(\dfrac{\partial f}{\partial x}\right)_{y,z} = 12xyz^2 + 4y$.

The subscripts tells us that y and z were held constant. To indicate that we obtained the next differential with respect to a *different* variable z, we write: $\dfrac{\partial}{\partial z}\left(\dfrac{\partial f}{\partial x}\right)$. We could rewrite this expression as $\dfrac{\partial^2 f}{\partial z \partial x}$, i.e. with the curly d terms nesting together, writing the first-used term on the right-hand side.

Therefore, $\dfrac{\partial}{\partial z}\left(\dfrac{\partial f}{\partial x}\right) = \dfrac{\partial^2 f}{\partial z \partial x} = 24xyz$. ■

As a very good generalization, the end result is not dictated by the order in which we perform the differentiation. In Worked Example 22.7, $\dfrac{\partial}{\partial x}\left(\dfrac{\partial f}{\partial z}\right) = \dfrac{\partial}{\partial z}\left(\dfrac{\partial f}{\partial x}\right)$: we can usually assume that obtaining a double partial differential is *commutative*. We call this aspect of their mathematics, the **Euler reciprocity relation**.

The notation here can get a little messy—longwinded even. In response, shorthand notations have arisen. Instead of saying $\partial/\partial x(\partial f/\partial y)$, we could say f_{xy}. We will not be using this new notation here.

Self-test 22.6

The ideal gas equation states, $p = nRT/V$.

22.6.1 Obtain the three partial differentials of p.

22.6.2 Obtain the six second partial differentials of p.

Comment on the applicability of the Euler reciprocity relation.

Second-order partial differential equations dominate fluid dynamics and dictate the behaviour of uncharged species as they diffuse to and from the interface between an electrode and the solution in which the solute is dissolved. Fick's two laws quantify the rate of flux J due to diffusion. (Migration and convection also bring material to the electrode surface, thereby supplementing the flux.)

Materials in solution diffuse from regions of high concentration to regions of lower concentration. **Fick's first law** postulates that the rate of diffusive flux J is directly proportional to the gradient: $J \propto$ a spatial derivative. In one dimension, we say

$$J = -D\left(\frac{\partial c}{\partial x}\right)_t$$

where c is concentration and x is distance from the electrode. The partial differential on the right-hand side describes the gradient in concentration. D is the diffusion coefficient, and acts as a constant of proportionality. In fact, the diffusion coefficient D quantifies the average speed of species as they diffuse through the fluid toward the electrode.

Fick's first law tells us nothing about the *timescales* of diffusion. We need **Fick's second law** if we wish to relate the spatial characteristics of diffusion to the temporal (i.e. time-based) behaviour. His second law is more complicated still. In one dimension:

$$\left(\frac{\partial c}{\partial t}\right)_x = D\left(\frac{\partial^2 c}{\partial x^2}\right)_t$$

where t is time. This equation predicts how diffusion causes the concentration gradient to change with time. Fourier's law of heat flow (or *heat conduction*), which describes the flow of heat through a conductor, is almost identical. Again, in one dimension:

$$\left(\frac{\partial T}{\partial t}\right) = \kappa\left(\frac{\partial^2 T}{\partial x^2}\right)$$

where T is temperature and the thermal conductivity κ is a property of the conducting material.

These equations are very complicated, particularly the second-order differentials. In consequence, they are generally horrendous to unravel. Their complexity helps explain why we find it so difficult to construct mathematical models that mimic electrochemical currents, because the magnitude of the current depends on *diffusion* of an analyte toward the electrode surface.

We can write almost identical equations in terms of pressure instead of concentration, which help us to predict changes in air pressure, itself helping us to predict weather patterns. But, again, the complexity of the equations requires powerful computer programmes, and helps explain why weather forecasts are relatively prone to error.

The **flux** J is the amount of material that reaches the surface of the electrode through a unit area *per unit time*.

Del ∇ and the Laplacian operator

By the end of this section, you will know:

- del ∇ is a partial differential in three dimensions.
- The second differential ∇^2 is called the Laplacian operator, and is particularly useful in quantum mechanics.
- The Laplacian operator is an abbreviation for a partial differential expression.

The Laplacian operator is named after the French mathematician Pierre-Simon, Marquis de **Laplace** (1749–1827).

One higher differential occurs so often in chemistry that we usually abbreviate it: **del** is a vector differential operator represented by the symbol ∇:

We give the name **nabla** to the symbol ∇.

$$\nabla = \frac{\partial}{\partial x} + \frac{\partial}{\partial y} + \frac{\partial}{\partial z} \tag{22.5}$$

This treatment of del is framed in terms of three dimensions. It is possible to treat the definitions and discussions more generally to encompass any *n*-dimensional space. (If *n* is 1, the differential is not partial.)

Equation (22.5) is useful, although chemists habitually employ its second differential, which we denote as ∇^2:

$$\nabla^2 = \frac{\partial^2}{\partial x^2} + \frac{\partial^2}{\partial y^2} + \frac{\partial^2}{\partial z^2} \tag{22.6}$$

Equation (22.6) defines the **Laplacian operator**. We call solutions of $\nabla^2 f = 0$, harmonic functions.

This kind of abbreviation makes many equations easier to comprehend, write, and remember. We also abbreviate its name: when speaking about ∇^2 we generally say 'del squared.'

Equation (22.6) describes the second partial differential of a function defined in terms of the familiar three-dimensional Cartesian coordinate system x, y, z. In fact, we need not restrict the Laplacian operator to Cartesian coordinates. For example, we can write ∇^2 in terms of 3D cylindrical polar coordinates:

$$\nabla^2 = \frac{1}{r} \frac{\partial}{\partial r} \left(r \frac{\partial}{\partial r} \right) + \frac{1}{r^2} \frac{\partial^2}{\partial \theta^2} + \frac{\partial^2}{\partial z^2} \tag{22.7}$$

or for 3D spherical polar coordinates:

$$\nabla^2 = \frac{1}{r^2} \frac{\partial}{\partial r} \left(r^2 \frac{\partial}{\partial r} \right) + \frac{1}{r^2 \sin\theta} \frac{\partial}{\partial \theta} \left(\sin\theta \frac{\partial}{\partial \theta} \right) + \frac{1}{r^2 \sin^2\theta} \frac{\partial^2}{\partial \phi^2}$$

if we use the product rule, then this becomes:

$$\nabla^2 = \frac{\partial^2}{\partial r^2} + \frac{2}{r} \frac{\partial}{\partial r} + \frac{1}{r^2} \frac{\partial^2}{\partial \theta^2} + \frac{\cos\theta}{r^2 \sin\theta} \frac{\partial}{\partial \theta} + \frac{1}{r^2 \sin^2\theta} \frac{\partial^2}{\partial \phi^2} \tag{22.8}$$

This is often simplified by writing it as:

$$\nabla^2 = \frac{\partial^2}{\partial r^2} + \frac{2}{r} \frac{\partial}{\partial r} + \frac{1^2}{r^2} \Lambda^2 \tag{22.9}$$

where Λ^2 is called the **Legendrian:**

$$\Lambda^2 = \frac{\partial^2}{\partial \theta^2} + \frac{\cos\theta}{\sin\theta} \frac{\partial}{\partial \theta} + \frac{1}{\sin^2\theta} \frac{\partial^2}{\partial \phi^2} \tag{22.10}$$

Depending on the way we apply del as an operator, we can make it describe:

- A *gradient* (or rate of change) of a function.

- The *divergence* (the extent to which something converges or diverges, see Chapter 28).

- The *rotational motion* at stationary points in a 3D fluid (the so-called *curl*, see Chapter 28).

The Laplacian operator and quantum chemistry

In a one-dimensional system, the Schrödinger wave equation yields the quantum-mechanical energy E of an electron orbiting a hydrogen atom:

$$E\psi = -\frac{\hbar^2}{2m}\frac{d^2\psi}{dx^2} + V(x)\psi$$

where m is the mass of the electron, h is the Planck constant, $\hbar = h/2\pi$, and ψ is the wavefunction. The first term on the right-hand side represents the kinetic energy of the electron and the second term V is the electron's potential energy; both terms are written in one direction—for brevity, along the x-axis.

In three dimensions, we rewrite the Schrödinger wave equation as

$$E\psi = -\frac{\hbar^2}{2m}\nabla^2\psi + V(x,y,z)\psi$$

Here, the value of ∇^2 represents the three-dimensional kinetic energy term of the Schrödinger wave equation. In this example, we say the Laplacian operator ∇^2 'operates over' three dimensions. For a one-electron system, the wavefunction can be written as a product of three functions each of which is only dependent on one variable: $\psi = R(r)\Theta(\theta)\Phi(\phi)$. This makes it much easier to find solutions of the Schrödinger wave equation.

■ **Worked Example 22.8**

A particle of mass m is free to move in three dimensions but at a *fixed* distance r from a central point. Show that the Schrödinger wave equation simplifies to the form:

$$E\psi = -\frac{\hbar^2}{2mr^2}\Lambda^2\psi.$$

Since r is constant, the first two terms in the Laplacian (eqn 22.9) must equal zero:

$$\frac{\partial^2\psi}{\partial r^2} + \frac{2}{r}\frac{\partial\psi}{\partial r} = 0.$$

Similarly, since the particle can only move on a spherical surface, potential energy $V = 0$. Therefore, from eqn (22.9), the Schrödinger wave equation becomes:

$$E\psi = -\frac{\hbar^2}{2mr^2}\Lambda^2\psi.$$

The motion of an electron and nucleus about the centre of mass is equivalent mathematically to the motion of a hypothetical particle, with reduced mass $\mu = \dfrac{m_{electron}m_{nucleus}}{m_{electron} + m_{nucleus}}$, about a fixed point. Therefore, the equations above can be applied to one-electron systems. ■

■ **Worked Example 22.9**

For a particle of mass m moving freely on the surface of a sphere, the wavefunction is: $\psi = N\sin\theta\, e^{i\phi}$. (This is actually a spherical harmonic function, Y_{l,m_l}.) Show that this wavefunction satisfies the equation: $\Lambda^2 Y_{l,m_l} = -l(l+1)Y_{l,m_l}$.

First, we should note that the argument of the exponential 'iϕ' is 'complex.' Complex numbers are discussed in more detail in Chapter 29. In this example, we use the definition, $i^2 = -1$.

Next, we calculate the individual differentials found in the expression for Λ^2:

$$\frac{\partial \psi}{\partial \theta} = N \cos \theta e^{i\phi} \qquad\qquad \frac{\partial^2 \psi}{\partial \theta^2} = -N \sin \theta e^{i\phi}$$

$$\frac{\partial \psi}{\partial \phi} = iN \sin \theta e^{i\phi} \qquad\qquad \frac{\partial^2 \psi}{\partial \phi^2} = i^2 N \sin \theta e^{i\phi} = -N \sin \theta e^{i\phi}$$

then, we substitute for these values:

$$\Lambda^2 \psi = \frac{\partial^2 \psi}{\partial \theta^2} + \frac{\cos \theta}{\sin \theta} \frac{\partial \psi}{\partial \theta} + \frac{1}{\sin^2 \theta} \frac{\partial^2 \psi}{\partial \phi^2}$$

$$\Lambda^2 \psi = -N \sin \theta e^{i\phi} + \frac{\cos \theta}{\sin \theta} \times N \cos \theta e^{i\phi} + \frac{1}{\sin^2 \theta} \times -N \sin \theta e^{i\phi}$$

Simplifying this expression:

$$\Lambda^2 \psi = \frac{N e^{i\phi}}{\sin \theta} \left(-\sin^2 \theta + \cos^2 \theta - 1\right)$$

Using the trigonometric identity, $\sin^2 \theta + \cos^2 \theta = 1$ (see Chapter 16),

$$\Lambda^2 \psi = \frac{N e^{i\phi}}{\sin \theta} \left(-\sin^2 \theta + \cos^2 \theta - \left(\sin^2 \theta + \cos^2 \theta\right)\right)$$

which simplifies to:

$$\Lambda^2 \psi = \frac{-2N \sin^2 \theta e^{i\phi}}{\sin \theta} = -2N \sin \theta e^{i\phi}$$

This can be written as: $\Lambda^2 \psi = -2\psi = l \times (l+1)\psi$.

Therefore, the wavefunction satisfies the equation for a spherical harmonic with $l = 1$. (It is actually $Y_{l,m_l} = Y_{1,1}$.) ■

Summary of key equations in the text

Partial differential relationships:

$$\left(\frac{\partial y}{\partial x}\right)_z = \frac{1}{(\partial x / \partial y)_z} \tag{22.1}$$

$$\left(\frac{\partial y}{\partial x}\right)_z = -\left(\frac{\partial y}{\partial z}\right)_x \left(\frac{\partial z}{\partial x}\right)_y \tag{22.2}$$

$$\left(\frac{\partial f}{\partial x}\right)_z = \left(\frac{\partial f}{\partial x}\right)_y + \left(\frac{\partial f}{\partial y}\right)_x \left(\frac{\partial y}{\partial x}\right)_z \tag{22.3}$$

The total differential:

$$df = \left(\frac{\partial f}{\partial x}\right) dx + \left(\frac{\partial f}{\partial y}\right) dy \tag{22.4}$$

The Laplacian operator, del:

$$\nabla = \frac{\partial}{\partial x} + \frac{\partial}{\partial y} + \frac{\partial}{\partial z} \tag{22.5}$$

$$\nabla^2 = \frac{\partial^2}{\partial x^2} + \frac{\partial^2}{\partial y^2} + \frac{\partial^2}{\partial z^2} \tag{22.6}$$

Del in cylindrical polar coordinates:

$$\nabla^2 = \frac{1}{r} \frac{\partial}{\partial r}\left(r \frac{\partial}{\partial r}\right) + \frac{1}{r^2} \frac{\partial^2}{\partial \theta^2} + \frac{\partial^2}{\partial z^2} \tag{22.7}$$

Del in spherical polar coordinates:

$$\nabla^2 = \frac{\partial^2}{\partial r^2} + \frac{2}{r}\frac{\partial}{\partial r} + \frac{1}{r^2}\frac{\partial^2}{\partial \theta^2} + \frac{\cos\theta}{r^2\sin\theta}\frac{\partial}{\partial \theta} + \frac{1}{r^2\sin^2\theta}\frac{\partial^2}{\partial \phi^2} \qquad (22.8)$$

$$\nabla^2 = \frac{\partial^2}{\partial r^2} + \frac{2}{r}\frac{\partial}{\partial r} + \frac{1}{r^2}\Lambda^2 \qquad (22.9)$$

where Λ^2 is called the Legendrian:

$$\Lambda^2 = \frac{\partial^2}{\partial \theta^2} + \frac{\cos\theta}{\sin\theta}\frac{\partial}{\partial \theta} + \frac{1}{\sin^2\theta}\frac{\partial^2}{\partial \phi^2} \qquad (22.10)$$

Additional problems

22.1 The volume of a one-component thermodynamic system is a function of pressure p, temperature T and amount of substance n. Write the total differential for volume, then compare the expression below with the total differential for volume:

$$dV = \alpha V dT - \kappa V\,dp + V_m\,dn$$

Then derive expressions for α, κ, and V_m.

22.2 The energy released by a current passing through a resistor is a function of potential, V, current, I, and time, t: $E = VIt$. Write the total differential of energy, E.

22.3 Consider the equation, $H = U + pV$. Differentiate this equation with respect to temperature, T at constant volume, V and hence derive an equation from thermodynamics. (p is pressure and U is internal energy.)

22.4 For an ideal gas, show that $\left(\dfrac{\partial^2 p}{\partial V \partial T}\right)_n = \left(\dfrac{\partial^2 p}{\partial T \partial V}\right)_n$.

22.5 The approximate internal energy, U, of a monatomic gas is, $U = \dfrac{3nRT}{2}$. Obtain the partial differentials, and show the Euler reciprocity relation holds.

22.6 From the definition of heat capacity C_p in Table 22.2 and the Clausius equality, $dS = dq/T$, obtain the equation $dS = \dfrac{C_p}{T}dT$.

22.7 Consider the equation $dU = TdS - pdV$, which comes from classical thermodynamics. All the terms have their usual meanings. Using this equation at constant volume (i.e. $dV = 0$), obtain a relation involving the heat capacity C_V. [Hint: start by writing the total differential of both dU and dS.]

22.8 Consider the Levich equation, which defines the current at the rotated-disc electrode, RDE:

$$I = \frac{0.62\,nFA\,c\,D^{\frac{2}{3}}\,\omega^{\frac{1}{2}}}{\sqrt[6]{\upsilon}}$$

where A is the area of the disc, D is the speed at which the analyte diffuses through solution, n is the number of electrons transferred in the redox reaction, and F is the Faraday constant. Show that $\dfrac{\partial^2 I}{\partial c\,\partial \omega} = \dfrac{\partial^2 I}{\partial \omega\,\partial c}$.

22.9 Starting with the relation, $dH = dU + pdV$, deduce a relation between the two heat capacities, C_p and C_V.

22.10 Show that the wavefunction $\psi = N\sin\theta\cos\theta\,e^{i\phi}$ satisfies the equation: $\Lambda^2 Y_{l,m_l} = -l(l+1)Y_{l,m_l}$.

23

Integration I

Reversing the process of differentiation

Introducing integration: integrating power functions by rule

By the end of this section, you will know:

- The process of integration performs the opposite function to differentiation.
- How to integrate a variety of simple polynomial functions by rule.

When we differentiate, we start with the equation of a line and obtain an expression to quantify the way it changes. Differential coefficients derived in this way allow us to calculate exactly a rate of change.

Now imagine we *start* with a mathematical expression describing a line's gradient—which is therefore a differential—yet we want to know about the line itself. In effect we want to perform the inverse process to differentiation. We call this reverse process, **integration**.

■ **Worked Example 23.1**

If a line has a gradient of x^3, what is the equation of the line it describes?

When we differentiate a simple polynomial expression using the rule in Chapter 17, we start with a function of the form $a\,x^n$. We decrease the power by '1', and multiply the factor 'a' by 'n', yielding $an\,x^{(n-1)}$. The function differentiated to obtain a gradient of x^3 must therefore have been x^4. But if we differentiated x^4 we would obtain $4x^3$, not $1 \times x^3$. We see how the original function must have been $\frac{1}{4}x^4 + c$. ■

In general, if $\dfrac{dy}{dx} = ax^n$ then

$$y = \frac{a\,x^{(n+1)}}{(n+1)} + c \qquad (23.1)$$

The additional term c is explained on p. 345 below.

Equation (23.1) cannot be used if $n = -1$. Such situations require eqn (23.5) on p. 344.

■ **Worked Example 23.2**

Integrate the expression $\dfrac{dy}{dx} = 5x^6$.

Before we employ the relationship in eqn (23.1), we must first introduce some nomenclature. First, we must decide which term represents the variable. The decision is easy in this example: the variable is clearly x. We say we integrate *with respect to x*. In terms of the symbols we employ, we write this methodology in two distinct ways:

- First, we write a sign to indicate that we wish to perform an integration: $\int$, We call this symbol an **integral sign** or a **script-S**. We then write the function we wish to integrate: $\int 5x^6$.

- We somehow need to indicate that we are interested in the variable 'x'. To this end, we write 'dx' after the integration sign and the function to be integrated. The d here is part of the operator, and the x is a variable. So we write $\int 5x^6\,dx$. In words, we often say, 'with respect to x' rather than just 'dee ex'.

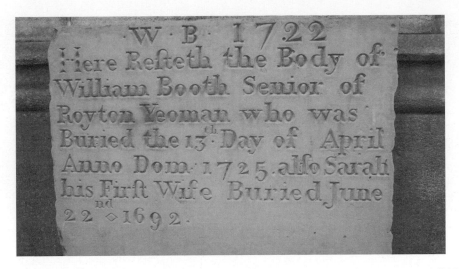

Figure 23.1 Leibniz chose the script-S because (in the handwriting of his day), a strong small 's' has this appearance. A mason carved this gravestone just after Leibniz's death: note the script-S in 'resteth' on the second line, 'also' on the sixth line and 'first' in the seventh. (Picture reproduced with the kind permission of the clergy and PCC of Oldham Parish Church, Oldham, Lancashire, UK.)

Aside

Probably the first person to develop the mathematics of integration was English mathematician, physicist, and philosopher Sir Isaac Newton (1643–1727). However, our current notation (dy/dx for a derivative and the **script-S** for an integral) derive from Leibniz (1646–1716). The calligraphic style of the 'script S' followed contemporary handwriting, and was chosen merely as an abbreviation for 'sum,' for reasons that will become apparent in the next chapter.

We now begin.

$$\frac{dy}{dx} = 5x^6 \quad \text{so} \quad y = \int 5x^6 dx$$

Integrating by rule (with eqn (23.1)) yields,

$$y = \frac{5x^{(6+1)}}{(6+1)} + c \quad \text{so} \quad y = \frac{5x^7}{7} + c \quad \blacksquare$$

We need to note:

- By convention, we generally do not convert the fraction '5/7' into a decimal, but leave it as a fraction.

- We occasionally tidy up the fraction by cancelling or rearranging.

- It is always a good idea to check the integration by differentiating the answer, to see if we get the initial expression.

■ **Worked Example 23.3**

Integrate the equation $\dfrac{1}{x^4}$ with respect to x.

Before we start, we must rewrite the expression in a more 'user-friendly' manner by removing the fraction. Using the laws of powers in Chapter 11, we write:

$$\frac{dy}{dx} = x^{-4}$$

We can now integrate by rule via eqn (23.1). We start by writing,

$$y = \int x^{-4} dx$$

so $\quad y = \dfrac{x^{-3}}{-3} + c$

which we could rewrite more tidily as $\quad y = -\dfrac{1}{3x^3} + c \quad \blacksquare$

■ **Worked Example 23.4**

Integrate the following expression: $\dfrac{dy}{dx} = \dfrac{1}{x^{1.2}} - 12x + 4$

In this example, for the first time we see a function comprising several terms. In just the same way as we **differentiate** an expression one term at a time, so we can **integrate** this expression one term at a time.

Strategy:

Anything raised to the power of zero has a value of '1', so $4x^0$ is actually 4×1, i.e. 4.

(i) It is easiest to rewrite the term '$1/x^{1.2}$' as $x^{-1.2}$.

(ii) Before we start we need to note how the final term '4' can be written as '$4x^0$'.

Solution:

$$\frac{dy}{dx} = x^{-1.2} - 12x + 4x^0 dx$$

$$y = \int \frac{dy}{dx} dx = \frac{x^{-0.2}}{-0.2} - \frac{12x^2}{2} + \frac{4x^1}{1} + c$$

so $y = -5x^{-0.2} - 6x^2 + 4x + c$, where c is a constant of integration, as explained on p. 345. ■

Self-test 23.1

Integrate the following expressions by rule using eqn (23.1):

23.1.1 $\dfrac{dy}{dx} = x^7$ **23.1.2** $\dfrac{dy}{dx} = 2x^5$

23.1.3 $\dfrac{dy}{dx} = x^{1.3}$ **23.1.4** $\dfrac{dy}{dx} = 3.7x^{8.2}$

23.1.5 $\dfrac{dy}{dx} = \dfrac{6}{x^3}$ **23.1.6** $\dfrac{dy}{dx} = \dfrac{1}{x^{4.2}} + 5$

23.1.7 $\dfrac{dy}{dx} = x^2 + 3x - 6$ **23.1.8** $\dfrac{dy}{dx} = \dfrac{4}{x^5} - \dfrac{3.5}{x^2}$

23.1.9 $\dfrac{dy}{dx} = \sqrt{x^7}$ **23.1.10** $\dfrac{dy}{dx} = \sqrt{x^6} - 2x^2$

Integrating functions other than simple powers

By the end of this section, you will know:

• How to integrate other functions by rule.

• The argument of an exponential, sine and cosine function remains unchanged following integration.

Just as we can differentiate functions other than polynomials, so we can integrate these same other functions. And we employ the same types of methodology.

■ **Worked Example 23.5**

What is the integral of e^{9x}?

The *differential* of an exponential is obtained by rewriting the exponential, without altering the argument in any way. We therefore start the integration of e^{9x} by writing just 'e^{9x}'.

We next add an additional factor, which is the differential of the argument. If we had differentiated e^{9x}, we would generate $9 \times e^{9x}$. To accommodate the additional factor of a, we need to divide the integral by '9'.

The integral of e^{9x} is therefore $\dfrac{1}{9}e^{9x}+c$. ■

The general rule for integrating exponentials is:

If $\quad \dfrac{dy}{dx}=e^{ax}\quad$ then

$$y=\frac{1}{a}e^{ax}+c \qquad\qquad (23.2)$$

where c is a constant of integration, as explained on p. 345.

■ **Worked Example 23.6**

Figure 23.2 shows the three p orbitals. In any one direction, we obtain an approximate probability P of finding an electron in a p-orbital by integrating the simplified function f

$$f=\exp\left(-\frac{kr}{a_0}\right)$$

where r is the distance of the electron from the atomic nucleus, k is a constant and a_0 is the atomic radius. What is integral of f with respect to r?

This function fits into the general equation in eqn (23.2) if we rewrite it saying, e^{ax} where $a=-k/a_0$. We can now insert terms into eqn (23.2):

If $\quad f=\exp\left(-\dfrac{kr}{a_0}\right)dr\quad$ then $\quad P=\int f\,dr=\left(-\dfrac{a_0}{k}\right)\exp\left(-\dfrac{kr}{a_0}\right)+c$

where c is a constant of integration as explained on p. 345. ■

> The function f has been simplified by removing an r^n term which should multiply the exponential.

> In this example, we have written exp rather than e, because the argument is bulky.

> **Remember:** Dividing anything by a *fraction* gives the same result as multiplying by the reciprocal of that fraction, so dividing by $-k/a_0$ is the same as multiplying by $-a_0/k$.

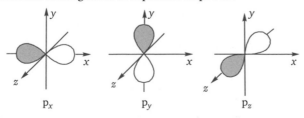

Figure 23.2 The shapes of the three p-orbitals.

■ **Worked Example 23.7**

Electrochemists often measure the **ac impedance** of a sample. They apply a voltage V across the sample and measure the resulting current. The magnitude of the voltage V alters periodically in a sinusoidal way, so

$V=\sin \omega t$

An **impedance** is an electrical resistance that changes periodically with time.

where t is time and ω is the frequency of the ac signal. The integral of the voltage yields mechanistic information. What is the integral of V with respect to time?

The rule for integrating sines and cosines are given respectively by eqn(23.3) and eqn (23.4). Notice how the argument remains intact in each case.

$$\text{If} \qquad \frac{dy}{dx} = \sin ax \quad \text{then} \quad y = -\frac{1}{a}\cos ax + c \qquad (23.3)$$

$$\text{and if} \qquad \frac{dy}{dx} = \cos ax \quad \text{then} \quad y = \frac{1}{a}\sin ax + c \qquad (23.4)$$

So, after inserting terms into eqn , we obtain:

$$\text{If} \quad V = \sin \omega t \quad \text{then} \quad \int V \, dt = -\frac{1}{\omega}\cos \omega t + c \quad \blacksquare$$

Integrating a reciprocal is more complicated:

$$\text{If} \qquad \frac{dy}{dx} = \frac{1}{x} \quad \text{then} \quad y = \ln x + c \qquad (23.5)$$

where c is a constant of integration, as defined on p. 345.

■ Worked Example 23.8

When the reaction of a chemical A proceeds via **first-order kinetics**, the rate of change of $[A]_t$ follows the equation:

$$\frac{d[A]}{dt} = -k[A]$$

Rearranging (including partial integration) yields:

$$-kt = \int \frac{1}{[A]} \, d[A]$$

Integrate the right-hand side of this equation to discern the relationship between t and $[A]_t$ during a first-order reaction.

Integration using the relationship in eqn (23.5) yields, 'ln $[A]$', so $-kt = \ln [A] + c$, which we often call the 'integrated first-order rate equation.' ■

Self-test 23.2

Integrate each of the following by rule:

23.2.1 $\dfrac{dy}{dx} = \cos 4x$

23.2.2 $\dfrac{dy}{dx} = 6\sin 4x$

23.2.3 $\dfrac{dy}{dx} = 5x^{-3}$

23.2.4 $\dfrac{dy}{dx} = \dfrac{3}{x}$

23.2.5 $\dfrac{dy}{dx} = \dfrac{1}{2}e^{5x}$

23.2.6 $\dfrac{dy}{dx} = 96.4$

23.2.7 $\dfrac{dy}{dx} = x^4 - \sin 2x$

23.2.8 $\dfrac{dy}{dx} = e^{dx} - e^{ex}$

23.2.9 $\dfrac{dy}{dx} = \dfrac{5}{x} + e^{32x}$

23.2.10 $\dfrac{dy}{dx} = b\cos ax - a\sin bx$

Constants of integration *c*

By the end of this section, you will know:

- Because a constant disappears when *differentiated*, the *integral* of a function needs a constant of integration, *c*.
- An integral with a constant of integration is called an *indefinite integral*.
- An integral without a constant of integration is called a *definite integral*.

Consider the following three functions: $y = x^2$, $y = x^2 + 3$ and $y = x^2 + 6$ (depicted in Fig. 23.3). The constant at the end of each expression (0, 3 or 6) shifts the curve up or down the page, but does not change the shape of the curve in any way. We now differentiate each:

$$y = x^2 \qquad \text{so} \qquad \frac{dy}{dx} = 2x$$

$$y = x^2 + 3 \qquad \text{so} \qquad \frac{dy}{dx} = 2x$$

$$y = x^2 + 6 \qquad \text{so} \qquad \frac{dy}{dx} = 2x$$

Because the shape of each curve is identical, the differential of each equation will be the same, and the constant disappears.

The loss of the constants places a severe limitation on the process of integration. If we want to integrate any of the three differentials above, we could not guess that the first had no constant (i.e. its constant was zero). Nor could we have guessed the value of the constant in the second was '3' and '6' in the third.

When we integrate a function, we need to accommodate this inherent uncertainty caused during differentiation. Accordingly, if we integrate $2x$ by rule, we obtain $y = x^2$. And because we know nothing about the constant in the original expression—whether

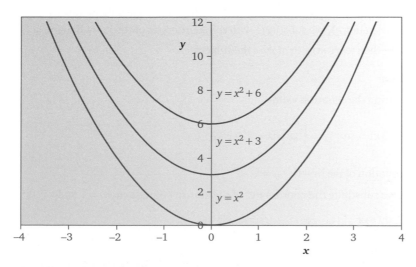

Figure 23.3 The gradient of the three functions: $y = x^2$, $y = x^2 + 3$ and $y = x^2 + 6$ are always the same, whatever the value of x.

it contained a constant at all and, if so, what its value was—we write the integral as, $y = x^2 + c$. We call the last term c a **constant of integration**. Unless additional information is available (as below), we do not know the value of c, so the integral is to some extent incomplete. We say such integrals are **indefinite**.

■ **Worked Example 23.9**

Integrate the expression $\dfrac{dy}{dx} = \dfrac{1}{x} + x^2 - \cos 2x$

We integrate this expression term by term, using eqns (23.1)–(23.5). But it also requires a constant of integration c:

Therefore, $y = \displaystyle\int \left(\dfrac{1}{x} + x^2 - \cos 2x \right) dx = \ln x + \ln x + \dfrac{x^3}{3} - \dfrac{1}{2}\sin 2x + c$ ■

Evaluating the constant of integration: definite integrals

By the end of this section, you will know:

- How to determine a value for the constant of integration c.
- The simplest method of determining c requires that we know the gradient of a line and a single point through which the line passes.

■ **Worked Example 23.10**

A line of gradient of $3x^2$ passes through the point $(1,11)$. What is the equation of the line?

Strategy:

(i) We first integrate the equation describing the gradient by rule with eqn (23.1).

(ii) Knowing the equation of the line (which at this stage will include a constant of integration c), we substitute with the coordinates of the known point.

(iii) We then rearrange to make c the subject.

Solution:

Integrating the equation yields,

$$\frac{dy}{dx} = 3x^2 \quad \text{so} \quad y = \int 3x^2 \, dx = x^3 + c$$

The equation of the line is: $y = x^3 + c$

Next, we substitute the coordinates of the known point, saying $x = 1$ and $y = 11$:

$$11 = 1^3 + c$$

so: $c = 10$

We deduce the equation of the line to be, $y = x^3 + 10$, which is depicted in Fig. 23.4. ■

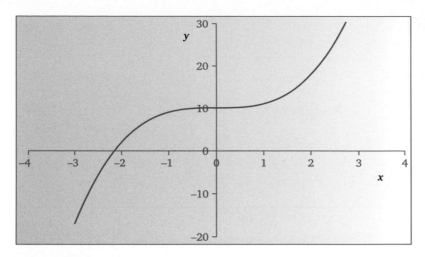

Figure 23.4 The line $y = x^3 + 10$ has the gradient $3x^2$ whatever the value of the x.

We need to note:

- We call the answer a **definite integral** when we know the value of the constant of integration.
- Usual practice tells us to leave the constant of integration as a fraction, without converting it to a decimal.

Self-test 23.3

Integrate the following differentials by rule using eqns (23.1)–(23.5), and in each case determine the constant of integration:

23.3.1 $\dfrac{dy}{dx} = x^3$ which goes through the point (0,9)

23.3.2 $\dfrac{dy}{dx} = e^{3x}$ which goes through the point (1,2)

23.3.3 $\dfrac{dy}{dx} = \sqrt{x}$ which goes through the point (7,3)

23.3.4 $\dfrac{dy}{dx} = \dfrac{1}{x}$ which goes through the point (4,5)

23.3.5 $\dfrac{dy}{dx} = x^3 - x^2$ which goes through the point (3,12)

23.3.6 $\dfrac{dy}{dx} = \dfrac{3}{x}$ which goes through the point (2,3)

Summary of key equations in the text

y	$\int y\, dx$	Equation number
a	$ax + c$	
$a x^n$	$\dfrac{ax^{(n+1)}}{(n+1)} + c$ but not if $n = -1$	(23.1)

(Continued)

e^{ax}	$\dfrac{1}{a}e^{ax}+c$	(23.2)
$\sin ax$	$-\dfrac{1}{a}\cos ax+c$	(23.3)
$\cos ax$	$\dfrac{1}{a}\sin ax+c$	(23.4)
$\dfrac{1}{x}$	$\ln x+c$	(23.5)
$\dfrac{a}{x}=a\times\dfrac{1}{x}$	$a\times\ln x+c$	

Additional problems

23.1 When an electrode is immersed in a solution of analyte and polarized, the time-dependent current I_t decreases with time t according to the **Cottrell equation**:

$$I_t = nFAc\sqrt{\frac{D}{\pi t}}$$

where other terms are constants. We define current as the rate of change of charge Q, i. e. $I=\dfrac{dQ}{dt}$.

Integrate the Cottrell equation to derive an expression relating Q and time t.

23.2 The enthalpy change ΔH associated with warming a solid from temperature T_1 to T_2 is a simple function of the heat capacity, C_p:

$$\Delta H = \int_{T_1}^{T_2} C_p\, dT$$

We can usually approximate the heat capacity C_p of a solid to a power series in T:

$$C_p = a + bT + \frac{c}{T^2}$$

Integrate this second expression to determine the enthalpy change ΔH when a substance is warmed.

23.3 (Following on from Additional problem 23.2): At very low temperatures, the heat capacity C_p of a solid is proportional to T^3, so we write $C_p = aT^3$.

Determine the temperature dependence of the enthalpy change ΔH when a substance is warmed at these low temperatures, i.e. integrate the expression, $C_p = aT^3$.

23.4 On a phase diagram for methane, and at temperatures just above the **triple point**, the slope of the solid–liquid phase boundary is given by the expression

$$\frac{dp}{dT} = 0.08446\,T^{0.85}$$

Integrate the right-hand side of this expression to obtain an expression relating p and T.

23.5 The attractive energy U between two particles depends on their separation r:

$$U = -\frac{A}{r^6}$$

where A is a jumble of constants. Integrate this expression with respect to r.

23.6 A line has the gradient $x^3 + x^2 - 1$ and goes through the point (2,20). Deduce the equation of the line.

23.7 One of the Maxwell expressions defines volume in terms of Gibbs function and pressure:

$$\left(\frac{\partial G}{\partial p}\right)_T = V.$$

Volume V is itself a function of pressure according to the ideal-gas equation, $pV = nRT$. Slight rearrangement and substituting for volume V gives:

$$dG = \frac{nRT}{p}\, dP$$

Derive a new equation, relating the change in Gibbs function ΔG for a change in pressure.

23.8 When the reaction of a chemical A proceeds via **second-order kinetics**, the rate of change of $[A]_t$ follows the equation:

$$\frac{d[A]}{dt} = -k[A]^2$$

Rearranging yields: $\dfrac{1}{[A]^2}\, d[A] = -k\, dt$

Integrate the expression $\displaystyle\int \frac{1}{[A]^2}\, d[A] = -k\int dt$

to discern the relationship between t and $[A]_t$ during a second-order reaction.

23.9 When the reaction of a chemical A proceeds via **third-order kinetics**, the rate of change of $[A]_t$ follows the equation.

$$\frac{d[A]}{dt} = -k[A]^3$$

Rearranging yields $\dfrac{1}{[A]^3} \times d[A] = -k\, dt$

Integrate the expression $\displaystyle\int \frac{1}{[A]^3}\, d[A] = -k\int dt$

to discern the relationship between t and $[A]_t$ during a third-order reaction.

23.10 A line goes through the point (3,202), and has the gradient $\dfrac{1}{x} + \exp(2x)$. What is the equation of the line?

24

Integration II

Separating the variables and integration with limits

Separating the variables

By the end of this section, you will know:

- It is possible to separate a differential expression into its two variables.
- Having separated the variables, it is common to perform two integration steps.

A differential expression will generally take the form

$$\frac{dy}{dx} = f(x)$$

where 'f(x)' is some (as yet undefined) function of x. We can treat the differential dy/dx as a fraction, which permits us to cross-multiply by 'dx'. If we do so, the expression will look like

$$dy = f(x)\, dx$$

We call this procedure **separating the variables**. We can integrate expressions like this in the usual way.

■ **Worked Example 24.1**

If $\dfrac{dy}{dx} = 4x^4$, separate the variables and then integrate with respect to x.

We first separate the variables, i.e. cross-multiply by 'dx':

$$dy = 4x^4 dx$$

We then integrate in the usual way:

$$\int dy = \int 4x^4\, dx$$

This equation looks odd because we see no function on its left-hand side. To get round this feature, we employ a dodge by saying that 'dy' is the same as '$1 \times dy$'. We then write:

$$\int 1\, dy = \int 4x^4\, dx$$

Integration is now straightforward. We integrate *both* sides, yielding:

so $$y = \frac{4x^5}{5} + c$$

The integration on the left-hand side works because the integral of 1 (i.e. $1 \times y^0$) is $1 \times y^1$.

This integration yields the same result as that obtained by integration in the usual way (see the previous chapter). ■

> Strictly, we must integrate *both* sides.

The power of this methodology becomes clearer when we realize how we can integrate *both* sides of an equation at the same time, even when neither side is a simple function of x or y.

■ **Worked Example 24.2**

Integrate the expression $y^2 \times \dfrac{dy}{dx} = 4x^6 + 2x^2 + 5$.

We first write the appropriate integration symbolism: $\int y^2\, dy = \int \left(4x^6 + 2x^2 + 5\right) dx$

We then integrate in the usual way: $\dfrac{y^3}{3} + c = \dfrac{4x^7}{7} + \dfrac{2x^3}{3} + 5x + c'$

c and c' are both constants of integration.

We usually wish to rearrange subsequently, e.g. combining these constants to form a new constant, c'' on only one side of the equation. In fact, many mathematicians simply write '$+ c$' on the right-hand side, and omit this last step. ■

Occasionally, we need to rearrange slightly *before* the integration. In such situations, we ensure that all instances of the controlled variable (including dx) are located on the right-hand side, and all instances of the observed variable (including dy) lie on the left-hand side.

■ **Worked Example 24.3**

A **phase diagram** depicts values of pressure and temperature at which two or more phases coexist at equilibrium. Figure 24.1 shows the phase diagram for carbon dioxide. We call the solid lines *phase boundaries*. A modified form of the Clapeyron equation quantifies the gradient of the bold phase boundary, which describes the boiling temperature T of liquid water as a function of the applied pressure p:

$$\frac{\mathrm{d}p}{\mathrm{d}T} = \frac{p\,\Delta H^{\ominus}_{(\text{vap})}}{RT^2}$$

where R is the gas constant and $\Delta H^{\ominus}_{(\text{vap})}$ is the molar enthalpy of boiling.

   Rearrange this equation, separate the variables and then integrate this expression to yield an equation describing the shape of the phase boundary.

Strategy:

(i) We rearrange this expression, so all instances of the controlled variable (in this case temperature, T) are located on the right-hand side, and all instances of the observed variable (in this case pressure, p) lie on the left-hand side.

(ii) We separate the variables.

(iii) We integrate.

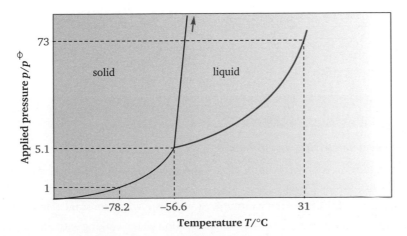

Figure 24.1 The phase diagram of carbon dioxide (not drawn to scale).

Solution:

(i) before we can integrate, we must **rearrange** because the left-hand side is expressed in terms of pressure, but a pressure term also appears on the right. We therefore cross-multiply with the p term:

$$\frac{1}{p}\frac{\mathrm{d}p}{\mathrm{d}T} = \frac{\Delta H^{\ominus}_{(vap)}}{RT^2}$$

(ii) We separate the variables:

$$\frac{1}{p}\,\mathrm{d}p = \frac{\Delta H^{\ominus}_{(vap)}}{RT^2}\,\mathrm{d}T$$

(iii) Finally, we **integrate**:

$$\int \frac{1}{p}\,\mathrm{d}p = \int \frac{\Delta H^{\ominus}_{(vap)}}{RT^2}\,\mathrm{d}T$$

We choose to place the term '$\Delta H^{\ominus}_{(vap)} \div R$' outside (i.e. to the left of) the right-hand integral because its value remains constant.

Because R and $\Delta H^{\ominus}_{(vap)}$ are both constants, while T is a variable, we will move the constants outside the integral sign:

$$\int \frac{1}{p}\,\mathrm{d}p = \frac{\Delta H^{\ominus}_{(vap)}}{R}\int \frac{1}{T^2}\,\mathrm{d}T$$

Integrating the reciprocal of p on the left-hand side yields a logarithm (see Chapter 23); and integrating the $1/T^2$ term on the right-hand side yields $-1/T$, so we obtain:

$$\ln p = -\frac{\Delta H^{\ominus}_{(vap)}}{R}\times\frac{1}{T} + c \quad \blacksquare$$

Self-test 24.1

In each case, separate the variables and then integrate the expression:

24.1.1 $\dfrac{\mathrm{d}y}{\mathrm{d}x} = zx^3$

24.1.2 $\dfrac{\mathrm{d}y}{\mathrm{d}x} = y^2 x^2$

24.1.3 $\dfrac{\mathrm{d}c}{\mathrm{d}e} = \dfrac{c^3}{e}$

24.1.4 $\dfrac{\mathrm{d}h}{\mathrm{d}g} = 4h^3 \sin g$

Integration with limits

By the end of this section, you will know:

- We employ the concept of 'limits' if we know the value of a differential at two values of the principal variable.
- Integrating with limits is an alternative way of removing a constant of integration.
- We usually do not write a constant of integration when working with limits.

In the previous chapter, we were able to evaluate a constant of integration if we knew a gradient (that is, a *differential*) and a single point. At other times, we may know the value of a variable under two separate conditions. We can utilize this situation to circumvent the need to write a constant of integration.

We call this procedure **using limits**.

Strategy:

(i) Perform an integration procedure.

(ii) Insert the upper value of the variable into this expression.

(iii) Insert the lower value of the variable into this expression.

(iv) Subtract the answer from part (iii) from the answer in part (ii).

Using the limits strategy, we cause the constant to disappear by subtracting our first answer from the second. The method allows us to evaluate an integral, even if the constant of integration is not at first apparent.

■ **Worked Example 24.4**

Evaluate the integral $\dfrac{dy}{dx} = x^3$ between the values of $x = 4$ and $x = 2$.

Strategy:

(i) We separate the variables, then integrate the expression in the normal way.

(ii) We insert first $x = 4$, then $x = 2$ into the resulting expression.

(iii) We subtract the second result from the first.

Solution:

(i) Integrating this differential according to eqn (23.1) yields, $\frac{1}{4} x^4 + c$. Because we want limits, we write the expression in square brackets, and write the limits vertically along the right-hand outer side:

$$\int dy = \int x^3 dx = \left[\frac{x^4}{4} + c \right]_2^4$$

(ii) We then insert the values of the limits, the upper value first:

$$\int dy = \left[\frac{4^4}{4} + c \right] - \left[\frac{2^4}{4} + c \right]$$

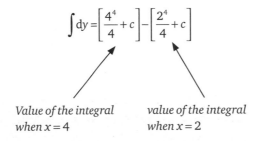

Value of the integral when x = 4   *value of the integral when x = 2*

(iii) The two c terms will cancel (which is one reason why we employ this procedure). In fact, we generally do not bother to write the constant c because it will cancel.

Having subtracted the constant c and inserted the limits, the integral is:

$$\int dy = 64 - 4 = 60$$

The numerical value of the integral is 60. ■

In words, we would normally rephrase the question to ask us to, 'integrate the function x^3 between the limits of 4 and 2'.

We call these two values of 2 and 4, the **limits**.

■ **Worked Example 24.5**

Evaluate the integral, $\int_0^{\pi/2} \sin 2x \, dx$.

Strategy:

(i) We integrate the function by rule, in this case employing eqn (23.3).

(ii) We insert first the upper limit of $\pi/2$ and the lower limit of 0, and subtract the result with the lower limit from the result with the upper limit.

(iii) Because this function involves trigonometry, we must employ *radians*.

Solution:

(i)   Inserting terms into eqn (23.3) yields:

$$\int_0^{\pi/2} \sin 2x \, dx = \left[-\frac{1}{2} \cos 2x \right]_0^{\pi/2}$$

Here, the 'dx' on the left-hand side means we integrate, 'with respect to x.'

(ii)   We next write the integral twice, and insert the upper limit in the first and the lower limit in the second:

$$\int_0^{\pi/2} \sin 2x \, dx = \left(-\frac{1}{2} \cos \left(2 \times \frac{\pi}{2} \right) \right) - \left(-\frac{1}{2} \cos (2 \times 0) \right)$$

The constant of c was not included because we know it will cancel. The cosine of $\pi/2$ is the same as the cosine of 180°, so its value is −1. The cosine of 0 is +1. Inserting these values yields:

$$\int_0^{\pi/2} \sin 2x \, dx = \left(-\frac{1}{2} \times (-1) \right) - \left(-\frac{1}{2} \times (1) \right)$$

Notice how we can say the value of the integral here is '1' rather than '1 + c'.

so $\int_0^{\pi/2} \sin 2x \, dx = 1$ ■

Self-test 24.2

Let y = the definite integrals shown. Evaluate each by integrating and inserting the limits.

24.2.1 $\int_1^2 x^4 - x^2 \, dx$ 24.2.2 $\int_1^3 e^{3x} \, dx$

24.2.3 $\int_{\pi/3}^{\pi} \cos 2x \, dx$ 24.2.4 $\int_0^5 x^3 - x2 + x - 6 \, dx$

24.2.5 $\int_{9.5}^{10} \frac{2}{x} \, dx$ 24.2.6 $\int_0^{\pi} \frac{\sin 6x}{3} \, dx$

Integration involving partial differentials

By the end of this section, you will know that even if the source differential is partial:

- We can separate variables.
- We can use limits.

We can integrate using partial differentials in just the same way as when using 'normal' differentials. The only real difference is the way that we need to keep a mental note, reminding ourselves that we have pegged one or more of the variables, so we cannot change them.

■ **Worked Example 24.6**

One of the Maxwell expressions defines volume in terms of Gibbs function and pressure: $\left(\dfrac{\partial G}{\partial p}\right)_T = V$.

Separate the variables in this expression to derive a new equation, relating the change in Gibbs function ΔG to a change in pressure. The subscripted T reminds us that temperature has been kept constant. Remember that any final expression is only valid at constant temperature.

Strategy:

1. We first separate the variables: $\partial G = V \, \partial p$.

2. We then recognize that volume V is itself a function of pressure according to the ideal-gas equation, $pV = nRT$. Substituting for volume V gives:

$$\partial G = \left(\frac{nRT}{p}\right) \partial p$$

Provided the temperature remains constant, we can then integrate. This has several consequences: the partial differentials ∂G and ∂p can be written as dG and dp. Secondly, we can regard the term nRT as a constant, and write it outside its respective integration sign.

We then write the appropriate integration symbolism. We say that:

- At pressure p_1, the Gibbs energy is G_1, and
- At pressure p_2, the Gibbs energy is G_2.

So we write:

$$\int_{G_1}^{G_2} 1 \, \mathrm{d}G = nRT \int_{p_1}^{p_2} \tfrac{1}{p} \, \mathrm{d}p$$

As before (p. 352), the '1' on the left-hand side is merely a mathematical dodge, and allows us to integrate *something*.

Integration using the template equations in Table 23.1, yields:

$$[G]_{G_1}^{G_2} = nRT \left[\ln p\right]_{p_1}^{p_2}$$

Inserting limits yields, $G_2 - G_1 = nRT \left[\ln p_2 - \ln p_1\right]$

$G_2 - G_1$ is the same as ΔG and the two logarithm terms can be combined using the second law of logarithms (see p. 187). We therefore generate the equation,

$$\Delta G = nRT \ln\left(\frac{p_2}{p_1}\right) \ ■$$

■ **Worked Example 24.7**

We define heat capacity C_p for a single, pure substance as $\left(\dfrac{\partial H}{\partial T}\right)_p$. Derive an integrated expression for ΔH.

Separating the variables yields

$$\partial H = C_p \, \partial T$$

We then insert the appropriate integration symbolism: $\int \partial H = \int C_p \, \partial T$.

This expression will yield a useable result, but we usually prefer to integrate between limits. In cases such as this, we don't actually have predetermined limits, so we 'invent' them: we say the enthalpy is H_1 at temperature T_1, and H_2 at T_2.

It does not matter which we write at the top of the integral signs and which at the bottom, provided they are respective values:

$$\int_{H_1}^{H_2} dH = \int_{T_1}^{T_2} C_p \, dT$$

Integrating then yields,

$$[H]_{H_1}^{H_2} = C_p [T]_{T_1}^{T_2}$$

We then insert the limits: $(H_2 - H_1) = C_p (T_2 - T_1)$. The term on the left-hand side is ΔH, so we write, $\Delta H = C_p (T_2 - T_1)$. ∎

This expression assumes that C_p is independent of temperature.

Additional problems

24.1   The relationship between Gibbs function G and temperature yields one of the so-called Maxwell relations:

$$\left(\frac{\partial G}{\partial T} \right)_p = -S$$

From the first law of thermodynamics, we substitute for $-S$ with $(G - H)/T$, allowing us to derive the partial differential:

$$\left(\frac{\partial (G/T)}{\partial T} \right)_p = -\frac{H}{T^2}$$

Integrate this expression. For limits, let G_2/T_2 relate to T_2 and let G_1/T_1 relate to T_1.

24.2   Consider a fixed amount of gas in a cylinder. The work w performed when the volume V of a gas is altered is given by the expression:

$$w = \int_{V_{(initial)}}^{V_{(final)}} p \, dV$$

Where p = pressure. Perform this integration to obtain an expression for w.

24.3   Current I is defined as the rate of change of charge Q

$$I = \frac{dQ}{dt}$$

where t is time. Separate the variables and integrate this expression to obtain an equation with Q as the subject.

24.4   When the reaction of a chemical A proceeds via **second-order kinetics**, the rate of change of $[A]_t$ follows the equation:

$$\frac{d[A]}{dt} = -k[A]^2$$

Rearrange, separate the variables, then integrate this expression to obtain the **integrated second-order rate equation.**

24.5   Acceleration a is defined as the rate of change of velocity v:

$$a = \frac{dv}{dt}$$

Integrate this expression using these limits. To obtain the constant of integration, the velocity at time $t = 0$ is u and the velocity at time $t = t$ is v.

24.6 Evaluate the following definite integral: $\int_2^4 \sqrt[3]{5x} \; dx$.

24.7 The speed at which electrons move through a semiconductor relates to its mobility μ. For many semiconductors, a gradient of a graph mobility μ of against temperature T, has the following form:

$$\frac{d\mu}{dT} = k\sqrt{T}$$

where k represents a collection of constants. Separate the variables and hence integrate this expression to obtain a relationship in which μ is the subject.

24.8 We define an object's velocity v as the rate at which its position changes:

$$v = \frac{dl}{dt}$$

where l is distance from the start position. Separate the variables and hence integrate this expression to obtain a relationship involving v, l and t.

24.9 When the reaction of a chemical A proceeds via **first-order kinetics**, the rate of change of $[A]_t$ follows the equation:

$$\frac{d[A]}{dt} = -k[A]$$

Rearrange, separate the variables, then integrate this expression to obtain the **integrated first-order rate equation**.

24.10 Evaluate the following definite integral: $\int_{\pi/12}^{\pi/6} \sin 4x \, dx$.

25

Integration III

Integration by parts, by substitution and integration tables

Integration by parts

By the end of this section, you will know:

- We can integrate the differentials produced by using the product rule (eqn (20.1)).
- Integration by parts allows us to integrate products that cannot be integrated more straightforwardly.
- Integration by parts is only really easy if one function is a simple polynomial.
- Occasionally, we need to integrate by parts more than once—a procedure we call successive reduction.

We saw in Chapter 20 how differentiating a product yields an answer in two halves:

$$\frac{d(u \times v)}{dx} = u \times \left(\frac{dv}{dx}\right) + v \times \left(\frac{du}{dx}\right) \tag{20.1}$$

where u and v are the two component functions of x within the product we are differentiating. In this section, we learn how to, in effect, reverse the product rule, and integrate expressions that (initially) look too complicated to solve. These expressions consist of two terms multiplied together. They must fit into the format described below.

Integrating the right-hand side of eqn (20.1) with respect to x is straightforward:

$$uv = \int u \left(\frac{dv}{dx}\right) dx + \int v \left(\frac{du}{dx}\right) dx \tag{25.1}$$

If we know either of the results on the right-hand side of eqn (20.1), we can straightforwardly determine the other. We therefore have the choice of calculating either of the two integrals. In practice, we will determine the easiest, or the one that is possible without recourse to complicated procedures.

If, for example, we decide that $\int u \left(\frac{dv}{dx}\right) dx$ is the easier function to find, then

$$\int u \left(\frac{dv}{dx}\right) dx = uv - \int v \left(\frac{du}{dx}\right) dx \tag{25.2}$$

This equation is the inverse of the product rule, eqn (20.1). We call its use, the rule of **integration by parts**. In words, we say this expression describes:

$$\begin{array}{ll} \text{Integration} \\ \text{by parts} \end{array} = \begin{array}{l} 1^{\text{st}} \text{ function} \times \\ \text{Integral of } 2^{\text{nd}} \\ \text{function} \end{array} - \begin{array}{l} \text{Integral of (integral} \\ \text{of the } 2^{\text{nd}} \text{ function} \times \\ \text{differential of the} \\ 1^{\text{st}} \text{ function)} \end{array}$$

The integration of a product therefore involves one integration problem and one differentiation problem … and a little bit of rearrangement.

When we integrate a product, we must first decide which term to call u and which (dv/dx). The choice can dictate whether we simplify the product sufficiently that integration is possible, or actually make the expression more difficult.

■ **Worked Example 25.1**

Integrate the product, $\int x\cos x\ dx$.

This expression comprises two completely different types of function.

Strategy:

(i) We decide which function to call u and which is dv/dx.

(ii) We insert terms into eqn (25.2).

Solution:

(i) We will call x 'u' and let 'dv/dx' equal $\cos x\ dx$:

$$u = x \qquad \frac{dv}{dx} = \cos x$$

$$\frac{du}{dx} = 1 \qquad v = \sin x$$

(ii) Substituting into eqn (25.2) gives, $\int x\cos x\ dx = x\sin x - \int \sin x\ \frac{d}{dx}(x)\ dx.$

so $\qquad \int x\cos x\,dx = x\sin x - \int \sin x\ dx$

and $\qquad \int x\cos x\ dx = x\sin x + \cos x + c$

where c is the inevitable constant of integration.

Therefore, instead of having to find the integral product, $\int x\cos x\ dx$, we only had to find the simpler integral, $\int \sin x\ dx$. ■

Without limits, integration by parts yields an indefinite integral.

The 'art' of integrating by parts lies in the ability to make the correct choice of u and v. In the example immediately above, $u = x$ and $dv/dx = \cos x$. The second term must be easy to integrate, and differentiating the first term must make the resultant integral easier to solve.

■ **Worked Example 25.2**

What would happen if we had tried to integrate the product, $\int x\cos x\,dx$ but made the opposite (unwise) choice of $u = \cos x$ and $dv/dx = x$?

Using the same procedure as that outlined in Worked Example 25.1, we obtain:

$$u = \cos x \qquad \frac{dv}{dx} = x$$

$$\frac{du}{dx} = -\sin x \qquad v = \frac{x^2}{2}$$

Inserting terms into eqn (25.2) yields,

$$\int x\cos x\ dx = \frac{1}{2}x^2\cos x - \frac{1}{2}\int x^2\sin x\ dx$$

The integral on the far right-hand side is now *more* difficult than the one we first sought to solve. ■

This example demonstrates the simple rule: if one of the components within the product is a polynomial, we should choose it as the function u in the expression in eqn (25.2).

In general, an integral with a polynomial of order n can be solved by repeated integration by parts, with n steps. Each application of the rule lowers the power of x by one. So integration of '$x^7 \sin x$' requires that we perform the integration-by-parts procedure 7 times: $-x^7 \cos x + 7x^6 \sin x + 42x^5 \cos x - 210x^4 \sin x - 840x^3 \cos x + 2520x^2 \sin x + 5040x \cos x - 5040 \sin x + c$.

■ Worked Example 25.3

Integrate the product, $x^2 e^x$.

First integration by parts:

Let $u = x^2$ and $dv/dx = e^x$. Therefore, $du/dx = 2x$, and $v = e^x$. Inserting terms into eqn (25.2) yields, $\int x^2 e^x \, dx = (x^2 \times e^x) - \int 2x \times e^x \, dx$. We cannot solve this second integral using the standard formulae in the previous chapters, because it too is a product. We need to integrate by parts a second time.

Second integration by parts:

We need to reduce it further: we say, $2x = u$ and $e^x = dv/dx$. Therefore, $du/dx = 2$, and $v = e^x$. This integral becomes, $\int 2x \ e^x \ dx = 2x e^x - 2 \int e^x \, dx$.

The overall integral is therefore, $\int x^2 \ e^x \ dx = x^2 e^x - 2x e^x + 2e^x + c$.

We might wish to factorize this result, saying, $\int x^2 \ e^x \ dx = e^x (x^2 - 2x + 2) + c$. ■

There is only one straightforward exception to the rule that the polynomial term should be selected as the function u to be differentiated and this occurs, when the product comprises a polynomial and a logarithm term (as shown in Worked Example 25.4).

Before attempting this integration, we should note that sometimes, it is a convenient 'dodge' to say a function has been multiplied by '1', and then treat the function as though it was a conventional product. In this way, we can integrate a function that is otherwise impossible, or much more difficult.

■ Worked Example 25.4

Integrate the *single* function $\ln x$.

Strategy:

(i) We will say this is a product of $1 \times \ln x$.

(ii) We integrate by parts in the usual way using eqn (25.2).

Solution:

(i) The reason why this problem is difficult and requires this new procedure follows because we can't integrate $\ln x$ directly. Therefore, it's imperative that it's $\ln x$ we choose as u; dv/dx is therefore 1.

(ii) $u = \ln x \qquad \dfrac{dv}{dx} = 1$

$\dfrac{du}{dx} = \dfrac{1}{x} \qquad v = x$

Inserting respective terms into eqn (25.2) gives:

$$\int \ln x \ dx = x \ln x - \int x \times \frac{1}{x} \, dx = \int \ln x \ dx = x \ln x - \int 1 \, dx$$

So the integral $\int \ln x \; dx = x \ln x - x + c$, where c is the usual constant of integration. This result is so unusual we should check it, differentiating using the product rule. The result is, $1 + \ln x - 1 = \ln x$. ∎

Occasionally, we need to be subtle, and look for patterns in our answers. This approach is particularly useful when integrating trigonometric functions, where repeated integration generates the pattern, $\sin \rightarrow \cos \rightarrow - \sin \rightarrow -\cos$. Simple exponentials such as e^x also generate themselves, and are susceptible to the same sort of approach.

■ **Worked Example 25.5**

Integrate the product, $e^x \cos x$.

Let $u = e^x$ and $dv/dx = \cos x$. Therefore, $du/dx = e^x$ and $v = \sin x$. Inserting terms into eqn (25.2) yields, $\int e^x \cos x \; dx = e^x \sin x - \int e^x \sin x \; dx$. It should be obvious that the integral portion here possesses exactly the same form as its source integral: the product of a trigonometric function and an exponential.

If we integrate the second integral on the right-hand side, the equation becomes:

$$\int e^x \cos x \; dx = e^x \sin x - \left(-e^x \cos x + \int e^x \cos x \; dx \right)$$

which simplifies to:

$$\int e^x \cos x \; dx = e^x \sin x + e^x \cos x - \int e^x \cos x \; dx$$

$$2 \int e^x \cos x \; dx = e^x (\sin x + \cos x)$$

so $\int e^x \cos x \; dx = \dfrac{e^x}{2} (\sin x + \cos x) + c$ ∎

> Notice how the last term on the right-hand side is the same as the term on the left-hand side.

In summary, integration by parts is straightforward when one term is a polynomial. Many of the other functions we encounter in chemistry require a degree of subtlety that is beyond the scope of this technique, so we need other methods.

Self-test 25.1

Integrate the following functions using the method of integration by parts.

25.1.1 $x \sin x$	**25.1.2** $x \ln x$
25.1.3 $x e^{ax}$	**25.1.4** $x^2 \sin x$
25.1.5 $x^3 \ln x$	

Integration by substitution

By the end of this section, you will know:

- If the function to be integrated has the form $f(g(x)) \times g'(x)$, then integration could be made possible by using the substitution $u = g(x)$ and changing the variable of integration.
- Backward substitution of the form $x = h(u)$ is sometimes also helpful.

Integration by parts is not always appropriate. In some cases, it is more effective to introduce a substitute variable and change the variable of integration. This is known as **integration by substitution**. We will recognize that there are some similarities to the chain rule for differentiation, which was discussed in Chapter 19.

To understand this method, we will start by remembering that integration is the reverse operation of differentiation. The chain rule tells us the differential of cos (x^5) is $-5x^4 \sin (x^5)$. Thus, if we are asked to evaluate $-\int 5x^4 \sin(x^5)dx$ then it is relatively easy for us to say that the answer is 'cos (x^5)'. We can make this result clearer using substitution.

■ **Worked Example 25.6**

Use the method of integration by substitution to evaluate $-\int 5x^4 \sin(x^5)dx$.

Let $u = x^5$. It follows that $\dfrac{du}{dx} = 5x^4$.

If we substitute these values into the integral: $-\int 5x^4 \sin x^5 dx = -\int \sin u \dfrac{du}{dx} dx$

This result can be simplified further to give: $-\int \sin u \, du = \cos u + c$.

Substituting for u: $-\int 5x^4 \sin(x^5)dx = \cos(x^5)+c$. ■

The general form for integration by substitution is:

$$\int f(g(x))\left(\frac{dg}{dx}\right) dx = \int f(g(x))g'(x) \, dx = \int f(u) \, du, \qquad (25.3)$$

where $u = g(x)$. The general strategy for this method is:

- Choose $u = g(x)$.

- Find $\dfrac{du}{dx} = \dfrac{dg}{dx} = g'(x)$.

If the second term in the integral is actually $g'(x)/a$, where a is a constant, then it may be necessary to multiply by $1/a$.

- Substitute u into $f(g(x))$ and replace $g'(x) \, dx$ by du.

- Evaluate $\int f(u) \, du$.

- Substitute back in for u in terms of x.

■ **Worked Example 25.7**

Use the method of integration by substitution to evaluate $\displaystyle\int \frac{x}{3x^2 +5} \, dx$.

Note:

When we look at this equation, we immediately see:

(i) The numerator in the function, x, is the differential of the denominator divided by 6.

(ii) This integral looks similar to: $\displaystyle\int \frac{1}{x} dx$, which we know is just $\ln x + c$.

Solution:

(i) Let $u = (3x^2 + 5)$.

(ii) $\dfrac{du}{dx} = 6x$ (Note that the second term in the integral is actually $g'(x)/6$.)

(iii) Substitute u into the integral: $\dfrac{1}{6}\displaystyle\int \frac{1}{u}\frac{du}{dx} dx$.

(iv) Evaluating this integral: $\dfrac{1}{6}\ln u + c$.

(v) Substituting back in for u: $\dfrac{1}{6}\ln(3x^2 + 5) + c$.

So, $\displaystyle\int \dfrac{x}{3x^2 + 5}\,dx = \dfrac{1}{6}\ln(3x^2 + 5) + c$. ∎

Limits: If we are interested in evaluating a definite integral, then we use:

$$\int_a^b f(g(x))g'(x)\,dx = \int_{g(a)}^{g(b)} f(u)\,du \tag{25.4}$$

We must convert the limits from '$x =$' to '$u =$'.

■ **Worked Example 25.8**

Use the method of integration by substitution to evaluate $\displaystyle\int_1^2 x\,e^{x^2}\,dx$.

(i) $u = x^2$

(ii) $\dfrac{du}{dx} = 2x$

(iii) If $x = 1$, $u = 1^2 = 1$

If $x = 2$, $u = 2^2 = 4$

(iv) Substitute u into the integral: $\dfrac{1}{2}\displaystyle\int_{u=1}^{u=4} e^u\,du$

Note that we have replaced $g'(x)\,dx$ by du directly and the limits have changed to the values for the u variable.

(v) Evaluate the integral: $\dfrac{1}{2}\left[e^u\right]_1^4 = \dfrac{1}{2}\left(e^4 - e^1\right) = 25.9$

Therefore, $\displaystyle\int_1^2 x\,e^{x^2}\,dx = 25.9$. ∎

Backward substitution:

Sometimes it is not possible to identify the du/dx term in the integral expression. For instance, if $\displaystyle\int \sqrt{1 - x^2}\,dx$ and we try to substitute $u = 1 - x^2$, then $du/dx = -2x$. However, this term is not present in the integral expression and it is not a constant. In these cases, it is necessary to:

- Use a substitution of the form $x = h(u)$, where $u = h^{-1}(x)$.
- $h(u)$ is often a trigonometric function.
- Replace dx by $(dx/du)\,du$.

■ **Worked Example 25.9**

Evaluate $\displaystyle\int_0^1 \sqrt{1 - x^2}\,dx$, using the backward substitution $x = \sin u$.

(i) $u = \sin^{-1} x$

(ii) $\dfrac{dx}{du} = \cos u$.   Therefore, $dx = \cos u\,du$.

(iii) If $x = 0$, $u = 0$.

If $x = 1$, $u = \pi/2$.

(iv) Substitute the expression for x in terms of u into the integral (note the change in the limits): $\int_0^{\frac{\pi}{2}} \sqrt{1 - \sin^2 u} \cos u \; du$

(v) Using the trigonometric identity, $\sin^2 u + \cos^2 u = 1$ (see Chapter 16):

$$\int_0^{\frac{\pi}{2}} \sqrt{\cos^2 u} \cos u \; du = \int_0^{\frac{\pi}{2}} \cos^2 u \; du$$

(vi) Using the trigonometric relationship, $\cos 2u = 2\cos^2 u - 1$ (see Chapter 16):

$$\int_0^{\frac{\pi}{2}} \left(\frac{\cos 2u + 1}{2} \right) du = \frac{1}{2} \times \left[\left(\frac{\sin 2u}{2} + u \right) \right]_0^{\frac{\pi}{2}}$$

(vii) Evaluating the integral: $\dfrac{1}{2} \left(\left(\dfrac{\sin\left(2 \times \frac{\pi}{2}\right)}{2} + \dfrac{\pi}{2} \right) - \left(\dfrac{\sin(2 \times 0)}{2} + 0 \right) \right)$

Hence, $\int_0^1 \sqrt{1 - x^2} \; dx = \dfrac{\pi}{4}$.

We have actually found the area of quarter of a circle, which is represented by the equation $x^2 + y^2 = 1$. ∎

Self-test 25.2

Tip: It is useful to check the answer by differentiating it, and seeing if it is identical to the function we were trying to integrate.

Use the method of integration by substitution to evaluate the following.

25.2.1 $\int x^5 \left(x^6 + 7 \right)^4 dx$ **25.2.2** $\int (\cos x) e^{\sin x} \; dx$

25.2.3 $\int_0^{\frac{\pi}{2}} \sin x \cos^3 x \; dx$

Integration via power series

By the end of this section, you will know:

- The integration of some mathematical functions, that are difficult to integrate, can be simplified by first writing them as a power series.
- We integrate such series term by term.

We can gain insights into some mathematical functions by rewriting them in the form of a **power series** in x. One of the most useful concerns exponential functions, which we write as an infinite series. For example,

$$e^{ax} = 1 + \frac{(ax)^1}{1!} + \frac{(ax)^2}{2!} + \; ... \; + \frac{(ax)^\infty}{\infty!} = 1 + \sum_{i=1}^{\infty} \frac{(ax)^n}{n!} \tag{25.5}$$

This is a **Maclaurin series**. The last term in the series will clearly have a value of zero, because anything divided by infinity = 0. Notice e^x also works for any argument, negative or positive.

We generally find that terms increase until perhaps the third or fourth term, as the numerator gets larger. But then, because the denominator gets larger more quickly than the numerator does, the terms get progressively smaller. In fact, the terms decrease rapidly. We say the power series **converges** because successive terms get progressively smaller. It is rare to need more than about ten power terms, and we can sometimes use far fewer, unless we require the answer to many decimal places.

Sine and cosine functions can be expressed by similar series approximations:

$$\sin x = x - \frac{x^3}{3!} + \frac{x^5}{5!} - \frac{x^7}{7!} + \dots \tag{25.6}$$

$$\cos x = 1 - \frac{x^2}{2!} + \frac{x^4}{4!} - \frac{x^6}{6!} + \dots \tag{25.7}$$

Self-test 25.3

25.3.1 The probability of finding an electron at a distance of r from the nucleus relates to the wavefunction ψ, which is given by the expression

$$\psi = Ae^{-Zr/a_o}$$

where a_o is the atomic radius, Z is the atomic number (a constant), and A is a collection of constants. Express this exponential as a power series including just five terms.

The great advantage of using reduction formulae concerns their ease of integration: we simply integrate each part of the expression, term by term.

■ **Worked Example 25.10**

Using eqn (25.5), show that the integral of e^{ax} is $\frac{1}{a}e^{ax}$.

Using eqn (25.5), the power to describe e^{ax} is,

$$e^{ax} = 1 + \frac{(ax)^1}{1} + \frac{(ax)^2}{2!} + \frac{(ax)^3}{3!} + \frac{(ax)^4}{4!} + \dots = 1 + \frac{ax}{1} + \frac{a^2x^2}{1 \times 2} + \frac{a^3x^3}{1 \times 2 \times 3} + \frac{a^4x^4}{1 \times 2 \times 3 \times 4} + \dots$$

The integral of this expression with respect to x is:

$$\int e^{ax}\, dx = x + \frac{1}{2} \times \frac{ax^2}{1} + \frac{1}{3} \times \frac{a^2x^3}{1 \times 2} + \frac{1}{4} \times \frac{a^3x^4}{1 \times 2 \times 3} + \frac{1}{5} \times \frac{a^4x^5}{1 \times 2 \times 3 \times 4} + \dots + c$$

where c is the inevitable constant of integration. Its value is $1/a$. We then tidy the denominators of each term:

$$\int e^{ax}\, dx = \frac{1}{a} + x + \frac{ax^2}{1 \times 2} + \frac{a^2x^3}{1 \times 2 \times 3} + \frac{a^3x^4}{1 \times 2 \times 3 \times 4} + \frac{a^4x^5}{1 \times 2 \times 3 \times 4 \times 5} + \dots$$

The denominators are now factorials again:

$$\int e^{ax}\, dx = \frac{1}{a} + x + \frac{ax^2}{2!} + \frac{a^2x^3}{3!} + \frac{a^3x^4}{4!} + \frac{a^4x^5}{5!} + \dots$$

Finally, we multiply top and bottom of each term by a (i.e. $a/a = 1$):

$$\int e^{ax}\, dx = \frac{1}{a} \times 1 + \frac{1}{a} \times ax + \frac{1}{a} \times \frac{a^2x^2}{2!} + \frac{1}{a} \times \frac{a^3x^3}{3!} + \frac{1}{a} \times \frac{a^4x^4}{4!} + \frac{1}{a} \times \frac{a^5x^5}{5!} + \dots$$

Factorizing out the $1/a$ terms, and tidying the numerator terms, yields,

$$\int e^{ax} \, dx = \frac{1}{a}\left(1 + ax + \frac{(ax)^2}{2!} + \frac{(ax)^3}{3!} + \frac{(ax)^4}{4!} + \frac{(ax)^5}{5!} + \cdots \right).$$

The term in the bracket is e^{ax}, so we summarize saying, $\int e^{ax} \, dx = \dfrac{1}{a}e^{ax}$. ■

Self-test 25.4

For each of the following, write a simple power series in x, and then integrate to yield an expression for $\int y \, dx$. Go as far as a term in x^4.

25.4.1 $y = e^{4x}$ **25.4.2** $y = x\, e^{8x}$

25.4.3 $y = 3x^2\, e^x$ **25.4.4** $y = x^2 \exp(zx/a_0)$

This form of reduction treatment is relatively simple, but can become tedious, particularly if the series does not converge quickly, necessitating a long series. It can be easier to obtain integrals from published sources, as described in the next section.

Published tables of integrals

By the end of this section, you will know:

- Many integrals are extremely complicated to solve.
- Published tables of integrals help the hard-pressed chemist.

We have seen already how some integrals can be extremely complicated to solve. They are difficult and, even if we knew a procedure to solve them, their solution could require a considerable length of time.

Help is at hand: chemists can take a simple way out by looking at published lists of definite integrals. For example, the *CRC handbook of chemistry and physics* ('The Rubber Handbook') contains a list of nearly a thousand integrals. The first integral in the list is familiar to us: the integral of ax^n in eqn (23.1). The last integral in the list, number 728, is the horrendous-looking,

$$\int \sin^n x \, dx = \frac{1}{2^{n-1}} \sum_{k=0}^{\frac{n-1}{2}} \binom{n}{k} \frac{\sin[n - 2k(\pi/2) - x]}{2k - n}$$

where n is odd; the answer to the integral differs for even values of n. Even chemists with years of mathematical training would be stuck without such lists. They are also useful for checking our answer if we need to integrate an unusual looking example.

Table 25.1 contains a few of the more useful standard integrals for which none of the answers is straightforward and/or intuitively obvious.

These published lists are never fail-safe:

- We have to copy them *exactly*.

- Most chemists have, at some point, seen an equation in such a list and mistaken it for the very similar but slightly different integral they require. This mistake is surprisingly easy to make if our integral is expressed in algebra other than x and y.

- Another potential problem concerns the way the published integrals of trigonometric functions relate to degrees in *radians*.

Table 25.1 Standard integrals.

Function	Integral $(+c)$	Domain
$\sec^2(ax)$	$\dfrac{1}{a}\tan(ax)$	$ax \neq (2n+1)\dfrac{\pi}{2}$
$\operatorname{cosec}^2(ax)$	$-\dfrac{1}{a}\cot(ax)$	$ax \neq n\pi$
$\dfrac{1}{x^2+a^2}$	$\dfrac{1}{a}\arctan\left(\dfrac{x}{a}\right)$	all real
$\dfrac{1}{(x-a)(x-b)}$	$\begin{cases} \dfrac{1}{(a-b)}\ln\left(\dfrac{a-x}{x-b}\right) \\[2mm] \dfrac{1}{(a-b)}\ln\left(\dfrac{x-a}{x-b}\right) \end{cases}$	$b < x < a$ $x < b < a$ or $x > a > b$
$\dfrac{1}{\sqrt{x^2+a^2}}$	$\ln\left(x+\sqrt{x^2+a^2}\right)$	all real
$\dfrac{1}{\sqrt{x^2-a^2}}$	$\begin{cases} \ln\left(x+\sqrt{x^2-a^2}\right) \\[2mm] \ln\left(-x-\sqrt{x^2-a^2}\right) \end{cases}$	$x > |a|$ $x < -|a|$
$\dfrac{1}{\sqrt{a^2-x^2}}$	$\arcsin\left(\dfrac{x}{a}\right)$	$-1 < \dfrac{x}{a} < 1$
$\ln(ax)$	$x\ln(ax) - x$	$ax > 0$
$\displaystyle\int_0^\infty \sqrt{x}\,e^{-x}\,dx$	$\dfrac{\sqrt{\pi}}{2}$	
$\displaystyle\int_0^\infty e^{-ax^2}\,dx$	$\dfrac{1}{2}\sqrt{\dfrac{\pi}{a}}$	
$\displaystyle\int_0^\infty \dfrac{\sin(x)}{x}\,dx$	$\dfrac{\pi}{2}$	
$\displaystyle\int_0^\infty \dfrac{\sin^2(x)}{x^2}\,dx$	$\dfrac{\pi}{2}$	
$\displaystyle\int_0^a \sin\left(\dfrac{n\pi x}{a}\right)\sin\left(\dfrac{m\pi x}{a}\right)dx$	0 if $m \neq n$ $\dfrac{a}{2}$ if $m = n$	
$\displaystyle\int_0^a \cos\left(\dfrac{n\pi x}{a}\right)\cos\left(\dfrac{m\pi x}{a}\right)dx$	0 if $m \neq n$ $\dfrac{a}{2}$ if $m = n$	
$\displaystyle\int_0^a \sin\left(\dfrac{n\pi x}{a}\right)\cos\left(\dfrac{n\pi x}{a}\right)dx$	0	

(Continued)

Table 25.1 (Continued)

Function	Integral $(+c)$	Domain
$\int_0^\infty x^n e^{-ax} dx$	$\dfrac{n!}{a^{(n+1)}}$	$a > 0$
$\int_0^\infty x^{2n} e^{-ax^2} dx = \dfrac{1}{2}\int_{-\infty}^\infty x^{2n} e^{-ax^2} dx$	$\dfrac{(2n-1)!!^*}{2(2a)^n}\sqrt{\dfrac{\pi}{a}}$	$a > 0, n = 0, 1, 2, \ldots$
$\int_0^\infty x^{2n+1} e^{-ax^2} dx$	$\dfrac{n!}{2a^{n+1}}$	$a > 0$

*A double factorial $(2n-1)!! = \dfrac{(2n)!}{2^n n!} = (2n-1) \times (2n-3) \times (2n-5) \times \ldots$

■ Worked Example 25.11

The average of the speed squared in the x-direction (v_x^2) for an ensemble of gas particles can be found by calculating the velocity distribution function:

$\langle v_x^2 \rangle = \displaystyle\int_{-\infty}^\infty v_x^2 f(v_x) dx$, where $f(v_x)$ is the Maxwell–Boltzmann velocity distribution:

$$f(v_x) = \left(\frac{m}{2\pi kT}\right)^{\frac{1}{2}} \exp\left(\frac{-mv_x^2}{2kT}\right)$$

Integrate this expression.

Let $A = \dfrac{m}{2kT}$ and substitute $f(v_x)$ into the integral. Integrate this expression:

$$\langle v_x^2 \rangle = \int_{-\infty}^\infty v_x^2 \times \sqrt{\frac{A}{\pi}} \exp(-Av_x^2) dx$$

This has the form of one of the standard integrals in Table 25.1:

$$\int_0^\infty x^{2n} e^{-ax^2} dx = \frac{1}{2}\int_{-\infty}^\infty x^{2n} e^{-ax^2} dx = \frac{(2n-1)!!}{2(2a)^n}\sqrt{\frac{\pi}{a}}, \text{ where } n = 1 \text{ and } a = A.$$

We will need to multiple the standard integral by 2 since the lower limit of our integral is $-\infty$ rather than 0.

Therefore, we can evaluate the integral as:

$$\langle v_x^2 \rangle = 2\sqrt{\frac{A}{\pi}} \frac{(2-1)!!}{2(2A)^1}\sqrt{\frac{\pi}{A}} = \frac{1}{2A}$$

Substituting in for A: $\langle v_x^2 \rangle = \dfrac{2kT}{2m} = \dfrac{kT}{m}$. ■

Choosing a method

Simple integrals, such as those discussed in Chapters 23 and 24, are relatively easy to solve. However, when evaluating more complicated integrals, the following approach is recommended to help when selecting a suitable method for integration:

- Check whether the integral is (or can be rearranged to give) a standard function found in published tables.
- If the integral is a product and one term of the product gives a constant when differentiated (one or more times), then try integration by parts.
- If the function being integrated has the form: $f(g(x)) \times g'(x)$, where f is a function that can be integrated, then we should try integration by substitution.
- If the function does not fit any of the forms above, then we see if we can find a backwards substitution that will convert the function to a form that can be integrated.

With practice (and some trial and error), we should be able to select the best method to solve the problem.

Summary of key equations in the text

Integration by parts:
$$\int u \left(\frac{dv}{dx} \right) dx = uv - \int v \left(\frac{du}{dx} \right) dx \text{ w} \tag{25.2}$$

Integration by substitution:
$$\int f(g(x)) g'(x) \, dx = \int f(u) \, du \tag{25.3}$$

$$\int_a^b f(g(x)) g'(x) \, dx = \int_{g(a)}^{g(b)} f(u) \, du \tag{25.4}$$

Maclaurin series:
$$e^{ax} = 1 + \frac{(ax)^1}{1!} + \frac{(ax)^2}{2!} + \dots + \frac{(ax)^n}{n!} + \dots \tag{25.5}$$

$$\sin x = x - \frac{x^3}{3!} + \frac{x^5}{5!} - \frac{x^7}{7!} + \dots \tag{25.6}$$

$$\cos x = 1 - \frac{x^2}{2!} + \frac{x^4}{4!} - \frac{x^6}{6!} + \dots \tag{25.7}$$

Additional problems

To simplify the mathematics, most of these examples are not chemistry related.

25.1 Using a power series, determine the value of e^3.

25.2 Use the Maclaurin series for $\sin x$ and $\cos x$ to show that $\int \cos x \, dx = \sin x$.

25.3 Integrate the product, $e^{-x} \sin x \, dx$.

25.4 The mathematical form $r^2 e^{-ar}$ occurs very often in quantum chemistry, particularly when integrating volumes (see next chapter). This function occurs so often because we multiply by the function by '$r^2 \sin \theta$' when converting a problem from Cartesian coordinates to spherical polar coordinates (see p. 257). So, if a function already includes an exponential function in r—as is common in quantum chemistry—we automatically generate $r^2 e^{-ar}$. For example, in spherical polar coordinates, the square of the wavefunction for a hydrogenic atom ψ is described by an equation of the form,

$$\psi^2 = N^2 \exp(-2r/a_0)$$

so $$\int \psi^2 \, dV = \int_0^{2\pi} \int_0^{\pi} \int_0^{\infty} N^2 \exp(-2r/a_0) \, r^2 \sin\theta \, dr \, d\theta \, d\phi$$

Integrate the r term: $\int r^2 \exp(-ar) \, dr$, where $a = 2/a_0$.

25.5 Following on from the integration shown in Additional problem 25.4, integrate the r term but this time include the limits: $\int_0^{\infty} r^2 \exp\left(-\dfrac{2r}{a_0}\right) dr$.

25.6 Integrate the product, $x \sin 3x$.

25.7 Integrate the product $x^2 \ln x$.

25.8 Show that if $x > a > b$, then $\displaystyle\int \frac{1}{(x-a)(x-b)} \, dx = \frac{1}{a-b} \ln\left(\frac{x-a}{x-b}\right)$. This integral is given in Table 25.1, which lists some of the standard integrals. It is known as the method of partial fractions and is a useful method to integrate quotients that can be factorized in this way.

25.9 To find the average distance of an electron from the nucleus in a hydrogen atom, it is necessary to find the expectation value of the operator that corresponds to the distance from the nucleus, r. Assume that the wavefunction is: $\psi = \left(\dfrac{1}{\pi a_0^3}\right)^{\!1/2} e^{-r/a_0}$ and that we have averaged over the angular spatial components, so the integral to solve is:

$$\int \psi^* r \psi \, dV = 4\pi \times \left(\frac{1}{\pi a_0^3}\right) \int_0^{\infty} r^3 e^{-2r/a_0} \, dr.$$

25.10 Use backwards substitution to find the following integral: $\displaystyle\int \frac{x}{\sqrt{x-4}} \, dx$ (Take $x = u^2 + 4$.)

26

Integration IV

Area and volume determination

Simple integration to determine a two-dimensional area

By the end of this section, you will know:

- Integrating an equation of a line between limits can yield an area.
- Integration affords the most accurate method of determining an area.

One of the easier methods of determining the area *A* beneath a curve drawn on a piece of graph paper is to count the squares between the axis and the curve. The method can be surprisingly accurate, especially if the squares are small. Indeed, the smaller the squares, the more accurate is the final area. The method is often accurate but very time consuming. And the scope for error increases as the complexity of the shape grows.

A somewhat simpler method, but also generally accurate, again relies on drawing: we draw the curve carefully on graph paper, and divide it into thin rectangles. Figure 26.1 shows such a curve: the area beneath the curve and between the two limits A and B is represented by a series of slender rectangles, each with the same width (call it 'δx'), but each with a differing height.

A moment's thought suggests that this method is also approximate. The rectangles align with the *x*-axis, but the top of each rectangle extends slightly beyond the curve. With care, the error associated with this method can be minimized: in practice, we ensure the curve cuts neatly across the top of each rectangle, such that the area (by eye) above the curve is the same as the beneath it, see Fig. 26.2. In this way, the errors per rectangle then cancel.

We obtain the area of a rectangle by multiplying its height by its width. The width of each rectangle here is δx, and the height is *y*. We see how the area per rectangle is $(y \times \delta x)$, and the area between the limits A and B and beneath the curve is then merely the sum of these rectangular areas: $\Sigma\,(y \times \delta x)$.

To minimize the errors still further, we ensure that each rectangle is infinitesimally narrow, so its width is no longer δx but d*x*. We say the area is:

$$\text{area} = \Sigma\,y\,\mathrm{d}x$$

In general practice, we usually write the function of the line in place of the letter '*y*'.

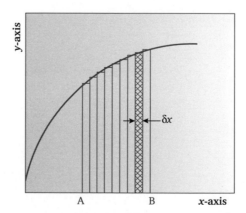

Figure 26.1 We can approximate the **area** beneath a curve to a series of slender rectangles of height *y* and width δx.

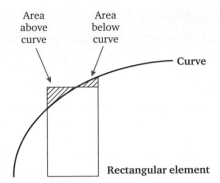

Figure 26.2 To minimize the errors, the area above the curve needs to be the same as the beneath it. The areas that need to cancel are shaded.

■ **Worked Example 26.1**

Calculate the area bounded by the line $y = x^2$ and the x-axis, and the vertical lines of $x = 3$ and $x = 6$ (the shaded portion in Fig. 26.3).

Strategy:

(i)   We integrate the equation of the curve $y = x^2$.

(ii) As limits, we insert numbers $x = 3$ and $x = 6$.

Solution:

(i)   The function is, $y = x^2$, so area $= \displaystyle\int_{3}^{6} x^2 \, dx$

(ii) Area $= \left[\dfrac{x^3}{3} \right]_{3}^{6} = \left(\dfrac{6^3}{3} - \dfrac{3^3}{3} \right) = \dfrac{216}{3} - \dfrac{27}{3} = 63.$

The area of the bound portion is 63 area units. ■

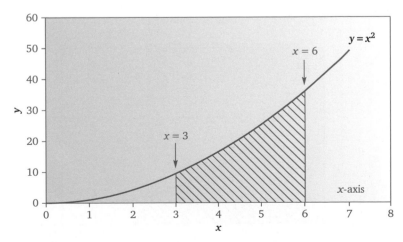

Figure 26.3 To determine the area of the shaded portion, we integrate the equation of the curve between the two limits $x = 6$ and $x = 3$.

■ Worked Example 26.2

The following equation defines the change in entropy ΔS of a substance as it warms up:

$$\Delta S = \text{area under curve} = \int_{T_1}^{T_2} \frac{C_p}{T} \, dT$$

where T_2 is the final, upper temperature and T_1 is the initial, lower temperature. C_p is the substance's heat capacity. To determine the value of ΔS, we plot a graph of C_p/T (as 'y') against T (as 'x'). Figure 26.4 shows this relationship diagrammatically.

If the heat capacity C_p for chloroform (**I**) is 70 J K^{-1} mol^{-1}, calculate the change in entropy when chloroform (**I**) is warmed from 220 to 240 K?

We start by rewriting the expression for ΔS:

$$\Delta S = \int_{T_1}^{T_2} \left(C_p \times \frac{1}{T} \right) dT$$

In this example, the value of C_p is constant. For this reason, we place it **outside the integral** (in practice, this means writing it 'out of the way' to the *left* of the integration sign):

$$\Delta S = C_p \int_{T_1}^{T_2} \frac{1}{T} \, dT$$

We now integrate the expression. Equation (23.5) in Chapter 23 tells us how to: $\int \frac{1}{x} \, dx = \ln x$. We write:

$$\Delta S = C_p [\ln T]_{T_1}^{T_2}$$

We next insert the limits:

$$\Delta S = C_p \{ \ln T_2 - \ln T_1 \}$$

Finally, we use the third law of logarithms (see p. 187) to group together the two logarithm terms:

$$\Delta S = C_p \ln \left(\frac{T_2}{T_1} \right)$$

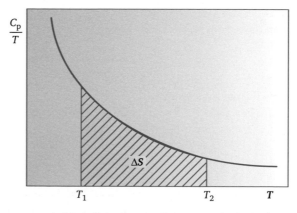

Figure 26.4 The entropy of warming chloroform is obtained as the area under a curve of C_p/T (as 'y') against T (as 'x').

Finally, to calculate the change in the entropy ΔS of chloroform (**I**) associated with warming, we insert values into the derived expression to obtain:

$$\Delta S = 70 \text{ J K}^{-1} \text{ mol}^{-1} \times \ln\left(\frac{240 \text{ K}}{220 \text{ K}}\right)$$

So the molar change in entropy for chloroform during warming from 220 to 240 K is $6.1 \text{ J K}^{-1} \text{ mol}^{-1}$. ■

Self-test 26.1

Determine the areas bounded by the following:

26.1.1   The x-axis, the curve $y = x^2 + 4$, and the lines $x = 3$ and $x = 5$.

26.1.2   The x-axis, the curve $y = \dfrac{1}{x^2}$ and the lines $x = 6$ and $x = 2$.

26.1.3   One mole of oxygen is warmed from 300 K to 350 K. $C_p(O_2) = (25.8 + 1.2 \times 10^{-2} T) \text{ J K}^{-1} \text{ mol}^{-1}$.
Calculate the associated rise in entropy ΔS by first integrating the function C_p/T, then inserting the two values of T as limits.

Integrating to obtain more complicated areas

By the end of this section, you will know:

- Sometimes a function or line crosses the x-axis.

- In such cases, the area we calculate using integration and just one pair of limits is often incorrect.

- It is necessary to use several sets of limits.

Sometimes a function or line crosses the x-axis. For example, the simple function $y = x^2 - 3x + 2$ crosses the x-axis twice, at $x = 1$ and at $x = 2$. In such cases, we need a more subtle approach when calculating an area using integration. Look at the graph of this function in Fig. 26.5: area (a) lies above the x-axis, while area (b) lies beneath it. The computed area will be the sum of these two. Unfortunately, the integration method will automatically compute area (a) as positive and area (b) as negative. So the computed sum will certainly be too small.

■ **Worked Example 26.3**

Find the area enclosed between the curve $y = x^2 - 3x + 2$, and the x-axis.
The line cuts the x-axis at $x = 1$ and at $x = 2$, so there are two enclosed areas: between $x = 0$ (the y-axis) and $x = 1$, and between $x = 1$ and $x = 2$.

Strategy:

(i)   We integrate the function.

(ii)   We insert *two separate pairs of limits,* one for each of the two enclosed areas.

(iii)   We sum the two areas, remembering to add the modulus of each, i.e. changing the sign of the negative portion.

Solution:

(i)   If $y = x^2 - 3x + 2$, $\int y\,dx = \dfrac{x^3}{3} - \dfrac{3x^2}{2} + 2x + c$

(ii)   Area (a) $= \left[\dfrac{x^3}{3} - \dfrac{3x^2}{2} + 2x\right]_0^1 = \left(\dfrac{1^3}{3} - \dfrac{3 \times 1^2}{2} + 2 \times 1\right) - (0) = \dfrac{5}{6}$

Area (b) $= \left[\dfrac{x^3}{3} - \dfrac{3x^2}{2} + 2x\right]_1^2 = \left(\dfrac{2^3}{3} - \dfrac{3 \times 2^2}{2} + 2 \times 2\right) - \left(\dfrac{1^3}{3} - \dfrac{3 \times 1^2}{2} + 2 \times 1\right)$

$= \dfrac{2}{3} - \dfrac{5}{6} = -\dfrac{1}{6}$

(iii)   Total area $= \dfrac{5}{6} + \dfrac{1}{6} = 1$ square unit

If we had integrated with only a single set of limits (using 2 and 0), we would have calculated an area of $\left(\dfrac{5}{6} - \dfrac{1}{6}\right) = \dfrac{2}{3}$, which is wrong. ∎

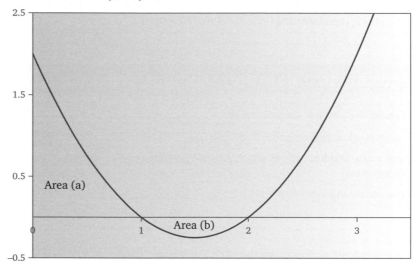

Figure 26.5 The curve $y = x^2 - 3x + 2$.

Self-test 26.2

Determine the areas enclosed by the following lines:

26.2.1 $y = 4x^3 - 12x^2 + 8x$, between the limits of 0 and 2.

26.2.2 $y = \cos x$, between the limits of 0 and π.

Solids of revolution

By the end of this section, you will know:

- How to calculate the volume of the solid created by rotating a line about either the x-axis or the y-axis.

- Just as it is possible to determine an area on a two-dimensional surface, so we can determine a *volume* by rotating a function about an axis.

Rotation about the x-axis: If we wish to rotate about the x-axis, we merely integrate a function of x with respect to x. The procedure generates a volume that we call a **solid of revolution**:

$$\pi\int_a^b (f(x))^2 \; dx = \pi\int_a^b y^2 \; dx \qquad (26.1)$$

We can use eqn (26.1) provided $f(x)$ is continuous throughout the interval, $a \le x \le b$.

■ **Worked Example 26.4**

Atomic orbitals with quantum number $l = 1$ (s orbitals) are spherical. If the radius of such a spherical orbital is r, what is the volume of an s orbital?

If the centre of the circle is the origin, the general equation of a circle of radius r is $x^2 + y^2 = r^2$. Consider the arc between $x = 0$ (where the circle strikes the y axis) and $x = r$ (where the circle cuts the x-axis): rotating this arc by $360°$ about the x-axis generates a hemisphere.

We obtain the volume of this hemisphere by integrating the function $x^2 + y^2 = r^2$ between the limits of $x = 0$ and $x = r$; see Fig. 26.6.

We obtain the volume V of the sphere as $2 \times \int_0^r \pi y^2 \; dx$. We need the factor of 2 to double the volume of the *hemi*sphere formed during integration, and thereby forming a complete sphere.

Substituting for y^2 yields, $\quad V = 2\pi \times \int_0^r (r^2 - x^2)\,dx$

Integration yields, $\quad V = 2\pi\left[r^2 x - \frac{1}{3}x^3 \right]_0^r$

> The factor of π appears *outside* the integral because it is a constant.

Inserting limits yields the familiar equation, $\quad V = 2\pi\left[r^3 - \frac{1}{3}r^3 \right] = \frac{4}{3}\pi r^3$ ■

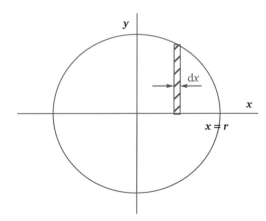

Figure 26.6 The equation $x^2 + y^2 = r^2$ defines a circle of radius r. Rotating an arc of this circle about the x-axis through $360°$ generates a perfect sphere.

Rotation about the y-axis:

The procedure is essentially the same if we wish to rotate about the y-axis (rather than the x-axis, as above): we merely integrate a function of x with respect to y rather than y with respect to x:

■ Worked Example 26.5

The potential energy of a particle E varies according to its distance r from another particle. To a good first approximation, the value of E follows a parabolic dependence. We define a parabola by the equation $y = 4x^2$. Determine the volume of this parabola between $y = 6$ and $y = 2$.

The volume we generate in this question has two flat, circular faces—one larger than the other. The two circular faces are parallel, horizontal, and joined by a smoothly curved surface. Figure 26.7 illustrates this bowl-shaped volume.

We obtain the volume V of this portion using the integral

$$\pi \int_2^6 \left(f(y)\right)^2 dy = \pi \int_2^6 x^2 dy.$$

Substituting for x^2 yields, $V = \pi \int_2^6 \dfrac{y}{4} dy$. The factor of ¼ is a constant, so we place it outside the integral.

Integration yields, $V = \dfrac{\pi}{4} \left[\dfrac{y^2}{2} \right]_2^6 = \dfrac{\pi}{8} \left[y^2 \right]_2^6$

Inserting limits gives, $V = \dfrac{\pi}{8} \left[6^2 - 2^2 \right]_2^6 = \dfrac{\pi}{8} [36 - 4]$

So, to 3 s.f., $V = 12.6$ volume units. ■

The examples above are perhaps artificially easy. Often, we must perform multiple integrations, which are described in the next section.

Self-test 26.3

Determine the following volumes:

26.3.1   The segment of a sphere (of radius 3) between $x = 1$ and $x = 2$.

26.3.2   A parabola is defined by the equation $y^2 = 3x$. Determine the volume of the solid formed by rotating this parabola about the x-axis between $x = 5$ and $x = 3$.

26.3.3   What is the area of overlap between the functions $y = x^2$ and $y = -2x + 8$?

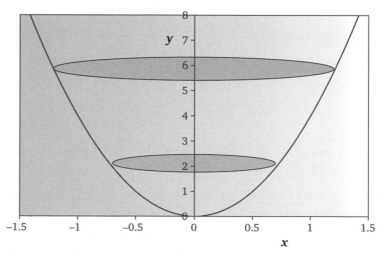

Figure 26.7 We form an object shaped like a soup-bowl from the parabola $y = 4x^2$ if we cut between $y = 6$ and $y = 2$.

Sometimes we need to calculate the volume of an **overlap**. For example, in the simple theory of covalent bonds, the LCAO (linear combination of atomic orbitals) model depends on orbitals overlapping in three-dimensional space. Electrons from either atom pair up within the volume of the overlap, which constitutes the desired bond. The mathematical forms of these orbitals can be quite fearsome; but the approach in the next section illustrates how we would determine either a 2D area or 3D volume.

Integration in three dimensions

By the end of this section, you will know:

- It is possible to integrate a function more than once, function by function.

- Multiple integration is particularly important when a function exists in more than one dimension.

The previous section introduced the idea of using integration to obtain a volume. But many of the integrations in that section were relatively simple, because the solids were formed by rotating about the *x*- or *y*-axis and we were able to obtain the volume via a single integration step. Usually, we obtain a volume by effectively integrating more than once—**multiple integration**.

If we want to determine the volume of a *three*-dimensional body, it is often necessary to integrate three times: once per dimension.

A triple (or *three-fold*) integral has the general form,

$$\text{volume} = \int_V f(x,y,z)\, dV = \int_{z_1}^{z_2} \int_{y_1}^{y_2} \int_{x_1}^{x_2} f(x,y,z)\, dx\, dy\, dz \qquad (26.2)$$

where *V*, as usual, means volume.

The volume element is a box of sides d*x*, d*y* and d*z*. Figure 26.8 demonstrates why the integral takes this form.

We need to note:

- This notation requires three integral signs and three 'd' terms. These terms behave much like nesting brackets:
    - The innermost integral sign relates to the innermost d term (here, they define *x*).
    - The middle integral sign relates to the middle d term (here, they define *y*).
    - The outermost integral sign relates to the outermost d term (here, they define *z*).
- We write an upper-case *V* against the integral sign, where the lower limit would normally appear.

We often find it more meaningful (and straightforward) to integrate using spherical polar coordinates (see p. 254). Figure 26.9 shows a schematic representation of the volume element d*V*. The height of the volume element is $r\, d\theta$ and the width is $r \sin \theta d\phi$. Therefore, the volume of the volume element is, $dV = r \sin \theta d\phi \times r\, d\theta \times dr = r^2 \sin \theta d\phi d\theta dr$.

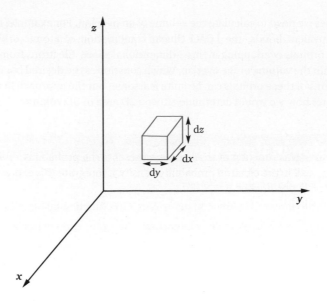

Figure 26.8 Integrating in three dimensions requires us to consider incremental changes in the x, y, and z directions. The volume element $dV = dx\,dy\,dz$.

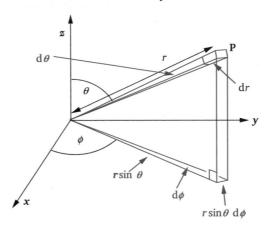

Figure 26.9 Schematic representation of a volume element dV depicted in spherical polar coordinates.

In spherical polar coordinates, the volume integral is

$$\int_V f(r,\theta,\phi)\mathrm{d}V = \int_{\phi_1}^{\phi_2}\int_{\theta_1}^{\theta_2}\int_{r_1}^{r_2} f(r,\theta,\phi)\, r^2 \sin\theta\, \mathrm{d}r\,\mathrm{d}\theta\,\mathrm{d}\phi \qquad (26.3)$$

where V is again volume.

Sometimes, we need to integrate over *all* of three-dimensional space. In other words, we wish to evaluate the value of a function *everywhere*. In fact, we merely rewrite eqns (26.2) and (26.3) with different limits:

- In Cartesian space, the volume integral becomes:

$$\text{volume} = \int f\, \mathrm{d}V = \int_{-\infty}^{\infty}\int_{-\infty}^{\infty}\int_{-\infty}^{\infty} f(x,y,z)\, \mathrm{d}x\,\mathrm{d}y\,\mathrm{d}z \qquad (26.4)$$

- and in spherical polar coordinates, the volume integral is,

$$\text{volume} = \int f\, \mathrm{d}V = \int_0^{2\pi}\int_0^{\pi}\int_0^{\infty} f(r,\theta,\phi)r^2 \sin\, \mathrm{d}r\,\mathrm{d}\theta\,\mathrm{d}\phi \qquad (26.5)$$

These integrals are particularly important in quantum mechanics, when evaluating atomic and molecular properties from wavefunctions obtained as solutions of the Schrödinger equation, because an orbital represents the probability of finding an electron within a given three-dimensional volume.

■ **Worked Example 26.6**

The square of the wavefunction of the hydrogen 1s orbital is $\psi_{1s}^2 = \dfrac{1}{\pi a_0^3} \exp\left(-\dfrac{2r}{a_0}\right)$.

The integral of this function over all space represents the probability of finding the electron (we call it 'the electron probability density'). Integrate this expression using spherical polar coordinates.

Expressed in spherical polar coordinates, the volume integral is,

$$\int \psi_{1s}^2 dV = \frac{1}{\pi a_0^3} \int_0^{2\pi} \int_0^{\pi} \int_0^{\infty} \exp\left(-\frac{2r}{a_0}\right) r^2 \sin\theta \, dr d\theta d\phi$$

Strategy:

(i)   We first split into its three component parts, accommodating r, θ, and ϕ.

   (This is only possible in this example because the components are cleanly separated. If the limits of one integral are dependent on another variable, then this integral must be performed first.)

(ii) We integrate each in turn.

(iii) We multiply together the results of each integral.

Solution:

(i) $\int \psi_{1s}^2 dV = \dfrac{1}{\pi a_0^3} \quad \underbrace{\int_0^{\infty} \exp\left(-\dfrac{2r}{a_0}\right) r^2 dr}_{r} \quad \underbrace{\int_0^{\pi} \sin\theta \, d\theta}_{\theta} \quad \underbrace{\int_0^{2\pi} d\phi}_{\phi}$

(ii) Integration with respect to r requires a published integral, in this case from Table 25.1 on p. 372:

$$\int_0^{\infty} x^n \exp(-ax)dx = \frac{n!}{a^{(n+1)}},$$

where $n!$ is the **factorial** of n defined on p. 29 of Chapter 2. In this case, $n = 2$, $a = 2/a_0$, so the integral has a value of $2/(2 \times a_0)^3$, which cancels to $\dfrac{a_0^3}{4}$.

Integrate with respect to ϕ: $\int_0^{2\pi} d\phi = 2\pi$.

Integrate with respect to θ: $\int_0^{\pi} \sin\theta \, d\theta = \left[-\cos\theta\right]_0^{\pi} = 2$.

(iii) The result of the three-fold integration is $2 \times 2\pi \times \dfrac{a_0^3}{4} = \pi a_0^3$.

The factor before the integral is $1/\left(\pi a_0^3\right)$, so the final result is, 1.

The final result is a probability; and a probability of 1 means a certainty. After a moment, this result becomes obvious: if the hydrogen atom has an electron, the chance of finding that electron in *all* space must be certain. ■

Sometimes, we perform one or two of the integration steps, but find it convenient to retain two or one of the integrals.

■ **Worked Example 26.7**

When we derive the Maxwell distribution of molecular speeds, we need to integrate the equation

$$\int n(v)\,\mathrm{d}V = \int A' \exp\left(-\frac{mv^2}{2kT}\right)\mathrm{d}V$$

where $n(v)$ is the number of particles having the velocity v; $\mathrm{d}V$ is the volume element of the velocity space; all other terms are constants.

Express this equation in terms of spherical polar coordinates, integrate it with respect to the three dimensions of θ and ϕ, but do not integrate with respect to v.

Strategy:

(i) We express the integral in polar coordinates.

(ii) We integrate in *two* steps, one for each of the dimensions θ and ϕ.

(iii) We multiply together the results of the two constituent integrals, but leave the $\mathrm{d}V$ term unaltered.

Solution:

(i) Expressed in spherical-polar coordinates, the integral becomes,

$$\int_0^{2\pi}\int_0^{\pi}\int_0^{\infty} n(v)\,v^2 \sin\theta\,\mathrm{d}v\,\mathrm{d}\theta\,\mathrm{d}\phi = \int_0^{2\pi}\int_0^{\pi}\int_0^{\infty} A'v^2 \exp\left(-\frac{mv^2}{2kT}\right)\sin\theta\,\mathrm{d}v\,\mathrm{d}\theta\,\mathrm{d}\phi$$

(ii) Integrate with respect to ϕ: we need the outermost integral: $\int_0^{2\pi}\mathrm{d}\phi = 2\pi$.

Integrate with respect to θ: we need the middle integral: $\int_0^{\pi}\sin\theta\,\mathrm{d}\theta = \left[-\cos\theta\right]_0^{\pi} = 2$.

(iii) The product of the first two integrals is clearly 4π. We multiply the remainder of the integral by 4π. Thus, we have derived the commonly found form of the Maxwell speed distribution:

$$n(v)\,\mathrm{d}v = 4\pi\,A'v^2 \exp\left(-\frac{mv^2}{2kT}\right)\mathrm{d}v.$$

In effect, this version of the equation refers to a distribution of molecular speeds v without considering the complication of molecular direction. Expressions of this type occur often in chemistry, for example as a simplified form of the Maxwell–Boltzmann equation. ■

Self-test 26.4

Integrate the following expressions over all space.

26.4.1 $V = \int_0^1\int_1^2\int_1^2\left(x^2 + yz\right)\mathrm{d}z\,\mathrm{d}y\,\mathrm{d}x$

26.4.2 What is the volume of the solid formed between the planes, $x = yz$ and $x = 0$, $y = 0, y = 1, z = 0$, and $z = 4$?

Summary of key equations in the text

Solid of revolution about the x-axis:

$$\pi \int_a^b (f(x))^2 \, \mathrm{d}x = \pi \int_a^b y^2 \mathrm{d}x \qquad (26.1)$$

Integration within a defined volume:

Cartesian: $\qquad \text{volume} = \int_V f(x,y,z) \, \mathrm{d}V = \int_{z_1}^{z_2} \int_{y_1}^{y_2} \int_{x_1}^{x_2} f(x,y,z) \, \mathrm{d}x \, \mathrm{d}y \, \mathrm{d}z \qquad (26.2)$

Spherical polar: $\quad \text{volume} = \int_V f(r,\theta,\phi) \mathrm{d}V = \int_{\phi_1}^{\phi_2} \int_{\theta_1}^{\theta_2} \int_{r_1}^{r_2} f(r,\theta,\phi) \, r^2 \sin\theta \, \mathrm{d}r \mathrm{d}\theta \mathrm{d}\phi \qquad (26.3)$

Integration over all space:

Cartesian: $\qquad \text{volume} = \int f \, \mathrm{d}V = \int_{-\infty}^{\infty} \int_{-\infty}^{\infty} \int_{-\infty}^{\infty} f(x,y,z) \, \mathrm{d}x \, \mathrm{d}y \, \mathrm{d}z \qquad (26.4)$

Spherical polar: $\quad \text{volume} = \int f \, \mathrm{d}V = \int_0^{2\pi} \int_0^{\pi} \int_0^{\infty} f(r,\theta,\phi) r^2 \sin\theta, \, \mathrm{d}r \, \mathrm{d}\theta \, \mathrm{d}\phi \qquad (26.5)$

Additional problems

To simplify the mathematics sufficiently, many of these examples below are not drawn from chemistry, or have been simplified very greatly.

26.1 Until recently, methylchloroform was used commonly as a degreasing agent, and as a solvent in typing-correction fluids. Calculate the increase in entropy associated with warming 3.2 moles of methylchloroform from 240 K to 330 K. The value of C_p is $(91.47 + 7.5 \times 10^{-2} \, T)$ in $\mathrm{J\,K^{-1}\,mol^{-1}}$.

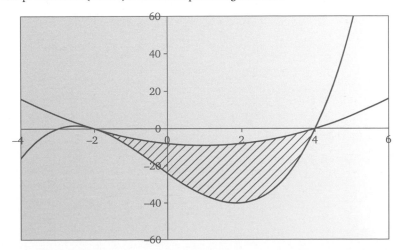

26.2 Consider the two simple functions $y = x^3 + x^2 - 14x - 24$ and $y = x^2 - 2x - 8$, which clearly overlap. What is the (shaded) area of overlap in the figure below?

26.3   Calculate the area bounded by the function, $y = 4x^2 - 2x + 5$, and the three lines, $x = 3, x = 1$, and the x-axis.

26.4   What is the volume occupied within the symmetrical cone formed when the function $y = 4x^3$ is rotated about the y-axis, and the plane defined by $y = 3$?

26.5   What is the area enclosed by the curve $y = 4x^2 + 12x + 8$, the straight lines $y = 3$ and $y = 8$, and the x-axis?

26.6   What is the volume created by rotating the exponential function $y = \exp(2x)$ around the x-axis, using the limits of $x = 2$ and $x = 4$?

26.7   This question concerns the wavefunction for the 1s orbital on a hydrogen atom. Normalize the wavefuntion, given that $\psi_{1s}^2 = N^2 \exp(-2r/a_0)$, where a_0 is the Bohr radius, and r the distance between the single electron and the nucleus.

26.8   Compute the triple integral of $f(x,y,z) = 8xyz$ over the solid between $z = 0$ and $z = 1$ and over the region $x = 0$ and $x = 1$, and $y = 2$ and $y = 3$.

26.9   Find the normalization constant for the wavefunction of the $Y_{1,0}$ spherical harmonic:
$$\psi = N\cos\theta.$$

26.10   Show that the following spherical harmonics are orthogonal: $\psi_1 = N_1\cos\theta$ and $\psi_2 = N_2\sin\theta e^{i\phi}$.

27

Matrices

Introducing matrices

By the end of this section, you will know:

- Matrices are symbolized in upright, bold type.
- A matrix is a rectangular quantity that obeys matrix algebra.
- A matrix has m rows and n columns.

A matrix is a means of compressing a large amount of information into a small amount of space. More formally, matrices are rectangular representations of data that obey the rules of mathematics called **matrix algebra**. They are sometimes called 'arrays.'

The following are all matrices:

$$\mathbf{P} = \begin{pmatrix} 1 & 3 & 7 \\ 4 & 3 & 2 \\ 0 & 5 & 9 \end{pmatrix} \qquad \mathbf{Q} = \begin{pmatrix} a \\ b \\ c \\ d \end{pmatrix} \qquad \mathbf{R} = \begin{pmatrix} 1 & 2 & 3 \end{pmatrix}$$

Notice how we use brackets to contain the information within each. We use the usual printing conventions, so the brackets are *curved* and the symbol for the matrix is upright and bold. When writing a symbol for a matrix by hand, rather than typing, it is best to write the letter and underline it, e.g. $\underline{A}$.

> **Care:** The plural of matrix is *matrices* and never *matrixs*.

The general form of a matrix is $\mathbf{A} = \begin{pmatrix} a_{11} & a_{12} & \cdots & a_{1n} \\ a_{21} & a_{22} & \cdots & a_{2n} \\ a_{31} & a_{32} & \cdots & a_{3n} \\ \cdots & \cdots & \ddots & \cdots \\ a_{m1} & a_{m2} & \cdots & a_{mn} \end{pmatrix}$

where each a_{mn} is the **matrix element** for the mth row and the nth column. The number of elements in the matrix is the product of m and n.

When describing the matrix, we say the **order** (or 'dimension') of the matrix is $(m \times n)$. Therefore, matrix $\mathbf{P}$ is a 3×3 matrix, matrix $\mathbf{Q}$ has an order of 4×1, and matrix $\mathbf{R}$ is of the order 1×3.

There are several types of matrix that are special:

- A *square* matrix has $m = n$. Matrix $\mathbf{P}$ above is a square matrix with $n = 3$.

- A *column* matrix has $n = 1$. Matrix $\mathbf{Q}$ above is a column matrix, with $m = 4$.

- A *row* matrix has $m = 1$. Matrix $\mathbf{R}$ above is a row matrix, with $n = 3$.

- An *identity* matrix has the property that, when multiplied by any other matrix, it returns the matrix unchanged. We denote such matrices as $\mathbf{I}$. Multiplying a matrix by $\mathbf{I}$ is equivalent to multiplying a number by 1. Matrix $\mathbf{S}$ below is a 4×4 identity matrix. This property of matrices can only be achieved with square matrices, so $\mathbf{I}$ is always square (i.e. $m = n$).

> The principal diagonal is written from top left to bottom right.

  The identity matrix is written with all its elements equated to zero, except along the principal diagonal, where each element has a value of 1.

- A *null* matrix has all its elements equal to zero. Matrix $\mathbf{T}$ below is a 3×3 null matrix. Null matrices can have any dimension, so they need not be square.

$$S = \begin{pmatrix} 1 & 0 & 0 & 0 \\ 0 & 1 & 0 & 0 \\ 0 & 0 & 1 & 0 \\ 0 & 0 & 0 & 1 \end{pmatrix} \qquad T = \begin{pmatrix} 0 & 0 & 0 \\ 0 & 0 & 0 \\ 0 & 0 & 0 \end{pmatrix}$$

Matrices are particularly useful in quantum mechanics, firstly because they can convey a huge amount of information in a very simple way. Secondly, they are useful because matrices are easily compatible with modern electronic computational devices. Matrix algebra therefore simplifies the mathematics required to solve the associated equations.

■ **Worked Example 27.1**

Consider the simple molecule butadiene (**I**), which has four carbon atoms in its backbone. (We will ignore all the hydrogens in this particular analysis.) Clearly, atom 1 is attached to atom 2, atom 2 to atom 3, and so on. Express this information in the form of an atom connectivity matrix.

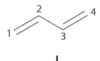

I

Strategy:

(i)   Because compound (**I**) has 4 atoms, we start by writing a 4×4 matrix.

(ii)   We use the element in row m and column n to indicate the connectivity of atom m, i.e. indicating which atoms are bonded to atom m. (In fact, it does not matter in this approach which numbers we assign to which atoms, provided we unambiguously assign rows and columns to the constituent atoms.)

As a simple example, we express the connectivity of atom 1 with atom 2, by writing the number of bonds between them, in element a_{12}. Since this is a double bond, $a_{12} = 2$. Since atom 2 is connected to atom 1 in the same way, we also inserted a value of 2 as the matrix element a_{21}.

(iii)   Atom 1 is not attached to atoms 3 or 4, so we insert a zero for matrix elements a_{13} and a_{14}.

(iv)   Within this mindset, the matrix element a_{11} would indicate 'the number of bonds between atom 1 and atom 1,' which is clearly not a meaningful concept. We instead insert the atomic number.

Solution:

The connectivity matrix describing compound (**I**) is therefore: $\begin{pmatrix} 6 & 2 & 0 & 0 \\ 2 & 6 & 1 & 0 \\ 0 & 1 & 6 & 2 \\ 0 & 0 & 2 & 6 \end{pmatrix}$. ■

Atom connectivity matrices are a useful way of summarizing key information about the atom type and bonding in a molecule in a more mathematical form. Given the matrix above, it would be easy for us to draw the butadiene molecule that it represents. We can use these matrices in computational drug-discovery methods.

> ### Self-test 27.1
>
> Write a matrix to describe the connectivity of the following compounds. It may help to number the atoms before starting.
>
> **27.1.1**   Cyclohexane (**II**)   **27.1.2**   2-methlypropanoic acid (**III**)
>
>
>
> **II**   **III**

The Hamiltonian, **H**, is a quantum-mechanical operator, which is associated with the total energy of the system. It is the sum of the kinetic energy and potential energy operators. It is used in the time-independent Schrödinger equation: $H\psi_v = E_n \psi_n$.

Common examples of matrices in quantum mechanics include the **orbital overlap matrix, S, and Hamiltonian matrix**, H. Such matrices are capable of great sophistication and power.

Hückel Theory:

We use the Hückel model to calculate the energy levels of delocalized π electrons in conjugated and aromatic molecules. Using this approach, a Hamiltonian matrix can be approximated for a molecule by letting each element in the matrix represent the interaction energy of the corresponding orbitals.

The model assumes that:

α and β are the so-called Hückel parameters. They have negative magnitudes and do not involve imaginary numbers (see Chapter 29). They have the dimensions of energy.

- $H_{ij} = \alpha$, if $i = j$ (i.e. same orbital).
- $H_{ij} = \beta$, if i and j are on neighbouring (i.e. bonded) atoms.
- $H_{ij} = 0$, for all other cases.

The α parameter is a Coulomb integral and can be thought of as the energy of an electron if it occupies orbital i. The β parameter is a resonance integral. Due to these approximations, the matrix formulated using this method is often called an 'effective Hamiltonian'.

> ### ■ Worked Example 27.2
>
> Consider the simple molecule butadiene (**I**), seen in Worked Example 27.1. Use the Hückel approximation to formulate an effective Hamiltonian matrix.
>
> **Strategy:**
>
> (i)   Because compound (**I**) has four p_z orbitals (one on each atom), we start by writing a 4×4 matrix.
>
> (ii)   We use the element in row i and column j to indicate the interaction energy of orbital i with orbital j. Following the rules listed above, the matrix element H_{11} is the energy of an electron when it occupies orbital 1, which is α. On the other hand, since orbital 1 and orbital 2 lie on neighbouring atoms, which are bonded together, the element H_{12} is β.

Solution:

The effective Hamiltonian matrix for compound (**I**) is therefore: $\begin{pmatrix} \alpha & \beta & 0 & 0 \\ \beta & \alpha & \beta & 0 \\ 0 & \beta & \alpha & \beta \\ 0 & 0 & \beta & \alpha \end{pmatrix}$. ■

■ **Worked Example 27.3**

Consider the simple heterocycle pyridine (**IV**), which comprises the same number of atoms as benzene, but with one carbon replaced with nitrogen. Again, we ignore all the hydrogens. The nitrogen has a different electronegativity to the carbon. Write an effective Hamiltonian matrix for (**IV**).

IV

To signify that a molecule contains a heteroatom, we multiply the α and β terms by simple factors when representing the interaction energies of the orbital on the nitrogen atom with itself or with the orbitals on the neighbouring carbon atoms. When an orbital is on nitrogen rather than on carbon, we use 1.5α rather than α because the atomic orbital on nitrogen is lower in energy than the atomic orbital on carbon. Carbon–nitrogen orbital interaction energies are characterized by values of $1.75\,\beta$, which reflects the difference in orbital size and electronegativity. The higher value of β reflects the stronger interaction that is formed when nitrogen is involved.

By convention, the nitrogen heteroatom is called atom 1. The effective Hamiltonian matrix for (**IV**) is:

$$\begin{array}{c c c c c c c} & \textit{1} & \textit{2} & \textit{3} & \textit{4} & \textit{5} & \textit{6} \\ \textit{1} & 1.5\alpha & 1.75\beta & 0 & 0 & 0 & 1.75\beta \\ \textit{2} & 1.75\beta & \alpha & \beta & 0 & 0 & 0 \\ \textit{3} & 0 & \beta & \alpha & \beta & 0 & 0 \\ \textit{4} & 0 & 0 & \beta & \alpha & \beta & 0 \\ \textit{5} & 0 & 0 & 0 & \beta & \alpha & \beta \\ \textit{6} & 1.75\beta & 0 & 0 & 0 & \beta & \alpha \end{array}$$ ■

The italic numbers *outside* the bracket, to the left and above, are included in this example for clarity alone: they relate to the atom positions, with the heterocyclic nitrogen as atom 1.

Self-test 27.2

Write an effective Hamiltonian matrix for the following compounds. The α coefficient for nitrogen is 1.5 and 1.4 for sulphur; the β coefficients are 1.75 and 1.60, respectively. Use the numbering system given below.

27.2.1 Methylethylamine (**V**)   **27.2.2** Thiophene (**VI**)

V   **VI**

Adding and subtracting matrices

By the end of this section, you will know:

- Adding and subtracting matrices requires that the matrices have the same order.
- The sum of two matrices is obtained as the sum of the respective elements.
- The difference of two matrices is obtained as the difference of the respective elements.

Having seen a few matrices, we now manipulate them.

If two matrices have the same order, we obtain their sum by adding together the respective elements. For example, we add two 2×2 matrices as follows:

$$\begin{pmatrix} a_{11} & a_{12} \\ a_{21} & a_{22} \end{pmatrix} + \begin{pmatrix} b_{11} & b_{12} \\ b_{21} & b_{22} \end{pmatrix} = \begin{pmatrix} a_{11}+b_{11} & a_{12}+b_{12} \\ a_{21}+b_{21} & a_{22}+b_{22} \end{pmatrix} \tag{27.1}$$

Extension to matrices of other orders is obvious. Addition in this way is wholly associative, so $(\mathbf{A}+\mathbf{B})+\mathbf{C} = \mathbf{A}+(\mathbf{B}+\mathbf{C})$.

■ **Worked Example 27.4**

Determine the sum of $\mathbf{A} = \begin{pmatrix} 1 & 2 & 3 & 4 \\ -2 & 3 & 5 & 6 \\ 0 & 4 & 8 & 9 \end{pmatrix}$ and $\mathbf{B} = \begin{pmatrix} 2 & -4 & 6 & 0 \\ 9 & 2 & 3 & 4 \\ 1 & -7 & 8 & 5 \end{pmatrix}$.

We will call the resultant sum $\mathbf{C}$. The element a_{11} in matrix $\mathbf{A}$ is 1 and the element a_{11} in matrix $\mathbf{B}$ is 2. Their sum is 3:

$$\begin{pmatrix} \textcircled{1} & 2 & 3 & 4 \\ -2 & 3 & 5 & 6 \\ 0 & 4 & 8 & 9 \end{pmatrix} + \begin{pmatrix} \textcircled{2} & -4 & 6 & 0 \\ 9 & 2 & 3 & 4 \\ 1 & -7 & 8 & 5 \end{pmatrix}$$

Similarly, the element a_{12} in matrix $\mathbf{A}$ is 2 and the element a_{12} in matrix $\mathbf{B}$ is -4. Clearly, their sum is -2.

$$\begin{pmatrix} 1 & \textcircled{2} & 3 & 4 \\ -2 & 3 & 5 & 6 \\ 0 & 4 & 8 & 9 \end{pmatrix} + \begin{pmatrix} 2 & \boxed{-4} & 6 & 0 \\ 9 & 2 & 3 & 4 \\ 1 & -7 & 8 & 5 \end{pmatrix}$$

In this way, we add together all the respective matrix elements to generate matrix $\mathbf{C}$:

$$\mathbf{C} = \begin{pmatrix} 3 & -2 & 9 & 4 \\ 7 & 5 & 8 & 10 \\ 1 & -3 & 16 & 14 \end{pmatrix} \blacksquare$$

The addition of matrices is completely commutative, so $\mathbf{A}+\mathbf{B}$ gives the same result as $\mathbf{B}+\mathbf{A}$.

We subtract matrices in exactly the same manner by subtracting the respective elements. For example, we subtract two 2×2 matrices as follows:

$$\begin{pmatrix} a_{11} & a_{12} \\ a_{21} & a_{22} \end{pmatrix} - \begin{pmatrix} b_{11} & b_{12} \\ b_{21} & b_{22} \end{pmatrix} = \begin{pmatrix} a_{11}-b_{11} & a_{12}-b_{12} \\ a_{21}-b_{21} & a_{22}-b_{22} \end{pmatrix} \tag{27.2}$$

Again, extension to matrices of other orders should be obvious.

■ **Worked Example 27.5**

If $D = \begin{pmatrix} 1 & -3 & 4 \\ 9 & 2 & 5 \\ 0 & -7 & 3 \end{pmatrix}$ and $E = \begin{pmatrix} 2 & 1 & 9 \\ 0 & 3 & -5 \\ 1 & 1 & 4 \end{pmatrix}$, determine the value of $D - E$.

The element a_{11} in matrix D is 1 and a_{11} in matrix E is 2. The subtraction '1 – 2' clearly yields –1.

$$\begin{pmatrix} \boxed{1} & -3 & 4 \\ 9 & 2 & 5 \\ 0 & -7 & 3 \end{pmatrix} - \begin{pmatrix} \boxed{2} & 1 & 9 \\ 0 & 3 & -5 \\ 1 & 1 & 4 \end{pmatrix}$$

In this way, we generate the new matrix $D - E = \begin{pmatrix} -1 & -4 & -5 \\ 9 & -1 & 10 \\ -1 & -8 & -1 \end{pmatrix}$. ■

Self-test 27.3

Perform the following matrix addition and subtraction problems.

27.3.1 $\begin{pmatrix} 1 & 2 \\ 4 & 6 \end{pmatrix} + \begin{pmatrix} 2 & -3 \\ 0 & 2 \end{pmatrix}$

27.3.2 $(3a \quad 4b \quad 8c \quad 5d) - (a \quad 3b \quad 9c \quad 5d)$

27.3.3 $\begin{pmatrix} 4 & -3 & 3 \\ 2 & 7 & -2 \\ -10 & 8 & 5 \end{pmatrix} - \begin{pmatrix} 2 & 9 & 0 \\ 3 & 8 & -2 \\ 2 & 7 & 2 \end{pmatrix}$

Multiplying matrices

By the end of this section, you will know:

- To multiply a matrix with a simple factor involves multiplying each element in turn by the factor.
- Matrices can only be multiplied if their orders are complementary.

Multiplying a matrix by a factor is sometimes referred to as the matrix's **scalar product**.

Multiplying a matrix by a number or factor is simple and straightforward: we merely multiply each element in the matrix by the factor.

■ **Worked Example 27.6**

What is the value of $3 \times \begin{pmatrix} 1 & 2 \\ 3 & 0 \\ 5 & -1 \end{pmatrix}$?

In simple examples like this, we multiply each matrix element by the external factor:

$$3 \times \begin{pmatrix} 1 & 2 \\ 3 & 0 \\ 5 & -1 \end{pmatrix} = \begin{pmatrix} 3 \times 1 & 3 \times 2 \\ 3 \times 3 & 3 \times 0 \\ 3 \times 5 & -3 \times 1 \end{pmatrix} = \begin{pmatrix} 3 & 6 \\ 9 & 0 \\ 15 & -3 \end{pmatrix}$$ ■

Scalar quantities have only a magnitude but no direction. Vector quantities have *both* a magnitude and a direction. We discuss this aspect in more detail in Chapter 28.

Self-test 27.4

Determine the following matrix scalar products.

27.4.1 $2\begin{pmatrix} 2 & 1 & -5 \\ 5 & 8 & 6 \\ -3 & 2 & 0 \end{pmatrix}$

27.4.2 $4.2\begin{pmatrix} 0 & 3 \\ 5 & 1 \end{pmatrix}$

27.4.3 $a\begin{pmatrix} a & 1 & c^2 & 2ac \\ 0 & b & b^2 & ab \end{pmatrix}$

27.4.4 $3x\begin{pmatrix} 2 & 5xy \\ 2x^{2.2} & x^2 \\ \dfrac{4.3}{x} & 0 \end{pmatrix}$

MULTIPLYING two matrices together is more time consuming than ADDITION or SUBTRACTION, but not particularly complicated. Two matrices **G** and **H** can only be MULTIPLIED together to generate matrix **K** if the number of columns in **G** is the same as the number of rows in **H**. If **G** has 5 rows and 3 columns, and **H** has 3 rows and 6 columns, the product of **G** and **H** will have 5 rows and 6 columns.

In general, the element $K_{ij} = \Sigma\, G_{ik} \times H_{kj}$. In words, we MULTIPLY together a row of **G** with a column of **H**.

If we multiply a matrix of order $m \times n$ with a matrix of order $n \times p$, the product will be of the order $m \times p$.

■ **Worked Example 27.7**

If $\mathbf{G} = \begin{pmatrix} 1 & 2 \\ 3 & 4 \\ -1 & 0 \\ 3 & 1 \end{pmatrix}$ and $\mathbf{H} = \begin{pmatrix} 1 & 3 & 1 \\ 2 & -1 & 0 \end{pmatrix}$, determine the product **G H**.

G has 2 columns and **H** has 2 rows, so we *can* multiply them together. The order of the final matrix K will be 4×3, because **G** has 4 rows and **H** has 3 columns. (We cannot multiply to obtain the product **H G** because **H** has 3 columns and **G** has 4 rows, meaning they are not complimentary or compatible.)

Strategy:

(i) We multiply the first row of **G** with the first column of **H**. We do so, element by element. In this Worked Example, there are two multiplications: 1×1 and 2×2. Their sum is $1 \times 1 + 2 \times 2 = 5$.

$$\begin{bmatrix} \boxed{1 \quad 2} \\ 3 & 4 \\ -1 & 0 \\ 3 & 1 \end{bmatrix} \times \begin{pmatrix} \boxed{1} & 3 & 1 \\ \boxed{2} & -1 & 0 \end{pmatrix} = \begin{pmatrix} 5 & ? & ? \\ ? & ? & ? \\ ? & ? & ? \\ ? & ? & ? \end{pmatrix}$$

Algebraically, we say $G_{11} \times H_{11} + G_{12} \times H_{12} = 5$. We write this result in the matrix on the right-hand side of the equation. Clearly, we do not yet know the values of the other 11 elements.

(ii) We can now multiply the *second* row of **G**, again with the first column of **H**.

$$\begin{pmatrix} 1 & 2 \\ \boxed{3 \quad 4} \\ -1 & 0 \\ 3 & 1 \end{pmatrix} \times \begin{pmatrix} \boxed{1} & 3 & 1 \\ \boxed{2} & -1 & 0 \end{pmatrix}.$$

The matrix **K** now looks like this: $\begin{pmatrix} 5 & ? & ? \\ 11 & ? & ? \\ ? & ? & ? \\ ? & ? & ? \end{pmatrix}$

As before, we obtain the terms by multiplying, element by element. The values are clearly 3×1 and 4×2, the sum of which is 11, so $G_{21} \times H_{11} + G_{22} \times H_{22} = 11$. We placed the answer of '11' in the first column and second row because we obtained it by multiplying the first row of H with the second row of G.

Using these methods, we obtain the final result,

$$K = \begin{pmatrix} 5 & 1 & 1 \\ 11 & 5 & 3 \\ -1 & -3 & -1 \\ 5 & 8 & 3 \end{pmatrix}$$

By following this procedure, we determine values for all twelve matrix elements. For example, element $K_{32} = -3$ is obtained as $(-3 \times 1) + (0 \times -1)$.

In summary, matrix multiplication can sometimes seem a little time consuming, but it is not conceptually difficult. ■

In this example, we readily multiplied G and H to obtain the product GH. We could not obtain the product HG, because the order of the two matrices was not compatible. Even when multiplying two square matrices, it is very rare for $AB = BA$. Matrix multiplication is not commutative.

Self-test 27.5

Perform the following matrix multiplication problems.

27.5.1 $\begin{pmatrix} 1 & 4 \\ 3 & 0 \end{pmatrix}\begin{pmatrix} 3 & 4 \\ 0 & 2 \end{pmatrix}$

27.5.2 $\begin{pmatrix} 2 & 0 & 3 \\ 1 & 2 & -1 \end{pmatrix}\begin{pmatrix} 1 \\ 2 \\ 4 \end{pmatrix}$

27.5.3 $\begin{pmatrix} a & 0 \\ a^2 & 1 \end{pmatrix}\begin{pmatrix} a \\ 2a \end{pmatrix}$

Matrices representing symmetry operators

By the end of this section, you will know:

- Symmetry operators (such as mirror planes and rotations) can be represented by matrices.
- The matrix R_{new} representing one symmetry operation (represented by R_1) followed by another (represented by R_2) can be obtained by using matrix multiplication. $R_{new} = R_2 R_1$

We have seen that matrix multiplication is relatively straightforward. One key application of this method is in the field of Group Theory, which is a mathematical approach to symmetry. The symmetry of a molecule is the key to a number of different properties:

- It determines whether a molecule is polar or chiral.
- It dictates which atomic orbitals contribute to a given molecular orbital.
- It identifies which vibrational modes are infrared or Raman active.
- It controls the selection rules for transitions between molecular states in spectroscopy.

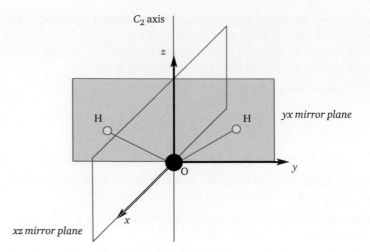

Figure 27.1 A water molecule has a C_2 rotation axis and two mirror planes

It is possible to identify different 'actions' that, after being applied to a molecule, leave it in a configuration that is indistinguishable from the original. These are known as symmetry operations. The most typical examples are reflections, rotations and inversions. For instance, if we look at the molecule of water shown in Fig. 27.1, it is clear that if the molecule lies in the y–z plane, then we can reflect the molecule in both the x–z mirror plane and the y–z mirror plane leaving the molecule looking unchanged. In addition, if we rotate the molecule about the z-axis through $180°$ then once again it will look identical to the original. We call this the C_2-axis. The mirror planes and rotation axes are known as symmetry elements. As we will see in the next Worked Example, matrices can be used to represent all of these symmetry operations.

■ **Worked Example 27.8**

Show that a reflection in the x–z plane can be represented by the matrix:

$$\begin{pmatrix} 1 & 0 & 0 \\ 0 & -1 & 0 \\ 0 & 0 & 1 \end{pmatrix}$$

In order to calculate the matrix, it is best to sketch a diagram to visualize the transformation. We will use Fig. 27.1. If we consider the coordinates of the hydrogen atom on the right-hand side to be (x_1, y_1, z_1), then we can see that after reflection in the x–z mirror plane:

- The x-coordinate will remain the same.
- The y-coordinate will become $-y_1$.
- The z-coordinate will remain the same.

Therefore, we can see that after reflection in the x–y mirror plane, a general point (x_1, y_1, z_1) will move to $(x_1, -y_1, z_1)$. If we call the final position (x_2, y_2, z_2) then we can summarize this action as:

$$x_2 = 1 \times x_1 + 0 \times y_1 + 0 \times z_1 \tag{i}$$

$$y_2 = 0 \times x_1 + (-1) \times y_1 + 0 \times z_1 \tag{ii}$$

$$z_2 = 0 \times x_1 + 0 \times y_1 + 1 \times z_1 \tag{iii}$$

We can represent this reflection by constructing a matrix such that $x_2 = R\,x_1$, where

the initial point is represented by the matrix: $x_1 = \begin{pmatrix} x_1 \\ y_1 \\ z_1 \end{pmatrix}$

and after reflection the position of the atom is given by: $x_2 = \begin{pmatrix} x_2 \\ y_2 \\ z_2 \end{pmatrix} = \begin{pmatrix} x_1 \\ -y_1 \\ z_1 \end{pmatrix}.$

We construct the first row of the matrix, by looking at the equation for x_2 (i) and writing down the coefficient in front of x_1 in column 1 (this is a 1), the coefficient in front of y_1 in column 2 (this is 0) and then the coefficient of z_1 in column 3 (this is also 0). We repeat this procedure for the second and third rows using the corresponding coefficients for y_2 and z_2, respectively. The matrix representing the reflection is:

$$R_{\text{reflection}} = \begin{pmatrix} 1 & 0 & 0 \\ 0 & -1 & 0 \\ 0 & 0 & 1 \end{pmatrix}.$$

We can easily check that it satisfies the equation $x_2 = R\,x_1$:

$$\begin{pmatrix} x_2 \\ y_2 \\ z_2 \end{pmatrix} = \begin{pmatrix} 1 & 0 & 0 \\ 0 & -1 & 0 \\ 0 & 0 & 1 \end{pmatrix}\begin{pmatrix} x_1 \\ y_1 \\ z_1 \end{pmatrix}$$

Now that we have a matrix representing this operation, it is very quick to calculate the effect of the operation on any given point. We can also use this method to investigate how a set of atomic orbitals transform under a symmetry operation. ∎

■ **Worked Example 27.9**

Show that a rotation about the z-axis through 180° can be represented by the matrix:
$\begin{pmatrix} -1 & 0 & 0 \\ 0 & -1 & 0 \\ 0 & 0 & 1 \end{pmatrix}$ and find the image of the point $(3, 4, 1)$ after this transformation.

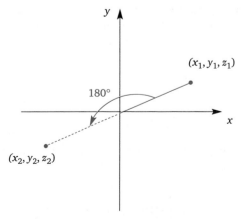

Figure 27.2 Rotation of a point (x_1, y_1, z_1) about the z-axis through 180° to point (x_2, y_2, z_2).

By inspecting Fig. 27.2, we can see that after a rotation of 180° about the z-axis plane:

- The x-coordinate will become $-x_1$.

- The y-coordinate will become $-y_1$.

- The z-coordinate will remain the same.

Therefore, after rotation about the z-axis, a general point (x_1, y_1, z_1) will move to $(-x_1, -y_1, z_1)$. If we call the final position (x_2, y_2, z_2) then we can summarize this action as:

$$x_2 = (-1) \times x_1 + 0 \times y_1 + 0 \times z_1$$

$$y_2 = 0 \times x_1 + (-1) \times y_1 + 0 \times z_1$$

$$z_2 = 0 \times x_1 + 0 \times y_1 + 1 \times z_1$$

The matrix representing the rotation is: $\mathbf{R}_{\text{rotation}} = \begin{pmatrix} -1 & 0 & 0 \\ 0 & -1 & 0 \\ 0 & 0 & 1 \end{pmatrix}$.

We can now find the image of the point (3, 4, 1) using matrix multiplication: $\mathbf{X}_2 = \mathbf{R}_{\text{rotation}} \mathbf{X}_1$.

$$\begin{pmatrix} x_2 \\ y_2 \\ z_2 \end{pmatrix} = \begin{pmatrix} -1 & 0 & 0 \\ 0 & -1 & 0 \\ 0 & 0 & 1 \end{pmatrix} \begin{pmatrix} 3 \\ 4 \\ 1 \end{pmatrix} = \begin{pmatrix} -3 \\ -4 \\ 1 \end{pmatrix} \blacksquare$$

Looking at the $\mathbf{R}$ matrices that represent the symmetry operations in Worked Examples 27.8 and 27.9, we should notice how the first column of the matrix is simply the image of the point (1, 0, 0) after the transformation. Similarly the second and third columns are the images of the points (0, 1, 0) and (0, 0, 1), respectively, after the transformation. This result is true for all linear transformations and provides a quick method to derive a transformation matrix. We can check that this method works for the Self-test 27.6 questions.

Matrix multiplication also makes it possible to evaluate a point after a series of operations. This is much quicker than drawing a sketch and evaluating the position of the image at each intermediate step. It is also likely to help eliminate errors.

■ Worked Example 27.10

Calculate the image of a point (x, y, z) after reflection in the x–z plane followed by rotation about the z-axis through 180°.

[Hint: Use the matrices found for these operations in Worked Examples 27.8 and 27.9.]

Now that we have the matrices representing the reflection and rotation operations, we can easily calculate the matrix representing the reflection *followed by* the rotation, $\mathbf{R} = \mathbf{R}_{\text{rotation}} \mathbf{R}_{\text{reflection}}$. (Note the order of the matrices—the transformation that is applied first goes on the right-hand side.):

$$\mathbf{R} = \begin{pmatrix} -1 & 0 & 0 \\ 0 & -1 & 0 \\ 0 & 0 & 1 \end{pmatrix} \begin{pmatrix} 1 & 0 & 0 \\ 0 & -1 & 0 \\ 0 & 0 & 1 \end{pmatrix} = \begin{pmatrix} -1 & 0 & 0 \\ 0 & 1 & 0 \\ 0 & 0 & 1 \end{pmatrix}$$

$$\begin{pmatrix} x_2 \\ y_2 \\ z_2 \end{pmatrix} = \begin{pmatrix} -1 & 0 & 0 \\ 0 & 1 & 0 \\ 0 & 0 & 1 \end{pmatrix} \begin{pmatrix} x \\ y \\ z \end{pmatrix} = \begin{pmatrix} -x \\ y \\ z \end{pmatrix}$$

We can see that after the application of these two matrices, a general point (x, y, z) will move to $(-x, y, z)$. This is equivalent to reflection in a y–z mirror plane. ∎

Self-test 27.6

Write down the matrices representing the following transformations that are applied to a general point (x, y, z):

27.6.1 Reflection in an x–y mirror plane

27.6.2 Rotation through 90° (in the anticlockwise direction) about the z-axis.

27.6.3 A two-step transformation: rotation through 90° (in the anticlockwise direction) followed by reflection in an xy mirror plane.

Using these matrix representations, it is possible to explore mathematically the symmetry of a molecule and, using sets of specialized tables (called 'character tables' that may be found in many physical chemistry textbooks), deduce many of the symmetry properties described at the start of this section.

The determinant of a matrix, det A

By the end of this section, you will know:

- The determinant of the matrix **A** is written as **det A** or $|A|$.
- Determinants are of crucial importance in quantum mechanics.
- The determinant of a simple 2×2 square matrix is obtained by inspection.
- The determinant of larger square matrices is obtained by reduction, which yields cofactors.

The **determinant** is a feature of matrices we have not yet discussed. It is a scalar property, so it has magnitude but not direction (see p. 420). The determinant is a rather abstract quantity, but we can relate it to the volume of a parallelepiped or the area of a triangle. We should not let the abstract nature of the determinant cause any worry, since it is an invaluable tool for determining key properties in quantum mechanics, as we see in the next section.

We denote the determinant of a matrix **A** as **det A**, $|A|$ or as Δ. In this text, we will usually use the first and second notations. Note that when using this second notation, we employ *vertical lines* rather than the *curved brackets* that surround a matrix.

In the following discussion, we will assume that all matrices are square (i.e. $m = n$) because it is not straightforward to obtain the determinant of a rectangular matrix. For a simple matrix **A**, we can obtain the determinant by inspection:

- **det A** for a 1×1 matrix (a_{11}) is the same as the single element a_{11}.

- **det A** for a 2×2 matrix $\begin{pmatrix} a_{11} & a_{12} \\ a_{21} & a_{22} \end{pmatrix}$ is obtained following eqn (27.3):

$$\det A = \begin{vmatrix} a & b \\ c & d \end{vmatrix} = ad - bc \qquad (27.3)$$

Take care: remember to use the rules of BODMAS here.

■ **Worked Example 27.11**

Obtain the determinant of the following 2×2 matrix: $\begin{pmatrix} 1 & 2 \\ 3 & 4 \end{pmatrix}$

Using the template in eqn (27.3), $\det A = (1 \times 4) - (2 \times 3) = 4 - 6 = -2$. ■

Obtaining the determinant of large matrices is considerably more time consuming. In effect, we split the matrix into smaller, handier, 'chunks,' and obtain the determinant of each. We call this process **reduction**.

Obtaining the determinant of larger matrices can seem quite difficult. For the 3×3 matrix $A = \begin{pmatrix} a_{11} & a_{12} & a_{13} \\ a_{21} & a_{22} & a_{23} \\ a_{31} & a_{32} & a_{33} \end{pmatrix}$, we use the template in eqn (27.4):

$$\det A = a_{11} \begin{vmatrix} a_{22} & a_{23} \\ a_{32} & a_{33} \end{vmatrix} - a_{12} \begin{vmatrix} a_{21} & a_{23} \\ a_{31} & a_{33} \end{vmatrix} + a_{13} \begin{vmatrix} a_{21} & a_{22} \\ a_{31} & a_{32} \end{vmatrix} \qquad (27.4)$$

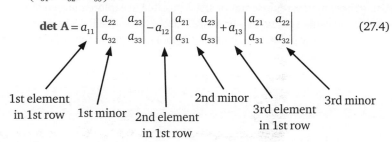

1st element in 1st row   1st minor   2nd element in 1st row   2nd minor   3rd element in 1st row   3rd minor

A square matrix of m rows and columns generates m pairs of terms. A moment's glance persuades us that eqn (27.4) follows a simple, regimented pattern.

There are three components to each term: a multiplication factor, an element from the first row of the matrix, a_{1i}, and a determinant. The multiplication factor is given by: $(-1)^{i+j}$ for element a_{ij}.

In the first term, we write down the multiplication factor associated with a_{11}, which is 1. Similarly, in the second and third terms, we calculate the factors -1 and 1 for a_{12} and a_{13}, respectively. In fact, it is easier to remember that the sign of the multiplication factor alternates as we write down each term associated with an element in the top row. It follows the clear pattern shown below, where '+' and '−' indicate that the term for that element should be multiplied by '+1' or '−1', respectively.

For a 2×2 matrix: $\begin{pmatrix} + & - \\ - & + \end{pmatrix}$

For a 3×3 matrix: $\begin{pmatrix} + & - & + \\ - & + & - \\ + & - & + \end{pmatrix}$

For a 4×4 matrix: $\begin{pmatrix} + & - & + & - \\ - & + & - & + \\ + & - & + & - \\ - & + & - & + \end{pmatrix}$

The third component in each term is a **minor**, M_{ij}. The minor of a_{mn} is the determinant of the matrix obtained by deleting the row m and the column n. We can see this effect diagrammatically: the element a_{mn} is indicated in blue and the lozenges encapsulate the remainder of the original matrix, which is the matrix used in the respective minor:

$$\begin{pmatrix} a_{11} & a_{12} & a_{13} \\ a_{21} & a_{22} & a_{23} \\ a_{31} & a_{32} & a_{33} \end{pmatrix} \qquad \begin{pmatrix} a_{11} & a_{12} & a_{13} \\ a_{21} & a_{22} & a_{23} \\ a_{31} & a_{32} & a_{33} \end{pmatrix} \qquad \begin{pmatrix} a_{11} & a_{12} & a_{13} \\ a_{21} & a_{22} & a_{23} \\ a_{31} & a_{32} & a_{33} \end{pmatrix}$$

Having written eqn (27.4), we then obtain the minors (which are just determinants of 2×2 matrices) in the usual way, using the template in eqn (27.3).

It is useful for us, at this point, to define a cofactor as the combination of the multiplication factor with the minor:

$$\text{cofactor} = (-1)^{i+j} M_{ij} \qquad (27.5)$$

We can see that the determinant in eqn (27.4) is simply:

the sum of (each element in the top row × its cofactor).

■ **Worked Example 27.12**

Establish the determinant of the matrix $P = \begin{vmatrix} 5 & 2 & 3 \\ 4 & 7 & 1 \\ 8 & 5 & 9 \end{vmatrix}$

Following the template in eqn (27.4), we write:

$$\det P = 5 \times \begin{vmatrix} 7 & 1 \\ 5 & 9 \end{vmatrix} - 2 \times \begin{vmatrix} 4 & 1 \\ 8 & 9 \end{vmatrix} + 3 \times \begin{vmatrix} 4 & 7 \\ 8 & 5 \end{vmatrix}$$

	Element	Cofactor	Determinant
First pair of terms	5	$\begin{vmatrix} 7 & 1 \\ 5 & 9 \end{vmatrix}$	$(7 \times 9) - (1 \times 5) = 63 - 5 = 58$
Second pair of terms	2	$-\begin{vmatrix} 4 & 1 \\ 8 & 9 \end{vmatrix}$	$(4 \times 9) - (1 \times 8) = 36 - 8 = 28$
Third pair of terms	3	$\begin{vmatrix} 4 & 7 \\ 8 & 5 \end{vmatrix}$	$(4 \times 5) - (7 \times 8) = 20 - 56 = -36$

Inserting these data into eqn yields

$$\det P = (5 \times 58) - (2 \times 28) + (3 \times -36) = 290 - 56 - 108 = 126. \quad ■$$

Reducing a 4×4 matrix generates four minors, each of which is the determinant of a 3×3 matrix. Unfortunately, being larger than 2×2, we must perform a separate process of reduction on each of these minors according to eqn (27.4). In fact, this process of reduction can become very tedious indeed if the sources matrix is large. But don't worry: computer programmes can reduce matrices of any size.

Quantum mechanicists often need to generate matrices having an order of a hundred or so.

Self-test 27.7

Obtain the determinant for each of the following matrices.

27.7.1 $\begin{pmatrix} 1 & 2 \\ 6 & 4 \end{pmatrix}$

27.7.2 $\begin{pmatrix} 1 & -2 & 7 \\ 3 & 0 & 2 \\ 2 & 1 & 6 \end{pmatrix}$

27.7.3 $\begin{pmatrix} a & 2s & 4s \\ -a & a & 4as \\ 0 & 3b & 0 \end{pmatrix}$

27.7.4 $\begin{pmatrix} 1 & 3 & 2 & 8 \\ 3 & 1 & 6 & 3 \\ 2 & 6 & 1 & 0 \\ 8 & 3 & 0 & 1 \end{pmatrix}$

Slater determinants:

One property of determinants that quantum chemists exploit is the fact that interchanging any two rows or columns of the matrix **A** changes the sign of the determinant of **A**. (We can confirm this for any of the matrices in this section.) Due to the Pauli exclusion principle, the wavefunctions describing a many-electron system must change sign (i.e. be antisymmetric) if two electrons are exchanged. These wavefunctions can look quite complicated when written out in full since each term consists of a hydrogen-like spatial orbital multiplied by a spin orbital. Instead, we write them in determinant form. These determinants were developed by John C. Slater in 1929, so we call them **Slater determinants**. The Slater determinant form makes it easier to interpret and construct a wavefunction. For instance, we can write the Slater determinant for the ground state of helium as:

$$\psi(1,2) = \frac{1}{\sqrt{2}} \begin{vmatrix} 1s(1)\alpha(1) & 1s(1)\beta(1) \\ 1s(2)\alpha(2) & 1s(2)\beta(2) \end{vmatrix}.$$

Each row contains the two spin orbitals, which are available to an electron in the ground state of helium. Both are occupied by the corresponding electron: electron 1 for row 1, electron 2 for row 2. It is relatively easy to extend this approach to describe the wavefunctions of atoms containing more than two electrons.

We involve determinants in the calculation of a number of other components of matrix algebra, as we see in the remaining sections.

The eigenvalues of a matrix

By the end of this section, you will know:

- An eigenvector of a square matrix A is a non-zero vector x, such that Ax = λx, where λ is the eigenvalue corresponding to the eigenvector x.

- The eigenvalues of a matrix are obtained by subtracting a constant amount from each element along the principal diagonal, and setting the determinant equal to zero.

Many properties in quantum mechanics can be determined by solving an eigenvalue equation. This is an equation of the form:

$$\mathbf{Ax} = \lambda\mathbf{x} \tag{27.6}$$

where **A** is a square matrix, **x** is a column vector, known as an **eigenfunction** of the matrix **A**, and λ is a number, which is called the eigenvalue corresponding to the eigenvector **x**. The matrix **A** is an operator that when applied to the function represented by **x** equals a number multiplied by the same vector **x**. **A** could be something as simple as a linear transformation that transforms the vector **x** into a multiple of **x**, i.e. the vector will change length but not direction.

In quantum mechanics, eigenvalue equations exist for a number of operators including those for the Hamiltonian, momentum and angular momentum. Of these, the time-independent Schrödinger equation is perhaps the most obvious example: $\hat{H}\psi_n = E_n\psi_n$.

In this case, $\hat{H}$ is the Hamiltonian operator matrix, which is associated with the total energy of the system. The wavefunctions, ψ_n, are the eigenfunctions of $\hat{H}$ and the allowed energies, E_n, are the corresponding eigenvalues.

When we try to 'solve' an eigenvalue problem, such as the Schrödinger equation, we are really trying to find the eigenvalues and corresponding eigenfunctions of the operator. The key to this process is to rewrite the equation as:

$$(\mathbf{A} - \lambda \mathbf{I})\mathbf{x} = 0 \qquad (27.7)$$

We can demonstrate eqn (27.7)= eqn (27.6) by multiplying out the bracket, and remembering that, by definition, $\mathbf{I}\,\mathbf{x} = \mathbf{x}$.

We should note that the concise equations given above actually represent the following simultaneous equations:

$$
\begin{aligned}
(a_{11} - \lambda)x_1 &+ a_{12}x_2 &+ a_{13}x_3 &\cdots & a_{1n}x_n &= 0 \\
a_{21}x_1 &+ (a_{22} - \lambda)x_2 &+ a_{23}x_3 &\cdots & a_{2n}x_n &= 0 \\
\vdots & \vdots & \vdots & \ddots & \vdots \\
a_{m1}x_1 &+ a_{m2}x_2 &+ a_{m3}x_3 &\cdots & (a_{mn} - \lambda)x_n &= 0
\end{aligned}
$$

They have the trivial solution $\mathbf{x} = 0$ for all values of λ, which we will ignore. The matrix $(\mathbf{A} - \lambda \mathbf{I})$ can only have non-zero solutions if the value of λ is chosen in such a way that the determinant of $(\mathbf{A} - \lambda \mathbf{I})$ is zero (i.e. $(\mathbf{A} - \lambda \mathbf{I})$ is a non-invertible matrix).

The **characteristic** (or **secular**) equation of the matrix $\mathbf{A}$ is defined as:

$$\mathbf{det}\ (\mathbf{A} - \lambda \mathbf{I}) = 0 \qquad (27.8)$$

which can be written in full as:

$$
\begin{vmatrix}
(a_{11} - \lambda) & a_{12} & a_{13} & \cdots & a_{1n} \\
a_{21} & (a_{22} - \lambda) & a_{23} & \cdots & a_{2n} \\
\vdots & \vdots & \vdots & \ddots & \vdots \\
a_{m1} & a_{m2} & a_{m3} & \cdots & (a_{mn} - \lambda)
\end{vmatrix} = 0.
$$

We call the left-hand side of this equation the **characteristic** (or **secular**) **determinant** of the matrix $\mathbf{A}$. It is a polynomial of order n in λ, and its roots are the values of λ. The values of λ for which this determinant has non-zero values are called the eigenvalues (or characteristic values) of the matrix $\mathbf{A}$.

The word **secular** here comes from a Latin source meaning 'age' or 'generation.' The term comes ultimately from astronomy, where the same style of equation appears in connection with slowly accumulating modifications of planetary orbits.

These values of λ are also called the **latent roots** or **proper roots** of the equation.

■ **Worked Example 27.13**

Solve the characteristic equation of the matrix $\mathbf{A} = \begin{pmatrix} -2 & 1 & 1 \\ -11 & 4 & 5 \\ -1 & 1 & 0 \end{pmatrix}$

Strategy:

(i) We first rewrite each element in the diagonal, adding '$-\lambda$'.

(ii) Write in determinant form.

(iii) Obtain the value of the determinant, and solve for λ.

Solution:

(i) and (ii) $\det \mathbf{A} = \begin{vmatrix} -2-\lambda & 1 & 1 \\ -11 & 4-\lambda & 5 \\ -1 & 1 & 0-\lambda \end{vmatrix}$

(iii) $\det \mathbf{A} = -(2+\lambda) \begin{vmatrix} 4-\lambda & 5 \\ 1 & -\lambda \end{vmatrix} - 1 \begin{vmatrix} -11 & 5 \\ -1 & -\lambda \end{vmatrix} + 1 \begin{vmatrix} -11 & 4-\lambda \\ -1 & 1 \end{vmatrix}$

Therefore, $\det \mathbf{A} = -(2+\lambda)\{-\lambda(4-\lambda)-5\} - 1\{11\lambda - (-5)\} + 1\{-11 - (-(4-\lambda))\}$

To determine the values of λ, we multiply, collect terms, and then factorize:

$$\det A = -(2+\lambda)\{\lambda^2 - 4\lambda - 5\} - 11\lambda - 5 - 11 + 4 - \lambda$$

$$= -(2\lambda^2 - 8\lambda - 10 + \lambda^3 - 4\lambda^2 - 5\lambda) - 12\lambda - 12$$

$$= -(\lambda^3 - 2\lambda^2 - 13\lambda - 10) - 12\lambda - 12$$

$$= -(\lambda^3 - 2\lambda^2 - 13\lambda - 10 + 12\lambda + 12)$$

$$= -(\lambda^3 - 2\lambda^2 - \lambda + 2)$$

$$= -(\lambda - 1)(\lambda^2 - \lambda - 2)$$

$$= -(\lambda - 1)(\lambda + 1)(\lambda - 2)$$

so $\lambda = 1, -1$ and 2.

We can check our answer is correct: the sum of the eigenvalues of A equals the sum of the elements on its leading diagonal. (We call this sum the **trace** of a square matrix.)
Sum of eigenvalues $= 1 + (-1) + 2 = 2$
Trace of $A = -2 + 4 = 2$
Thus, we can be fairly confident that we have correctly solved the characteristic equation. ∎

Tip: A useful check of the answer is that the sum of the eigenvalues of matrix A equals the sum of the elements on its leading diagonal.

This example is purely numerical, but the methodology is very commonly utilized in quantum-mechanical calculations.

■ **Worked Example 27.14**

In the molecular orbital theory for a π-electron system, the energy states of the π-electrons are described by the matrix eigenvalue equation, $Hc = Ec$, where the matrix H represents the effective Hamiltonian for a π-electron in the system, the eigenvalues E of H are the orbital energies of the π-electrons, and the eigenvectors c represent the corresponding molecular orbitals.

Write the effective Hamiltonian matrix for cyclobutadiene (**VII**) and solve to obtain the energy E.

VII

To start, we write the effective Hamiltonian matrix as $H = \begin{pmatrix} \alpha & \beta & 0 & \beta \\ \beta & \alpha & \beta & 0 \\ 0 & \beta & \alpha & \beta \\ \beta & 0 & \beta & \alpha \end{pmatrix}$.

When written as the characteristic equation, and in determinate form, we obtain:

$$\det(H - EI) = \begin{vmatrix} \alpha - E & \beta & 0 & \beta \\ \beta & \alpha - E & \beta & 0 \\ 0 & \beta & \alpha - E & \beta \\ \beta & 0 & \beta & \alpha - E \end{vmatrix} = 0$$

The determinant is $(\alpha - E)^2 (\alpha - E + 2\beta)(\alpha - E - 2\beta) = 0$.

To have a value of zero, the eigenvalues are therefore

$$E_1 = \alpha + 2\beta, E_2 = E_3 = \alpha, \quad \text{and} \quad E_4 = \alpha - 2\beta.$$

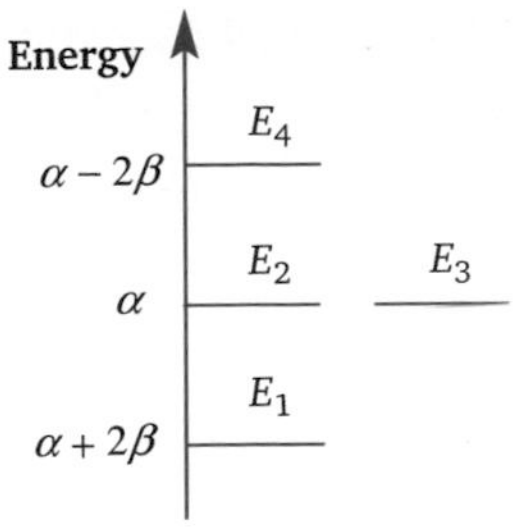

Figure 27.3 A simple molecular-orbital energy diagram of cyclobutadiene.

Again, we can check our answer is correct since:

Sum of eigenvalues $= \alpha + 2\beta + \alpha + \alpha + \alpha - 2\beta = 4\alpha = $ Trace of H

Figure 27.3 shows these molecular-orbitals energies diagrammatically. E_2 and E_3 are the same and therefore energetically degenerate. ■

In some cases, we are interested in finding both the eigenvalues and the eigenvectors of a matrix. To find the eigenvalues and eigenvectors of a matrix A, we must:

(i)   Solve the characteristic equation, $\det(A - \lambda I) = 0$ to find any eigenvalue λ.

(ii)   Substitute each eigenvalue λ found in step (i), into the eigenvalue equation.

(iii)   Solve the simultaneous equations to find the corresponding eigenfunction.

In normal mode analysis of molecular vibrations, it can be shown that Newton's equations of motion for a given system of oscillating atoms satisfy the following equation:

$$\ddot{x} = -\omega^2 x \qquad (27.9)$$

where $\ddot{x}$ can be written in the form $\ddot{x} = Ax$. $\ddot{x}$ is the (mass-weighted) Hessian matrix, representing the second derivative of the energy, and A is called the dynamic matrix. By substituting in for $\ddot{x}$, we can see that eqn (27.9) can be written in the form of a typical eigenvalue equation (27.6):

$$Ax = \lambda x$$

where the eigenvalue, λ, is the negative of the square of the normal mode angular frequency, ω, associated with the corresponding normal mode eigenvector, x, which gives us the coefficients for each of the atomic displacements involved in this vibrational mode.

■ **Worked Example 27.15**

Find the normal mode angular frequency and corresponding eigenvectors of the dynamic matrix for a diatomic molecule that is given below:

$$A = \begin{pmatrix} -4 & 2 \\ 2 & -4 \end{pmatrix}$$

Strategy:

(i)   First, we write the characteristic equation, $\det(A - \lambda I) = 0$:

$$\det(A - \lambda I) = \begin{vmatrix} -4 - \lambda & 2 \\ 2 & -4 - \lambda \end{vmatrix} = 0$$

We expand the determinant using eqn (27.3).

$(-4 - \lambda)(-4 - \lambda) - 2 \times 2 = 0$

It is relatively simple to find the roots of this equation since:

$(-4 - \lambda)^2 = 2^2$

Therefore, $-4 - \lambda = \pm 2$

and $\quad \lambda = -6 \quad$ or $\quad -2$.

(ii) When $\lambda = -6$, the eigenvector equations are:

$$\begin{pmatrix} -4-(-6) & 2 \\ 2 & -4-(-6) \end{pmatrix} \begin{pmatrix} x_1 \\ x_2 \end{pmatrix} = \begin{pmatrix} 0 \\ 0 \end{pmatrix}$$

If we multiply these matrices, we can see that:

$$2x_1 + 2x_2 = 0$$

$$2x_1 + 2x_2 = 0$$

These equations are the same and reduce to: $x_1 = -x_2$.

Therefore, we can see that $\begin{pmatrix} 1 \\ -1 \end{pmatrix}$ is one possible eigenvector.

(Any scalar multiple of the eigenvector selected will also be an eigenvector with the same corresponding eigenvalue. We have just selected the simplest form that we could find.)

When $\lambda = -2$, the eigenvector equations are:

$$\begin{pmatrix} -4-(-2) & 2 \\ 2 & -4-(-2) \end{pmatrix} \begin{pmatrix} x_1 \\ x_2 \end{pmatrix} = \begin{pmatrix} 0 \\ 0 \end{pmatrix}$$

If we multiply these matrices, we can see that:

$$-2x_1 + 2x_2 = 0$$

$$2x_1 - 2x_2 = 0$$

As before, these equations are the same and reduce to: $x_1 = x_2$.

Therefore, we can see that $\begin{pmatrix} 1 \\ 1 \end{pmatrix}$ is one possible eigenvector with eigenvalue -2.

The superscripted T here means transpose; see next section

We know that the normal mode angular frequencies can be found from the eigenvalue: $-\lambda = \omega^2$. Therefore, for the eigenvector $(1 \ -1)^T$, with corresponding eigenvalue $\lambda = -2$, $\omega = \sqrt{2}$. For the eigenvector $(1 \ 1)^T$, with corresponding eigenvalue $\lambda = -6$, $\omega = \sqrt{6}$. ∎

Self-test 27.8

Write an effective Hamiltonian matrix to describe the allyl radical (**VIII**), $CH_2CHCH_2{}^\bullet$, and hence solve to obtain the energies E of and corresponding molecular orbitals for the π-electrons.

$$\textbf{VIII} \quad CH_2 - \overset{\bullet}{C}H - CH_2$$

The transpose of a matrix A^T

By the end of this section, you will know:

- The transpose of a matrix is denoted as A^T.
- We obtain the transpose of a matrix by exchanging the elements of rows a_{nm} with the corresponding elements for columns, a_{mn}.
- The elements along the diagonal of A remain completely unchanged in A^T.

We denote the **transpose** of a matrix A as A^T. We obtain the transpose by simply exchanging rows for columns, so if a matrix element is a_{nm}, it becomes element a_{mn} in the transpose.

■ **Worked Example 27.16**

Determine the transpose of the matrix $A = \begin{pmatrix} 5 & 2 & 3 \\ 4 & 7 & 1 \\ 8 & 5 & 9 \end{pmatrix}$

In practice, we obtain the transpose A^T by considering the mirror image of the matrix A along the diagonal top-left–bottom-right:

$\begin{pmatrix} 5 & 2 & 3 \\ 4 & 7 & 1 \\ 8 & 5 & 9 \end{pmatrix}$ The mirror plane employed when obtaining the transpose A^T

The transpose A^T is therefore $\begin{pmatrix} 5 & 4 & 8 \\ 2 & 7 & 5 \\ 3 & 1 & 9 \end{pmatrix}$

Notice how the elements along the diagonal remain completely unchanged. ■

In the case of a square matrix (i.e. $m = n$), the transpose acts as a simple test of whether a matrix is symmetric: for all symmetric matrices, $A = A^T$. For example,

$$A = \begin{pmatrix} 1 & 2 \\ 2 & 4 \end{pmatrix} \quad \text{and} \quad A^T = \begin{pmatrix} 1 & 2 \\ 2 & 4 \end{pmatrix} = A$$

Self-test 27.9

Obtain the transpose A^T for each of the matrices in Self-test 27.7

The inverse of a matrix A^{-1}

By the end of this section, you will know:

- The inverse of a simple matrix A is denoted as A^{-1}.
- If we multiply A by A^{-1} we get the identity matrix.
- We obtain the inverse of a matrix A either by calculating the determinant and the transpose of the matrix of cofactors or by using Gaussian elimination.

We denote the **inverse** of a matrix A as A^{-1}. Every matrix has an inverse, provided the value of the determinant is not zero. If the determinant *is* zero, we say the matrix is **singular** (or **degenerate**), and we cannot then write an inverse.

A square matrix having a non-zero determinant (and hence an inverse) is said to be **non-singular** or **invertible**. In the remainder of this chapter, we will concentrate on the algebra of non-singular matrices, because we will need matrices capable of transforming to an inverse matrix.

It will probably have been noticed in the first half of this chapter that we discussed matrix ADDITION, SUBTRACTION and MULTIPLICATION, but not DIVISION. In fact, we never talk about matrix DIVISION, and talk instead about MULTIPLYING by the inverse matrix, A^{-1}.

This choice of nomenclature has an important consequence. In normal algebra, we often cancel by dividing a term by itself, to yield 1. In matrix algebra, as we cannot divide, we do not talk in terms of A/A, but say instead $A \times A^{-1} = I$. This product is important. Any matrix multiplied by its inverse yields the identity matrix I:

$$A \times A^{-1} = I \tag{27.10}$$

■ **Worked Example 27.17**

Show the product of the matrix $G = \begin{pmatrix} 1 & 2 \\ 3 & 4 \end{pmatrix}$ and its inverse $G^{-1} = -\frac{1}{2}\begin{pmatrix} 4 & -2 \\ -3 & 1 \end{pmatrix}$ is the identity matrix I.

We obtain the product using the multiplication methods above. For example, the first element a_{11} of the product matrix is $-\frac{1}{2}[(1 \times 4) + (2 \times -3)] = -\frac{1}{2} \times (4 - 6)$ which is 1.

The product of these two matrixes is the identity matrix $I = \begin{pmatrix} 1 & 0 \\ 0 & 1 \end{pmatrix}$. In other words, this result proves that matrix G is the inverse of matrix G^{-1}. ■

We need to know how to derive an inverse matrix.

- The inverse of a 1×1 matrix (a_{11}) is the reciprocal of its sole element, $1/a_{11}$.

- The inverse of a simple 2×2 matrix is obtained using the template:

$$\text{if} \quad A = \begin{pmatrix} a & b \\ c & d \end{pmatrix} \quad \text{then} \quad A^{-1} = \frac{1}{(ad - cb)}\begin{pmatrix} d & -b \\ -c & a \end{pmatrix} \tag{27.11}$$

- We can obtain the inverse of larger matrices by using the determinant and the transpose of the matrix of cofactors of A:

$$A^{-1} = \frac{1}{\det A} C^{T} \tag{27.12}$$

where c_{ij} are the cofactors of a_{ij}.

- Alternatively, the inverse of large matrices may be obtained by using the Gaussian elimination method. Details of this method can be found in more advanced texts.

■ **Worked Example 27.18**

Establish the inverse of the matrix Q: $\begin{pmatrix} 2 & 5 \\ 3 & 4 \end{pmatrix}$

Using the template in eqn (27.11), the inverse matrix

$$Q^{-1} = \frac{1}{8 - 15}\begin{pmatrix} 4 & -5 \\ -3 & 2 \end{pmatrix} = \frac{1}{-7}\begin{pmatrix} 4 & -5 \\ -3 & 2 \end{pmatrix} = \begin{pmatrix} -\frac{4}{7} & \frac{5}{7} \\ \frac{3}{7} & -\frac{2}{7} \end{pmatrix}. ■$$

Obtaining the **inverse** of a 3×3 (or larger) matrix A involves a multistep process:

(i)   Obtain the determinant of A, using eqn (27.4).

(ii) Highlight the matrices required to calculate the cofactors.

(iii) Obtain the cofactors of the source matrix.

(iv) Rewrite the matrix, replacing each element with its cofactor. We will call this matrix C.

(v)   Obtain the transpose of this matrix (C^{T}), i.e. the rows become the columns and the columns become the rows.

(vi) Multiply the matrix C^{T} by $1/\det A$.

■ **Worked Example 27.19**

Obtain the inverse of the matrix of $A = \begin{pmatrix} 1 & 4 & 3 \\ 2 & 1 & 0 \\ 1 & 3 & 2 \end{pmatrix}$

We follow the strategy listed above.

Solution:

(i) Using eqn (27.4) to calculate the determinant, we reduce the matrix:

$$\det A = \begin{vmatrix} 1 & 4 & 3 \\ 2 & 1 & 0 \\ 1 & 3 & 2 \end{vmatrix} = 1\begin{vmatrix} 1 & 0 \\ 3 & 2 \end{vmatrix} - 4\begin{vmatrix} 2 & 0 \\ 1 & 2 \end{vmatrix} + 3\begin{vmatrix} 2 & 1 \\ 1 & 3 \end{vmatrix}$$

The determinant is:

$$\det A = 1 \times (1 \times 2 - 3 \times 0) - 4 (2 \times 2 - 0 \times 1) + 3 (2 \times 3 - 1 \times 1) = 2 - 16 + 15 = 1.$$

(ii) We obtain the cofactors of the source matrix. We have already learnt how to obtain cofactors in the section above on determinants. There, we only obtained n cofactors for each $n \times n$ matrix. When writing an inverse, we must obtain all $n \times n$ cofactors. Therefore, this 3×3 matrix has 9 cofactors.

First, we highlight the matrix required to calculate the cofactor, for the element in *italic* type. From the top row:

From the *middle* row:

And from the *bottom* row:

This process is seen to be identical to that used when obtaining determinants.

(iii) Using the highlighted matrices, we calculate the corresponding cofactors. This step is easy when using the template for a determinant in eqn (27.3). We must remember to include the multiplication factor $(-1)^{i+j}$, as explained in eqn (27.5).

(iv) We then insert the cofactors into a new 3×3 matrix.

Inserting the appropriate cofactors (shown to the left of each bracket) yields:

$$\begin{pmatrix} +(2) & -(4) & +(5) \\ -(-1) & +(-1) & -(-1) \\ +(-3) & -(-6) & +(-7) \end{pmatrix} = \begin{pmatrix} 2 & -4 & 5 \\ 1 & -1 & 1 \\ -3 & 6 & -7 \end{pmatrix}$$

(v) We obtain the transpose of this last matrix, as

$$\begin{pmatrix} 2 & 1 & -3 \\ -4 & -1 & 6 \\ 5 & 1 & -7 \end{pmatrix}.$$

(vi) From eqn (27.12), the inverse $A^{-1} = \dfrac{1}{1} \begin{pmatrix} 2 & 1 & -3 \\ -4 & -1 & 6 \\ 5 & 1 & -7 \end{pmatrix}$,

which is clearly $\begin{pmatrix} 2 & 1 & -3 \\ -4 & -1 & 6 \\ 5 & 1 & -7 \end{pmatrix}$. ∎

Obtaining the inverse of a large matrix can be quite painful. Luckily, modern computers can perform algebra of this type without apparent effort.

Self-test 27.10

Obtain the inverse A^{-1} for each of the following matrices.

27.10.1 $\begin{pmatrix} 1 & 2 \\ 6 & 4 \end{pmatrix}$
  27.10.2 $\begin{pmatrix} 2 & 1 & 3 \\ 0 & 1 & 2 \\ 0 & 0 & 3 \end{pmatrix}$

27.10.3 $\begin{pmatrix} 1 & 2 & 3 \\ 4 & 3 & 2 \\ 0 & 2 & 1 \end{pmatrix}$
  27.10.4 $\begin{pmatrix} a & b & bc \\ a^2 & c & b^2 \\ ab & ac & c \end{pmatrix}$

Using matrices to solve simultaneous linear equations

By the end of this section, you will know:

- Matrices afford a simple means of solving simultaneous linear equations.
- A coefficient matrix will have the same number of columns as variables.

In Chapter 10, we introduced a straightforward algebraic method of solving simultaneous linear equations. The method has many advantages:

- It is conceptually simple.
- It is generally easy and quick to use.

- Simple substitution of the values back into the source equations gives a ready indication of whether the answers are correct.

But the algebraic solution of simultaneous linear equations has one major drawback. In Chapter 10, we discussed the solution of *two* simultaneous linear equations. The method is readily extended to solve any number of simultaneous linear equations, but becomes increasingly cumbersome as the number of variables increases.

In this section, we see how matrices allow for the simple solution of simultaneous linear equations of any number of variables. The first step is to summarize the data in such a way that we generate a series of linear equations. These equations must be unique, by which we mean one cannot be reduced to yield another (see p. 148). We say the equations are **inhomogeneous**.

The linear equations will look something like this:

$$
\begin{aligned}
a_{11}x_1 &+a_{12}x_2 &+a_{13}x_3 &\cdots &a_{1n}x_n &= b_1 \\
a_{21}x_1 &+a_{22}x_2 &+a_{23}x_3 &\cdots &a_{2n}x_n &= b_2 \\
\vdots & \vdots & \vdots & \vdots & \ddots & \vdots \\
a_{m1}x_1 &+a_{m2}x_2 &+a_{m3}x_3 &\cdots &a_{mn}x_n &= b_m
\end{aligned}
$$

where the a values are coefficients, and the x values are the unknowns.

These data can be written in the form of simple matrices:

$$
\begin{pmatrix}
a_{11} & a_{12} & a_{13} & \cdots & a_{1n} \\
a_{21} & a_{22} & a_{23} & \cdots & a_{2n} \\
\vdots & \vdots & \vdots & \ddots & \vdots \\
a_{m1} & a_{m2} & a_{m3} & \cdots & a_{mn}
\end{pmatrix}
\begin{pmatrix}
x_1 \\ x_2 \\ \vdots \\ x_n
\end{pmatrix}
=
\begin{pmatrix}
b_1 \\ b_2 \\ \vdots \\ b_n
\end{pmatrix}
$$

> These equations are described as **linear** because they do not involve power terms.

Writing the data in this format yields the **coefficient matrix**, so-called because it contains the coefficients of the unknowns. We can write the coefficient matrix in abbreviated format as, $\mathbf{A}\,\mathbf{x} = \mathbf{b}$. n simultaneous equations yield a matrix of order $n \times n$. Computers can routinely solve up to 10 000 simultaneous equations if they are presented in a matrix format.

Simple algebraic manipulation of both sides yields the relation $\mathbf{x} = \mathbf{A}^{-1}\mathbf{b}$. So multiplying the matrix $\mathbf{b}$ by the inverse of matrix $\mathbf{A}$ straightforwardly yields a matrix comprising the values of $\mathbf{x}$ that solve the simultaneous equations.

■ **Worked Example 27.20**

A compound contains only sulphur and a new, unknown, element—call it E. A compound of E is analysed by mass spectrometry. Two molecular fragments are found: ES_2 has a mass of 375 amu, and ES_3 has a mass of 407 amu. Use matrices to obtain the mass of the new element E.

Let x be the atomic mass of E and y be the atomic mass of sulphur. We generate the following pair of linear equations:

$x + 2y = 375$

$x + 3y = 407$

This example may be slightly artificial insofar as we could obtain the answer by simple inspection alone. But the answer we obtain using matrix algebra will confirm the answer obtained by inspection.

Strategy:

(i) We insert the coefficients in a simple 2×2 coefficient matrix (**A**).

(ii) Adjacent to the right-hand side of this first matrix, we write a second: a 1×2 matrix of x and y (**x**).

(iii) We write a third matrix (**b**) comprising the two numbers to the right of the equals signs in the source equations: $\begin{pmatrix} 1 & 2 \\ 1 & 3 \end{pmatrix}\begin{pmatrix} x \\ y \end{pmatrix} = \begin{pmatrix} 375 \\ 407 \end{pmatrix}$. We denote it as, $\mathbf{A} \times \mathbf{x} = \mathbf{b}$.

(iv) We rearrange to obtain: $\mathbf{x} = \mathbf{A}^{-1}\,\mathbf{b}$, so we need to determine the inverse of **A**. Using the template in eqn (27.11), the inverse matrix $\mathbf{A}^{-1}$ is

$$\frac{1}{1}\begin{pmatrix} 3 & -2 \\ -1 & 1 \end{pmatrix} = \begin{pmatrix} 3 & -2 \\ -1 & 1 \end{pmatrix}.$$

(v) Matrix algebra yields the matrix **x** that contains the values of x and y that solve the two source equations.

In this example, MULTIPLYING this matrix with matrix **b** yields the desired data:

$$\begin{pmatrix} 3 & -2 \\ -1 & 1 \end{pmatrix}\begin{pmatrix} 375 \\ 407 \end{pmatrix} = \begin{pmatrix} (3 \times 375) - (2 \times 407) \\ (-1 \times 375) + (1 \times 407) \end{pmatrix} = \begin{pmatrix} 311 \\ 32 \end{pmatrix}$$

Solution:

The mass of the new element E is therefore 311 amu, and the mass of sulphur is 32 amu. ∎

The real power of this technique becomes apparent when we have more than two linear equations to solve. Unfortunately, obtaining the inverse of such matrices is far more complicated.

■ **Worked Example 27.21**

Use matrices to solve the following three equations, i.e. to obtain values of x, y and z.

$x + 4y + 3z = 3$

$2x + y = 2$

$x + 3y + 2z = 4$

Strategy:

(i) We generate a coefficient matrix (**A**) by inserting the coefficients in a simple 3×3 matrix. We insert a coefficient of 0 if a term does not appear (e.g. there is no z term in the second equation here).

(ii) To the right-hand side of this first matrix, we write a second: a 1×3 matrix of x, y and z (**x**).

(iii) A third matrix (**b**) comprises the three numbers to the right of the equals signs in the source equations.

We write a value of 0 for matrix element a_{23} in the left-hand matrix because there are no z terms in source equation 2.

$$\begin{pmatrix} 1 & 4 & 3 \\ 2 & 1 & 0 \\ 1 & 3 & 2 \end{pmatrix}\begin{pmatrix} x \\ y \\ z \end{pmatrix} = \begin{pmatrix} 3 \\ 2 \\ 4 \end{pmatrix}$$

We have already obtained the inverse of this 3×3 matrix in Worked Example 27.19:

$\det \mathbf{A} = 1.$ Therefore, $\quad \mathbf{A}^{-1} = \dfrac{1}{1}\begin{pmatrix} 2 & 1 & -3 \\ -4 & -1 & 6 \\ 5 & 1 & -7 \end{pmatrix} = \begin{pmatrix} 2 & 1 & -3 \\ -4 & -1 & 6 \\ 5 & 1 & -7 \end{pmatrix}$

Therefore, $\begin{pmatrix} x \\ y \\ z \end{pmatrix} = \mathbf{A}^{-1}\begin{pmatrix} 3 \\ 2 \\ 4 \end{pmatrix}$

so $\begin{pmatrix} x \\ y \\ z \end{pmatrix} = \begin{pmatrix} 2 & 1 & -3 \\ -4 & -1 & 6 \\ 5 & 1 & -7 \end{pmatrix}\begin{pmatrix} 3 \\ 2 \\ 4 \end{pmatrix} = \begin{pmatrix} (2\times3)+(1\times2)+(-3\times4) \\ (-4\times3)+(-1\times2)+(6\times4) \\ (5\times3)+(1\times2)+(-7\times4) \end{pmatrix} = \begin{pmatrix} -4 \\ 10 \\ -11 \end{pmatrix}$

The final bracket on the right-hand side is the **x**. Therefore, $x = -4, y = 10$ and $z = -11$. Inserting these numbers into the source equations confirms this result. ■

Self-test 27.11

27.11.1 Solve the following three simultaneous equations using this matrix method.

$$x - 2y + 8z = 5$$

$$2x + 15z = 3y + 6$$

$$8y + 22 = 4x + 30z$$

Summary of key equations in the text

Addition: $\begin{pmatrix} a_{11} & a_{12} \\ a_{21} & a_{22} \end{pmatrix} + \begin{pmatrix} b_{11} & b_{12} \\ b_{21} & b_{22} \end{pmatrix} = \begin{pmatrix} a_{11}+b_{11} & a_{12}+b_{12} \\ a_{21}+b_{21} & a_{22}+b_{22} \end{pmatrix}$ (27.1)

Subtraction: $\begin{pmatrix} a_{11} & a_{12} \\ a_{21} & a_{22} \end{pmatrix} - \begin{pmatrix} b_{11} & b_{12} \\ b_{21} & b_{22} \end{pmatrix} = \begin{pmatrix} a_{11}-b_{11} & a_{12}-b_{12} \\ a_{21}-b_{21} & a_{22}-b_{22} \end{pmatrix}$ (27.2)

Determinant:

(2×2 matrix) $\qquad \det \mathbf{A} = \begin{vmatrix} a & b \\ c & d \end{vmatrix} = ad - bc$ (27.3)

(3×3 matrix) $\quad \det \mathbf{A} = a_{11}\begin{vmatrix} a_{22} & a_{23} \\ a_{32} & a_{33} \end{vmatrix} - a_{12}\begin{vmatrix} a_{21} & a_{23} \\ a_{31} & a_{33} \end{vmatrix} + a_{13}\begin{vmatrix} a_{21} & a_{22} \\ a_{31} & a_{32} \end{vmatrix}$ (27.4)

$$\text{cofactor} = (-1)^{i+j} M_{ij} \qquad (27.5)$$

Eigenvalue equation: $\qquad \mathbf{A}\mathbf{x} = \lambda\mathbf{x}$ (27.6)

Characteristic equation: $\qquad \det(\mathbf{A} - \lambda\mathbf{I}) = 0$ (27.8)

Equation of motion: $\qquad \ddot{\mathbf{x}} = (\mathbf{A}\mathbf{x}) = -\omega^2\mathbf{x}$ (27.9)

Inverse: $A \times A^{-1} = I$ (27.10)

$(2 \times 2 \text{ matrix})$ if $A = \begin{pmatrix} a & b \\ c & d \end{pmatrix}$ then $A^{-1} = \dfrac{1}{(ad - cb)} \begin{pmatrix} d & -b \\ -c & a \end{pmatrix}$ (27.11)

$(3 \times 3 \text{ matrix})$ $A^{-1} = \dfrac{1}{\det A} C^{T}$ (27.12)

where c_{ij} are the cofactors of a_{ij}.

Additional problems

27.1 Write the effective Hamiltonian matrix for the cyclopentadienyl radical anion.

27.2 What is the value of $\begin{pmatrix} 0 & 0 & 5 \\ 2 & 4 & -9 \\ 1 & 4 & 6 \\ 3 & 3 & 1 \end{pmatrix} - \begin{pmatrix} 0 & 3 & -3 \\ 3 & 2 & 1 \\ 2 & -2 & 0 \\ 6 & 7 & 2 \end{pmatrix}$?

27.3 Consider the following two linear equations:

$$2x - 3y = 5$$
and $$x + 5y = 9$$

Write a coefficient matrix, and hence solve to ascertain the unique values of x and y.

27.4 Show that the matrices $B = \begin{pmatrix} -15 & 2 & -3 \\ 0 & 1 & 0.5 \\ 2 & 0 & 0.5 \end{pmatrix}$ and $C = \begin{pmatrix} 1 & -2 & 8 \\ 2 & -3 & 15 \\ -4 & 8 & -30 \end{pmatrix}$ are inverses.

27.5 What is the determinant of the matrix, $\begin{pmatrix} a & 5b \\ 3ab^2 & 4a \end{pmatrix}$?

27.6 What is the determinant of the matrix $A = \begin{pmatrix} 1 & 2 & 5 \\ 2 & 2 & 0 \\ 0 & 4 & 3 \end{pmatrix}$?

27.7 Determine a value for:

$$\begin{pmatrix} 1 & 9 & 6 & 2 & 2 & 0 \\ 6 & 2 & 3 & -1 & 8 & 1 \\ 1 & 0 & 6 & 31 & 13 & 1 \end{pmatrix} + \begin{pmatrix} -4 & 1 & 0 & 3 & -2 & 2 \\ 2 & 1 & 0 & 0 & 2 & 3 \\ 0 & 7 & 7 & 2 & 1 & 0 \end{pmatrix}$$

27.8 Write the effective Hamiltonian matrix for naphthalene.

27.9 What is the scalar product, $3.2a \begin{pmatrix} a & 2 \\ a^2 & 3e \end{pmatrix}$?

27.10 Write the effective Hamiltonian matrix for pyrrole, taking the α and β coefficients for nitrogen as 1.5 and 1.75, respectively. Let the nitrogen be atom 1 in the ring.

28

Vectors

Introducing vectors and scalars

By the end of this section, you will know:

- A scalar has no direction; merely a magnitude.
- Vectors are symbolized in bold type.
- Vectors have both magnitude and direction.
- A vector can be expressed in terms of plane polar coordinates, as a magnitude and a direction.
- In Cartesian coordinates, we can define a vector in terms of a unit vectors, $\hat{\imath}$, $\hat{\jmath}$ and $\hat{k}$.
- The magnitude of the vector in Cartesian space is obtained using Pythagoras' theorem and the coefficients of the unit vectors.

In chemistry as well as in everyday life, most things have a straightforward magnitude: a flask, two seconds, three metres, four moles or five grammes. These trivial examples are alike insofar as each is a **scalar** quantity: each has a magnitude but no direction. We cannot think of an amount of substance having a direction, but it can have a magnitude.

It is often useful to know more than a magnitude alone. Newcomers to an unknown town asking the way to a hotel will want to know more than the distance to a hotel, but its direction also. They want information in the form of a **vector**. Many quantities in chemistry are vectors: for example, velocity has both magnitude and direction. In this sense, other vector quantities include dipole moment, force, electric field, linear and angular momentum, and force.

The word vector itself means 'carrier' in Latin, and comes from the verb *vehere*, which means *to carry*. We use vectors often in chemistry (and other physical sciences) to express quantities that cannot be described completely by a scalar. In print, we usually denote a vector with **bold** print, e.g. **a**. Since it is not easy to indicate bold print when writing, we indicate the presence of a vector by underlining, as a̲, or including a tilde beneath each vector, as a̰ . Occasionally, especially in physics, we write a vector with an arrow pointing right a̅ , or add a hat, â. We will only use the bold notation here.

This symbolism has one important consequence. Many physicochemical variables are vectors, such as velocity, force and electric field. We usually write the symbols of these variables in normal type, as v, F and $Ɛ$, respectively. In this chapter only, we will use **v**, **F** and **E**, to emphasize their vector character.

Because a vector has both magnitude and direction, we need to describe both when defining a vector. Probably the simplest, conceptually, involves the use of plane polar coordinates. Given a starting point, we can describe any vector using both a length and an angle; see Fig. 28.1.

In molecular biology, *vector* denotes a DNA sequence used to transfer foreign genetic material into another cell. We will not here be using the vector in this biological sense.

We may also see **r** described as the *radius* or *radial coordinate*; and the angle θ as the *colatitude*, *zenith angle*, *normal angle*, or *polar angle*.

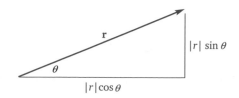

Figure 28.1 A vector of magnitude $|\,\mathbf{r}\,|$ and angle θ.

In Fig. 28.1, $|r|$ is the vector's *magnitude* and θ is its *inclination angle*, which defines its *direction*. In the language of mathematics, vectors comprise an **ordered pair** consisting of a magnitude and direction. These data are often, but not always, cited within angled brackets, e.g. $< |r|, \theta >$.

Using the trigonometry from Chapter 16, we can see that the vector in Fig. 28.1 has moved in the horizontal direction by an amount of,

$$|r| \cos \theta \qquad (28.1)$$

At the same time, we have moved vertically by an amount

$$|r| \sin \theta \qquad (28.2)$$

We call these distances the *x-component* and the *y-component*, respectively.

For many purposes, we will find it easier to describe vectors in Cartesian space. We can do so in many ways. For example, we can describe the motion in the form of a matrix, either writing the *x*-component at the top and the *y*-component beneath it, $\begin{pmatrix} x \\ y \end{pmatrix}$ or on a line, as (x,y).

Vectors need not have two components. We can have 2, 3, n or an infinite number of components. For our purposes, three-dimensional motion is the most common situation we need to describe. It is obvious how to extend the notation above to three or many more dimensions.

In an alternative approach, we artificially separate a vector into a series of movements along the three Cartesian axes. But instead of lengths along these axes, we think in terms of multiples of **unit vectors**:

- We call the unit vector in the *x*-direction $\hat{\imath}$.

- We call the unit vector in the *y*-direction $\hat{\jmath}$.

- We call the unit vector in the *z*-direction $\hat{k}$.

If a vector **A** describes three-dimensional motion in Cartesian space, we can represent it in terms of these unit vectors $\hat{\imath}$, $\hat{\jmath}$ and $\hat{k}$, as:

$$\mathbf{A} = A_x \hat{\imath} + A_y \hat{\jmath} + A_z \hat{k} \qquad (28.3)$$

where A_x, A_y and A_z are the factors describing the number of unit vectors travelled. We therefore represent a vector such as $(3,7,2)$ by writing, $3\hat{\imath} + 7\hat{\jmath} + 2\hat{k}$.

The magnitude of a vector is found by taking the **modulus** of this vector, which is indicated by the short vertical lines either side of the vector, A. This property is only interested in the magnitude of the vector, as shown by the fact that the modulus of $|A| = |-A|$. We obtain the *magnitude* of the vector via Pythagoras' theorem in Chapter 16, as,

Magnitude of $$\mathbf{A} = |A| = \sqrt{A_x^2 + A_y^2 + A_z^2} \qquad (28.4)$$

For the vector $A = 3\hat{\imath} + 7\hat{\jmath} + 2\hat{k}$, the magnitude is $|A| = \sqrt{3^2 + 7^2 + 2^2} = 7.87$. The magnitude of the vector B, where $B = -3\hat{\imath} - 7\hat{\jmath} - 2\hat{k} = -A$, is also 7.87.

> The unit vector is a vector that has been *normalized* to ensure its magnitude is 1.

Self-test 28.1

What is the magnitude of the following vectors?

28.1.1 $(2,3,5)$ **28.1.2** $(6,0,12.6)$

What is the length in the *x*- and *y*-directions of the following vectors:

28.1.3 $(3,40°)$ **28.1.4** $(6.2,-150°)$

Scalar multiplication of vectors

By the end of this section, you will know:

- The multiplication of a vector by a scalar yields another vector having exactly the same direction but a different magnitude.

If we want λ multiples of a vector **p**, the total vector is λ**p**. We often call this process **scalar multiplication**, because the multiplication factor, being merely a constant, is a scalar. Notice how the vector obtained by scalar multiplication changes in *magnitude*, but its *direction* does not change. In effect, the vector becomes longer or shorter.

A **scalar** quantity has magnitude but no direction. Scalar quantities include mass, amount of material or luminous intensity. Time is not a straightforward scalar, for it has direction (forward and backwards).

■ **Worked Example 28.1**

A molecule of acetylene (**I**) moves according to the vector $q = (2,5)$. What is the single vector describing the motion after (**I**) has repeated this vector motion 3 times?

$$H-C\equiv C-H$$
$$\text{I}$$

We obtain the scalar product of this vector in exactly the same way as obtaining the scalar product of any other matrix (see p. 395). In matrix form, we obtain the scalar product as,

$$3q = 3\begin{pmatrix} 2 \\ 5 \end{pmatrix} = \begin{pmatrix} 3\times 2 \\ 3\times 5 \end{pmatrix} = \begin{pmatrix} 6 \\ 15 \end{pmatrix}$$

The vector we obtain is parallel with the source vector **q** because it is in the same direction. It is merely 3 times longer. ■

A vector can have units.

■ **Worked Example 28.2**

A molecule of water has mass 18 atomic mass unit (amu) and moves with a velocity described by the vector $v = 2\hat{\imath} - \hat{\jmath}$, where the factors before $\hat{\imath}$ and $\hat{\jmath}$ are in m s^{-1}. Calculate its momentum, p.

We define momentum as $p = \text{mass} \times \text{velocity}$.

Therefore, $p = 18(2\hat{\imath} - \hat{\jmath}) = 36\hat{\imath} - 18\hat{\jmath}$ amu m s^{-1}.

1 amu has a mass of $1.66 - 10^{-27}$ kg. The momentum p is therefore, $1.66 \times 10^{-27} \times (36\hat{\imath} - 18\hat{\jmath}) = (5.976\hat{\imath} - 2.989\hat{\jmath}) \times 10^{-26}$ N s. ■

N s is the SI unit of momentum.

Self-test 28.2

Multiply the following vectors.

28.2.1 $1.5 \times (2,4)$ 28.2.2 $2.73 \times (5,2.2)$

The addition and subtraction of vectors

By the end of this section, you will know:

- We add vectors by adding together their respective elements.
- We subtract vectors by subtracting their respective elements.

Addition:

Vector addition is entirely *commutative*, so $a + b = b + a$. It is also *associative*, so it does not matter in which order we do the addition: $a + (b + c) = (a + b) + c$.

Adding vectors together is easy. We first resolve the vectors into the respective components in the $\hat{i}$, $\hat{j}$ and $\hat{k}$ directions, cf. eqn (28.3). We then sum them as a linear combination in each direction, so the magnitude of the final vector in the $\hat{i}$ direction is $\Sigma(A_x)$, the magnitude in the $\hat{j}$ direction is $\Sigma(A_y)$, and the magnitude in the $\hat{k}$ direction is $\Sigma(A_z)$.

> ### ■ Worked Example 28.3
>
> Add together the two vectors v_1 and v_2: $v_1 = (2,1)$ and $v_2 = (3,4)$

We can perform the sum in several ways:

Simple addition:

We merely sum the x-components, as $2 + 3 = 5$, and the y-components as, $1 + 4 = 5$. Therefore, $v_1 + v_2 = (5,5)$.

Addition using vectors:

We write the x- and y-components vertically:

$$\begin{pmatrix} x \\ y \end{pmatrix} = \begin{pmatrix} 2+3 \\ 1+4 \end{pmatrix} = \begin{pmatrix} 5 \\ 5 \end{pmatrix}.$$

Graphically:

In effect, by summing these two vectors, we say the motion represents five steps horizontally per three steps vertically. So it is also possible to do the sum graphically.

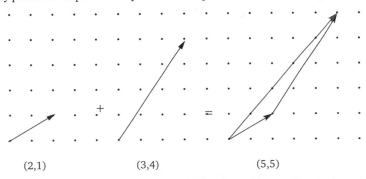

(2,1)　　　　　　　(3,4)　　　　　　　(5,5)

The new vector in blue does not have the same direction as either of the two source vectors. It is not parallel with them. The final vector will only be parallel with one of the source vectors by chance, i.e. if several vectors, when summed, have the same x, y, z profile as a multiple of one of the source vectors.

Graphical addition of vectors is so time consuming, however, and so prone to inaccuracies that we generally avoid this method. It does, however, help us gain a *physical* idea of the final vector. ■

> ### ■ Worked Example 28.4
>
> The unit cell of sodium chloride has a simple cubic structure depicted in Fig. 28.2. If the lattice parameters are $|a| = |b| = |c| = 282$ pm. What is the length of the bold vector from the sodium to the chloride ion?

In the x-direction, the vector length is $3 \times |a|$
In the y-direction, the vector length is $1 \times |b|$
In the z-direction, the vector length is $1 \times |c|$

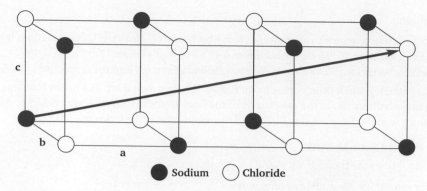

Figure 28.2 Three unit cells of sodium chloride. We obtain the length of the bold vector as a linear combination, according to eqn (28.4).

Let us represent $| \mathbf{a} | = a$ and remember that in this case $| \mathbf{b} | = | \mathbf{c} | = a$.

From eqn (28.4), the length of the vector is, $\sqrt{(3a)^2 + a^2 + a^2} = \sqrt{11a^2} = 3.32a$ bond lengths $= 3.32 \times 282$ pm $= 935.29$ pm (to 2 d.p.). ■

Vector subtraction:

In common with normal subtraction, vector subtraction is not commutative: $(\mathbf{a} - \mathbf{b}) \neq (\mathbf{b} - \mathbf{a})$.

To effect a simple subtraction, we first resolve the vectors into the respective components in the $\hat{\mathbf{i}}$, $\hat{\mathbf{j}}$ and $\hat{\mathbf{k}}$ directions, cf. eqn (28.3). We then subtract the respective components in each direction, so the magnitude of the final vector in the $\hat{\mathbf{i}}$ direction is $(a_x - b_x)$, the magnitude in the $\hat{\mathbf{j}}$ direction is $(a_y - b_y)$, and the magnitude in the $\hat{\mathbf{k}}$ direction is $(a_z - b_z)$.

■ **Worked Example 28.5**

If $\mathbf{c} = 3\hat{\mathbf{i}} - \hat{\mathbf{j}} + \hat{\mathbf{k}}$, and $\mathbf{d} = \hat{\mathbf{i}} + 2\hat{\mathbf{j}} - \hat{\mathbf{k}}$, what is the vector produced as, $\mathbf{c} - \mathbf{d}$?

In the x-direction: $3\hat{\mathbf{i}} - \hat{\mathbf{i}} = 2\hat{\mathbf{i}}$

In the y-direction: $-\hat{\mathbf{j}} - 2\hat{\mathbf{j}} = -3\hat{\mathbf{j}}$

In the y-direction: $\hat{\mathbf{k}} - (-\hat{\mathbf{k}}) = 2\hat{\mathbf{k}}$

So the new vector is, $\mathbf{c} - \mathbf{d} = 2\hat{\mathbf{i}} - 3\hat{\mathbf{j}} + 2\hat{\mathbf{k}}$. ■

Self-test 28.3

Consider the vectors, $\mathbf{a} = (3,2,1)$, $\mathbf{b} = (5,0,4)$ and $\mathbf{c} = (3,9,1)$. Combine them as follows, and calculate their magnitude:

28.3.1 $4\mathbf{a} + 2\mathbf{b}$ **28.3.2** $3\mathbf{a} - 2\mathbf{b} + 4\mathbf{c}$

Relative motion

By the end of this section, you will know:

• The motion of a species A relative to another species B is obtained via vector subtraction, as $\mathbf{x}_A - \mathbf{x}_B$ (where $\mathbf{x}$ is a vector describing a physicochemical parameter associated with the motion).

Using vector algebra, it is easy to describe the motion of object A with respect to object B, when we know how each is moving with respect to a reference object O.

Think of two vectors **a** and **b**, which are operating in different directions. An external observer can straightforwardly see the directions of each, but what is the direction of one *relative* to the other? It's the same problem we experience when sitting in a plane, and see another plane flying past us at a different speed, altitude and direction.

■ **Worked Example 28.6**

Two molecules of the ozone-depleting CFC trichlorotrifluoroethane (**II**), occupy the same flask. Being gaseous, the molecules are moving fast. Molecule A moves with a velocity described by the vector (5,3), call it v_A; and the molecule B moves with a velocity vector (−2,9), call it v_B. What is the relative motion of molecule A when viewed from molecule B?

II

Strategy:

(i) Because *both* molecules are moving, we need to effectively stop molecule B (the 'observer' molecule).

(ii) To stop the molecule B, we subtract its motion vector v_B from the vectors for both molecules A and B. This subtraction has the effect of turning the observer vector into a stationary point—it becomes an *objective* reference point.

(iii) So, in vector notation, we obtain the motion of molecule A relative to B as,

$$v_A - v_B \qquad (28.5)$$

When split into scalar components, we rewrite this equation as:

A relative to B in the x-direction $= v_{A,x} - v_{B,x}$

A relative to B in the y-direction $= v_{A,y} - v_{B,y}$

Solution:

In the x-direction: $5 - (-2) = 7$

In the y-direction: $3 - 9 = -6$.

So the vector describing the motion of A relative to B is (7,−6). Figure 28.3 shows these vectors diagrammatically. To denote the way vector **b** has been subtracted, its arrow has been reversed (its head now points downward). The bold arrow describes the vector. ■

This equation is not valid at speeds approaching the speed of light, e.g. for electrons moving through a good conductor such as silver or gold. In such cases, we need Einstein's special theory of relativity.

Self-test 28.4

What is the relative motion of A (moving with velocity vector **a**) relative to B (moving with velocity vector **b**)?

28.4.1   **a** = (2,7) **b** = (−5,3)   **28.4.2**   **a** = (4.2,3.1) **b** = (−6.2,−2.2)

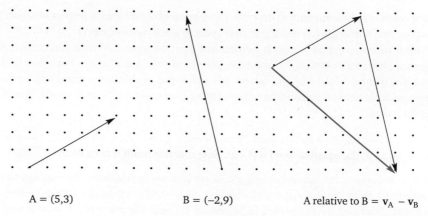

A = (5,3)   B = (−2,9)   A relative to B = $\mathbf{v}_A - \mathbf{v}_B$

Figure 28.3 We obtain the motion of A relative to B as the vector subtraction, $\mathbf{v}_A - \mathbf{v}_B$.

Multiplying vectors, I: the scalar (dot) product

By the end of this section, you will know:

- Vectors may be multiplied together in two ways: as a cross-product and as a dot product.
- The dot product is also called the scalar product.
- The dot product of two vectors has magnitude but no direction, so is a scalar.

There are two straightforward ways in which we can multiply together two vectors. In the simplest, the result has magnitude but no direction. It is therefore a scalar and not a vector, so we call it the scalar product. Most mathematical treatments prefer to call it the **dot product**.

Occasionally, we call the dot product the *inner product*. In this context, we would usually employ the notation $<a, b>$. We will be using neither this name nor its notation.

We define the dot product of the vectors **a** and **b** as,

$$\mathbf{a} \cdot \mathbf{b} = |\,\mathbf{a}\,|\,|\,\mathbf{b}\,|\cos\theta \tag{28.6}$$

In words, we would say aloud, '**a** dot **b**'. The bold **a** and **b** on the left-hand side are the source vectors. The term $|\,\mathbf{a}\,|\,|\,\mathbf{b}\,|$ on the right-hand side is the modulus of **a** multiplied by the modulus of **b**. The dot product is commutative, so $\mathbf{a} \cdot \mathbf{b} = \mathbf{b} \cdot \mathbf{a}$. This commutation still applies if one of the vectors has been multiplied by a factor k, because the eventual product is still a scalar, $(k\,\mathbf{a}) \cdot \mathbf{b} = k\,(\mathbf{a} \cdot \mathbf{b})$.

The angle θ is the angle between the directions of **a** and **b**. It follows immediately from the definition of the dot product that $\mathbf{a} \cdot \mathbf{b} = 0$ if **a** is *perpendicular* to **b**, because $\cos 90° = 0$. This condition means we can only consider two vectors as being orthogonal if (and only if) their dot product is zero. This condition assumes that both vectors have a finite non-zero magnitude. This property provides a simple method to test the condition of orthogonality.

By resolving the vectors **a** and **b** into their x-, y- and z-components, we can rewrite eqn (28.6) as,

$$\mathbf{a} \cdot \mathbf{b} = a_x b_x + a_y b_y + a_z b_z \tag{28.7}$$

The words orthogonal and perpendicular both mean, at right angles (i.e. 90°).

These terms arise because $a_x\hat{\imath}\cdot b_x\hat{\imath}=a_xb_x\hat{\imath}\cdot\hat{\imath}=a_xb_x\times\mid 1\mid\cos 0°=a_xb_x$. There are no terms such as a_xb_y, because the x- and y-axes exist at right angles. Also, (from the definition of the dot product) $\hat{\imath}\cdot\hat{\jmath}=\mid 1\times 1\mid\cos 90°=0$.

Accordingly, in general,

$$\mathbf{a}\cdot\mathbf{b}=\sum_{i=1}^{n}a_ib_i$$

As a further consequence of the angle term, because $\cos 0°=1$, we can obtain a dot product between any non-zero vector and itself. For a unit vector, we obtain $\hat{\imath}\cdot\hat{\imath}=1$, etc. Furthermore, for any vector $\mathbf{a}$, the square $\mathbf{a}\cdot\mathbf{a}=\mid\mathbf{a}\mid^2=a^2$. Note that this result is not a vector.

■ **Worked Example 28.7**

A force F of 10 N applied at an angle of 60° causes a block of mass m to move a distance of 10 m along a smooth plane surface. Find the work done w, if $w=\mathbf{F}\cdot\Delta x$.

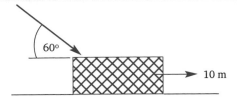

Strategy:

For the purposes of the mathematics here, F and Δx are force and displacement vectors. We therefore use F and Δx as the source vectors.

Solution:

Expanding the expression of work, we have:

$$w=\mathbf{F}\cdot\Delta x$$

where, F = 10 N, Δx = 10 m and $\cos\theta=\cos 60°=0.5$
Therefore, $w=10\times 10\times 0.5=50$ J. ■

Sometimes we know the coordinates of two source vectors $\mathbf{a}$ and $\mathbf{b}$, and want to ascertain the angle between them. We rearrange eqn (28.6), and obtain:

$$\theta=\cos^{-1}\left(\frac{a_xb_x+a_yb_y+a_zb_z}{\mid\mathbf{a}\mid\mid\mathbf{b}\mid}\right) \tag{28.8}$$

where the numerator in the bracket comes from eqn (28.7).

■ **Worked Example 28.8**

When using the wall-jet electrode, we squirt two narrow jets of solution at a stationary, planar electrode—the 'wall'. If one jet has a momentum vector ($\mathbf{u}=(2,3,4)$) and the other a momentum vector, ($\mathbf{v}=(1,-2,3)$), calculate the angle between the two jets of solution.

Strategy:

(i) The two jets of solution acts as vectors.

(ii) We resolve the two vectors into their component unit vectors. Being momentum vectors, the result will have units of kg m s^{-1}.

(iii) We determine the magnitudes of the two source vectors using the Pythagoras theorem, in eqn (28.4), and hence the product $|\mathbf{u}||\mathbf{v}|$. The result will have units of $(kg\ m\ s^{-1})^2$.

(iv) We insert these data into the rearranged form of the dot-product equation: eqn (28.8).

Solution:

(i) $\mathbf{u} = 2\hat{\imath} + 3\hat{\jmath} + 4\hat{k}$ and $\mathbf{v} = \hat{\imath} - 2\hat{\jmath} + 3\hat{k}$

From eqn (28.7), $\mathbf{u} \cdot \mathbf{v} = (2 \times 1) + (3 \times (-2)) + (4 \times 3) = 2 - 6 + 12 = 8\ kg\ m\ s^{-1}$.

(ii) $|\mathbf{u}| = \sqrt{2^2 + 3^2 + 4^2} = \sqrt{29} = 5.385$ and $|\mathbf{v}| = \sqrt{1^2 + (-2)^2 + 3^2} = \sqrt{14} = 3.742$

so $|\mathbf{u}||\mathbf{v}| = 20.149\ kg^2\ m^2\ s^{-2}$.

(iii) Inserting terms into eqn (28.8), we obtain: $\cos\theta = \dfrac{8}{20.149} = 0.397$

so $\theta = 66.6°$. ∎

Self-test 28.5

28.5.1 Calculate the dot product of the two vectors, $\mathbf{r} = 6\hat{\imath} + 5\hat{\jmath}$ and $\mathbf{s} = 2\hat{\imath} - 8\hat{\jmath}$.

28.5.2 Two force vectors of magnitude 6 N and 4 N, respectively, operate at an angle of 40°. What is the scalar product of these two vectors?

28.5.3 Find the angle between the $\mathbf{p} = 2\hat{\imath} + \hat{\jmath} - \hat{k}$ and $\mathbf{q} = \hat{\imath} - \hat{k}$.

28.5.4 Find the angle between $\mathbf{a} = 2\hat{\imath} - 3\hat{\jmath} + \hat{k}$ and $\mathbf{b} = 4\hat{\imath} + \hat{\jmath} - 3\hat{k}$.

28.5.5 A molecule of adrenaline (**III**) moves in a straight line. The velocity vector (3,2) describes its motion. Calculate its kinetic energy. Assume the velocity components in the $\hat{\imath}$ and $\hat{\jmath}$ directions have units of m s⁻¹.

III

Multiplying vectors, II: the vector (cross-) product

By the end of this section, you will know:

- The cross-product of any two vectors is itself a vector.
- The cross-product is also called the *vector product*.

The second way we multiply vectors together involves the so-called **cross-product**, which is also known as the vector product and the Gibbs vector product, after the American genius J. Willard Gibbs.

We will denote the cross-product with a simple multiplication sign, e.g. $\mathbf{a} \times \mathbf{b}$. Other texts use other notations, such as $\mathbf{a} \wedge \mathbf{b}$. We will not use this style here.

The result of a cross-product of two vectors has both magnitude *and* direction, and is a straightforward vector in three-dimensional Cartesian space. The cross-product can only be applied to two vectors and not a scalar and a vector, nor to two scalars.

Mathematically, we define the cross-product of the two vectors **a** and **b** as,

$$\mathbf{a} \times \mathbf{b} = (\,|\,\mathbf{a}\,|\,|\,\mathbf{b}\,|\,\sin\theta)\,\hat{\mathbf{n}} \qquad\qquad (28.9)$$

where **a** and **b** on the left-hand side are the source vectors, and the $|\,\mathbf{a}\,|\,|\,\mathbf{b}\,|$ term on the right-hand side is the directionless (i.e. scalar) product of the magnitudes of **a** and **b**.

The vertical lines either side of **a** and **b** imply their modulus. As the angle between two vectors cannot be greater than 180°, the cross-product is always positive. The magnitude of the cross-product can be interpreted as the positive area of the parallelogram having **a** and **b** as sides (see Fig. 28.4).

The vectors **a** and **b** define a plane in 3D space. The vector $\hat{\mathbf{n}}$ in eqn (28.9) is the unit vector perpendicular to the plane defined by **a** and **b**. The cross-product of the vectors **a** and **b** therefore operates at right angles to either of the source vectors. Figure 28.5 demonstrates the so-called *right-hand rule*, which helps us discern the direction of $\hat{\mathbf{n}}$. Here, we hold the right hand with the thumb, forefinger and middle finger at mutual right angles (see Fig. 28.5). The forefinger gives the direction of **a**, the middle finger gives the direction of **b**, and, the product vector $\hat{\mathbf{n}}$ follows the direction of the thumb.

There are several important consequences of the cross-product definition in eqn (28.9):

1. The unit vectors $\hat{\imath}$, $\hat{\jmath}$, and $\hat{\mathbf{k}}$ satisfy the following equalities:

    $$\hat{\imath} \times \hat{\jmath} = \hat{\mathbf{k}} \quad \hat{\jmath} \times \hat{\mathbf{k}} = \hat{\imath} \quad \hat{\mathbf{k}} \times \hat{\imath} = \hat{\jmath}$$

2. It matters which of the two vectors we consider first: we say that a cross-product is *anticommutative*. In consequence, $\mathbf{b} \times \mathbf{a} = -(\mathbf{a} \times \mathbf{b})$. This result implies that following identities:

    $$\hat{\jmath} \times \hat{\imath} = -\hat{\mathbf{k}} \quad \hat{\mathbf{k}} \times \hat{\jmath} = -\hat{\imath} \quad \hat{\imath} \times \hat{\mathbf{k}} = -\hat{\jmath}$$

3. If two vectors are parallel, then they are at an angle of $\theta = 0$ to each other. Because $\sin 0 = 0$, we deduce the following relationships:

    $$\hat{\imath} \times \hat{\imath} = \hat{\jmath} \times \hat{\jmath} = \hat{\mathbf{k}} \times \hat{\mathbf{k}} = 0$$

The third consequence allows us to multiply two vectors together by considering each vector as a combination of $\hat{\imath}$, $\hat{\jmath}$ and $\hat{\mathbf{k}}$ components in a bracket, and multiplying out the brackets (see Chapter 7). It's often useful to rewrite eqn (28.9) having first resolved **a** and **b** into their component parts:

$$\mathbf{a} \times \mathbf{b} = (a_y b_z - a_z b_y)\hat{\imath} - (a_x b_z - a_z b_x)\hat{\jmath} + (a_x b_y - a_y b_x)\hat{\mathbf{k}} \qquad (28.10)$$

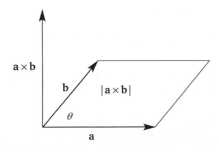

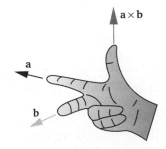

Figure 28.4 The magnitude of the cross-product can be seen as the (positive) area of the parallelogram having **a** and **b** as sides.

Figure 28.5 We obtain the direction of the cross-product vector using the right-hand rule.

where, for example, a_y is the component of the vector **a** along the y-axis. Notice how each of the round brackets on the right-hand side displays a symmetrical pattern. It can be easier to see if we write it as:

$$\mathbf{a} \times \mathbf{b} = \begin{pmatrix} a_x \\ a_y \\ a_z \end{pmatrix} \times \begin{pmatrix} b_x \\ b_y \\ b_z \end{pmatrix} = \begin{pmatrix} a_y b_z - a_z b_y \\ -(a_x b_z - a_z b_x) \\ a_x b_y - a_y b_x \end{pmatrix}$$

We find the entry for each row by covering up that row and cross-multiplying the terms in the other rows (not forgetting the minus sign for the $\hat{\jmath}$ component).

These results come from matrix algebra, and indicate how the cross-product is similar to the calculation of a determinant. Actually, we could represent eqn (28.10) by writing the determinant of the matrix,

$$\begin{pmatrix} \hat{\imath} & \hat{\jmath} & \hat{k} \\ a_x & a_y & a_z \\ b_x & b_y & b_z \end{pmatrix}$$

■ **Worked Example 28.9**

Find the cross-product of the two vectors, $\mathbf{a} = 2\hat{\imath} + 3\hat{\jmath} - \hat{k}$ and $\mathbf{b} = -3\hat{\imath} - 2\hat{\jmath}$.

Inserting terms directly into eqn (28.10) yields the cross-product of two vectors:

$$\mathbf{a} \times \mathbf{b} = (3 \times 0 - (-1)(-2))\,\hat{\imath} - (2 \times 0 - (-1) \times (-3))\,\hat{\jmath} + (2 \times (-2) - 3 \times (-3))\hat{k}$$

so $\mathbf{a} \times \mathbf{b} = (0-2)\,\hat{\imath} - (0-3)\,\hat{\jmath} + (-4 - (-9))\,\hat{k}$

i.e. $\mathbf{a} \times \mathbf{b} = -2\,\hat{\imath} + 3\hat{\jmath} + 5\hat{k}$

Alternatively, we could perform this same calculation using matrices:

$$\mathbf{a} \times \mathbf{b} = \begin{pmatrix} 2 \\ 3 \\ -1 \end{pmatrix} \times \begin{pmatrix} -3 \\ -2 \\ 0 \end{pmatrix} = \begin{pmatrix} -2 \\ -(-3) \\ 5 \end{pmatrix} = \begin{pmatrix} -2 \\ 3 \\ 5 \end{pmatrix} ■$$

Self-test 28.6

28.6.1   Find the cross-product $\mathbf{u} \times \mathbf{v}$, if $\mathbf{u} = 2\hat{\imath} + \hat{\jmath} - 3\hat{k}$ and $\mathbf{v} = 4\,\hat{\jmath} + 5\hat{k}$

28.6.2   Find the cross-product $\mathbf{a} \times \mathbf{b}$, if $\mathbf{a} = (1,2,3)$ and $\mathbf{b} = (1,1,1)$.

Vector derivatives: the gradient, curl and divergence of a vector

By the end of this section, you will know:

• The gradient of a scalar field always points in the direction in which the function grows fastest.

• The gradient of a vector has a value of zero at the function's local maxima or minima.

- The magnitude of the curl gives the maximum magnitude of vector rotation.
- We define a vector's curl as the vector field having magnitude equal to the maximum 'circulation' at each point and to be oriented perpendicularly to this plane of circulation for each point.
- The divergence of a vector field is a measure of the flux density.
- A positive divergence indicates that the vector field is expanding (i.e. there is a flux source).
- A negative divergence indicates that the vector field is contracting (ie. there is a flux sink).

The gradient of a vector:

In this context, *gradient* means a derivative. In vector calculus, the *gradient* of a scalar field is a 'vector field.' A scalar field is a distribution of scalar values on a 2- and 3-dimensional space. We represent a scalar field by a scalar function of 2- or 3-spatial coordinates, such as $T(x, y, z)$. The vector field of a scalar function points in the direction of the greatest rate of increase of the scalar field. Its magnitude represents the vector's greatest rate of change.

These definitions sound a little theoretical so consider, for example, a flask of reagent in which the concentration is defined by a scalar field c. At each point (x, y, z) within the flask, the concentration is given by the function $c(x, y, z)$, which indicates that the concentration depends on the position within the flask. This situation is common in chemistry, when:

- The reaction is inhomogeneous, for example when a solution-phase reagent reacts with a gas or solid e.g. an electrode.
- As an extension of the criterion above, any system in which one reagent is diffusing through a solution.

At every point within the flask, the gradient of c shows the direction in which the concentration rises most quickly. The magnitude of the gradient will determine how fast the concentration rises in that direction. This same argument holds for other variables, particularly temperature and chemical potential.

We often use vector fields in physical chemistry to model, for example, the speed and direction of a fluid as it moves through 3D space, e.g. in a chemical reactor or in electrochemical hydrodynamics. Such vectors are also necessary when modelling the strength and direction of a force, such as the magnetic or electrostatic force, as it changes from point to point.

The properties of the gradient are:

- It always points towards the direction in which the function grows the most.
- It has a value of zero at a local maximum or minimum of the function.

We symbolize the gradient (or gradient vector field) of a scalar function $f(x_1, x_2, ..., x_n)$ using the operator del, ∇. The gradient $= \nabla f$. It should be noted that the del symbol, ∇, used here is bold to indicate that the result is a vector. Other mathematicians employ other notations, such as $\mathbf{grad}(f)$. If writing the gradient rather than printing, some people write an arrow above the del: $\vec{\nabla} f$. We will not use these forms here.

A **vector field** is a construction in vector calculus that associates a vector with every point in a subset of 3D space.

We define the gradient of a scalar field as

$$\nabla f = \left(\frac{\partial f}{\partial x}\right)\hat{\imath} + \left(\frac{\partial f}{\partial y}\right)\hat{\jmath} + \left(\frac{\partial f}{\partial z}\right)\hat{k} \tag{28.11}$$

We can expand this definition to accommodate cylindrical or spherical polar coordinates (cf. p. 256). If ∇ is the operator in Cartesian coordinates, $\dfrac{\partial}{\partial x}\hat{\imath} + \dfrac{\partial}{\partial y}\hat{\jmath} + \dfrac{\partial}{\partial z}\hat{k}$,

• ∇ in *plane polar coordinates* is

$$\nabla = e_r \frac{\partial}{\partial r} + \frac{1}{r}e_\theta \frac{\partial}{\partial \theta}$$

• ∇ in *cylindrical polar coordinates* is

$$\nabla = e_r \frac{\partial}{\partial r} + e_\theta \frac{1}{r}\frac{\partial}{\partial \theta} + e_z \frac{\partial}{\partial z}$$

• ∇ in *spherical polar coordinates* is

$$\nabla = e_r \frac{\partial}{\partial r} + e_\theta \frac{1}{r}\frac{\partial}{\partial \theta} + e_\phi \frac{1}{r\sin\theta}\frac{\partial}{\partial \phi}$$

where e_r, e_θ, e_ϕ and e_z are the unit vectors that relate to the r, θ, ϕ and z dimensions for each system.

The directional derivative (or *slope*) of a scalar field, f, in a direction specified by a unit vector $\hat{\mathbf{I}}$ is: $\mathbf{grad}f \cdot \hat{\mathbf{I}}$, or $\dfrac{df}{dl} = \nabla f \cdot \hat{\mathbf{I}}$. To obtain the gradient at any point in 3D space (x,y,z), we insert the coordinates of the point into the expression obtained via eqn (28.11).

■ **Worked Example 28.10**

A solution of reagent is pumped at pressure through a tube reactor. If the speed is $v = 3x^2y - y^3z^2$, what is the gradient of the speed at the point $(1,-2,1)$?

Strategy:

(i) We differentiate the velocity vector v three times, once for each of the terms in eqn (28.11).

(ii) We insert the coordinates of the point $(1,-2,1)$ into the equation from part (i).

Solution:

(i) The three differentials are:

$$\left(\frac{\partial v}{\partial x}\right) = \left(\frac{\partial(3x^2y - y^3z^2)}{\partial x}\right) = 6xy$$

$$\left(\frac{\partial v}{\partial y}\right) = \left(\frac{\partial(3x^2y - y^3z^2)}{\partial y}\right) = 3x^2 - 3y^2z^2$$

$$\left(\frac{\partial v}{\partial z}\right) = \left(\frac{\partial(3x^2y - y^3z^2)}{\partial z}\right) = -2y^3z$$

So $\nabla v = 6xy\,\hat{\imath} + (3x^2 - 3y^2z^2)\,\hat{\jmath} + (-2y^3z)\hat{k}$

(ii) Inserting the coordinates of the point yields,

$$\nabla v = 6(1)\times(-2))\,\hat{\imath} + (3(1)^2 - 3(-2)^2 \times(1)^2)\,\hat{\jmath} + (-2(-2)^3 \times(1))\hat{k}$$

so $\nabla v = -12\hat{\imath} - 9\,\hat{\jmath} + 16\hat{k}$ ■

> **Self-test 28.7**
>
> Find the gradient of the following vectors at the point indicated:
>
> **28.7.1** $F = 3xy^2z - 4x^2yz^3$ at the point $(3,1,2)$
>
> **28.7.2** $\chi = 2xz^4 - x^2y$, at the point $(2,3,4)$

The curl of a vector:

We will adopt the notation, **curl u**. The curl of the vector **u** is sometimes symbolized using the del operator (from p. 334), as $\nabla \times \mathbf{u}$—note the cross. The curl therefore has the form of a cross-product, meaning the curl is also itself a vector.

The curl is a vector operator. It shows a vector field's tendency to rotate about a point (hence its name). We can think of it as the circulation per unit area, circulation density, or rate of rotation (i.e. amount of twisting at a single point). For example, imagine shrinking a whirlpool, making it progressively smaller, yet keeping the force the same. We will eventually have a situation involving lots of power within a small area—which represents a large curl. If we widen the whirlpool while keeping the force the same as before, we will experience a smaller curl. Zero circulation equates to a zero curl (and we call any vector field having a zero curl either *irrotational* or **conservative**).

So the curl quantifies a vector field's rate of rotation. We define a vector's curl as the vector field having magnitude equal to the maximum 'circulation' at each point and to be oriented perpendicularly to this plane of circulation for each point. We need the curl when dealing with fluid mechanics and elasticity theory. It also underpins the theory of electromagnetism, where it arises in two of the four Maxwell equations, e.g.

$$\mathbf{curl\ E} = \nabla \times \mathbf{E} = -\left(\frac{\partial \mathbf{B}}{\partial t}\right)$$

where **B** is magnetic field, **E** is electric field and t is time.

The length and direction of the curl vector characterize the rotation at any point in 3D space:

- The direction of the curl is the axis of rotation, as determined by the right-hand rule (see p. 429) in which a person's fingers indicate the direction of circulation.

- The magnitude of the curl gives the magnitude of rotation.

For example, if the vector field represents the flow velocity of a moving fluid, then the curl is the *circulation density* of the fluid. The curl is therefore very important in electrochemistry when considering solution flow over hydrodynamic electrodes.

The curl is a form of differentiation for vector fields. At any point in 3D space, it is proportional to the on-axis force ('torque') experienced at that point. If **A** has the form $A_x\hat{\imath} + A_y\hat{\jmath} + A_z\hat{k}$, we obtain its curl via the matrix:

$$\mathbf{curl\ A} = \nabla \times \mathbf{A} = \begin{vmatrix} \hat{\imath} & \hat{\jmath} & \hat{k} \\ \dfrac{\partial}{\partial x} & \dfrac{\partial}{\partial y} & \dfrac{\partial}{\partial z} \\ A_x & A_y & A_z \end{vmatrix}$$

This form helps explain why ∇ is used for the curl, because ∇ is a gradient operator. Notice the symmetry within this matrix: the first column relates to the x-axis, the second to the y-axis and the third to the z-axis. The curl of this vector can be expressed in terms of the individual $\hat{\imath}$, $\hat{\jmath}$ and $\hat{k}$ components, as shown in Additional problem 28.4 .

If **curl F** = 0, then **F** is a conservative field, which means that all line integrals between any pair of points are independent of the path.

We obtain the curl in terms of Cartesian coordinates:

$$\mathbf{curl\,A} = \nabla \times \mathbf{A} = \left(\frac{\partial A_z}{\partial y} - \frac{\partial A_y}{\partial z}\right)\hat{\mathbf{i}} + \left(\frac{\partial A_x}{\partial z} - \frac{\partial A_z}{\partial x}\right)\hat{\mathbf{j}} + \left(\frac{\partial A_y}{\partial x} - \frac{\partial A_x}{\partial y}\right)\hat{\mathbf{k}} \qquad (28.12)$$

Again, notice the symmetry of the terms within each bracket.

■ **Worked Example 28.11**

An electrochemist immerses an electrode in a flow cell. Its coordinates are (1,–1,1). The current at the electrode relates to the maximum velocity of solution, so we need the curl. If the velocity of the solution is characterized by the vector, $\mathbf{v} = xz^3\,\hat{\mathbf{i}} - 2x^2 yz\,\hat{\mathbf{j}} + 2yz^4\,\hat{\mathbf{k}}$, what is the curl at the electrode?

Strategy:

(i) We differentiate the equation defining the velocity, according to eqn (28.12).

(ii) We insert the coordinates of the electrode.

Solution:

(i) If velocity is $\mathbf{v}$, then $\mathbf{v} = \begin{pmatrix} xz^3 \\ -2x^2 yz \\ 2yz^4 \end{pmatrix}$, i.e. $\mathbf{v} = xz^3\hat{\mathbf{i}} - 2x^2 yz\,\hat{\mathbf{j}} + 2yz^4\,\hat{\mathbf{k}}$.

The constituent terms are:

$$\frac{\partial v_z}{\partial y} = \frac{\partial(2yz^4)}{\partial y} = 2z^4 \qquad \frac{\partial v_y}{\partial z} = \frac{\partial(-2x^2 yz)}{\partial z} = -2x^2 y$$

$$\frac{\partial v_x}{\partial z} = \frac{\partial(xz^3)}{\partial z} = 3xz^2 \qquad \frac{\partial v_z}{\partial x} = \frac{\partial(2yz^4)}{\partial x} = 0$$

$$\frac{\partial v_y}{\partial x} = \frac{\partial(-2x^2 yz)}{\partial x} = -4xyz \qquad \frac{\partial v_x}{\partial y} = \frac{\partial(xz^3)}{\partial y} = 0$$

Inserting terms into eqn (28.12) yields,

$$\mathbf{curl\,v} = (2z^4 - (-2x^2 y))\hat{\mathbf{i}} + (3xz^2 - 0)\,\hat{\mathbf{j}} + (-4xyz - 0)\hat{\mathbf{k}}$$

$$\mathbf{curl\,v} = (2z^4 + 2x^2 y)\hat{\mathbf{i}} + 3xz^2\,\hat{\mathbf{j}} - 4xyz\hat{\mathbf{k}}$$

(ii) Inserting the coordinates of the electrode, (1, –1, 1):

$$\mathbf{curl\,v} = (2-2)\hat{\mathbf{i}} + 3\,\hat{\mathbf{j}} + 4\hat{\mathbf{k}} = 3\,\hat{\mathbf{j}} + 4\hat{\mathbf{k}}. \ ■$$

Self-test 28.8

Find the curl of the following equations, at the points indicated.

28.8.1 $\mathbf{G} = 2xyz\,\hat{\mathbf{i}} + 3x^2 z^3\hat{\mathbf{j}} - 4y^3 z\hat{\mathbf{k}}$ at $(3,4,1)$

28.8.2 $\beta = 2x^2 y^3 z\,\hat{\mathbf{i}} - yz^3\,\hat{\mathbf{j}} + 4xy^2\,\hat{\mathbf{k}}$ at $(0,2,5)$

Conservative fields:

A conservative field $\mathbf{F}$ is a **vector field** which shows the following properties (and any one of the following conditions implies all of the others):

- The line integral $\int_A^B \mathbf{F} \cdot \mathbf{dr}$ is independent of the path from A to B, for all paths, assuming that each path lies completely in the region.

- The line integrals around all closed curves in the region are zero (i.e. the circulation is zero).

- **curl F** = 0, at every point of the region.

- There must be a scalar potential U such that $\mathbf{F} = -\mathbf{grad}\, U$.

The most obvious example of a conservative field is gravity. It is easiest to prove that the curl of the gravitational field is zero using the spherical polar coordinate form of the curl operator, which we have not included here. However, having derived this result and based on the conditions above, we would conclude that the work done to move something from A to B in the gravitational vector field g is independent of the path. This does indeed match with our real-life experience.

The divergence of a vector:

The divergence of a vector field measures the amount of flux entering or leaving at a particular point. If the divergence is non-zero, then there must either be a source (if the value is positive) or a sink (if the value is negative) at that point. We write the divergence as the dot product of the operator ∇ and a vector field $\mathbf{F}(x,y,z)$, which in Cartesian coordinates gives us:

$$\mathrm{div}\, \mathbf{F} = \nabla \cdot \mathbf{F} = \left(\left(\frac{\partial}{\partial x}\right)\hat{\imath} + \left(\frac{\partial}{\partial y}\right)\hat{\jmath} + \left(\frac{\partial}{\partial z}\right)\hat{k} \right) \cdot \left(F_x\, \hat{\imath} + F_y\, \hat{\jmath} + F_z\, \hat{k} \right)$$

or more simply

$$\mathrm{div}\, \mathbf{F} = \nabla \cdot \mathbf{F} = \left(\frac{\partial F_x}{\partial x}\right) + \left(\frac{\partial F_y}{\partial y}\right) + \left(\frac{\partial F_z}{\partial z}\right) \tag{28.13}$$

The divergence of a vector field is a scalar. We can also define this property in terms of cylindrical or spherical polar coordinates:

- For a vector field $\mathbf{F}(r, \theta, z) = F_r \mathbf{e}_r + F_\theta \mathbf{e}_\theta + F_z \mathbf{e}_z$, the divergence in *cylindrical polar coordinates* is

$$\nabla \cdot \mathbf{F} = \frac{\partial F_r}{\partial r} + \frac{1}{r}F_r + \frac{1}{r}\frac{\partial F_\theta}{\partial \theta} + \frac{\partial F_z}{\partial z}.$$

- For a vector field $\mathbf{F}(r, \theta, \phi) = F_r \mathbf{e}_r + F_\theta \mathbf{e}_\theta + F_\phi \mathbf{e}_\phi$, the divergence in *spherical polar coordinates* is

$$\nabla \cdot \mathbf{F} = \frac{\partial F_r}{\partial r} + \frac{1}{r}\left(\frac{\partial F_\theta}{\partial \theta} + 2F_r\right) + \frac{1}{r\sin\theta}\left(\frac{\partial F_\phi}{\partial \phi} + F_\theta \cos\theta\right).$$

■ **Worked Example 28.12**

The heat flow rate in a solution is given by the vector field $\mathbf{J} = 3xy^2\,\hat{\imath} + 2y^3z\,\hat{\jmath} + 4xyz^2\,\hat{k}$. Calculate the heat source density, S, given that $S = \mathrm{div}\, \mathbf{J}$, and evaluate S at the point $(3, 2, 1)$.

Strategy:

(i) We differentiate the components of **J** required for eqn (28.13).

(ii) We substitute the resultant three terms into eqn (28.13).

(iii) We insert the values for x, y and z.

Solution:

(i) The three differentials are:

$$\left(\frac{\partial J_x}{\partial x}\right) = \left(\frac{\partial\left(3xy^2\right)}{\partial x}\right) = 3y^2$$

$$\left(\frac{\partial J_y}{\partial y}\right) = \left(\frac{\partial\left(2y^3z\right)}{\partial y}\right) = 6y^2z$$

$$\left(\frac{\partial J_z}{\partial z}\right) = \left(\frac{\partial\left(4xyz^2\right)}{\partial z}\right) = 8xyz$$

(ii) The divergence of **J** is:

$$S = \nabla \cdot \mathbf{J} = 3y^2 + 6y^2z + 8xyz$$

(iii) At the point (3, 2, 1),

$$S = (3 \times 2^2) + (6 \times 2^2 \times 1) + (8 \times 3 \times 2 \times 1) = 84\,\text{Wm}^{-3}.$$

Since the divergence of **J** is positive at (3, 2, 1), we can see that this is indeed a source of heat flow. ■

Self-test 28.9

Find the divergence of the vector fields, given in Self-test 28.8, at the points indicated.

Summary of key equations in the text

Vector in terms of unit vectors:	$\mathbf{A} = A_x\hat{\mathbf{i}} + A_y\hat{\mathbf{j}} + A_z\hat{\mathbf{k}}$	(28.3)
Magnitude of a vector:	$\lvert\mathbf{A}\rvert = \sqrt{A_x^2 + A_y^2 + A_z^2}$	(28.4)
Relative motion of A relative to B:	$\mathbf{v}_A - \mathbf{v}_B$	(28.5)
The scalar (dot) product:	$\mathbf{a} \cdot \mathbf{b} = \lvert\mathbf{a}\rvert\,\lvert\mathbf{b}\rvert\cos\theta$	(28.6)
	$\mathbf{a} \cdot \mathbf{b} = a_x b_x + a_y b_y + a_z b_z$	(28.7)
The vector (cross-) product:	$\mathbf{a} \times \mathbf{b} = (\lvert a\rvert\,\lvert b\rvert\sin\theta)\hat{\mathbf{n}}$	(28.9)
In 3D:	$\mathbf{a} \times \mathbf{b} = (a_y b_z - a_z b_y)\hat{\mathbf{i}} - (a_x b_z - a_z b_x)\hat{\mathbf{j}} + (a_x b_y - a_y b_x)\hat{\mathbf{k}}$	(28.10)
Vector gradient	$\nabla f = \left(\frac{\partial f}{\partial x}\right)\hat{\mathbf{i}} + \left(\frac{\partial f}{\partial y}\right)\hat{\mathbf{j}} + \left(\frac{\partial f}{\partial z}\right)\hat{\mathbf{k}}$	(28.11)
Directional derivative:	$\nabla f \cdot \hat{\mathbf{I}}$	
Vector curl:	$\mathbf{curl\,A} = \nabla \times \mathbf{A} = \left(\frac{\partial A_z}{\partial y} - \frac{\partial A_y}{\partial z}\right)\hat{\mathbf{i}} + \left(\frac{\partial A_x}{\partial z} - \frac{\partial A_z}{\partial x}\right)\hat{\mathbf{j}} + \left(\frac{\partial A_y}{\partial x} - \frac{\partial A_x}{\partial y}\right)\hat{\mathbf{k}}$	(28.12)
Divergence:	$\mathbf{div\,F} = \nabla \cdot \mathbf{F} = \left(\frac{\partial F_x}{\partial x}\right) + \left(\frac{\partial F_y}{\partial y}\right) + \left(\frac{\partial F_z}{\partial z}\right).$	(28.13)

Additional problems

28.1 What is the work w done by a force F described by the vector $2t\,\hat{\imath}+4\,\hat{\jmath}$ on a particle of velocity $\mathbf{v}=5\hat{\imath}-t\,\hat{\jmath}$, in a time interval of $0<t<2$ s. [Hint: $w=\int_0^2 \mathbf{F}\cdot\mathbf{v}\,dt.$]

28.2 Using $\mathbf{H}=4\hat{\imath}+7\hat{\jmath}-3\hat{k}$ as an example vector, show that the cross-product of a vector with itself is zero.

28.3 What is the work needed to lift a rotary evaporator of mass 10 kg from the point (2,3,3) to the point (5,4,7)?

Take the acceleration due to gravity g as 9.8 m s^{-2}.

28.4 Derive eqn (28.12) from the definition of the $\operatorname{curl}\mathbf{A}=\nabla\times\mathbf{A}=\begin{vmatrix}\hat{\imath} & \hat{\jmath} & \hat{k}\\ \dfrac{\partial}{\partial x} & \dfrac{\partial}{\partial y} & \dfrac{\partial}{\partial z}\\ A_x & A_y & A_z\end{vmatrix}.$

28.5 What is the power exerted by a force $\mathbf{F}=5t^2\hat{\imath}+2t\hat{\jmath}$ on a particle of velocity $\mathbf{v}=t\hat{\imath}-2t^2\hat{\jmath}$ at time t?

28.6 A force $\mathbf{F}=3\hat{\imath}+2\hat{\jmath}+4\hat{k}$ acts through the point with position vector $\mathbf{r}=2\hat{\imath}-\hat{\jmath}+3\hat{k}$. What is its torque about a perpendicular axis through this point?

28.7 A molecule moves inside a mass spectrometer at a velocity v of 40.2 m s^{-1}. Its trajectory is 40° to the detector. Resolve the velocity into its component parts.

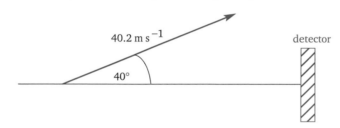

28.8 Given that the scalar field $f(x,y,z)=A/(x^2+y^2+z^2)^{1/2}$, find the derivative of f in the direction of the unit vector $\hat{d}=\dfrac{\left(\hat{\imath}+\hat{\jmath}+\hat{k}\right)}{\sqrt{3}}$ at (1,2,2).

28.9 A rigid rotor consisting of two atoms separated by a fixed bond length is rotated about its centre of mass. This is equivalent to the rotation of a single particle of reduced mass μ about the centre of mass coordinates on a ring of radius, r, where r is the bond length. The angular momentum is defined by $\mathbf{l}=\mathbf{r}\times\mathbf{p}$, where $\mathbf{r}$ is the position vector relative to the centre of mass and $\mathbf{p}$ the momentum vector of the particle.

Use the definition of the cross-product to show that the kinetic energy of this rigid rotor is

$$E=\frac{l^2}{2I},\ \text{where}\ I=\mu r^2\ \text{is the moment of inertia.}$$

28.10 Find the volume of a parallelepiped, which can be calculated using the triple scalar product: $\mathbf{a}\cdot\mathbf{b}\times\mathbf{c}$, when $\mathbf{a}=\hat{\imath}+\hat{\jmath}+3\hat{k}$, $\mathbf{b}=2\hat{\imath}-4\hat{\jmath}+\hat{k}$ and $\mathbf{c}=3\hat{\imath}+2\hat{\jmath}+\hat{k}$.

29

Complex numbers

Introducing complex numbers

By the end of this section, you will know:

- Numbers fall into three categories: real, imaginary and complex.
- The square root of −1 is called i.
- Numbers involving only i and its multiples are termed *imaginary*.
- Numbers not involving i and its multiples are termed *real*.
- Numbers comprising both real and imaginary numbers are termed *complex*.

All the numbers we have considered in this book so far, whether positive or negative, have a basis in fact. They can be visualized readily. As an example, the process of squaring a number is usually regarded as very simple, so $4^2 = 16$, $3^2 = 9$, $a \times a = a^2$, $3a \times 3a = 9a^2$, and so forth. Using exactly the same methodology, $(-4)^2 = +16$, $(-3)^2 = +9$, $(-a)^2 = +a^2$. It is possible to propose a square root to each of these numbers.

Clearly, the examples in the previous paragraph are so simple they are trivial. But what is the square root of *minus 1*? A moment's thought suggests there is no valid answer, because the square of a negative number is *always* positive. It is therefore simply not possible to square either a negative number or a positive number in order to generate a negative number: $+1^2$ equals $+1$ and $(-1)^2$ also equals $+1$.

Chemists, and physical scientists in general, have identified a large number of situations where it would be useful to pretend that it is possible to have a square root of −1. Accordingly, they *define* an answer:

$$i = \sqrt{-1} \qquad (29.1)$$

Because there is no genuine answer to the problem of $\sqrt{-1}$, we say that i is **imaginary**. The imaginary number $i = \sqrt{-1}$ behaves exactly like any other number in algebra, without any anomalies. i is merely a symbol. We could have chosen any other letter.

We say that numbers that have a genuine square root are **real**. Algebraic terms such as a, b, or $5c$, are all assumed to be real. In this context, zero is also a real number.

As a third category, there are numbers that comprise both a real and an imaginary component. For example, consider the general number z:

$$z = x + iy \qquad (29.2)$$

where x is real and the coefficient y (which relates to the imaginary part) is also a real number. We call such composites **complex numbers**. In physical science, $y = 0$ in the overwhelming majority of cases. In other words, there is only a real component. 5 and b are real numbers; 2i and yi are imaginary numbers; but $3 + 9i$ and $-5 - 2i$ are both complex. By convention, we usually give a complex number the symbol z.

Two complex numbers are equal if and only if:

- Their real parts are equal, and
- Their imaginary parts are equal.

We need to note:

Chemists employ a subtle use of adjective:

- **Complex** can mean either an association compound, e.g. $[Ag(NH_3)_4]^+$ or a number incorporating $\sqrt{-1}$, which we call i.

Squares and square roots were discussed in Chapter 11.

- **Complicated** means difficult, not straightforward. It can also mean situations where mathematical modelling is impossible, and therefore necessitates the use of approximate solutions.

Aside:
Complex numbers were invented when it was discovered that solving some cubic equations required intermediate calculations containing the square roots of negative numbers, even when the final solutions were real numbers. It is now clear that many, many situations in the physical sciences require complex numbers. Examples include electronics, quantum mechanics, magnetism and angular momentum. In practice, complex numbers allow us to simplify the mathematics of such physical parameters as well as completing the number system.

> **Self-test 29.1**
>
> Categorize each of the following: are they real, imaginary or complex?
>
> **29.1.1** 39   **29.1.2** 4i
>
> **29.1.3** 21 + 7i   **29.1.4** $a + 7i$
>
> **29.1.5** $bi + 8i$ **29.1.6** 3.142

Simplifying complex numbers

By the end of this section, you will know:

- Complex numbers may be added and subtracted like any other algebraic sums. Some complex numbers can be easily simplified. If the imaginary part is expressed as the square root of a negative number, we merely take the root of the number;

So while the words 'real' and 'imaginary' were meaningful when complex numbers were first envisaged, the term is no longer useful. The use of such numbers in later applications shows that nature has no preference for 'real' numbers; indeed, many *real* descriptions of *real* things actually require complex numbers. The 'imaginary' part is every bit as physical as the 'real' part. For example, when a time-dependent voltage is applied across a capacitor, the resistance changes with time. If the frequency is ω, the resistance is given by the equation, $z_c = \dfrac{1}{i \omega C}$ where C is the value of the capacitance. Similarly, the time-dependent resistance of an inductor L is, $Z_L = i \omega L$.

■ **Worked Example 29.1**

What is the value of $\sqrt{-25}$?

$\sqrt{-25}$ is the same as $\sqrt{25} \times \sqrt{-1}$. Therefore:

$$\sqrt{-25} = \sqrt{25} \times \sqrt{-1}$$

$$\sqrt{-25} = 5 \times \sqrt{-1}$$

$$\sqrt{-25} = 5i \quad ■$$

If a number comprises both real components and the square root of a negative number, we leave the real component and perform the same algebra on the imaginary component as that in Worked Example 29.1.

■ **Worked Example 29.2**

What is the value of $2 - \sqrt{-64}$?

The real component needs no further algebraic attention. We shall leave it as 2.

The imaginary component of $\sqrt{-64}$ is the same as $\sqrt{+64} \times \sqrt{-1}$. The first root has the value of $+8$. $\sqrt{-64}$ is therefore the same as 8i. The value of the complex number is therefore $2 - 8i$. ■

> **Self-test 29.2**
>
> Write these complex numbers in the form, $a + bi$, where a and b are real numbers.
>
> **29.2.1** $\sqrt{-9}$ **29.2.2** $\sqrt{-36}$
>
> **29.2.3** $11 - \sqrt{-100}$ **29.2.4** $14 - \sqrt{-36}$
>
> **29.2.5** $8 - \sqrt{-12}$ **29.2.6** $2 - \sqrt{-50}$

The simple algebra of complex numbers

By the end of this section, you will know:

- Complex numbers may be added and subtracted like any other algebraic sum or difference.

Adding and subtracting complex numbers

Adding and subtracting complex numbers is very simple. We merely add (or subtract) like terms, so we add together (or subtract) the real components and then add (or subtract) the imaginary components.

> ### ■ Worked Example 29.3
>
> An electrochemist makes a circuit comprising a capacitor and an inductor. The frequency-dependent resistance (or *impedance*) of the circuit is $z(\text{capacitor}) = 12 - 4\,\text{i}$ and $z(\text{inductor}) = 17 + 6\text{i}$. What is the total impedance?

A time-or frequency-dependent resistance is called an impedance, which has the symbol z.

The sum of the real components is $12 + 17 = 29$. The sum of the imaginary components is $(-4) + 6 = 2$, so the overall impedance is $z = 29 + 2\text{i}$. ■

In general, then, adding the two numbers $z_1 = x_1 + y_1\text{i}$ and $z_2 = x_2 + y_2\text{i}$ follows the pattern:

$$z_1 + z_2 = (x_1 + x_2) + (y_1 + y_2)\,\text{i} \tag{29.3}$$

Subtracting with complex numbers is just as easy as addition.

> ### ■ Worked Example 29.4
>
> The electrochemist measures the total impedance of a different related circuit as $z(\text{total}) = 40 + 12\text{i}$. If the impedance of the capacitor is $z(\text{capacitor}) = 22 + 3\text{i}$, what is the impedance of the inductor?

The impedance of the inductor is the difference between $z(\text{total})$ and $z(\text{capacitor})$. Therefore:

$$z(\text{inductor}) = z(\text{total}) - z(\text{capacitor})$$

Inserting numbers: $(40 + 12\text{i}) - (22 + 3\text{i}) = (40 - 22) + (12 - 3)\text{i} = 18 + 9\text{i}$ ■

In general, then, subtracting the two numbers $z_1 = x_1 + y_1\text{i}$ and $z_2 = x_2 + y_2\text{i}$ follows the pattern:

$$z_1 - z_2 = (x_1 - x_2) + (y_1 - y_2)\,\text{i} \tag{29.4}$$

Self-test 29.3

Perform the following addition and subtraction problems.

29.3.1 $4 + \text{i}$ add $5 + 2\text{i}$ **29.3.2** $5.3 + 3.2\text{i}$ subtract $2.7 - 4.1\text{i}$

29.3.3 $(45.87 + 6.33\text{i}) - (0.34 + 11.60\text{i})$ **29.3.4** $(5a + 11b\text{i}) + (a + 15b\text{i})$

29.3.5 $(a + 23b + 3c + d\text{i}) - (-2a + 3b + 5c - 3d\text{i}) + (10a + 4b + 6c + 13d\text{i})$

29.3.6 $\sqrt{-48} + \left(-1 - \sqrt{-75}\right)$ **29.3.7** $\left(12 - \sqrt{-72}\right) + \left(7 + \sqrt{-98}\right)$

Multiplying complex numbers

By the end of this section, you will know that:

- The product of two real numbers is real.
- The product of two *imaginary* numbers is real.
- When multiplying together two complex numbers, we use the same procedure as multiplying together two brackets (see p. 90 ff.).

Multiplying complex numbers is trickier than adding or subtracting, because of the imaginary components. When the numbers have only real components, the procedure is so easy it's trivial: for example, $2a \times 4b = 8ab$. We covered this algebra in Chapter 5.

When we wish to multiply two imaginary numbers, we merely multiply the two coefficients and then include the factor, $i \times i = -1$:

■ **Worked Example 29.5**

What is the product of 2i and 10i?

2i is the same as $2 \times i$; 10i is the same as $10 \times i$. Since multiplication is associative (see p. 34), we can say $2i \times 10i = (2 \times 10) \times (i \times i)$.
Therefore: $2i \times 10i = 20 \times i^2 = 10 \times -1 = -20$. ■

When we multiply together two complex numbers, we treat each as a bracket. We already know how to multiply together brackets (see p. 90 ff.).

■ **Worked Example 29.6**

What is the product of $4 + 2i$ and $5 + 3i$?

First we multiply, term by term, the whole of the *right*-hand bracket with the real number 4 from the *left*-hand bracket:

$$(4+2i) \times (5+3i)$$

The results of the two multiplication steps are:

    Step **1**: 20   and   Step **2**: 12i

Then we multiply the whole of the *right*-hand bracket, this time with the imaginary component from the *left*-hand bracket, i.e. with 2i:

$$(4+2i) \times (5+3i)$$

The results of the two multiplication steps are:

    Step **3**: 10i   Step **4**: $6i^2$

In order, the terms **1–4** are: $20 + 12i + 10i + 6i^2$. Summing yields $20 + 22i + 6i^2$. But $i^2 = -1$, so $6i^2 = -6$. The final sum is therefore $14 + 22i$. ■

In general, then, multiplying the two numbers $z_1 = x_1 + y_1 i$ and $z_2 = x_2 + y_2 i$ follows the pattern:

$$z_1 \times z_2 = (x_1 x_2 - y_1 y_2) + (x_1 y_2 + x_2 y_1)\, i \qquad (29.5)$$

We can look back and confirm that Worked Example 29.6 follows this template exactly.

Self-test 29.4

Multiply together the following complex numbers.

29.4.1 $(15 + 3i)\,(-2 + i)$ **29.4.2** $(-1 + 3i)\,(4 - 2i)$

29.4.3 $(3 + 2i)\,(3 - 2i)$ **29.4.4** $(4.2 + 7.3i)\,(9.2 - 11.6i)$

29.4.5 $(a + bi)\,(a - bi)$ **29.4.6** $(2a + 3bi)\,(3a - 7bi)$

29.4.7 $(a^2 + 2b^2 i)\,(3a^3 - 4bi)$ **29.4.8** $\left(\sqrt{a} + 5\sqrt{b}i\right)\left(2\sqrt{a} - 4\sqrt{b}i\right)$

Multiplying complex numbers is easier using the so-called **polar forms**.

■ **Worked Example 29.7**

Multiply the two complex numbers A and B, where $A = r_1(\cos\theta + i\sin\theta)$ and $B = r_2 (\cos\varphi + i\sin\varphi)$.

Simple multiplication of A and B yields:

$$A \times B = r_1 r_2 (\cos\theta\cos\varphi - \sin\theta\sin\varphi) + i\, r_1 r_2 (\sin\theta\cos\varphi + \sin\varphi\cos\theta)$$

By using straightforward trigonometric formulae (see Chapter 16), this answer simplifies to:

$$A \times B = r_1 r_2 [\cos(\theta + \varphi) + i\sin(\theta + \varphi)].$$

So the modulus of the product, $r_1 r_2$, is simply the product of the moduli of A and B i.e. r_1 and r_2. But notice how the argument of the product, $\theta + \varphi$, is the *sum* of the arguments of A and B, i.e. θ and φ.

This gives a simple rule for multiplying complex numbers in polar form.

- We multiply the moduli.

- We add together the arguments. ■

Self-test 29.5

Multiply together the following complex numbers.

29.5.1 $3(\cos\theta + i\sin\theta) \times 4(\cos 2b + i\sin 2b)$

29.5.2 $a^2\left(\cos\dfrac{\pi}{2} + i\sin\dfrac{\pi}{2}\right) \times a\left(\cos\dfrac{\pi}{4} + i\sin\dfrac{\pi}{4}\right)$

The complex conjugate

By the end of this chapter, you will know:

- The complex conjugate of the general complex number $a + bi$ is $a - bi$.

- Complex conjugates are represented with a star: the conjugate of z is z^*.

- The product of z and z^* is a real number.

Look at the answer to problem 29.4.5 in Self-test 29.4. The answer is $a^2 + b^2$, a perfect square. Also notice how the answer is wholly real, for it does not contain any imaginary components.

In general, if a complex number is $a + bi$, we say the almost identical complex number $a - bi$ is its **complex conjugate**. The only difference is the change in sign. If the sign was originally negative then the complex conjugate would involve a positive sign.

■ **Worked Example 29.8**

In quantum electrodynamics, the gauge covariant derivative D_μ is $\partial_\mu + eA_\mu i$. Write its complex conjugate.

To obtain the complex conjugate, we rewrite the expression almost exactly. The only difference is that we replace i with –i. In this Worked Example, such a transposition changes the plus sign into a minus.

The complex conjugate of $D_\mu = \partial_\mu + eA_\mu i$ is $D_\mu{}^* = \partial_\mu - eA_\mu i$ ■

Self-test 29.6

Write the complex conjugate for the first complex number in each of the examples in Self-test 29.4.

Often, the complex conjugate is abbreviated with an asterisk: if the complex number is z, then its conjugate is z^*. In Worked Example 29.8, the complex conjugate is therefore written as $D_\mu{}^*$.

The complex conjugate has one particularly useful aspect. The product of a complex number and its conjugate yields a number that is wholly *real*; there is no imaginary component.

■ **Worked Example 29.9**

φ is a wavefunction. Determine the value of $\varphi \, \varphi^*$ if $\varphi = a + bi$.

If $\varphi = a + bi$ then its conjugate will be $\varphi^* = a - bi$.
Multiplying these two yields: $a^2 + abi - abi - b^2 i^2$. The second and third terms cancel completely, leaving $a^2 + - b^2 i^2$. By definition, $i^2 = -1$, so $\varphi \, \varphi^* = a^2 + b^2$, which has no imaginary component. ■

The answer to Worked Example 29.9 is not only real but also a perfect square. The perfect square was introduced in Chapter 7, where we saw that $(a^2 - b^2)$ factorizes to yield $(a + b)(a - b)$. An appreciation of complex number theory suggests that $(a^2 + b^2)$ can be factorized as a perfect square:

$$(a^2 + b^2) = (a + bi)(a - bi) \tag{29.6}$$

Aside:
The complex conjugate is important in quantum chemistry, because the simplest test of whether two wavefunctions φ_i and φ_j^* are 'orthogonal' is that $\int \varphi_i \varphi_j^* \, d\tau = 0$, where τ means 'all space.' In saying the product of their integral vanishes, we mean the two orbitals do not overlap – which precludes any possibility of a covalent bond.

Complex fractions

By the end of this section, you will know that to divide complex numbers:

- We first write the division as a fraction, then multiply both numerator and denominator by the complex conjugate of the denominator.
- The final result will be a new complex number with both real and imaginary parts.

When dividing a complex number by another complex number, we must first multiply both numerator and denominator by the complex conjugate of the denominator. By this means, the new denominator becomes a real number with no imaginary component.

■ **Worked Example 29.10**

What is $(2 + 4i) \div (3 + 10i)$?

Strategy:

(i) Write the complex conjugate of the denominator.

(ii) Multiply both the top and the bottom of the fraction by the complex conjugate (i.e. in effect we multiply by 1.)

(iii) We rearrange.

Solution:

(i) The complex conjugate of $(3 + 10i)$ is $(3 - 10i)$.

(ii) Multiply both top and bottom of the fraction by the conjugate:

$$\frac{(2+4i)(3-10i)}{(3+10i)(3-10i)} = \frac{2\times3+4\times10+(4\times3-2\times10)i}{3^2+10^2}$$

Notice how the denominator on the right-hand side is wholly real.

> Remember to use the laws of BODMAS when determining the value of the numerator.

(iii) We multiply the products on the top line, then group the real and imaginary terms in the numerator, as we did on p. 442.

Rearrangement yields $\frac{46}{109} - i\left(\frac{8}{109}\right)$. We cannot readily simplify this result by cancelling. Cancelling is permissible in other examples. ■

In general, then, dividing the two numbers $z_1 = x_1 + y_1i$ by $z_2 = x_2 + y_2i$ follows the pattern:

$$\frac{z_1}{z_2} = \frac{x_1x_2+y_1y_2}{x_2^2+y_2^2} + i\left(\frac{x_2y_1-x_1y_2}{x_2^2+y_2^2}\right) \tag{29.7}$$

Self-test 29.7

Perform the following division problems.

29.7.1 $(2+3i) \div i$ 29.7.2 $(4+5i) \div 2i$

29.7.3 $\dfrac{2+7i}{6+i}$ 29.7.4 $\dfrac{4-9i}{5-2i}$

This methodology also shows us a way of moving a complex number from the denominator of a fraction to the numerator (which is conceptually far easier).

■ **Worked Example 29.11**

The impedance of a capacitor is given on p. 441. Rewrite this complex expression without i in the denominator.

Strategy:

(i) We multiply both the top and bottom of the fraction by i.

(ii) We then rearrange.

Solution:

(i) $z = \dfrac{1}{i\omega C} \times \dfrac{i}{i}$

Multiplying by i/i in this way is the same as multiplying by 1. So we have not actually changed the equation at all.

(ii) The denominator is now $i^2 \omega C$, which is the same as $-\omega C$. The numerator is now i.

We rewrite as: $z = \dfrac{i}{-\omega C}$.

Finally, to reposition the minus sign, we take it from the denominator, and rewrite the equation as either $z = \dfrac{-i}{\omega C}$ or, better, $-\dfrac{i}{\omega C}$.

de Moivre's theorem and complex numbers involving exponentials

By the end of this section, you will know:

- de Moivre's theorem combines real and complex trigonometric function, as $(\cos \theta + i \sin \theta)^n = \cos n\theta + i \sin n\theta$
- To interconvert between exponential and trigonometric functions, Euler's formula says: $e^{ikx} = \cos k\,\theta + i \sin k\,\theta$

Many complex numbers are expressed in polar coordinates (see Chapter 16). When expressed as a vector, a complex has both magnitude r and a direction θ. Having introduced the concept of direction θ, it is useful to think in terms of complex trigonometric functions.

The theorem of Abraham de Moivre (1667–1754) combines real and complex trigonometric function:

$$(\cos \theta + i \sin \theta)^n = \cos n\theta + i \sin n\theta \qquad (29.8)$$

where the value of θ is the same in both of the trigonometric terms.

■ **Worked Example 29.12**

Calculate a value for $\left(\cos\dfrac{\pi}{4} + i \sin\dfrac{\pi}{4} \right)^4$.

Using de Moivre's theorem in eqn (29.8),

$$\left(\cos\frac{\pi}{4} + i\sin\frac{\pi}{4} \right)^4 = \left(\cos\left(4 \times \frac{\pi}{4}\right) + i \sin\left(4 \times \frac{\pi}{4}\right) \right)$$

Clearly, cancelling is possible on the right-hand side, yielding:

$$\left(\cos\frac{\pi}{4} + i \sin\frac{\pi}{4} \right)^4 = (\cos \pi + i \sin \pi)$$

$\sin \pi = 0$ and $\cos \pi = -1$. So $\left(\cos\dfrac{\pi}{4} + i \sin\dfrac{\pi}{4} \right)^4 = -1$. Note how this result is real. ■

Soon after de Moivre formulated his theorem, the Swiss mathematician Leonhard Euler (1707–1783) adapted it to reveal a relationship between complex trigonometric numbers and exponential functions:

$$e^{i\theta} = \cos\theta + i\sin\theta \qquad (29.9)$$

This relationship is particularly useful in quantum mechanics, because so many wavefunctions have complex exponential terms. Confusingly, a great many people today call eqn (29.9) 'de Moivre's theorem'. The combination of sine, i and cosine together occurs so often that we often see $\cos(n\theta) + i\sin(n\theta)$ abbreviated as $cis(n\theta)$.

Combining eqns (29.8) and (29.9) yields the more useful relationship:

$$e^{ik\theta} = (\cos\theta + i\sin\theta)^k \qquad (29.10)$$

which can itself be rewritten as:

$$e^{ik\theta} = (\cos k\theta + i\sin k\theta) \qquad (29.11)$$

This relationship is only really valid if the modulus is 1. If not, then

$$re^{ik\theta} = r(\cos k\theta + i\sin k\theta) \qquad (29.12)$$

As a special case of this theorem, substituting $\theta = \pi$ radians, yields the equation $e^{i\pi} + 1 = 0$.

■ **Worked Example 29.13**

Using de Moivre's theorem, multiply the two complex numbers A and B, where

$$A = r_1 e^{i\theta} \text{ and } B = r_2 e^{i\varphi}.$$

Using the exponential form of these complex numbers and the appropriate power law (see Chapter 11) the solution can easily be found:

$$A B = r_1 e^{i\theta} \times r_2 e^{i\varphi}$$
$$= r_1 r_2 e^{i(\theta+\varphi)}$$
$$= r_1 r_2 [\cos(\theta+\varphi) + i\sin(\theta+\varphi)]$$

This problem is identical to the Worked Example 29.7 on p. 444. This identity can be seen by converting the exponentials to trigonometric functions using de Moivre's theorem, eqn (29.9).

Therefore, $A = r_1(\cos\theta + i\sin\theta)$

and $B = r_2(\cos\varphi + i\sin\varphi)$ ■

■ **Worked Example 29.14**

The radial wavefunction for a 2p orbital of hydrogen with $m_l = \pm 1$ has the following form,

$$\psi_{2p} = \mp\frac{1}{\sqrt{2}} \, r \, \sin\theta \, e^{\pm i\phi} f(r)$$

The symbol $\pm$ means 'plus or minus.' The related symbol $\mp$ means essentially the same, except the part of an equation labelled $\mp$ must always take the opposite sign to the part labelled $\pm$.

Rewrite the exponential component as a complex trigonometric function, and hence rewrite the entire equation.

The exponential function is $e^{\pm i\phi}$. At this stage of the answer, we will ignore the $\mp$ symbol because it relates to the $\pm$ symbol immediately after the equals sign (i.e. it does not stand on its own). Using the relation in eqn (29.11), we write the exponential as:
$e^{i\phi} = \cos\phi + i\sin\phi$.

As a result of the $\mp$ and $\pm$ signs, the overall expression has two forms:

$$\psi_{2p} = -\frac{1}{\sqrt{2}} \ r \ \sin\theta \ f(r)(\cos\phi + i \ \sin\phi) \text{ and}$$

$$\psi_{2p} = \frac{1}{\sqrt{2}} \ r \ \sin\theta \ f(r)(\cos(-\phi) + i \ \sin(-\phi)).$$

The second expression can be simplified slightly:

$$\psi_{2p} = \frac{1}{\sqrt{2}} \ r \ \sin\theta \ f(r)(\cos\phi - i \ \sin\phi). \ \blacksquare$$

Self-test 29.8

Use Euler's formula to rewrite the following:

29.8.1 $4e^{i\theta}$ **29.8.2** $b(\cos d\varphi + i \sin d\varphi)$

29.8.3 $(\cos \varphi + i \sin \varphi)^4$ **29.8.4** $c^2 e^{ifx}$

The graphical representation of complex numbers

By the end of this section, you will know:

- An Argand diagram is a Cartesian graph plotted using the imaginary component of a complex number as y against the real component as x.
- Argand diagrams are popular in electrochemistry, and are used to elucidate the composition of their electrochemical systems.

Complex numbers can be viewed as points on a straightforward two-dimensional Cartesian coordinate system.

Such graphs are produced by plotting the imaginary component of a complex number (as y) against the real component (as x); see Fig. 29.1. We call such a system 'the complex plane' and the resultant graph is often termed an **Argand diagram** after Jean-Robert

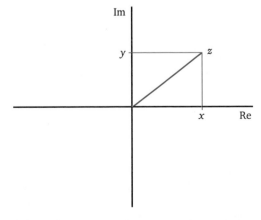

Figure 29.1 An Argand diagram. The imaginary number $z = x + iy$ can be represented on a graph by plotting it as a point (x, y). The real (Re) component is plotted on the x-axis and the imaginary (Im) component on the y-axis.

The representation of a complex number by its Cartesian coordinates is called the **Cartesian form** or **rectangular form** or **algebraic form** of that complex number.

Argand (1768–1822) who first devised them. In effect, we are saying a direct visual correspondence exists between the Cartesian $x-y$ plane (see Chapters 8 and 9) and the complex numbers, $x + iy$. In an Argand diagram, each complex number z represents a unique point. The point can also be envisaged as a position vector.

The length of the solid line on Fig. 29.1 represents the magnitude of the complex number z. By direct analogy with Pythagoras' theorem (p. 254), this length relates to the magnitudes of the real and imaginary components of z:

$$|z| = \sqrt{x^2 + y^2} \tag{29.13}$$

The vertical lines either side of z are termed **modulus lines**, and tell us to consider only the *magnitude* of z and not its *sign*; clearly, a line on a graph cannot have a negative length.

■ **Worked Example 29.15**

The appearance of the Argand diagram is useful for discerning the components within the circuit under analysis. The impedance of a straightforward resistor R is real (call it z_R), and is independent of frequency. Conversely, the impedance z of a capacitor C is imaginary, and varies with frequency ω according to the equation on p. 441. The overall impedance of the circuit therefore contains both real and imaginary components, and is complex.

Because the impedance is complex, it can be represented on an Argand diagram. Electrochemists believe the interface between an electrode and a solution behaves like a capacitor and resistor placed in parallel. How do electrochemists detect the presence of this 'RC element'?

The voltage applied to the circuit has a frequency ω. Because the impedance z_C varies with frequency ω, the overall complex impedance of a circuit will also vary with frequency.

Electrochemists draw a point on an Argand diagram to represent the complex impedance at each frequency. These points are then joined together. And now comes the clever bit: the shapes generated in the Argand diagram can be used to distinguish between different arrangements of electrical components. For example, when a resistor and capacitor are placed in *series*, the Argand diagram shows a vertical line displaced from the origin by an amount equal to z_R; see Fig. 29.2(a).

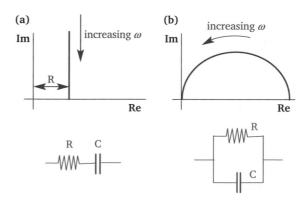

Figure 29.2 Argand diagrams representing the complex impedance can be used to distinguish between different arrangements of electrical components: resistor and capacitor (a) in *series* and (b) in *parallel*.

Conversely, when the circuit components are placed in *parallel*, the impedance is calculated as:

$$\frac{1}{z(\text{overall})} = \frac{1}{z_C} + \frac{1}{z_R}$$

And, owing to this mathematical relation, when a resistor is placed in parallel with a capacitor, the Argand diagram describes a semicircular arc. This arc is simply made up of the impedances of the circuit at individual, fixed frequencies, as indicated by the curved arrow on Fig. 29.2(b).

So, when electrochemists draw an Argand diagram and see a semicircular arc, they can be fairly sure that a resistor and a capacitor are placed in *parallel*. (Within the electrochemical community, such an Argand diagram is often termed a 'Nyquist plot.') ■

Self-test 29.9

29.9.1 A resistor and a capacitor are arranged in parallel. Derive an expression for $z(\text{overall})$, by inserting the equations $z_R = R$ and $z_C = 1/i\omega C$.

[Hint: use the equation in Worked Example 29.14, and rearrange to show $z(\text{total}) = R/(1 + i\omega CR)$.]

Complex numbers and Fourier-transform operations

By the end of this section, you will know:

* A Fourier transform is a mathematical means of interconverting between two interconnected variables.

Consider the propagation of electromagnetic waves of light. It is possible to analyse and measure the propagation in terms of the distance covered, or in terms of temporal measurements, i.e. time. For a simple sinusoidal wave, interconversion between these two different descriptions of the physical system—time and distance—is simple: the length of the wave, λ, and the frequency, v, with which the wave propagates are connected by the simple equation $c = \lambda v$. Knowledge of one variable allows the other to be obtained.

In a broadly similar way, mathematical methods based on ideas by Joseph Fourier (1768–1830) enable us to transform functions based on *length* into complementary functions that we express in terms of *time*. For a so-called **Fourier transform** to work, the two domains of time and length must be continuous and unbounded.

Many forms of molecular spectroscopy employ methods based on the mathematics of Fourier transforms, especially nuclear magnetic resonance (NMR), electron-spin resonance (ESR), infrared (IR) and Raman vibration spectroscopy, and mass spectrometry.

Experimentally, the spectrometer observes how the spectrum intensity varies according to energy. Fourier transform analysis of the energy dependence of the spectrum (which relates to the frequency and hence the time domain) enables the accurate computation of the spectrum of the source, which is construed in terms of wave*length*.

Chemists are interested in several possible pairs of interconnected data that are available via other Fourier-transformed techniques:

Superposition means the linear combination (addition) of simple sine and cosine waves.

- If the wavefunction $\psi(x)$ describes quantum-mechanical wave motion, then $\psi(x)$ gives the particle's *position*. The Fourier transform of $\psi(x)$ yields a particle's *momentum*.

- In molecular spectroscopy, measurements relate to the time domain of electromagnetic radiation. When the observed variable t relates to *time*, the transform variable f represents ordinary *frequencies*.

There are several common conventions for performing the Fourier transform of a periodic function, x. The simplest employs exponential functions of imaginary numbers:

$$X(\omega) = \int_{-\infty}^{\infty} x(t)e^{-i\omega t}dt \tag{29.14}$$

We have used the term 'Fourier transform' here to mean the process whereby a formula 'transforms' one function into the other. In fact the term is very much wider.

Here, the Fourier-transform function $X(\omega)$ defines the *frequencies* at which a molecular response is non-zero. (For chemists, this 'molecular response' usually relates to molecular spectroscopy.) The function $x(t)$ yields the *times* at which the same signal is non-zero. All frequencies are expressed in terms of angular frequency, $\omega = 2\pi f$, for which the units are radians per second (see Chapter 16). Frequency f is termed the *transform variable* and is simply a series of integers, which are real numbers.

When suitable conditions hold, the signal in the time domain can be calculated using the inverse transform of eqn (29.14):

$$x(t) = \frac{1}{2\pi}\int_{-\infty}^{\infty} X(\omega)e^{i\omega t}d\omega \tag{29.15}$$

Again, all values of t must represent real numbers.

Summary of key equations in the text

Definition of i:	$i = \sqrt{-1}$	(29.1)
Definition of a complex number:	$z = x + yi$	(29.2)
Adding complex numbers:	$z_1 + z_2 = (x_1 + x_2) + (y_1 + y_2)i$	(29.3)
Subtracting complex numbers:	$z_1 - z_2 = (x_1 - x_2) + (y_1 - y_2)i$	(29.4)
The product of 2 complex numbers:	$z_1 \times z_2 = (x_1 x_2 - y_1 y_2) + (x_1 y_2 + x_2 y_1)i$	(29.5)
Perfect square involving complex numbers:	$(a^2 + b^2) = (a + bi)(a - bi)$	(29.6)
Complex fractions;	$\dfrac{z_1}{z_2} = \dfrac{x_1 x_2 + y_1 y_2}{x_2^2 + y_2^2} + i\left(\dfrac{x_2 y_1 - x_1 y_2}{x_2^2 + y_2^2}\right)$	(29.7)
de Moivre's theorem:	$(\cos\theta + i\sin\theta)^n = \cos n\theta + i\sin n\theta$	(29.8)
Euler's theorem:	$e^{i\theta} = (\cos\theta + i\sin\theta)$	(29.9)
	$e^{ik\theta} = (\cos k\theta + i\sin k\theta)$	(29.11)
	$re^{ik\theta} = r(\cos k\theta + i\sin k\theta)$	(29.12)
Coordinates of i on an Argand diagram:	$\lvert z \rvert = \sqrt{x^2 + y^2}$	(29.13)
Fourier transforms:	$X(\omega) = \int_{-\infty}^{\infty} x(t)e^{-i\omega t}dt$	(29.14)
	$x(t) = \dfrac{1}{2\pi}\int_{-\infty}^{\infty} X(\omega)e^{i\omega t}d\omega$	(29.15)

Additional problems

Problems 29.1–3 relate to operations in NMR, which are often described according to combinations of the following Pauli spin matrices:

$$I_x = \frac{1}{2}\begin{pmatrix} 0 & 1 \\ 1 & 0 \end{pmatrix} \qquad I_y = \frac{1}{2i}\begin{pmatrix} 0 & 1 \\ -1 & 0 \end{pmatrix} \qquad I_z = \frac{1}{2}\begin{pmatrix} 1 & 0 \\ 0 & -1 \end{pmatrix}$$

29.1 Prove that $I_x I_y - I_y I_x = i I_z$.

29.2 Calculate the identity: $I_y I_z - I_z I_y = ?$

29.3 Calculate the identity: $I_z I_x - I_x I_z = ?$

29.4 An electrochemical circuit is set up. Three components are arranged in parallel: a capacitor C and the two resistors R_1 and R_2. What is the total impedance?

29.5 What is the conjugate of the complex number $12 + 6i$.

29.6 Consider the perfect square $(9a^2 + 49b^2)$: using complex numbers, factorize this expression.

29.7 One solution to the Schrödinger equation for a particle, that is free to move parallel to the x-axis with zero potential energy, is: $\psi = Ae^{ikx} + Be^{-ikx}$. Rewrite the equation in terms of trigonometric functions.

29.8 The **wavefunction** of a molecular orbital Ψ will have an expression of the kind, $\Psi = (f + ig)$, and its complex conjugate will have the form $\Psi^* = (f - ig)$. The probability of finding an electron within this orbital is a function of the product $\Psi\Psi^*$. Multiply together the two wavefunctions Ψ and Ψ^*, i.e. multiply the brackets in the following problem:

$$\Psi\Psi^* = (f + ig)(f - ig)$$

29.9 Using the quantum-mechanical form for the momentum operator in one dimension:
$$p = -i\hbar\frac{d}{dx}.$$
Construct an operator for the kinetic energy, E_{KE}.

29.10 The **wavefunction** of a particle trapped in a one-dimensional box with infinitely high sides is:
$$\psi = Ae^{ikx} \pm Ae^{-ikx}$$

where the first term is the wavefunction due to a particle moving with momentum $\hbar k$ in the positive x-direction and the second term is the wavefunction when the particle is moving in the opposite direction with momentum $-\hbar k$.

Use de Moivre's theorem to express the wavefunction in terms of sine and cosine.

30

Dimensional analysis

Dimensional analysis

By the end of this section, you will know:

- Dimensional analysis represents a way of looking at the units of an equation.
- An equation is not correct until its units are also correct.
- The units on both sides of an equation must be the same, if the equation is correct.
- Therefore, analysing the units in an equation is a simple and efficient way of testing whether an equation is correct.

We often call the manipulation of units **dimensional analysis**. Strictly, the analysis of dimensions is independent of the units employed. For example, the units of speed or velocity may be metres per second, miles per hour, or even millimetres per year, depending on the context. But the dimensions of speed must nevertheless always start with a distance $[l]$ divided by a time $[t]$: speed $= [l] \div [t]$.

Dimensional analysis has two particularly useful aspects:

1. It can be used to determine the units of a variable in an equation.
2. Using the usual rules of algebra, we can use it to determine whether an equation is correct, dimensionally. This latter aspect relies on the simple truth: the units in an equation should always balance. All equations in any scientific discipline are either dimensionally correct or wrong.

■ **Worked Example 30.1**

Using the definition of concentration

$$\text{concentration} = \left(\frac{\text{amount of material}}{\text{volume of solvent}} \right)$$

determine the SI units of concentration c.

> Note: a concentration cited with units of mol m^{-3} will give different numerical values of concentration to the more usual mol dm^{-3}. For example, the concentration 1 mol dm^{-3} is the same as 1000 mol m^{-3}.

We obtain the concentration of a solution by dividing the amount of material n in solution by the volume of the solvent V. Because we obtain concentration as the fraction $n \div$ volume, we obtain the units of concentration by dividing the units of the numerator (the top row) by the units of the denominator (the bottom row).

The unit for amount of material is the mole. We have already seen that volume is obtained as l^3, so the SI unit of volume is m^3. Concentration therefore has the units of mol $\div$ m^3. Chapter 11 tells us that DIVIDING by m^3 is the same as MULTIPLYING by m^{-3}. The SI unit of concentration is 'mol m^{-3}'. ■

■ **Worked Example 30.2**

Ohm's law says that passing a current of magnitude I through a wire of resistance R induces a voltage of magnitude V according to the equation

$$V = IR$$

If the current has the unit of amps A and the resistance has the unit of Ohms, Ω, what are the units of the right-hand side of this equation?

The terms are I and R with the respective units of A and Ω. The units of the right-hand side are therefore 'A Ω'. In other words, a volt (the unit of electrical potential) is an amp ohm. ■

■ Worked Example 30.3

We obtain the conductivity κ of an ionic compound in solution as the conductance G multiplied by the cell constant $(l \div A)$:

$$\kappa = G \times \left(\frac{l}{A}\right)$$

What are the units of (a) the cell constant $(l \div A)$ and (b) the conductivity κ?

(a) The cell constant

From the definition above, the cell constant represents the ratio of separation to area. Both length and area have the unit of metre.

Inserting units into this definition yields: $\left(\dfrac{m}{m^2}\right) = \dfrac{1}{m}$

so the unit of the cell constant is $1/m$. Using the laws of indices from Chapter 11, this unit is best written as m^{-1}.

(b) The unit of conductivity

The unit of conductance G is the Siemen S. The unit of the conductivity κ is therefore equal to the units of G multiplied by the units of cell constant $(l \div A)$:
The units of $\kappa = S \times m^{-1} = S\ m^{-1}$. ■

> The unit on the far left is a Greek Kappa κ and not a Latin K, although they probably look the same in most fonts.

> Notice how the units are printed in upright type, in contrast to variables, which we generally print in *italic* type.

■ Worked Example 30.4

We define the molar conductivity Λ of an ionic compound in solution Λ as the conductivity κ divided by its concentration c.

$$\Lambda = \frac{\kappa}{c}$$

What are the units of Λ?

We insert units: units of $\Lambda = \dfrac{S\ m^{-1}}{mol\ m^{-3}}$. Cancelling and subsequent rearrangement then yields $S\ m^2\ mol^{-1}$. ■

Self-test 30.1

Derive the units for each of the following.

30.1.1 The first-order constant of reaction k_1, given that rate $= k_1$ [reactant]. The units of 'rate' are 'mol $m^{-3}\ s^{-1}$'.

30.1.2 During combustion within a car engine, the rate of burning is given by the expression rate $= k_2$ $[O_2] \times$ [hydrocarbon]. Determine the units of the second-order rate constant k_2. The rate of reaction again has the units of 'mol $m^{-3}\ s^{-1}$'.

Using units to determine dimensional correctness

By the end of this section, you will know:

- An equation must be dimensionally correct if it is itself to be considered correct.

- Dimensional analysis is a powerful way of telling if an equation is correct.

- Dimensional analysis provides a means of making an 'intelligent guess' about which parts of an equation are missing.

An equation must be dimensionally correct. If the dimensions are incorrect, then so is the equation. For example, we cannot say 1 mile + 1 inch = 2 miles. While the numbers might add up to 2, the sum is simply nonsense with the units in place.

■ **Worked Example 30.5**

Show the following version of the Nernst equation is incorrect.

$$E = E^{\ominus} + \frac{RT}{n}\ln\left(\frac{[O]}{[R]}\right)$$

Aside

We cannot take the logarithm of anything except a number. Therefore, a logarithm term will never have units because it is merely a dimensionless number.

As before, we start by inserting the units of each term. n the number of electrons in the fully balanced redox equation is a stoichiometric number and therefore dimensionless. The logarithm term is automatically dimensionless, because a logarithm is merely a number. We indicate this lack of a dimension by writing '[1]'.

$$[V] = [V] + \frac{[J\,K^{-1}\,mol^{-1}] \times [K]}{[1]} \times [1]$$

While the K^{-1} and K terms readily cancel, the remainder is clearly an irresolvable mess. We conclude that this version of the Nernst equation is incorrect, so we should not use it.

The correct version of the Nernst equation has the Faraday constant in the denominator of the fraction, as RT/nF. The units of the Faraday constant are C mol^{-1}. Inserting units into the correct Nernst equation yields:

$$[V] = [V] + \frac{[J\,K^{-1}\,mol^{-1}] \times [K]}{[1] \times [C\,mol^{-1}]} \times [1]$$

so $[V] = [V] + [J\,C^{-1}]$

and 1 J per Coulomb is one definition of a volt.

Accordingly, we obtain $[V] = [V] + [V]$. Each term in the equation has the same units, thereby demonstrating the equation is dimensionally correct.

While the equation is dimensionally correct, it is wise to note that it could still be wrong. The n term has no dimensions, so this analysis would have yielded the same result if we had forgotten the n term, for example, or used n^2. ■

■ **Worked Example 30.6**

A student in an examination suddenly cannot remember the full details of the Clausius–Clapeyron equation, but does know the basic structure:

$$\ln\left(\frac{p_2}{p_1}\right) = -\Delta H^{\ominus}_{evaporation}\left(\frac{1}{T_2} - \frac{1}{T_1}\right)$$

By looking at the dimensions alone, guess which term is missing.

As before, we start by rewriting the equation, inserting the dimensions of each term. We can dispense with the minus sign. The logarithm is automatically dimensionless:

$$[1] = [J\,mol^{-1}]\left[\frac{1}{K} - \frac{1}{K}\right]$$

Because the whole of the left-hand side is dimensionless, we are aiming for the right-hand side to be dimensionless also. We simplify the bracket by noting that each of the two terms is a reciprocal temperature. The result of the subtraction can only be a reciprocal temperature also. In fact, whatever the actual values of T_1 and T_2, dimensionally, the bracket will have the units of 1/K. So we can simplify further:

$$[1] = [\text{J mol}^{-1}]\left[\frac{1}{K}\right]$$

The right-hand side of the equation has the units of J K^{-1} mol^{-1}.

So the simplest way to correct this mistaken version of the Clausius–Clapeyron equation is to divide the right-hand side with a parameter or constant that has the same units of J K^{-1} mol^{-1}. There are three common choices in the chemistry of phase equilibria: heat capacity C, entropy ΔS and the gas constant R.

The student should remember, perhaps after a moment's thought, that R is the correct choice. No competent student would 'remember' a version of the Clausius–Clapeyron equation with C or ΔS in the denominator.

The Clausius–Clapeyron equation is therefore: $\ln\left(\dfrac{p_2}{p_1}\right) = -\dfrac{\Delta H^{\ominus}_{\text{evaporation}}}{R}\left(\dfrac{1}{T_2} - \dfrac{1}{T_1}\right).$ ∎

Self-test 30.2

Show the following equations are dimensionally correct.

30.2.1   The Kirchhoff equation: $\Delta H_2 = \Delta H_1 + C_p\,(T_2 - T_1)$

30.2.2   Blagden's Law:

$$\Delta T = K_{(\text{cryoscopic})} \times \left(\frac{\text{mass of solute}}{\text{molar mass of solute}}\right) \times \frac{1000}{\text{mass of solvent}}$$

where $K_{(\text{cryoscopic})}$ has the units of K kg mol^{-1}.

Use dimensional analysis to make an 'intelligent guess' which parts of following equations are missing.

30.2.3 $\Delta G^{\ominus} = \Delta H^{\ominus} - \Delta S^{\ominus}$

30.2.4   The Gibbs–Helmholtz equation: $\dfrac{\Delta G_2}{T_2} = \Delta G_1 + \Delta H\left(\dfrac{1}{T_2} - \dfrac{1}{T_1}\right)$

SI units: the concepts

By the end of this section, you will know:

- The SI system comprises seven base units.
- We can make any unit using a suitable assortment of these seven alone.
- We employ shorthand units when the SI units appear complicated.

It's easy to draw a line of length 1 metre. We simply need a ruler. But why does a metre have a length of exactly one metre; who says what length a 'metre' should take? The most recent definition comes from the *Système Internationale* (SI) in Paris, which defines a metre as 1 650 763.73 wavelengths of the light emitted *in vacuo* by krypton-86. This definition is sensible insofar as any well-equipped laboratory in the world can reproduce it.

The SI system was designed in such a way that all units and definitions are self-consistent. The system comprises seven so-called **base units** (or 'fundamental units'), as defined in Table 30.1. We say that metre, m, is the base unit for a length because we cannot define m in terms of anything simpler.

In the SI system, we bolt together multiples of these seven base units to assemble the units of any physicochemical parameter. For example, since an area, A, can be the product of two lengths, within the SI system, we express the units of A in terms of m^2.

Table 30.1 The seven fundamental SI physical quantities and their units.

Physical Quantity	Symbol*	SI Unit	Abbreviation
Length	l	metre	m
Mass	m	kilogram	kg
Time	t	second	s
Electrical current	I	ampère	A
Thermodynamic temperature	T	kelvin	K
Amount of material	n	mole	mol
Luminous intensity	I_V	candela	cd

*Notice the use of type here: (1) We italicize the **symbol** for each quantity because it is a variable. (2) We print the **abbreviation** for the unit with an upright typeface because it is not a variable. (3) The **SI units** all start with a lower-case letter.

■ **Worked Example 30.7**

Using SI units, what is the unit of volume, V?

The simplest volume to visualize is a cube. The volume of a cube is the product of the lengths l of its three sides. Therefore, volume has the SI unit of $l \times l \times l = l^3$.

The SI unit of volume is l^3. ■

We see the enormous power of the SI system when forming units from the seven base units. For example, we define velocity as a length travelled per unit time: m s^{-1}. We call this pairing a **compound unit**. Here, just as we form compounds by combining elements, so we now formulate compound units from the seven SI base units.

We say the compound unit for velocity (m s^{-1}) is 'irreducible' because we cannot express it more simply. There is no way by which we can express either the length or the time components in a way that is simpler.

Self-test 30.3

Derive the SI unit(s) for each of the following.

30.3.1 Molecular velocity v, where v = distance ÷ time

30.3.2 Acceleration a, where a = change in velocity ÷ time

30.3.3 Wavenumber $\bar{v}$, where $\bar{v} = 1/\lambda$

Sometimes a compound unit becomes quite complicated and requires an abbreviation.

■ **Worked Example 30.8**

Derive the compound units of energy using the Einstein equation, $E = mc^2$.
We start by inserting terms into the equation:

$$\text{Units of } E = [\text{kg}] \times [\text{m s}^{-1}] \times [\text{m s}^{-1}]$$

Combining the terms yields, $\text{kg m}^2 \text{ s}^{-2}$

This combination of units is so cumbersome that we generally use shorthand instead. The shorthand of choice is the joule, J. Indeed, we would only ever consider this collection of units if we needed to simplify another expression. ■

Table 30.2 lists several other compound units.

Table 30.2 Selection of a few physicochemical parameters that comprise combinations of the seven SI fundamental quantities.

Quantity	Symbol	SI units	non-SI unit
Acceleration	a	m s^{-2}	–
Density	ρ	kg m^{-3}	–
Energy	E	$\text{kg m}^2 \text{ s}^{-2}$	J
Force	F	kg m s^{-2}	N
Potential	ϕ or E	$\text{kg m}^2 \text{ s}^{-3} \text{ A}^{-1}$	V
Pressure	p	$\text{kg m}^{-1} \text{ s}^{-2}$	Pa

Occasionally, the derivation of a unit requires two or more steps.

■ **Worked Example 30.9**

Starting with the ideal-gas equation, what are the units of pressure? [Hint: use the expression for energy in Table 30.2.]

Rearranging the ideal-gas equation to make p the subject yields $p = (nRT)/V$.

Inserting units for each term gives: $\dfrac{[\text{mol}] \times [\text{J K}^{-1} \text{ mol}^{-1}] \times [\text{K}]}{[\text{m}^3]}$

We greatly simplify the expression by cancelling the mol and mol^{-1} terms, and the K and K^{-1} terms, to yield the surprising result J m^{-3}.

But J is not an SI unit, so we must substitute for J. To do so, we recall that joule is the SI unit of energy, so we substitute using the expression for energy in Table 30.2, saying $p = [\text{kg m}^2 \text{ s}^{-2}] \times [\text{m}^{-3}]$. Further cancelling gives $p = \text{kg m}^{-1} \text{ s}^{-2}$.

We could have obtained this expression in several different ways. For example, pressure is defined as (force ÷ area); and from Newton's second law of motion, force = (mass × acceleration). Therefore, $p = (\text{mass} \times \text{acceleration})/\text{area}$. ■

Self-test 30.4

Derive the SI units for each of the following using the appropriate equation.

30.4.1 Force F, using Newton's Second Law, $F = ma$.

30.4.2 Show that the expression, $p = (\text{mass} \times \text{acceleration})/\text{area}$, gives the same result as that in Worked Example 30.9.

30.4.3 The gas constant R, using the ideal gas equation $pV = nRT$.

Interconversion between units

By the end of this section, you will know:

- Chemists often prefer to use non-SI units, for convenience or habit.
- We can readily change between SI units and non-SI units, often using standard factors.

Probably a majority of the SI units we commonly employ are straightforward. But many generate numbers that are either routinely too large or too small. For this reason, chemists have preferred non-SI units.

For example, bond lengths are always in the range 100–200 pm, i.e. $1–3 \times 10^{-10}$ m. Many crystallographers prefer not to use picometres but use the old-fashioned unit the Ångström, where $1 \text{ Å} = 10^{-10}$ m. Bond lengths expressed in the units of Ångström require no standard factors.

As a second example, the SI unit of concentration is mol m^{-3}. The volume term in this compound unit is simply vast—the size of a typical bath. It is difficult to get laboratory glassware this size, so chemists generally express concentration in terms of the amount of material in a much smaller volume, the litre (1 litre = 1 dm^3).

■ **Worked Example 30.10**

How many litres are contained within an SI volume of one cubic metre?

We start by noting how it's common to see a litre written as 1 dm^3. Remembering the standard prefixes from page 3, we say that one decimetre is a tenth of one metre, i.e. 10 cm. This should be obvious, because 1 m = 100 cm and 'deci' means 'one tenth.' Then, to solve this problem involves a 'dodge.'

Strategy:

(i) Note the relationship between the factors. In this case, by definition, $1 \text{ dm} = \frac{1}{10}$ m; so, by simple rearrangement, 10 dm = 1 m.

(ii) We also note that $1 \text{ m}^3 = (1 \text{ m})^3$.

(iii) We then write the identity, $(1 \text{ m})^3 = (1 \text{ m})^3$.

(iv) Substitute on the right-hand side with the relationship from part (i), i.e. 1 m = 10 dm. We obtain, $(1 \text{ m})^3 = (10 \text{ dm})^3$.

> In this context, **identity** means that both sides of the equation are exactly the same.

Multiplying the bracket yields, $1 \text{ m}^3 = 10^3 \text{ dm}^3$, so a cubic metre contains exactly a thousand cubic litres. ■

Self-test 30.5

In each case, how should we manipulate the data to make the variable dimensionless?

30.5.1 If $1 \text{ °C} = \frac{5}{9} \text{ °F}$, what is 1 °F, when written in terms of °C?

30.5.2 If 1 inch = 2.5 cm, what is the relationship between a cubic inch and a cubic centimetre?

Quantity calculus: unit manipulation for graphs and tables

By the end of this section, you will know:

- We can only plot dimensionless data on a graph. The axis label must also be dimensionless.
- We can only tabulate dimensionless data. The table header must also be dimensionless.
- Data for plotting or tabulation need to be made dimensionless.

Related to dimensional analysis is **quantity calculus**. Despite its name, it does not refer to differentiation or integration. It is particularly useful when setting out the label rows in a table or labelling a graph axis. Quantity calculus tells us how to handle physical quantities and their units using the normal rules of algebra.

We define any physical quantity in terms of a numerical value and its units, as discussed in Chapter 1.

Labelling axes on a graph:

When we plot a graph, it is essential we label the axes. Otherwise, we will not know what the axes refer to. In consequence, the trends and correlations we draw from the graph may be completely bogus.

When plotting anything on a graph, we need to remember that we cannot plot energies, pressures, temperatures, or indeed any other physicochemical variable. We can only plot *numbers*. We therefore need to convert the variable into a number before we plot the data.

■ Worked Example 30.11

We need to plot a graph on which the abscissa variable is the change in Gibbs function, ΔG. How should we label the axis? A sample datum is $\Delta G = -15$ kJ mol^{-1}.

We must convert the abscissa variable into a dimensionless number before we plot anything on the *x*-axis. In this example, the abscissa variable has the units of energy (kJ mol^{-1}). The key to any conversion is to recognize how each datum is also an equation.

> Remember, datum is the singular of data.

Strategy:

(i) Take the algebraic expression for a simple datum.

(ii) Rearrange the expression to generate a dimensionless *number* by dividing by the units—in this case 'kJ mol^{-1}'.

Solution:

For the Worked Example, the sample datum is $\Delta G = -15$ kJ mol^{-1}. We divide both sides of this equation by the units:

$$\text{If} \quad \Delta G = -15 \text{ kJ mol}^{-1} \quad \text{then} \quad \Delta G / \text{kJ mol}^{-1} = -15$$

Before the rearrangement, both sides of the equation have the units of energy. After the rearrangement, both sides of the equation are dimensionless. Because −15 is a number, it can be readily plotted or tabulated, as desired.

> All we did here was to divide both sides of the equation by the units 'kJ mol^{-1}'.

To plot this datum:

- We align the point against 15 on the *x*-axis in the usual way.
- We label the *x*-axis, we write, 'ΔG/kJ mol^{-1}'.
- Look at the following examples:

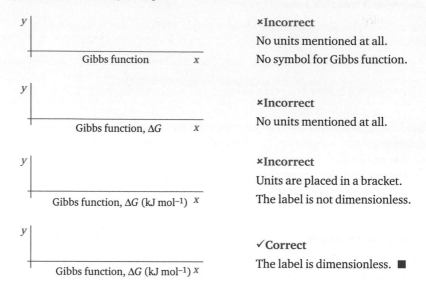

×**Incorrect**
No units mentioned at all.
No symbol for Gibbs function.

×**Incorrect**
No units mentioned at all.

×**Incorrect**
Units are placed in a bracket.
The label is not dimensionless.

✓**Correct**
The label is dimensionless. ∎

Many chemists prefer to go further. Because the kJ term in the example above is in effect '$10^3 \times J$', they decide that all factors—whatever the format—should also participate in the rearrangement.

To illustrate this point, we continue with the previous example. $\Delta G = -15$ kJ mol^{-1} can be re-expressed as $\Delta G = -15$ J mol$^{-1} \times 1000$. We can rearrange this in two stages:

1. We rearrange the factor: $\Delta G = -15$ J mol$^{-1} \times 1000$ so $10^{-3}\,\Delta G = -15$ J mol^{-1}
2. We rearrange the units: $10^{-3}\,\Delta G = -15$ J mol^{-1} so $10^{-3}\,\Delta G$/J mol$^{-1} = -15$
3. Beneath the *x*-axis, we write the label '$10^{-3}\,\Delta G$/J mol^{-1}'.

Labelling a table header:

The label at the head of a table should also be completely dimensionless and written without a factor. We formulate the header in much the same way as we made up an axis label for a graph. It's a good idea to take one datum, treat it as an equation, and then treat the result as a table label.

We do this because it would be tedious to write the factor against every column entry. We obviate the need for a factor by incorporating the inverse of the factor into the column header.

■ **Worked Example 30.12**

We need to tabulate some data. In one of the columns, the lowest value is 'Applied voltage $= 2 \times 10^4$ V' and the highest is 'Applied voltage $= 9 \times 10^4$ V'. When writing the table, how should we label the column heading?

Strategy:

(i) The data are applied voltages, so we start by simply writing the words 'Applied voltage.'

(ii) Each datum has a factor of '10^4' in common. We multiply both sides of the equation by the inverse of '$\times 10^4$', which is clearly $\times 10^{-4}$.' Therefore, if 'Applied voltage $= 2 \times 10^4$ V' then '10^{-4} Applied voltage $= 2$ V'.

(iii) The data have the unit 'V' in common. Again, we need to avoid repetition of these 'V' terms. We again incorporate the unit into the column header. If '10^{-4} Applied voltage $= 9$ V' then '10^{-4} Applied voltage/V $= 9$'.

The column label in the table is therefore the left-hand side of this equation. We say, '10^{-4} Applied voltage/V.'

In summary:

(i) We write a description of the entry: Applied voltage

(ii) We multiply by the inverse of the factor: 10^{-4} Applied voltage

(iii) We divide by the units: 10^{-4} Applied voltage/V.

Look at the following examples:

Applied voltage	
2×10^4 V	
3×10^4 V	
5×10^4 V	
9×10^4 V	

✗ Incorrect

• None of the entries is dimensionless.

• Highly repetitious.

Applied voltage/V	
2×10^4	
3×10^4	
5×10^4	
9×10^4	

✗ Incorrect

• Still highly repetitious.

• Takes much longer to write.

10^{-4} Applied voltage/V	
2	
3	
5	
9	

✓ Correct

• All entries are now dimensionless.

■

It is a very common error to read this answer, and conclude that we are to *divide* each entry by 10^{-4}. In fact, each of these entries *has already been multiplied* by a factor of 10^{-4}.

Many people employ a variant of this method. They keep together the factor and the unit. In Worked Example 30.12, the column header would then read 'Applied voltage/10^4 V'. Labelling in this way has several advantages: it is easier to extract data from the table with the correct factor (rather than its inverse). Many people also find it more intuitive.

This alternative method does, however, deviate from the stated rationale, that we divide by the units.

Self-test 30.6

In each case, how should we manipulate the data to make the variable dimensionless?

30.6.1 Entropy change, $\Delta S = 32 \text{ J K}^{-1} \text{ mol}^{-1}$

30.6.2 Energy released, $E = 15 \text{ kJ mol}^{-1}$

30.6.3 Mass, $m = 12 \text{ g}$

30.6.4 Temperature increase, $\Delta T = 1.2 \text{ K}$

30.6.5 Pressure, $p = 3.2 \times 10^5 \text{ Pa}$

30.6.6 Amount, $n = 2.67 \text{ mmol}$

Additional problems

30.1 A chemist manipulates the applied pressure in an autoclave to study a reaction rate. The pressures are $1 \times 10^5 \text{ Pa}$, $3 \times 10^5 \text{ Pa}$, $5 \times 10^5 \text{ Pa}$, $10 \times 10^5 \text{ Pa}$, and $30 \times 10^5 \text{ Pa}$. When drawing up a table of this data, what should be the column label?

30.2 We need to plot a graph. The ordinate variable is pressure in units of bars. How should the y-axis be labelled?

30.3 Starting from $\Delta G^{\ominus} = \Delta H^{\ominus} - T\Delta S^{\ominus}$, deduce the units of $\Delta S^{\ominus}$.

30.4 Until relatively recently, standard pressure was defined as one atmosphere = 101 325 Pa. Standard pressure $p^{\ominus}$ is now defined as 1 bar = 10^5 Pa. What is (i) The value of 1 atm in terms of bars, and (ii) The value of 1 bar in terms of atmospheres?

30.5 What are the units of the coulomb, C? [Hint: use the definition of current.]

30.6 A chemist needs to work out the units of molar mass.

30.7 What is the SI unit of potential? [Hint: use the Nernst equation, remembering that the units of the Faraday constant are $C \text{ mol}^{-1}$.]

30.8 The gas constant R may be calculated using the relation, $R = N_A k_B$, where N_A is the Avogadro constant = $6.022 \times 1023 \text{ mol}^{-1}$, and k_B is the Boltzmann constant. What are the units of k_B?

30.9 What are the units of pressure in the Clapeyron equation?

30.10 Determine the SI units of heat capacity C_p from its definition, $C_p = \left(\dfrac{dH}{dT} \right)_p$.

Answers to self-test questions

Chapter 1:
The display of numbers: Standard factors, scientific notation, significant figures and decimal places

Self-test 1.1

	Variable	number	factor	unit(s)
1.1.1	mass	2.65	k (10^3)	g
1.1.2	frequency	94.5	M (10^6)	Hz
1.1.3	wavelength	500	n (10^{-9})	m
1.1.4	current	0.3	m (10^{-3})	A
1.1.5	mass of beaker	340	—	g
1.1.6	mass of large car	1.4	M (10^6)	g
1.1.7	energy liberated	34	M (10^6)	$J\,mol^{-1}$
1.1.8	length of bond	130	p (10^{-12})	m
1.1.9	potential difference	550	m (10^{-3})	V
1.1.10	speed	34	—	$m\,s^{-1}$

Self-test 1.2

1.2.1	energy = 12.3 kJ mol^{-1}	1.2.2	frequency = 500 MHz or 0.5 GHz
1.2.3	length of road = 3.4 km	1.2.4	voltage = 30 kV
1.2.5	mass of truck = 36 Mg (which, incidentally, is the same as 36 tonnes)	1.2.6	amount of material = 1.2 Mmol
1.2.7	energy evolved = 2. 034 MJ mol^{-1}	1.2.8	wavelength λ = 0.98 µm or 980 nm
1.2.9	bond length = 156 pm	1.2.10	diameter of a fly's eye = 10 µm

Self-test 1.3

1.3.1	energy = 1.23×10^4 J mol^{-1}	1.3.2	Frequency = 5.0×10^8 Hz
1.3.3	length of road = 3.4×10^3 m	1.3.4	voltage = 3.0×10^4 V
1.3.5	mass of truck = 3.6×10^7 g	1.3.6	cost = £1.2×10^6
1.3.7	energetic content = 2. 034×10^6 J	1.3.8	wavelength λ = 9.8×10^{-7} m
1.3.9	bond length = 1.56×10^{-10} m	1.3.10	the diameter of a fly's eye = 1.0×10^{-5} m
1.3.11	$F = 9.65 \times 10^5$ C mol^{-1}	1.3.12	$q = 1.6 \times 10^{-19}$ C

Self-test 1.4

Before we start, we need to convert some of the data into decimals:

- $178 \text{ day} = \dfrac{178}{365.25} \text{year} = 0.487\ 337\ 440 \text{ year}$

- $5 \text{ hours} = \dfrac{5}{(24 \times 365.25)} \text{year} = 0.000\ 570\ 385\ 581 \text{ year}$

- $15\text{minutes} = \dfrac{15}{(60 \times 24 \times 365.25)}\,\text{year} = 0.000\,028\,519\,279\,\text{year}$

So, to a ludicrous number of s.f., the student's age is 18.487 936 340 years.

1.4.1	1 s.f.: age = 2×10, i.e. 20 years	1.4.2	2 s.f.: 18 years
1.4.3	3 s.f.: 18.5 years	1.4.4	4 s.f.: 18.49 years
1.4.5	5 s.f.: 18.488 years		

Self-test 1.5

1.5.1	energy change = -135 kJ	1.5.2	$emf = 1.432$ V
1.5.3	volume = $2.0\ \text{m}^3$	1.5.4	amount of material = 3.22 mol
1.5.5	mass = 1 M g or 1 000 000 g	1.5.6	1 metre = 1 650 764 wavelengths

Self-test 1.6

1.6.1 $V = 120 \times 151 \times 146.5 = 2\,654\,580\ \text{pm}^3$. Because a is expressed to 2 s.f., $V = 2.7 \times 10^6\ \text{pm}^3$.

1.6.2 $n = 0.250 \times 0.05\ \text{mol} = 0.0125\ \text{mol}$. Because the concentration was expressed to 1 s.f., $n = 0.01\ \text{mol}$.

1.6.3 rate $= 9.3 \times 10^{-2} \times 0.3 = 0.0279\ \text{mol dm}^{-3}\,\text{s}^{-1}$. Because the concentration was expressed to 1 s.f., rate $= 3 \times 10^{-2}\ \text{mol dm}^{-3}\,\text{s}^{-1}$.

1.6.4 $c = 3.2 \times 10^{-4} \div 0.5 = 6.4 \times 10^{-4}\ \text{mol dm}^{-3}$. Because the volume was expressed to 1 s.f., $c = 6 \times 10^{-4}\ \text{mol dm}^{-3}$.

1.6.5 $n = 4 \div 422 = 9.4787 \times 10^{-3}\ \text{mol}$. Because the mass was expressed to 1 s.f., $n = 9 \times 10^{-3}\ \text{mol} = 9\ \text{mmol}$.

1.6.6 $Q = 87.3 \times 10^{-3} \div 0.32 = 0.272\,812\,5\ \text{C cm}^2$. Because the area was expressed to expressed to 2 s.f., $Q = 0.27\ \text{C cm}^2$.

1.6.7 $p = \left(\dfrac{0.13 \times 8.314 \times 298}{14.2}\right)\text{Pa} = 22.681\,997\ \text{Pa}$. Because the amount of gas n was expressed to expressed to 2 s.f., $p = 23\ \text{Pa}$.

1.6.8 $\Delta G^{\ominus} = -8.314 \times 298 \times 4.0 = -9910.288\ \text{J mol}^{-1}$. Because K was expressed to 2 s.f., $\Delta G^{\ominus} = -9.9\ \text{kJ mol}^{-1}$.

Self-test 1.7

1.7.1 Rewriting: mass $= (12.0 + 1001 - 130.62)\ \text{g} = 882.38\ \text{g}$. To 0 d.p., this mass is 882 g.

1.7.2 Charge = 94 836.43 C. This is 94 836 C to 0 d.p.

1.7.3 Amount of material = 9.8879 mol. This is 9.9 mol to 1 d.p.

1.7.4 Rewriting: time $= 60.4\ \text{s} + 0.000\,012\ \text{s} + 0.033\,96\ \text{s} + 4.0\ \text{s} = 64.433\,972\,2\ \text{s}$. To 1 d.p., this answer is 64.4 s.

Chapter 2:
Algebra I: Introducing notation, nomenclature, symbols and operators

Self-test 2.1

2.1.1	$\Delta X = 10 - 22 = -12$	2.1.2	$\Delta X = 33 - 5.2 = 27.8$
2.1.3	$\Delta X = 9.37 - 9.34 = 0.03$	2.1.4	$\Delta G = 8 - 3.6\ \text{kJ mol}^{-1} = 4.4\ \text{kJ mol}^{-1}$
2.1.5	$\Delta(Abs) = 0.6 - 1.1 = -0.5$	2.1.6	$\Delta(emf) = 1.50 - 1.45\ \text{V} = 0.05\ \text{V}$

Self-test 2.2

2.2.1 $\sum_{i=1}^{4} N_i M_i = N_K M_K + N_{Fe} M_{Fe} + N_C M_C + N_N M_N + N_H M_H + N_O M_O = (4 \times 39) + (1 \times 56) + (6 \times 12) + (6 \times 14) + (6 \times 1) + (3 \times 16) = 422$ g mol^{-1}

2.2.2 $\sum_{i=1}^{3} N_i M_i = N_{Cu} M_{Cu} + N_S M_S + N_O M_O = (1 \times 64) + (1 \times 32) + (4 \times 16) = 160$ g mol^{-1}

2.2.3 $\sum_{i=1}^{2} N_i M_i = N_H M_H + N_O M_O = (2 \times 1) + (2 \times 16) = 34$ g mol^{-1}

2.2.4 $\sum_{i=1}^{3} N_i M_i = N_C M_C + N_H M_H + N_O M_O = (7 \times 2) + (6 \times 1) + (2 \times 16) = 122$ g mol^{-1}

2.2.5 $\sum_{i=1}^{3} N_i M_i = N_C M_C + N_N M_N + N_H M_H = (5 \times 12) + (1 \times 14) + (5 \times 1) = 79$ g mol^{-1}

2.2.6 $\sum_{i=1}^{3} N_i M_i = N_C M_C + N_O M_O + N_H M_H = (9 \times 12) + (4 \times 16) + (8 \times 1) = 180$ g mol^{-1}

Self-test 2.3

2.3.1 Scheme 2.1: Compound (**VIII**) C_3H_6O has a molar mass of 58 g mol^{-1} and compound (**IX**) $C_2H_6O_2$ has a molar mass of 62 g mol^{-1}. Water H_2O has a molar mass of 18 g mol^{-1}. We obtain the mass of product via the expression:

$$\sum_{i=1}^{2} v_i \left(\sum_{j=1}^{m} N_{ij} M_j \right)$$

where v_i is the stoichiometric number, N_j is the amount of moles of species j and N_j is the molar mass of the species j.

Mass of product = $(0.35 \times 58) + (0.35 \times 62) - (0.35 \times 18)$ g

Mass of product = $(20.3 + 21.7 - 6.3) = 35.7$ g

> Note the way we use i in one sum and j in the second. This practice avoids ambiguity.

2.3.2 We start by writing $\sum_{i=1}^{5} \left(Cr_i \times S_i \right)$ where Cr_i is the number of credits per unit, and S_i is the student's percentage mark per unit. Inserting numbers yields:

> Concerning the use of percentages, see Chapter 5.

$$\text{Total number of marks} = \left(10 \times \frac{45}{100}\right) + \left(20 \times \frac{56}{100}\right) + \left(20 \times \frac{70}{100}\right) + \left(20 \times \frac{62}{100}\right) + \left(30 \times \frac{40}{100}\right)$$

$$\qquad\qquad\qquad \text{Maths} \qquad \text{Inorganic} \qquad \text{Organic} \qquad \text{Physical} \qquad \text{Laboratory}$$

So $\dfrac{450 + 1120 + 1400 + 1240 + 1200}{100} = \dfrac{5410}{100} = 54.1\%$

Self-test 2.4

2.4.1 $1 \times 2 \times 3 \times 4 \times 5 = 120$

2.4.2 $\dfrac{1}{2^2} \times \dfrac{1}{3^2} \times \dfrac{1}{4^2} = \dfrac{1}{4} \times \dfrac{1}{9} \times \dfrac{1}{16} = \dfrac{1}{(4 \times 9 \times 16)} = \dfrac{1}{576} = 0.001736$

2.4.3 $5^3 \times 6^3 \times 7^3 = 125 \times 216 \times 343 = 9{,}261{,}000$

2.4.4 $\sqrt{1} \times \sqrt{2} \times \sqrt{3} = (1) \times (1.414) \times (1.732) = 2.45$

> Note the use of brackets in the term, '$(4 \times 9 \times 16)$'. These brackets illustrate the need for care when inputting data into a pocket calculator.

2.4.5 $\prod_{i=5}^{10} i$ 2.4.6 $\prod_{i=9}^{12} \dfrac{1}{i}$

2.4.7 $\prod_{i=3}^{6} i^2$ 2.4.8 $\prod_{i=1}^{5} i^2$

2.4.9 $\prod_{i=1}^{6} i$ 2.4.10 $\prod_{i=1}^{14} i$

Chapter 3:
Algebra II: The correct order to perform a series of operations: BODMAS

Self-test 3.1

3.1.1 $x = 36$ (the order does not matter here, so the calculation is *associative*)

3.1.2 $x = (6 \times 7) - (8 \times 2) = 42 - 16 = 26$

3.1.3 $x = (2 + 3) \times 5 = (5) \times 5 = 25$ 3.1.4 $x = 4 + 25 - 81 = -52$

3.1.5 $x = \dfrac{24}{14} = 1.714$ 3.1.6 $x = \dfrac{(54) - 2}{3 + (10)} = \dfrac{52}{13} = 4$

3.1.7 $x = 5 - \sqrt{4 - (12) + 44} = 5 \times \sqrt{36} = 5 \times 6 = 30$

3.1.8 $x = \left(\dfrac{12}{6}\right)^2 - 56 = (2)^2 - 56 = 4 - 56 = -52$

Chapter 4:
Algebra III: Simplification and elementary rearrangements

Self-test 4.1

4.1.1	$7a$	4.1.2	$33b$
4.1.3	$23c$	4.1.4	$12g + 3h + 7i$
4.1.5	$16g$	4.1.6	$14f + 8e - 7d$
4.1.7	$10f - h$ (which is the same as '$-1 \times h + 10f$')	4.1.8	$4C + 10H + S + O$ (which is the same as the empirical formula, $C_4H_{10}OS$)

Self-test 4.2

4.2.1   9 has been ADDED to the left-hand side, so we so we perform the inverse operation and SUBTRACT 9 from both sides: $x = 12 - 9 = 3$.

4.2.2   We first collect terms: $x - 4v = 4v$. Next, since $4v$ has been SUBTRACTED from the left-hand side, we so we perform the inverse operation and ADD $4v$ to both sides: $x = 4v + 4v$. We then tidy up by collecting terms: $x = 8v$.

4.2.3   Because 12 has been SUBTRACTED from x, we perform the inverse operation, and ADD 12 to both sides: so $a + 12 = x$.

4.2.4 $4c$ has been ADDED to x, so we perform the inverse operation, and SUBTRACT $4c$ from both sides: so $a - 4c = x$.

4.2.5 $m_{(two)}$ has been SUBTRACTED from $m_{(one)}$, so we perform the inverse operation and ADD $m_{(two)}$ to both sides: so, $4p + m_{(two)} = m_{(one)}$

4.2.6   We first group terms: $p = x + 5b$. $5b$ has been ADDED to x, so we SUBTRACT $5b$ from both sides: so, $p - 5b = x$.

Self-test 4.3

4.2.1 $18 \div 6 = 3$, so $y = 3c$

4.2.2 $2 \div 4 = \dfrac{1}{2}$, so $\;y = \dfrac{d}{2}\;$ (or $\;y = \dfrac{1}{2}d$)

4.2.3 $d \div d = 1$, so $y = \frac{1}{4}$

4.2.4 c and d both appear on top and bottom, so $\;y = \dfrac{b}{e} \times \dfrac{c}{c} \times \dfrac{d}{d}$, yielding, $\;y = \dfrac{b}{e}$

4.2.5   Rewriting slightly: $y = \dfrac{6}{3d} \times \dfrac{4}{20} = \dfrac{2}{d} \times \dfrac{1}{5}$, so $= \dfrac{2}{5d}$ or $\dfrac{2}{5}d$

4.2.6 $y = \dfrac{6a \times a \times b}{6a}$. The $6a$ terms on top and bottom cancel, leaving $y = ab$

Self-test 4.4

4.4.1 $\quad x = \dfrac{y}{m}$

4.4.2 $\quad \dfrac{24x}{24} = \dfrac{5y}{24}$ so $x = \dfrac{5y}{24}$

4.4.3 $\quad x = \dfrac{3z}{y^2}$

4.4.4 $\quad x = \dfrac{12}{ab}$

4.4.5 $\quad x = \dfrac{mgh}{55}$

4.4.6 $\quad x = \dfrac{55h}{mg}$

4.4.7 $\quad 11x = g,$ so $x = \dfrac{g}{11}$

4.4.8 $\quad 8x = 34,$ so $x = \dfrac{34}{8},$ i.e. $x = \dfrac{17}{4}$

4.4.9 $\quad 6f = 13x,$ so $x = \dfrac{6f}{13}$

4.4.10 $\quad p - 2q = 6x,$ so $x = \dfrac{p-2q}{6}$

Self-test 4.5

4.5.1 $\quad y = x^2 \rightarrow x = \sqrt{y}$

4.5.2 $\quad 4y = x^3 \rightarrow x = \sqrt[3]{4y} = (4y)^{\frac{1}{3}}$

4.5.3 $\quad by = x^4 \rightarrow \sqrt[4]{by} = (by)^{\frac{1}{4}}$

4.5.4 $\quad y = \sqrt[3]{x} \rightarrow x = y^3$

4.5.5 $\quad ay = \sqrt{x} \rightarrow x = (ay)^2 = a^2 y^2$

4.5.6 $\quad 4y = \sqrt[4]{x} \rightarrow x = (4y)^4 = 256 y^4$

Chapter 5:
Algebra IV: Fractions and percentages

Self-test 5.1

5.1.1 $\quad \dfrac{3}{9} = \dfrac{3 \times 1}{3 \times 3} = 1 \times \dfrac{1}{3} = \dfrac{1}{3}$

5.1.2 $\quad \dfrac{3}{81} = \dfrac{3 \times 1}{3 \times 27} = 1 \times \dfrac{1}{27} = \dfrac{1}{27}$

5.1.3 $\quad$ We notice how the numbers on top and bottom end in 5 and 0, so we can cancel by

the factor of '5': $\dfrac{25}{200} = \dfrac{5 \times 5}{5 \times 40} = \dfrac{5}{40}$.

This new fraction can also be cancelled for the same reason: $\dfrac{5}{40} = \dfrac{5 \times 1}{5 \times 8} = \dfrac{1}{8}$

5.1.4 $\quad$ We notice how the numbers on top and bottom end in even numbers, so we can
successively cancel by dividing top and bottom by a factor of '2':

$$\dfrac{252}{48} = \dfrac{2 \times 126}{2 \times 24} = \dfrac{126}{24} \rightarrow \dfrac{126}{24} = \dfrac{2 \times 63}{2 \times 12} = \dfrac{63}{12} = \dfrac{3 \times 21}{3 \times 4} = \dfrac{21}{4} = 5\tfrac{1}{4}$$

(If the digits in a number add up to a number divisible by '3', then the number itself
is divisible by '3'.)

> Notice the way we often cancel progressively, as we see a pattern emerge.

5.1.5 $\quad$ Many people prefer to cancel the algebraic terms separately from the numbers:

letters first $\dfrac{12abc}{3acd} = \dfrac{12b}{3d}$; then numbers: $= \dfrac{12b}{3d} = \dfrac{3 \times 4b}{3 \times 1d} = 1 \times \dfrac{4b}{d} = \dfrac{4b}{d}$

5.1.6 $\quad$ We cancel the a^2 term by recognizing that $a^2 = a \times a$: $\dfrac{6a}{3a^2} = \dfrac{6a}{3a \times a}$.

Again, we cancel letters first: $\dfrac{6a}{3a \times a} = \dfrac{6}{3a}$, then cancel the numbers: $\dfrac{6}{3a} = \dfrac{3 \times 2}{3a} = \dfrac{2}{a}$.

> We discuss the cancelling of polynomials in Chapter 11.

Self-test 5.2

5.2.1   The fraction is proper but vulgar: $\dfrac{3}{9} = 0.33\dot{3}$.

5.2.2   The fraction is proper: $\dfrac{6}{24} = \dfrac{1}{4} = 0.25$.

5.2.3 $\dfrac{5}{21} = 0.238$: proper but vulgar, because the decimal is irrational.

5.2.4 $\dfrac{6}{13} = 0.462$: proper but vulgar.

5.2.5   The fraction $\dfrac{125}{24}$ is improper because the numerator is larger than the denominator. Indeed, the fraction could be rewritten as $5\dfrac{5}{24}$. Its value as a decimal is $5.208\,\dot{3}$, so the fraction is also vulgar.

5.2.6   The fraction $\dfrac{16}{5}$ is also improper: it could be rewritten as $3\dfrac{1}{5}$. Its decimal is 3.2, so the fraction is not vulgar.

Self-test 5.3

5.3.1 $\dfrac{1}{2} \times \dfrac{1}{3} = \dfrac{1 \times 1}{2 \times 3} = \dfrac{1}{6}$

5.3.2 $\dfrac{3}{22} \times \dfrac{14}{3} = \dfrac{3 \times 14}{22 \times 3} = \dfrac{42}{66}$. Cancelling yields, $\dfrac{7}{11}$

5.3.3 $\dfrac{1}{2} \times \dfrac{1}{3} \times \dfrac{1}{4} \times \dfrac{1}{5} = \dfrac{1 \times 1 \times 1 \times 1}{2 \times 3 \times 4 \times 5} = \dfrac{1}{120}$

5.3.4 $\dfrac{1}{2} \times \dfrac{11}{10} \times \dfrac{2}{7} \times \dfrac{3}{31} = \dfrac{1 \times 11 \times 2 \times 3}{2 \times 10 \times 7 \times 31} = \dfrac{66}{4340}$. Cancelling yields $\dfrac{33}{2170}$.

5.3.5 $\dfrac{1}{a} \times \dfrac{1}{3b} \times \dfrac{1}{2c} = \dfrac{1 \times 1 \times 1}{2a \times 3b \times 2c} = \dfrac{1}{6abc}$

5.3.6 $\dfrac{1}{b} \times \dfrac{c}{12a} \times \dfrac{2d}{a} = \dfrac{1 \times c \times 2d}{12a^2b} = \dfrac{2cd}{12a^2b}$. Cancelling yields $\dfrac{cd}{6a^2b}$.

Self-test 5.4

5.4.1 $\dfrac{1}{2} + \dfrac{1}{2} = \dfrac{1+1}{2} = \dfrac{2}{2} = 1$

5.4.2 $\dfrac{1}{22} + \dfrac{4}{3} = \dfrac{3 + 4 \times 22}{22 \times 3} = \dfrac{3 + 88}{66} = \dfrac{91}{66}$. This result can be simplified to $1\dfrac{25}{66}$.

5.4.3 $\dfrac{2}{5} + \dfrac{1}{15} = \dfrac{(2 \times 15) + (1 \times 5)}{5 \times 15} = \dfrac{30 + 5}{75} = \dfrac{35}{75}$. This last result can be cancelled to form $\dfrac{7}{15}$. An alternative approach would recognize that $\dfrac{2}{5}$ is the same as $\dfrac{6}{15}$. We then say, $\dfrac{6}{15} + \dfrac{1}{15} = \dfrac{6+1}{15} = \dfrac{7}{15}$

5.4.4 $\dfrac{1}{a} + \dfrac{1}{b} = \dfrac{b+a}{ab}$

5.4.5 $\dfrac{1}{2} + \dfrac{1}{2} + \dfrac{1}{2} = \dfrac{1+1+1}{2} = \dfrac{3}{2}$. This result can be simplified to $1\dfrac{1}{2}$.

5.4.6 $\dfrac{3}{22} + \dfrac{4}{3} + \dfrac{1}{5} = \dfrac{3 \times (3 \times 5) + 4 \times (22 \times 5) + 1 \times (22 \times 3)}{22 \times 3 \times 5} = \dfrac{45 + 440 + 66}{330} = \dfrac{551}{330}$

Self-test 5.5

5.5.1 $\dfrac{1}{2} - \dfrac{1}{6} = \dfrac{6-2}{2 \times 6} = \dfrac{4}{12} = \dfrac{1}{3}$. Alternatively, note how $\dfrac{1}{2} = \dfrac{3}{6}$. We then redo the calculation, saying: $\dfrac{3}{6} - \dfrac{1}{6} = \dfrac{3-1}{6} = \dfrac{2}{6}$, which cancels to $\dfrac{1}{3}$.

5.5.2 $\dfrac{2}{3} - \dfrac{1}{6} = \dfrac{(2 \times 6) - (1 \times 3)}{3 \times 6} = \dfrac{9}{18} = \dfrac{1}{2}$. It might be easier to recognize how $\dfrac{2}{3} = \dfrac{4}{6}$, then

       say: $\dfrac{4}{6} - \dfrac{1}{6} = \dfrac{4-1}{6} = \dfrac{3}{6} = \dfrac{1}{2}$.

5.5.3 $\dfrac{11}{12} - \dfrac{3}{8} = \dfrac{(11 \times 8) - (3 \times 12)}{12 \times 8} = \dfrac{88 - 36}{96} = \dfrac{52}{96} = \dfrac{13}{24}$

5.5.4 $\dfrac{4}{13} - \dfrac{1}{22} = \dfrac{(4 \times 22) - (1 \times 13)}{13 \times 22} = \dfrac{88 - 13}{286} = \dfrac{75}{286}$

5.5.5 $\dfrac{1}{2a} - \dfrac{b}{3a} = \dfrac{3a - 2ab}{3a \times 2a} = \dfrac{3a - 2ab}{6a^2}$. We can then factorize and cancel (see Chapters 7

       and 4, respectively), which yields: $\dfrac{3 - 2b}{6a}$.

       We could have done this calculation slightly differently, by first noting the similarity

       between the denominators, and saying: $\dfrac{1}{2a} - \dfrac{b}{3a} = \dfrac{3 - 2b}{(3 \times 2)a} = \dfrac{3 - 2b}{6a}$, which yields

       the same answer.

5.5.6   We first multiply the right-hand fraction by '1' – in this case, by $\dfrac{a}{a}$:

$$\dfrac{2}{a^2} - \dfrac{3}{a} = \dfrac{2}{a^2} - \dfrac{3 \times a}{a \times a} = \dfrac{2}{a^2} - \dfrac{3a}{a^2}$$

       We then place over a common denominator: $\dfrac{2}{a^2} - \dfrac{3a}{a^2} = \dfrac{2 - 3a}{a^2}$.

Self-test 5.6

5.6.1 $\dfrac{1}{2} \div \dfrac{1}{6} = \dfrac{1}{2} \times \dfrac{6}{1} = \dfrac{6}{2} = 3$

5.6.2 $\dfrac{2}{5} \div \dfrac{2}{7} = \dfrac{2}{5} \times \dfrac{7}{2} = \dfrac{7}{5}$

5.6.3 $\dfrac{3}{7} \div \dfrac{1}{12} = \dfrac{3}{7} \times \dfrac{12}{1} = \dfrac{36}{7}$. This result can be simplified to $5\dfrac{1}{7}$.

5.6.4 $\dfrac{1}{7} \div \dfrac{5}{12} = \dfrac{1}{7} \times \dfrac{12}{5} = \dfrac{12}{7 \times 5} = \dfrac{12}{35}$

5.6.5 $\dfrac{1}{a} \div \dfrac{1}{ab} = \dfrac{1}{a} \times \dfrac{ab}{1} = \dfrac{ab}{a} = b$

5.6.6 $\dfrac{b^2 c}{a^2} \div \dfrac{bc^2}{a} = \dfrac{b^2 c}{a^2} \times \dfrac{a}{bc^2} = \dfrac{ab^2 c}{a^2 bc^2} = \dfrac{b}{ac}$

Self-test 5.7

5.7.1   Using eqn (5.8), percentage $= \dfrac{0.35 \text{ mol}}{0.6 \text{ mol}} \times 100 = 58\%$.

5.7.2   Using eqn (5.8), percentage $= \dfrac{\pounds 14.00}{\pounds 12.60} = 1.11$, so the percentage is 111%.

       The increase is therefore 11%.

5.7.3   Using eqn (5.8), percentage $= \dfrac{0.23 \text{ mol dm}^{-3}}{0.60 \text{ mol dm}^{-3}} \times 100 = 38\%$ of the original

       concentration. It has been diluted by $(100 - 38)\% = 62\%$.

5.7.4   Using eqn (5.8), percentage $= \dfrac{8.54 \times 10^3 (\text{mol dm}^{-3})^{-1} \text{ s}^{-1}}{3.1 \times 10^4 (\text{mol dm}^{-3})^{-1} \text{ s}^{-1}} \times 100 = 27.5\%$.

       The rate has decreased to 27.5% of its former value, so it has decreased by

       $(100 - 27.5) = 72.5\%$.

5.7.5   Using eqn (5.10), $72\% = 72 \times \dfrac{0.75 \text{ mol dm}^{-3}}{100} = 0.54 \text{ mol dm}^{-3}$.

Note how the rate of reaction is very sensitive to temperature, which explains why kinetics experiments should always be performed in a constant-temperature ('thermostatted') bath.

5.7.6   Using eqn (5.10), $0.3\% = 0.3 \times \dfrac{50g}{100} = 0.15g.$

5.7.7   Using eqn (5.10), $17.5\% = 17.5 \times \dfrac{£12\,000}{100} = £2100.$

5.7.8   Using eqn (5.10), $43\% = 43 \times \dfrac{12\,\text{mol}}{100} = 5.2\,\text{mol}.$

Chapter 6:
Algebra V: Rearranging equations according to the rules of algebra

In each case, the right-pointing arrow merely points to the result of the next stage of rearrangement.

Self-test 6.1

6.1.1   Combining terms: $y = 8x + 1$. SUBTRACTING the '1': $y - 1 = 8x$. DIVIDING by 8 yields, $x = \dfrac{y-1}{8}$

6.1.2 $y = 3x - 7 \;\rightarrow\; y + 7 = 3x \;\rightarrow\; x = \dfrac{y+7}{3}$

6.1.3 $y = 5 - 4x \;\rightarrow\; y - 5 = -4x \;\rightarrow\; x = \dfrac{y-5}{-4} = \dfrac{5-y}{4}$

6.1.4 $y = \dfrac{x4}{7} \;\rightarrow\; 7y = x - 4 \;\rightarrow\; x = 7y + 4$

6.1.5 $y = 8\,(x - 1) \;\rightarrow\; \dfrac{y}{8} = x - 1 \;\rightarrow\; x = \dfrac{y}{8} + 1$

6.1.6 $y = 8x\,(b - 1) \;\rightarrow\; \dfrac{y}{(b-1)} = 8x \;\rightarrow\; x = \dfrac{y}{8(b-1)}$

In fact, we could have performed these two steps at the *same* time

6.1.7 $y = \dfrac{d}{x} \;\rightarrow\; xy = d \;\rightarrow\; x = \dfrac{d}{y}$

6.1.8 $y = \dfrac{3}{x2} \;\rightarrow\; y\,(x - 2) = 3 \;\rightarrow\; x - 2 = \dfrac{3}{y} \;\rightarrow\; x = \dfrac{3}{y} + 2$

6.1.9 $y = \dfrac{ax}{4d} \;\rightarrow\; 4dy = a - x \;\rightarrow\; 4dy - a = -x \;\rightarrow\; x = a - 4dy$

The last step is represents **MULTIPLYING** both sides of the equation by '−1'

6.1.10 $y = \dfrac{x-9}{c+d} \;\rightarrow\; y(c + d) = x - 9 \;\rightarrow\; x = y(c + d) + 9$

6.1.11 $C = \dfrac{5}{9} \times (F - 32) \;\rightarrow\; \dfrac{9}{5} C = F - 32 \;\rightarrow\; F = \dfrac{9}{5} C + 32$

Self-test 6.2

6.2.1 $y = x^2 \;\rightarrow\; x = \sqrt{y}$

6.2.2 $y = -4x^2 \;\rightarrow\; -\dfrac{y}{4} = x^2 \;\rightarrow\; x = \sqrt{-\dfrac{y}{4}} = \dfrac{\sqrt{-y}}{2}$

6.2.3 $y = x^2 + 7 \;\rightarrow\; y - 7 = x^2 \;\rightarrow\; x = \sqrt{y-7}$

6.2.4 $y = 4\,(c - x^2) \;\rightarrow\; \dfrac{y}{4} - c - x^2 \;\rightarrow\; \dfrac{y}{4} + x^2 = c \;\rightarrow\; x^2 = c - \dfrac{y}{4}$

$\rightarrow\; x = \sqrt{c - \dfrac{y}{4}}$

6.2.5 $y = c(x^2 + 1)$ $\rightarrow$ $\dfrac{y}{c} = x^2 + 1$ $\rightarrow$ $\dfrac{y}{c} - 1 = x^2$ $\rightarrow$ $x = \sqrt{\dfrac{y}{c} - 1}$

6.2.6 $y = (x - a)^2$ $\rightarrow$ $\sqrt{y} = x - a$ $\rightarrow$ $x = \sqrt{y} + a$

6.2.7 $y = \sqrt{x - 9}$ $\rightarrow$ $y^2 = x - 9$ $\rightarrow$ $x = y^2 + 9$

6.2.8 $y = 5 \times \sqrt{x - v}$ $\rightarrow$ $\dfrac{y}{5} = \sqrt{x - v}$ $\rightarrow$ $\left(\dfrac{y}{5}\right)^2 = x - v$ $\rightarrow$ $x = \left(\dfrac{y}{5}\right)^2 + v$

     If we want to write this result without brackets, we say $\dfrac{y^2}{25} + v$.

6.2.9 $y = \sqrt{a - x}$ $\rightarrow$ $y^2 = a - x$ $\rightarrow$ $x + y^2 = a$ $\rightarrow$ $x = a - y^2$

6.2.10 $y = \left(\dfrac{x + 1}{5}\right)^2$ $\rightarrow$ $\sqrt{y} = \dfrac{x + 1}{5}$ $\rightarrow$ $55\sqrt{y} = x + 1$ $\rightarrow$ $x = 5\sqrt{y} - 1$

6.2.11 $\delta = \dfrac{1.61\, v^{\frac{1}{6}} D^{\frac{1}{3}}}{\sqrt{\omega}}$ $\rightarrow$ $\delta\sqrt{\omega} = 1.61 v^{\frac{1}{6}} D^{\frac{1}{3}}$ $\rightarrow$ $\sqrt{\omega} = \dfrac{1.61\, v^{\frac{1}{6}} D^{\frac{1}{3}}}{\delta} \rightarrow$

     $\omega = \left(\dfrac{1.61\, v^{\frac{1}{6}} D^{\frac{1}{3}}}{\delta}\right)^2 = \dfrac{2.59\, v^{\frac{1}{3}} D^{\frac{2}{3}}}{\delta^2}$

Chapter 7:
Algebra V: Simplifying equations: brackets and factorizing

Self-test 7.1

7.1.1 $pa + pb$

7.1.2 $bpx - bpy$

7.1.3 $ac + 2a + c + 2$

7.1.4 $a^2 \times a = a^3$, so $a^3 - ab + a^2 b^2 - b^3$

7.1.5   It's easier if we perform the **MULTIPLICATION** process in two separate stages:
       $g\,[HJ + 3H + 2J + 6]$, then:
       $gHJ + 3gH + 2gJ + 6g$

7.1.6 $J^2 - 4$

7.1.7 $\dfrac{3}{2} - \dfrac{a}{2c} + 12v - \dfrac{4av}{c}$

7.1.8 $\dfrac{e}{c} + \dfrac{5}{c} - \dfrac{e}{d} - \dfrac{5}{d}$ This answer can itself be factorized: $\dfrac{e + 5}{c} - \dfrac{e + 5}{d} = (e + 5)$
       $\left(\dfrac{1}{c} - \dfrac{1}{d}\right) = (e + 5)\left(\dfrac{d - c}{cd}\right)$

Self-test 7.2

Remember that '$-$' $\times$ '$-$' $=$ '$+$'.

7.2.1 $a^2 + 2ab + b^2$

7.2.2 $x^2 - 2xy + y^2$

7.2.3 $x^2 - 2cx + c^2$

7.2.4 $x^2 + 2xy + y^2 = (-1)^2 (x + y)^2 = (x + y)^2$

7.2.5 $4x^2 - 4xy + y^2$

7.2.6 $a^2 x^2 - 2abxy + b^2 y^2$

7.2.7 $25 - 10y^2 + y^4$

7.2.8 $1 - 8y + 16y^2$

Self-test 7.3

7.3.1 $a\,(1 + b)$

7.3.2 $2\,(x - y)$

7.3.3 $a\,(b + 2c + d)$

7.3.4 $2a\,(2 + c + 3b)$

7.3.5 $a\,(a + b)$

7.3.6 $a\,(a + 4)$

7.3.7 $a\,(a+3+6b)$

7.3.8 $x\,(y-x+2)$

7.3.9 $x\,(x^2+6x+4)$

7.3.10   (i)   We multiply out the brackets: $x^2+8x+16+4x^2-12x+9-25-x$

        (ii)   We collect together the like terms: $5x^2-5x$

        (iii)   We factorize, saying $5x\,(x-1)$

Self-test 7.4

7.4.1 $(x-b)\,(x+b)$ 7.4.2 $\left(c-\sqrt{p}\right)\left(c+\sqrt{p}\right)$

7.4.3 $(\alpha-\beta)\,(\alpha+\beta)$ 7.4.4 $\left(\omega-4\sqrt{\pi}\right)\left(\omega+4\sqrt{\pi}\right)$

7.4.5 $(x-2)\,(x+2)$ 7.4.6 $(a-9)\,(a+9)$

7.4.7 $(d-100)\,(d+100)$ 7.4.8 $\left(a-\sqrt{5.32}\right)\left(a+\sqrt{5.32}\right)$

Writing this last answer as $(a-2.31)\,(a+2.31)$ is correct, but less precise.

Self-test 7.5

7.5.1 $(x+3)\,(x+1)$ 7.5.2 $(x+2)\,(x+3)$

7.5.3 $(x+4)\,(x+1)$ 7.5.4 $(x+6)\,(x+1)$

7.5.5 $(x+4)\,(x+5)$ 7.5.6 $(x+20)\,(x+5)$

7.5.7 $(x+3)\,(x+3)$ 7.5.8 $(x+8)\,(x+7)$

7.5.9 $(x+2)\,(x+5)$ 7.5.10 $(x+3)\,(x+5)$

Self-test 7.6

7.6.1 $(x+4)\,(2x-3)=0$

7.6.2 $(2x+2)\,(2x+3)=2\,(x+1)\,(2x+3)=0$

7.6.3 $(x+1)\,(10x-1)=0$

7.6.4 $(x-2)\,(5x-1)=0$

7.6.5 $\left(x-\dfrac{1}{2}\right)\left(\dfrac{1}{2}x+2\right)=\dfrac{1}{4}(2x-1)(x+4)=0$

7.6.6 $\dfrac{2}{5}(x-1)(5x+1)=2(x-1)\left(x+\dfrac{1}{5}\right)=0$

7.6.7 $(x+0.123)\,(x+0.456)=0$

7.6.8 $(x+0.593)\,(x-0.843)=0$

Chapter 8:
Graphs I: Pictorial representations of functions

Self-test 8.1

8.1.1   Quantitative   8.1.2   Qualitative

8.1.3   Qualitative   8.1.4   Quantitative

8.1.5   Qualitative (i.e. we are   8.1.6   Quantitative

        later than we should be,

        but later by an *unspecified*

        amount of time)

Self-test 8.2

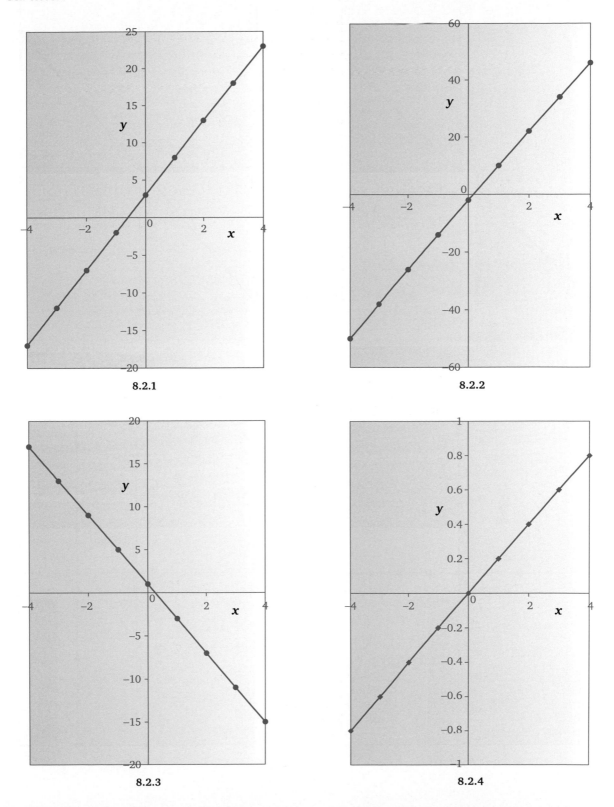

8.2.1

8.2.2

8.2.3

8.2.4

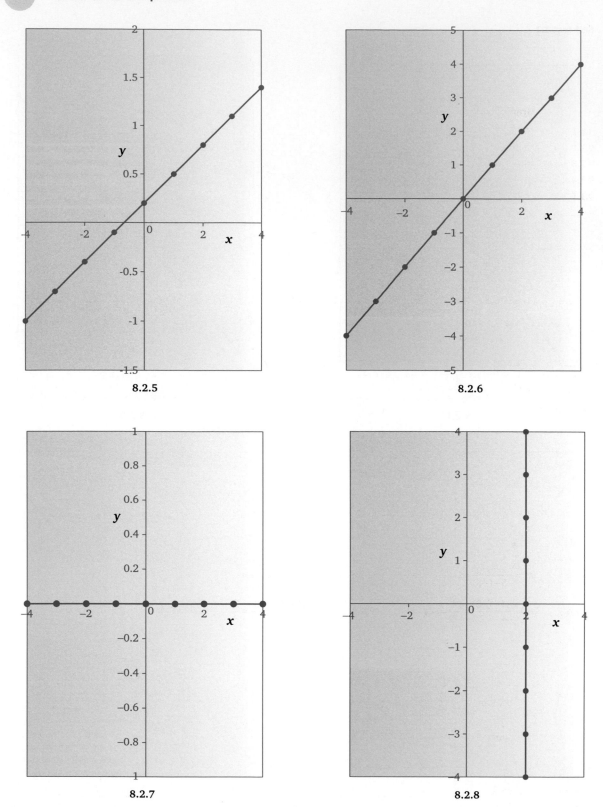

8.2.5

8.2.6

8.2.7

8.2.8

Chapter 9:
Graphs II: The equation of a straight-line graph

Self-test 9.1

	Gradient, m	Intercept, c
9.1.1	2	4
9.1.2	3.5	−2
9.1.3	4	8.4
9.1.4	4	3
9.1.5	1	22
9.1.6	−4	−4.3

Self-test 9.2

9.2.1 $\quad y = 2x + 5$

9.2.2 $\quad y = \dfrac{5}{3}x = -10$

9.2.3 $\quad y = -2x - 3$

9.2.4 $\quad y = \dfrac{4}{7.1}x$ (which could be written as $y = \dfrac{4}{7.1}x + 0$)

9.2.5 $\quad y = \dfrac{19}{30.2}x + \dfrac{10.4}{30.2}$

9.2.6 $\quad y = x - 3$

9.2.7 $\quad y = -\dfrac{2}{3.17}x + \dfrac{1.22}{3.17}$

9.2.8 $\quad y = \dfrac{1}{10^2}x - 10$

Self-test 9.3

9.3.1 $\quad$ 3 $\qquad$ 9.3.2 $\quad$ 0 $\qquad$ 9.3.3 $\quad$ −0.5

Self-test 9.4

9.4.1 $\quad m = \dfrac{4-2}{2-1} = \dfrac{2}{1} = 2$

9.4.2 $\quad m = \dfrac{22-17}{5-10} = \dfrac{5}{-5} = -1$

9.4.3 $\quad m = \dfrac{14-5}{6-3} = \dfrac{9}{3} = 3$

9.4.4 $\quad m = \dfrac{7-3}{0-(-2)} = \dfrac{4}{2} = 2$

9.4.5 $\quad m = \dfrac{-5-(-3)}{-5-(-3)} = 1$

9.4.6 $\quad m = \dfrac{5-(-3)}{4-2} = \dfrac{8}{2} = 4$

9.4.7 $\quad m = \dfrac{-4-(-8)}{-4-(-2)} = \dfrac{4}{-2} = -2$

9.4.8 $\quad m = \dfrac{12-4}{7-(-5)} = \dfrac{8}{12} = \dfrac{2}{3}$

Self-test 9.5

9.5.1 $\quad$ Intercept $= 4 \quad$ gradient $= \dfrac{8-4}{6-0} = \dfrac{4}{6} = \dfrac{2}{3} \qquad y = \dfrac{2}{3}x + 4$

9.5.2 $\quad$ Intercept $= 6 \quad$ gradient $= \dfrac{12-6}{30-0} = \dfrac{6}{30} = \dfrac{1}{5} \qquad y = 0.2x + 6$

9.5.3 $\quad$ Intercept $= 40 \quad$ gradient $= \dfrac{40-10}{0-15} = \dfrac{30}{-15} = -2 \quad y = -2x + 40$

Self-test 9.6

9.6.1 $\quad y = 3x - 1$

9.6.2 $\quad y = 10x - 36$

9.6.3 $\quad y = -4x + 12$

9.6.4 $\quad y = 2.5x - 5$

9.6.5 $\quad y = -3.5x - 14$

Self-test 9.7

9.7.1   gradient $=\dfrac{4-2}{2-1}=2$ line $=y=2x+0$ i.e. $y=2x$

9.7.2   gradient $=\dfrac{-11-(-2)}{3-0}=\dfrac{-9}{3}=-3$ line $=y=-3x-2$

9.7.3   gradient $=\dfrac{50-12}{28-9}=\dfrac{38}{19}=2$ line $=y=2x-6$

9.7.4   gradient $=\dfrac{10-9}{6-7}=\dfrac{1}{-1}=-1$ line $=y=-x+16$

9.7.5   gradient $=\dfrac{-4-2}{-3-1}=\dfrac{-6}{-4}=\dfrac{3}{2}$ line $=y=\dfrac{3}{2}x+\tfrac{1}{2}$

9.7.5   gradient $=\dfrac{-6-(-3)}{-2-(-3)}=\dfrac{-3}{1}=-3$ line $=y=-3x-12$

Chapter 10:
Algebra VII: Solving simultaneous linear equations

Self-test 10.1

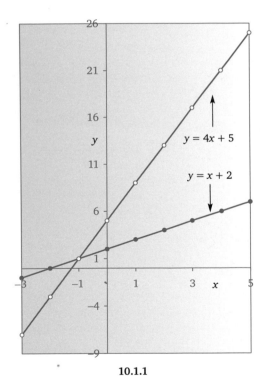

10.1.1
The lines intersect at $(-1,1)$

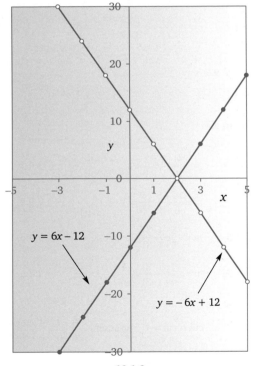

10.1.2
The lines intersect at $(2,0)$

Self-test 10.2

10.2.1 $x=1$ and $y=7$ 10.2.2 $x=-4$ and $y=-2$

10.2.3 $x=-1$ and $y=3$ 10.2.4 $x=0$ and $y=5$

10.2.5 $x=-4$ and $y=-5$ 10.2.6 $x=-3$ and $y=-2$

Self-test 10.3

10.3.1	$x = 8$ and $y = 21$	10.3.2	$x = -4$ and $y = -2$
10.3.3	$x = 1$ and $y = 2$	10.3.4	$x = 7$ and $y = 9$
10.3.5	$x = 11$ and $y = 5$	10.3.6	$x = 2.5$ and $y = 3$

Self-test 10.4

10.4.1	$x = 3$ and $y = 4$	10.4.2	$x = 2$ and $y = 5$
10.4.3	$x = 3$ and $y = 14$	10.4.4	$x = 2$ and $y = -1$
10.4.5	$x = 0.5$ and $y = 1$	10.4.6	$x = -0.4$ and $y = 12$

Chapter 11:
Powers I: Introducing indices and powers

Self-test 11.1

11.1.1	T^7	11.1.2	c^5
11.1.3	a^3	11.1.4	q^9
11.1.5	f^1	11.1.6	g^5

Self-test 11.2

11.2.1	$g^2 h^5$	11.2.2	$c^2 d^2 g^2 h^6$
11.2.3	$c^1 d^1 e^1 f^2$	11.2.4	$x^4 h^2 I^3$
11.2.5	$a^{(2+1+1)} b^2 = a^4 b^2$	11.2.6	$x^{(3+2+1)} y^{(1+1)} = x^6 y^2$
11.2.7	$d^{(1+1+3)} e^{(4+1+2)} = d^5 e^7$	11.2.8	$x^{(2+2+1)} y^{(6+1)} = x^5 y^7$

Self-test 11.3

11.3.1	a^{-3}	11.3.2	c^{-2}
11.3.3	$1 \div b^3 = b^{-3}$	11.3.4	$1 \div d^4 = d^{-4}$
11.3.5	$z^{-3.3}$	11.3.6	$1 \div p^{13.3} = p^{-13.3}$
11.3.7	$1 \div d^5 = d^{-5}$	11.3.8	$1 \div (j^4 k^5 h^3) = h^{-3} j^{-4} k^{-5}$

Self-test 11.4

11.4.1	$1/2$	11.4.2	$1/100$
11.4.3	$1/(100 \times 100) = 1/10{,}000$	11.4.4	$1/(6 \times 6 \times 6) = 1/216$ (or 216^{-1})
11.4.5	$1/5^5 = 1/3125$ (or 3125^{-1})	11.4.6	$1/13^3 = 1/2197$ (or 2197^{-1})

Self-test 11.5

11.5.1	$10^{(2+4)} = 10^6$	11.5.2	$10^{(3+5)} = 10^8$
11.5.3	$10^{(0+2)} = 10^2$	11.5.4	$10^{(20+40)} = 10^{60}$
11.5.5	$6^{(3+12)} = 6^{15}$	11.5.6	$7^{(2+4)} = 7^6$
11.5.7	$b^{(9+2)} = b^{11}$	11.5.8	$z^{(15+2)} = z^{17}$
11.5.9	$b^{(4.1+7.2+3.8)} = b^{15.1}$	11.5.10	$k^{(6.22+8.12)} = k^{14.34}$

Self-test 11.6

11.6.1	$10^{(2-4)} = 10^{-2}$	11.6.2	$10^{(3-5)} = 10^{-2}$
11.6.3	$10^{(0-12)} = 10^{-12}$	11.6.4	$10^{(-2-4)} = 10^{-6}$
11.6.5	$4^{(3-12)} = 4^{-9}$	11.6.6	$1.5^{(2-4)} = 1.5^{-2}$
11.6.7	$b^{(9-(-2))} = b^{11}$	11.6.8	$z^{(5-2.5)} = z^{2.5}$
11.6.9	$z^{(1.3-2.7)} = z^{-1.4}$	11.6.10	$c^{(3.15-2.93)} = c^{0.22}$

Self-test 11.7

11.7.1	$10^{(2 \times 5)} = 10^{10}$		11.7.2	$10^{(3 \times 10)} = 10^{30}$
11.7.3	$a^{(7 \times 3)} = a^{21}$		11.7.4	$a^{(3 \times 7)} = a^{21}$
11.7.5	$p^{(4.4 \times 1.2)} = p^{5.28}$		11.7.6	$7^{(3.3 \times 7.8)} = 7^{25.74}$

Self-test 11.8

11.8.1	$27^{\frac{1}{3}} = (3)$		11.8.2	$36^{\frac{1}{2}} = (6)$
11.8.3	$p^{\frac{1}{4}}$		11.8.4	$(7b)^{\frac{1}{9}}$
11.8.5	$12^{\frac{1}{j}}$		11.8.6	$k^{\frac{1}{i}}$

Chapter 12:
Powers II: Exponentials and logarithms

Self-test 12.1

12.1.1	20.09		12.1.2	24.53
12.1.3	1.203×10^6		12.1.4	0.04076

12.1.5 $\quad K = \exp\left(-\dfrac{-12\,000}{8.314 \times 298}\right) = \exp\left(+\dfrac{12\,000}{2478}\right) = \exp(4.843) = 126.9$

12.1.6 $\quad K = \exp\left(-\dfrac{-40\,200}{8.314 \times 298}\right) = \exp\left(+\dfrac{40200}{2478}\right) = \exp(16.226) = 1.111 \times 10^7$

Self-test 12.2

12.2.1 $\quad e^{\ln x} = e^7 \;\rightarrow\; x = e^7$

12.2.2 $\quad e^{\ln x^2} = e^y \;\rightarrow\; x^2 = e^y \;\rightarrow\; x = \sqrt{e^y} = \left(e^y\right)^{\frac{1}{2}} = \exp\left(\dfrac{y}{2}\right)$

12.2.3 $\quad e^{\ln xt} = e^y \;\rightarrow\; xt = e^y \;\rightarrow\; x = \dfrac{e^y}{t}$

12.2.4 $\quad \ln(e^x) = \ln 3 \;\rightarrow\; x = \ln 3$

12.2.5 $\quad \ln(e^{-6x}) = \ln h \;\rightarrow\; -6x = \ln h \;\rightarrow\; x = -\dfrac{\ln h}{6}$

12.2.6 $\quad \ln(e^{(x^2)}) = \ln y \;\rightarrow\; x^2 = \ln y \;\rightarrow\; x = \sqrt{\ln y}$

Self-test 12.3

12.3.1   3.91

12.3.2   1.531

12.3.3   The calculator display will say 'error,' because it's impossible to take the logarithm of a negative number.

12.3.4 $\log 340 = 2.531 = 1 + \log 34$ (As given in the answer to Self-test 12.3.2.)
Similarly, $\log 3400 = 2 + \log 34$.
As we see in the next section, this result is due to the way, $340 = 10 \times 34$
and using the first log law:
$\log 340 = \log(10 \times 34) = \log 10 + \log 34 = 1 + \log 34$
$\log 3400 = \log(100 \times 34) = \log 100 + \log 34 = 2 + \log 34$

12.3.5   0, by definition: $1 = e^0$, $\ln 1 = \ln(e^0) = 0$.

12.3.6   3

12.3.7   This calculation is not possible. The domain of the $\ln(ax)$ is $ax > 0$. On a calculator, the display would say 'error' because it is impossible to display an index so small that e^x can have a value of 0.

12.3.8   1, by definition.

Self-test 12.4

12.4.1 $\ln(5 \times 3) = \ln 15$ 12.4.2 $\ln(5 \times 8 \times 2) = \ln 80$

12.4.3 $\log(20 \times 7) = \log 140$ 12.4.4 $\log_p 7ab$

12.4.5 $\log ghj$ 12.4.6 $\log nmp$

12.4.7 $\log_6 qr$ 12.4.8 $\log_n (6 \times 4 \times 2 \times ft) = \log_n (48\,ft)$

Self-test 12. 5

In each case, we start by combining the two terms using the second law of logarithms, eqn (12.4). We then cancel the terms within the bracketed argument.

12.5.1 $\ln\left(\dfrac{6}{3}\right) = \ln 2$ 12.5.2 $\ln\left(\dfrac{5f}{5}\right) = \ln f$

12.5.3 $\log\left(\dfrac{12}{4}\right) = \log 3$ 12.5.4 $\log\left(\dfrac{y^2}{y}\right) = \log y$

12.5.5 $\log\left(\dfrac{6g}{3g}\right) = \log 2$ 12.5.6 $\log\left(\dfrac{h}{p}\right)$

Self-test 12.6

12.6.1 $5\log a = \log a^5$

12.6.2 $\log b + \log (b^2)^2 = \log (b \times (b^2)^2) = \log b^5$

12.6.3 $\log c^{3/2} + \log c = \log (c^{3/2} \times c) = \log c^{5/2}$

12.6.4 $\ln (c^2)^3 = \ln c^6$

12.6.5 $\ln y^6 + \ln (4y)^2 = \ln (y^6 \times 4^2 y^2) = \ln 16\,y^8$

Self-test 12.7

12.7.1   We start by writing, $x = 65 \times 41$. Taking the logarithm of both sides gives:
$\ln x = \ln 65 + \ln 41$. $\ln 65 = 4.1744$ and $\ln 41 = 3.7136$. Therefore, $\ln x = 7.888$.
So $x = \exp(7.888) = 2\,665.11$. To 4 s.f., $65 \times 41 = 2665$.

12.7.2   We start by writing, $x = 12^{3.5}$. We then take the logarithm of both sides: $\ln x = \ln 12^{3.5}$.
The third law of logarithms allows us to rewrite this phrase, as: $\ln x = 3.5 \times \ln 12$.
$\ln 12 = 2.4849$, so $\ln x = 3.5 \times 2.4849 = 8.69717$.
We then take the antilog of both sides: $x = \exp(8.69717) = 5986$.
The value of $12^{3.5}$ is 5986 (to 4 s.f.).

12.7.3   We start by writing, $x = \sqrt[5.8]{1265}$. We then take the logarithm of both sides:
$\ln x = \ln \left(\sqrt[5.8]{1265}\right)$. The third law of logarithms, allows us to rewrite this phrase, as:
$\ln x = \dfrac{1}{5.8} \ln 1265$.
$\ln 1265 = 7.1428$, so $\ln x = \dfrac{1}{5.8} \times 7.1428 = 1.2315$.
We then take the antilog of both sides: $x = \exp(1.2315) = 3.426$.
The value of $\sqrt[5.8]{1265}$ is 3.43 (or stated another way, $3.43^{5.8} = 1265$).

Chapter 13:
Powers III: Obtaining linear graphs from non-linear functions

Self-test 13.1

		$y =$	$m \times$	$x +$	c
13.1.1	The Nernst equation:	$E_{Cu^{2+},Cu}$	$= \dfrac{RT}{2F}$	$\ln[Cu^{2+}]$	$+ E^{\ominus}_{Cu^{2+},Cu}$
13.1.2	The Clausius–Clapeyron equation:	$\ln p$	$= -\dfrac{\Delta H^{\ominus}_{(vap)}}{R}$	$\dfrac{1}{T}$	$+ c$
13.1.3	The second-order rate equation:	$\dfrac{1}{c_t}$	$= k$	t	$+ \dfrac{1}{c_{t=start}}$

13.1.4   Boyle's law: p $= c$ $\dfrac{1}{V}$

(i.e. the constant $= 0$)

13.1.5   The van't Hoff isochore: $\ln K$ $= -\dfrac{\Delta H^{\ominus}}{R}$ $\dfrac{1}{T}$ $+ c$

Self-test 13.2

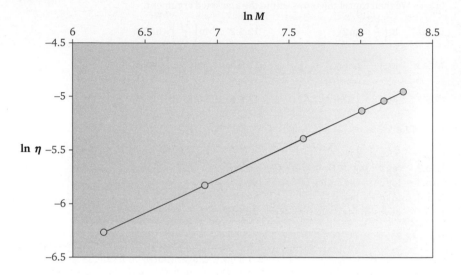

Look at the graph above of $\ln \eta$ (as 'y') against $\ln M$ (as 'x'), which is clearly linear. Its gradient is 0.63 and its intercept is -10.2. Accordingly, $\alpha = 0.63$ and $K = \exp(-10.2)$ so $K = 3.8 \times 10^{-5}$. Therefore, for polystyrene in benzene, the **Mark–Houwink equation** has the form, $\eta = 3.8 \times 10^{-5} M^{0.63}$.

Self-test 13.3

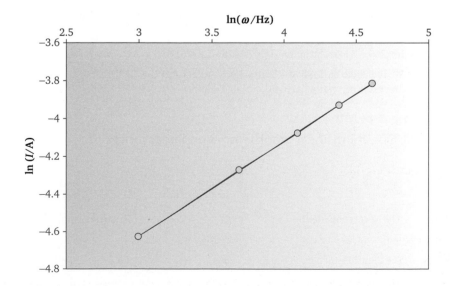

Look at the graph above of $\ln I$ (as 'y') against $\ln \omega$ (as 'x'), which is clearly is linear. Its gradient is 0.50, so the relationship is $I = k\omega^{\frac{1}{2}}$, which we often write as $I = k\sqrt{\omega}$. (This square-root relationship underpins the **Levich** equation.) The value of $k = e^{-6.12} = 2.19 \times 10^{-3}$ A Hz$^{-\frac{1}{2}}$.

Chapter 14:
Statistics I: Averages and simple data analysis

Self-test 14.1

    14.1.1 $\bar{m} = 11.97g$ **14.1.2** $\bar{c} = 691.8$ ppm   **14.1.3** $\overline{\Delta H} = 32.16$ KJ mol^{-1}

Self-test 14.2

    14.2.1   The mass 9.21 mg occurs three times, so 9.21 mg is the mode.
    14.2.2   The concentration 215 ppm occurs four times, so 215 ppm is the mode.
    14.2.3   The concentration 33 μg dm^{-3} occurs four times, so 33 μg dm^{-3} is the mode.

Self-test 14.3

    14.3.1   The data set comprises a total of 9 items, so the median value if the fifth. When placed in order, the median is therefore 1.93 cm.
    14.3.2   The data set comprises 6 data, so the median is the mean of items 3 and 4. The median is therefore the mean of 474 and 475, i.e. 474.5 kJ mol^{-1}.

Self-test 14.4

    14.4.1 $s = 0.033$ cm (Each data set comprises fewer than 10 entries, so the correct standard deviation term to use is the *sample* standard deviation, s.)
    14.4.2 $s = 1.47$ kJ mol^{-1}

Self-test 14.5

	N	x	$x_{(next)}$	$x_{(minimum)}$	$x_{(maximum)}$	$Q_{(exp)}$	Confidence to reject?
14.5.1	5	**0.92**	0.72	0.65	0.92	0.741	$95 < \% < 99$
14.5.2	4	**226**	219	214	226	0.583	Less than 90 %
14.5.3	6	**32.2**	34.3	32.2	35.7	0.600	$90 < \% < 95$

Chapter 15:
Statistics II: Treatment and assessment of errors

Self-test 15.1

    15.1.1   2.42 : 1 (not safe)   **15.1.2**   27.8 : 1 (safe)
    15.1.3   125 : 1 (safe)   **15.1.4**   29.7 : 1 (safe)
    15.1.5   92 : 1 (safe),
             because 12 mV = 0.012 V
    15.1.6   68 : 1 (safe)

Self-test 15.2

    15.2.1   Using eqn (15.3), the minimum error = (0.5 mm ÷ 112 mm) = 4.46×10^{-3} (to 3 s.f.), or 0.45%.
    15.2.2   Using eqn (15.3), the minimum error = (0.5 °C ÷ 24.9 °C) = 0.020, or 0.2%.
    15.2.3   The temperature divisions are 0.1 °C, so the error is 0.05 °C; and the balance has divisions of 1 mg, so the error is 0.0005 g.

Using eqn (15.4),

$$\left(\text{minimum error}\right)^2 = \left(\frac{0.05\ ^\circ\text{C}}{0.7\ ^\circ\text{C}}\right)^2 + \left(\frac{0.0005\ \text{g}}{3.503\ \text{g}}\right)^2$$

$$\left(\text{minimum error}\right)^2 = 5.1 \times 10^{-3} + 2.04 \times 10^{-8}$$

$$\left(\text{minimum error}\right)^2 = 5.10 \times 10^{-3}$$

so, the minimum error $= \sqrt{5.10 \times 10^{-3}} = 0.0714$

The minimum error represents about 7%. The overwhelming majority of this error comes from the reading of temperature. The chemist should use a superior thermometer, such as a Beckmann thermometer, or maybe a thermocouple of suitable sensitivity.

15.2.4 The balance has divisions of 1 mg, so the error is 0.0005 g; we already know the innate error of the flask and the instrument; and the ruler measuring path length has divisions of 0.5 mm, so the error is 0.25 mm.

We insert these values into eqn (15.4):

$$\left(\text{minimum error}\right)^2 = \left(\frac{0.0005\ \text{g}}{0.504\text{g}}\right)^2 + \left(\frac{0.5\ \text{cm}^3}{100\ \text{cm}^3}\right)^2 + \left(\frac{0.25\ \text{mm}}{10\ \text{mm}}\right)^2 + \left(\frac{0.05}{0.74}\right)^2$$

$$\left(\text{minimum error}\right)^2 = 9.842 \times 10^{-7} + 2.500 \times 10^{-5} + 6.25 \times 10^{-4} + 4.565 \times 10^{-3}$$

$$\left(\text{minimum error}\right)^2 = 5.216 \times 10^{-3}$$

so minimum error $= \sqrt{5.216 \times 10^{-3}} = 0.0722$

The minimum error is 7.2%. The principal error is the absorbance reading. The chemist should use a superior instrument; for example, the innate error in the absorbance reading with a good, modern instrument can be as low as 0.001.

Self-test 15.3

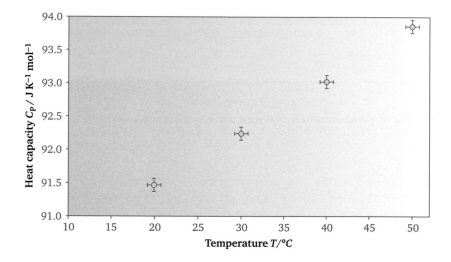

Self-test 15.4

The graph shows some slight scatter, and with a correlation coefficient r of -0.9910. The negative sign is a direct consequence of the negative slope. The value of r^2 is therefore $(-0.9910)^2 = 0.982$.

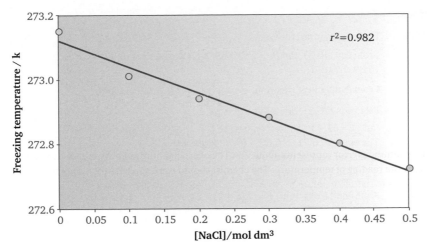

Chapter 16:
Trigonometry

Self-test 16.1

16.1.1	0.809	**16.1.2**	0.839
16.1.3	0.213	**16.1.4**	0.5
16.1.5	0.707	**16.1.6**	5.67

Self-test 16.2

16.2.1 $\theta = \sin^{-1}(0.3)$ → $\theta = 17.4°$

16.2.2 $\theta = \cos^{-1}(0.92)$ → $\theta = 23.1°$

16.2.3 $2\cos\theta = 0.4$ → $\cos\theta = (0.4 \div 2) = 0.2$ → $\theta = \cos^{-1}(0.2)$, so $\theta = 78.5°$

16.2.4 $\frac{1}{2}\sin\theta = 0.45$ → $\sin\theta = (2 \times 0.45) = 0.9$ → $\theta = \sin^{-1}(0.9)$, so $\theta = 64.2°$

16.2.5 $\cos\theta + 2 = 2.1$ → $\cos\theta = 0.1$ → $\theta = \cos^{-1}(0.1)$, so $\theta = 84.3°$

16.2.6 $\sin^2\theta = 0.9$ → $\sin\theta = -\sqrt{0.9} = -0.948$ → $\theta = \sin^{-1}(-0.948)$, so $\theta = -71.6°$.

16.2.7 $2\tan^2\theta = 9$ → $\tan^2\theta = (9 \div 2) = 4.5$ → $\tan\theta = \pm\sqrt{4.5} = \pm2.12$.
→ $\theta = \tan^{-1}(\pm2.12) = \pm64.8°$

16.2.8 $4\sin^3\theta = 0.76$ → $\sin^3\theta = (0.76 \div 4) = 0.19$ → $\sin\theta = \sqrt[3]{0.19} = 0.575$
→ $\theta = \sin^{-1}(0.575) = 35.1°$

Note that we would normally express these angles in terms of π, i.e. in *radians*.

Self-test 16.3

16.3.1	$\pi/3$ rad	**16.3.2**	$\dfrac{5\pi}{6}$ rad
16.3.3	$0.88\,\pi$ rad $= 2.79$ rad	**16.3.4**	$0.094 \times 2\,\pi$ rad $= 0.593$ rad

Self-test 16.4

16.4.1 $c^2 = 2^2 + 5^2$ → $4 + 25 = 29$ → $c = \sqrt{29} = 5.39$ cm

16.4.2 $c^2 = 7^2 + 13^2$ → $49 + 169 = 218$ → $c = \sqrt{218} = 14.8$ km

16.4.3 $150^2 = 130^2 + a^2 \rightarrow 22\,500 = 16\,900 + b^2 \rightarrow b^2 = 22\,500 - 16\,900 \rightarrow b^2 = 5\,600 \rightarrow$
$b = \sqrt{5\,600} \rightarrow b = 74.8$ pm. (It is not profitable to consider the negative root.)

16.4.4 $c^2 = 150^2 + 140^2$ → $22\,500 + 19\,600 = 42\,100$ →
$c = \sqrt{42\,100} = 205$ pm

Self-test 16.5

16.5.1   From Pythagoras' theorem, $r = \sqrt{2^2 + 5^2} = \sqrt{29} = 5.39$ to 3 s.f.

From eqn (16.3), $\tan \theta = (5/2) = 68.2°$ So the point is $(5.39, 68.2°) = (5.39, 1.19 \text{ rad})$.

16.5.2   From Pythagoras' theorem, $r = \sqrt{3^2 + 6^2} = \sqrt{45} = 6.71$ to 3 s.f.

From eqn (16.3), $\tan (180° - \theta) = \dfrac{6}{3} = 2$.

$\tan^{-1} 2 = 63.43°$. Therefore, $\theta = 116.6°$.
The point is $(6.71, 116.6°) = (6.71, 2.03 \text{ rad})$.

16.5.3   From Pythagoras' theorem,

$$r = \sqrt{4.2^2 + 1.9^2} = \sqrt{21.25} = 4.61 \text{ to 3 s.f.}$$

From eqn (16.3), $\tan^{-1} \theta = (1.9/4.2) = 24.3°$.
So the point is $(4.61, 24.3°) = (4.61, 0.425 \text{ rad})$.

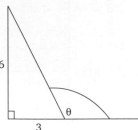

Self-test 16.6

16.6.1 $c^2 = 4^2 + 7^2 - (2 \times 4 \times 7) \times \cos 73°$
$\rightarrow c^2 = 16 + 49 - 56 \cos 73°$
$\rightarrow c^2 = 65 - (56 \times 0.292) \rightarrow c^2 = 65 - 16.4$
$\rightarrow c^2 = 48.63 \rightarrow c = 6.97$

16.6.2 $8^2 = 5^2 + 4^2 - (2 \times 5 \times 4 \cos \theta) \quad \rightarrow \quad 64 = (25 + 16 - 40 \cos \theta)$

$\rightarrow 64 = 41 - 40 \cos \theta \rightarrow (64 - 41) = -40 \cos \theta \rightarrow -\dfrac{23}{40} = \cos \theta$
$\rightarrow \theta = \cos^{-1} (-0.578) \rightarrow \theta = 125°$

Chapter 17:
Differentiation I: Rates of change, tangents, and differentiation

Self-test 17.1

17.1.1	Gradient $= 6$	**17.1.2**	Gradient $= 48$
17.1.3	Gradient $= 9$	**17.1.4**	Gradient $= 18$

Self-test 17.2

17.2.1 $\dfrac{dy}{dx} = \dfrac{(x + \delta x)^2 + 3 - (x^2 + 3)}{(x + \delta x) - x} = \dfrac{x^2 + 2x\delta x + \delta x^2 - x^2}{\delta x} = 2x + \delta x = 2x \quad$ when $\delta x = 0$.

17.2.2 $\dfrac{dy}{dx} = \dfrac{(x^3 + 3x^2 \delta x + 3x\delta x^2 + \delta x^3) - x^3}{(x + \delta x) - x} = \dfrac{3x^2 \delta x + 3x\delta x^2 + \delta x^3}{\delta x}$

$= 3x^2 + 3x\delta x + \delta x^2 = 3x^2 \text{ when } \delta x = 0.$

17.2.3 $\dfrac{dy}{dx} = \dfrac{(4(x + \delta x)^3) - 4x^3}{(x + \delta x) - x} = \dfrac{4(x^3 + 3x^2 \delta x + 3x\delta x^2 + \delta x^3) + 4x^3}{\delta x}$

$= \dfrac{12x^2 \delta x + 12x\delta x^2 + 4\delta x^3}{\delta x} = 12x^2 + 12x\delta x + 4\delta x^2 = 12x^2 \text{ when } \delta x = 0$

17.2.4 $\dfrac{dy}{dx} = \dfrac{(x + \delta x)^3 + 4 - (x^3 + 4)}{(x + \delta x) - x} = \dfrac{(x^3 + 3x^2 \delta x + 3x\delta x^2 + \delta x^3) + 4 - (x^3 + 4)}{\delta x}$

$= \dfrac{3x^2 \delta x + 3x\delta x^2 + \delta x^3}{\delta x} = 3x^2 + 3x\delta x + \delta x^2 = 3x^2 \quad \text{when } \delta x = 0 .$

Notice how the additional constant term '4' vanishes.

Self-test 17.3

17.3.1 $\dfrac{dy}{dx} = 6x^5$ 17.3.2 $\dfrac{dy}{dx} = 12x^{11}$

17.3.3 $\dfrac{dy}{dx} = 5x^4$ 17.3.4 $\dfrac{dy}{dx} = 3x^2$

17.3.5 $\dfrac{dy}{dx} = 1$ 17.3.6 $\dfrac{dh}{dT} = 5$

17.3.7 $\dfrac{dw}{dp} = -11d^{-12} = -\dfrac{11}{d^{12}}$ 17.3.8 $\dfrac{dc}{dq} = -7q^{-8} = -\dfrac{1}{q^8}$

17.3.9 $\dfrac{dh}{dk} = -3k^{-4} = -\dfrac{3}{k^4}$ 17.3.10 $\dfrac{dz}{dp} = 2.73\,p^{1.73}$

Self-test 17.4

17.4.1 $\dfrac{dy}{dx} = 3 \times 3\,x^{3-1} = 9x^2$ 17.4.2 $\dfrac{dy}{dx} = 5 \times 14\,x^{14-1} = 70x^{13}$

17.4.3 $\dfrac{dy}{dx} = 20x^4$ 17.4.4 $\dfrac{dy}{dx} = -8.4x^{-8} = -\dfrac{8.4}{x^8}$

17.4.5 $\dfrac{dy}{dx} = 25.8\,x^5$ 17.4.6 $\dfrac{dy}{dx} = -\left(4 \times 10^6\right)x^{-5} = -\dfrac{4 \times 10^6}{x^5}$

Self-test 17.5

17.5.1 $\dfrac{dy}{dx} = 9x^2$ 17.5.2 $\dfrac{dy}{dx} = 4x$

17.5.3 $\sqrt{x} = x^{\frac{1}{2}}$, so $\dfrac{dy}{dx} = 1.6x^{-\frac{1}{2}} = \dfrac{1.6}{\sqrt{x}}$ 17.5.4 $\dfrac{dy}{dx} = 1.2 \times 10^6 x$

17.5.5 $\sqrt[3]{x} = x^{\frac{1}{3}}$, so $\dfrac{dy}{dx} = 1.55x^{-\frac{2}{3}} = \dfrac{1.55}{x^{\frac{2}{3}}}$ 17.6.6 $\dfrac{dy}{dx} = 92x^{8.2}$

Self-test 17.6

17.6.1 $\dfrac{dy}{dx} = 3x^2$ 17.6.2 $\dfrac{dy}{dx} = 4x$

17.6.3 $\dfrac{dy}{dx} = -\frac{1}{2}x^{-\frac{1}{2}} = -\dfrac{1}{2\sqrt{x}}$ 17.6.4 $\dfrac{dy}{dx} = -3x^{-4} = -\dfrac{3}{x^4}$

17.6.5 $\dfrac{dy}{dx} = \frac{1}{2} \times 2 \times x = x$ 17.6.6 $\dfrac{dy}{dx} = 32x^3 - \dfrac{6}{x^3}$

Self-test 17.7

17.7.1 $\dfrac{dy}{dx} = 4x^3 + 2x$ 17.7.2 $\dfrac{dy}{dx} = 3x^2 - 1$

17.7.3 $\dfrac{dy}{dx} = 7x^6 - 12x$ 17.7.4 $\dfrac{dy}{dx} = \frac{1}{2}x^{-\frac{1}{2}} + 12 = \dfrac{1}{2\sqrt{x}} + 12$

Chapter 18:
Differentiation II: Differentiating other functions

Self-test 18.1

18.1.1 $\dfrac{dy}{dx} = 5\,e^{5x}$ 18.1.2 $\dfrac{dy}{dx} = 3.4e^{3.4x}$

18.1.3 $\dfrac{dy}{dx} = 7fe^{fx}$ 18.1.4 $\dfrac{dy}{dx} = 39.06\,e^{4.2x}$

18.1.5 $\dfrac{dy}{dx} = d^2\,e^{dx}$, where the factor of d^2 comes from '$d \times d\,e^{dx}$,

18.1.6 $\dfrac{dy}{dx} = (3.492 \times 10^{-6})\, e^{(4 \times 10^{-7})x}$

18.1.7 $\dfrac{dy}{dx} = e^x + e^{-x}$ **18.1.8** $\dfrac{dy}{dx} = 12x^3 - 14x + 5e^{5x}$

Self-test 18.2

18.2.1 $\dfrac{dy}{dx} = \dfrac{1}{x}$ **18.2.2** $\dfrac{dy}{dx} = \dfrac{1}{x}$

18.2.3 $\dfrac{dy}{dx} = \dfrac{5.4}{x}$ **18.2.4** $\dfrac{dy}{dx} = -\dfrac{1}{x}$

18.2.5 $\dfrac{dy}{dx} = \dfrac{1}{x}$ **18.2.6** $\dfrac{dy}{dx} = \dfrac{ab}{x}$

18.2.7 $\dfrac{dy}{dx} = 4 - \dfrac{3}{x}$ **18.2.8** $\dfrac{dy}{dx} = 4x^3 + \dfrac{3}{x^4} + \dfrac{1}{x}$

18.2.9 $\dfrac{dy}{dx} = 1000 + \dfrac{1}{x}$ **18.2.10** $\dfrac{dy}{dx} = \dfrac{1}{x} - \dfrac{c}{x}$

Self-test 18.3

18.3.1 $\dfrac{dy}{dx} = 4\cos 4x$ **18.3.2** $\dfrac{dy}{dx} = 12\cos 3x$

18.3.3 $\dfrac{dy}{dx} = -97.2\cos(8.1x)$ **18.3.4** $\dfrac{dy}{dx} = -44\sin(44x)$

18.3.5 $\dfrac{dy}{dx} = 49.14\sin(-7.8x)$

18.3.6 $\dfrac{dy}{dx} = -d^2\sin(dx)$, where the factor of d^2 comes from '$d \times d$' $\cos(dx)$

18.3.7 $\dfrac{dy}{dx} = \cos x - \sin x$

18.3.8   We first rewrite: $\dfrac{\cos x}{2} = \text{'}\dfrac{1}{2} \times \cos x\text{'}$ and $\dfrac{\sin 3x}{4} = \text{'}\dfrac{1}{4} \times \sin 3x\text{'}$

therefore $\dfrac{dy}{dx} = -\dfrac{\sin x}{2} - \dfrac{3\cos 3x}{4}$

(it's probably better not to write the second half as $0.75 \cos 3x$)

Self-test 18.4

18.4.1 $\dfrac{dy}{dx} = -4 \times 5\sin 5x$ so gradient $= -20\sin(5\pi/6) = -20 \times 0.5 = -10$

18.4.2 $\dfrac{dy}{dx} = 2 \times 7\cos 7x$, so gradient $= 14\cos(7\pi/8) = -12.934$

Chapter 19
Differentiation III: Differentiating functions of functions: the chain rule

Self-test 19.1

	Problem	First function	Second function
19.1.1	$y = (x^2 + 2)^2$	$x^2 + 2$	power (square)
19.1.2	$y = (\ln x)^2$	$\ln x$	power (square)
19.1.3	$y = \sin(e^{3x})$	e^{3x}	sine
19.1.4	$y = \exp(x^4)$	polynomial (x^4)	exponential
19.1.5	$y = \ln(x^3)$	polynomial (x^3)	logarithm (ln)

19.1.6	$y = \sin^5 x \equiv (\sin x)^5$	sine (x)	power (fifth)
19.1.7	$y = \cos(x^4 - x^2)$	polynomial $(x^4 - x^2)$	cosine
19.1.8	$y = 3\ln\left(\sqrt{x}\right)$	polynomial $\left(\sqrt{x}\right)$	$3 \times$ logarithm $(3\ln)$
19.1.9	$y = \ln(3x^{3/2})$	polynomial $(3x^{3/2})$	logarithm $(\ln)$
19.1.10	$y = \exp\left(\dfrac{3}{x^2} - \dfrac{2}{x^3}\right)$	polynomial $\left(\dfrac{3}{x^2} - \dfrac{2}{x^3}\right)$	exponential

Self-test 19.2

19.2.1 $\dfrac{dy}{dx} = 4x\,(x^2 + 2)$

19.2.2 $\dfrac{dy}{dx} = \dfrac{2}{x}\ln x$

19.2.3 $\dfrac{dy}{dx} = 3e^{3x} \times \cos\left(e^{3x}\right)$

19.2.4 $\dfrac{dy}{dx} = 4x^3\,e(x^4)$ or $4x^3\exp(x^4)$

19.2.5 $\dfrac{dy}{dx} = \dfrac{3x^2}{x^3} = \dfrac{3}{x}$

19.2.6 $\dfrac{dy}{dx} = 5\cos x\,(\sin x)^4 = 5\cos x\,\sin^4 x$

19.2.7 $\dfrac{dy}{dx} = -(4x^3 - 2x)\sin(x^4 - x^2)$

19.2.8 Remember that $\sqrt{x} = x^{\frac{1}{2}}$ and $\dfrac{1}{\sqrt{x}} = x^{-\frac{1}{2}}$

$\dfrac{dy}{dx} = 3 \times \dfrac{1}{2} \times \dfrac{x^{-\frac{1}{2}}}{x^{\frac{1}{2}}}$ The fraction $\dfrac{x^{-\frac{1}{2}}}{x^{\frac{1}{2}}} \equiv \dfrac{1}{\sqrt{x}} \times \dfrac{1}{\sqrt{x}} = \dfrac{1}{x}$,

so the differential becomes $\dfrac{3}{2} \times \dfrac{1}{x}$, i.e. $\dfrac{3}{2x}$

19.2.9 $u = 3x^{\frac{3}{2}}$ and $y = \ln u$. $u = \dfrac{dy}{du} = \dfrac{1}{u}$ and $\dfrac{du}{dx} = \dfrac{3}{2} \times 3 \times x^{\frac{1}{2}}$

The chain rule gives, $\dfrac{dy}{dx} = \dfrac{1}{u} \times \dfrac{3}{2} \times 3 \times x^{\frac{1}{2}} = \dfrac{1}{3x^{\frac{3}{2}}} \times \dfrac{9x^{\frac{1}{2}}}{2} = \dfrac{3}{2}\dfrac{x^{\frac{1}{2}}}{x^{\frac{3}{2}}} = \dfrac{3}{2x}$

19.2.10 $\dfrac{dy}{dx} = \left(-\dfrac{6}{x^3} + \dfrac{6}{x^4}\right) \times \exp\left(\dfrac{3}{x^2} - \dfrac{2}{x^3}\right)$

Chapter 20
Differentiation IV: The product rule and the quotient rule

Self-test 20.1

20.1.1 We start by writing the logarithm function within brackets, in order to separate it from the power term: $y = x^4\,(\ln x)$. There is no further means of simplification, so we see how there are two separate functions, so y here is a product.

20.1.2 Using the second law of powers (Chapter 11), the two powers can cancel: $y = x^{(3-5)} = x^{-2}$, so y is a simple function.

20.1.3 It is not possible to cancel this expression or to simplify it further. Accordingly, it is a quotient.

20.1.4 Using the first law of powers (see Chapter 11), we can rewrite the expression as $y = 4 \times x^{(4+5)} = 4x^9$. Accordingly, y is a simple, single function.

20.1.5 The two functions $\sin x$ and $\cos x$ are separate, so y is a product.

20.1.6 We start by writing the sin function in brackets to separate it from the power term, so $y = x^3\,(\sin 3x^2)$. There is no further means of simplification, so we see how there are two separate functions, so y here is a product.

20.1.7 We start by recognizing two functions, both sin and ln. There is no further means of simplification, so we see how there are two separate functions, so y here is a product.

20.1.8 It is not possible to cancel this expression or to simplify it further. Accordingly, it is a quotient.

Self-test 20.2

u	$\dfrac{du}{dx}$	v	$\dfrac{dv}{dx}$	$\dfrac{dy}{dx}$ obtained via the product rule
				(The first expression employs square brackets [ ] to indicate the two differential terms. The subsequent expressions use curly brackets { } to indicate factorizing.)
20.2.1 $2x^4$	$8x^3$	$\sin 3x$	$3\cos 3x$	$2x^4 \times [3\cos 3x] + \sin 3x \times [8x^3]$ $= 2x^3\{3x\cos 3x + 4\sin 3x\}$
20.2.2 $\ln 2x$	$\dfrac{1}{x}$	$\exp(2x)$	$2\exp(2x)$	$\ln 2x \times [2\exp(2x)] + \exp(2x) \times \left[\dfrac{1}{x}\right]$ $= \exp(2x)\left\{2\ln 2x + \dfrac{1}{x}\right\}$
20.2.3 x^5	$5x^4$	$\ln x$	$\dfrac{1}{x}$	$x^5 \times \left[\dfrac{1}{x}\right] + \ln x \times [5x^4] = x^4\{1 + 5\ln x\}$
20.2.4 $2x^{-4}$	$-8x^{-5}$	$\cos 4x$	$-4\sin 4x$	$2x^{-4} \times [-4\sin 4x] + \cos 4x \times [-8x^{-5}]$ $= -8x^{-5}\{x\sin 4x + \cos 4x\}$
20.2.5 $\sin^2 x$	via the chain rule: $2\sin x\cos x$	$\cos^2 x$	via the chain rule: $-2\sin x\cos x$	$\sin^2 x \times [-2\sin x\cos x] + \cos^2 x \times$ $[2\sin x\cos x] = 2\sin x\ \cos x \times$ $\left\{\cos^2 x - \sin^2 x\right\}$
20.2.6 $\ln x^2$	via the chain rule: $\dfrac{1}{x^2} \times 2x = \dfrac{2}{x}$	$\sin^3 2x$	via the chain rule: $3 \times 2\sin^2$ $2x\cos 2x$	$\ln x^2 \times [6\sin 2x\cos 2x] + \sin^3 2x \times$ $\left[\dfrac{2}{x}\right] = 2\sin^2 2x\left\{3\ln x^2\cos 2x + \dfrac{\sin 2x}{x}\right\}$

Self-test 20.3

u	$\dfrac{du}{dx}$	v	$\dfrac{dv}{dx}$	$\dfrac{dy}{dx}$ obtained via the product rule
				(The first expression employs square brackets [ ] to indicate the two differential terms. The subsequent expressions use curly brackets { } to indicate factorizing.)
20.3.1 x^2	$2x$	$\sin x$	$\cos x$	$\dfrac{\sin x[2x] - x^2[\cos x]}{(\sin x)^2} = \dfrac{x\{2\sin x - x\cos x\}}{\sin^2 x}$
20.3.2 $\ln x$	$\dfrac{1}{x}$	x^4	$4x^3$	$\dfrac{x^4\left[\dfrac{1}{x}\right] - \ln x[4x^3]}{\left(x^4\right)^2} = \dfrac{x^3 - 4x^3\ln x}{x^8}$ $= \dfrac{x^3\{1 - 4\ln x\}}{x^8} = \dfrac{1 - 4\ln x}{x^5}$

20.3.3 $\quad$ exp $2x \quad 2\exp 2x \quad 3x^3 \quad 9x^2 \quad \dfrac{3x^3\left[2\exp(2x)\right]-\exp 2x\left[9x^2\right]}{\left(3x^3\right)^2} =$

$$\dfrac{3x^2\exp 2x\{2x-3\}}{9x^6} = \dfrac{\exp 2x\{2x-3\}}{3x^4}$$

Chapter 21
Differentiation V: Maxima and minima on graphs: second differentials

Self-test 21.1

21.1.1 The function x^4 is a simple, single-term polynomial function, and will therefore have one turning point.

21.1.2 The equation is a simple quadratic, and will therefore have a single turning point.

21.1.3 This long expression involves a series of polynomials. The highest power is '4', i.e. x^4, so we expect up to $(4-1) = 3$ turning points.

21.1.4 Another long expression involving a series of polynomials. The highest power is '5', i.e. x^5, so we expect up to $(5-1) = 4$ turning points.

Self-test 21.2

21.2.1 $\dfrac{dy}{dx} = 2x+1$. At the turning point, $0 = 2x+1$, so $x = -\frac{1}{2}$.

Back-substitution yields $y = (-\frac{1}{2})^2 + -\frac{1}{2} + 5$, so $y = 4.75$.

21.2.2 $\dfrac{dy}{dx} = 6x-12$. At the turning point, $0 = 6x-12$, so $x = 2$.

Back-substitution yields $y = 3 \times (2)^2 - 12 \times 2 - 7$, so $y = -19$.

21.2.3 $\dfrac{dy}{dx} = -10x+4$. At the turning point, $0 = -10x+4$, so $x = 0.4$.

Back-substitution yields $y = -5 \times (0.4)^2 + 4 \times (0.4) + 2$, so $y = 2.8$.

21.2.4 $\dfrac{dy}{dx} = 6.4x-9.1$. At the turning point, $0 = 6.4x-9.1$, so $x = 1.421$.

Back-substitution yields $y = 3.2 \times (1.421)^2 - 9.1 \times (1.421) = 6.462 - 12.931$, so $y = -6.469$.

21.2.5 $\dfrac{dy}{dx} = x^2+10x+24$, which factorizes to $\dfrac{dy}{dx} = (x+6)(x+4)$.

At the turning point, $0 = (x+6)(x+4)$, and $x = -6$ or -4. Back-substitution allows us to compute the turning points as: $(-6,-29)$ and $(-4,-30.3)$

21.2.6 $\dfrac{dy}{dx} = 3x^2+18x+24$, which factorizes to $\dfrac{dy}{dx} = 3(x+2)(x+4)$.

At the turning point, therefore, $0 = (x+2)(x+4)$. Back-substitution yields the turning points as: $(-2,-20)$ and $(-4,-16)$.

Self-test 21.3

21.3.1 $\dfrac{dy}{dx} = 3x^2$, turning point $= (0,2)$, $\dfrac{d^2y}{dx^2} = 6x$, i.e. 0 at $x = 0$, so an inflection

21.3.2 $\dfrac{dy}{dx} = 6x+6$, turning point $= (-1,-3)$, $\dfrac{d^2y}{dx^2} = 6$, i.e. positive, so a minimum

21.3.3 $\dfrac{dy}{dx} = 12x^2 + 8x = 4x(3x+2)$ $\quad$ and $\quad$ $\dfrac{d^2y}{dx^2} = 24x+8$

Turning points occur either when $4x = 0$, i.e. when $x = 0$, or when $(3x+2) = 0$

i.e. when $x = -\dfrac{2}{3}$.

First turning point $= (0, 12)$, so $\dfrac{d^2y}{dx^2} = 8$ when $x = 0$, i.e. positive, so a minimum.

Second turning point $= (-\dfrac{2}{3}, 12.59)$, so $\dfrac{d^2y}{dx^2} = -8$, when $x = -\dfrac{2}{3}$, i.e. negative, so a maximum.

Chapter 22
Differentiation VI: Partial differentiation and polar coordinates

Self-test 22.1

22.1.1 $\left(\dfrac{\partial T}{\partial p}\right)_V$ 22.1.2 $\left(\dfrac{\partial V}{\partial p}\right)_T$ 22.1.3 $\left(\dfrac{\partial T}{\partial V}\right)_p$

Self-test 22.2

22.2.1 $\left(\dfrac{\partial x}{\partial c}\right)_{a,b,d}$ 22.2.2 $\left(\dfrac{\partial a}{\partial B}\right)_{t,u}$

Self-test 22.3

22.3.1 $\left(\dfrac{\partial f}{\partial x}\right)_{y,z} = 8x + 5y^2$ $\left(\dfrac{\partial f}{\partial y}\right)_{x,z} = 10xy,$ $\left(\dfrac{\partial f}{\partial z}\right)_{x,y} = -6$

22.3.2 $\left(\dfrac{\partial f}{\partial x}\right)_{y,z} = 16.96x^{4.3}$ $\left(\dfrac{\partial f}{\partial y}\right)_{x,z} = -\dfrac{5}{y^2 z}$ $\left(\dfrac{\partial f}{\partial z}\right)_{x,y} = -\dfrac{5}{yz^2} - \dfrac{4}{z}$

Self-test 22.4

22.4.1   Using the partial differential relationship in eqn (22.2),

$$\left(\frac{\partial p}{\partial T}\right)_V = -\left(\frac{\partial p}{\partial V}\right)_T \left(\frac{\partial V}{\partial T}\right)_p$$

from the definitions given:

$$\left(\frac{\partial V}{\partial T}\right)_p = \alpha V, \quad \text{and} \quad \left(\frac{\partial V}{\partial p}\right)_T = -\kappa V,$$

Applying eqn (22.1) to the equation for $(\partial V/\partial p)_T$, we find:

$$\left(\frac{\partial p}{\partial V}\right)_T = \frac{1}{(\partial V/\partial p)_T} = \frac{-1}{\kappa V}.$$

According to the partial derivative relationship given in eqn (22.2),

$$\left(\frac{\partial p}{\partial T}\right)_V = -\left(\frac{-1}{\kappa V}\right) \times \alpha V \quad \text{so,} \quad \left(\frac{\partial p}{\partial T}\right)_V = \frac{\alpha}{\kappa}$$

Self-test 22.5

22.5.1 $dE = \left(\dfrac{\partial E}{\partial v}\right)_m dv + \left(\dfrac{\partial E}{\partial m}\right)_v dm$

22.5.2 $dS = \left(\dfrac{\partial S}{\partial C_p}\right)_T dC_p + \left(\dfrac{\partial S}{\partial T}\right)_{C_p} dT$

22.5.3 $dT = \left(\dfrac{\partial T}{\partial n}\right)_{p,V} dn + \left(\dfrac{\partial T}{\partial p}\right)_{n,V} dp + \left(\dfrac{\partial T}{\partial V}\right)_{n,p} dV$

Self-test 22.6

22.6.1 $\left(\dfrac{\partial p}{\partial V}\right)_{n,T} = -\dfrac{nRT}{V^2}$ $\left(\dfrac{\partial p}{\partial n}\right)_{V,T} = \dfrac{RT}{V}$ $\left(\dfrac{\partial p}{\partial T}\right)_{V,n} = \dfrac{nR}{V}$

22.6.2 $\dfrac{\partial}{\partial T}\left(\dfrac{\partial p}{\partial V}\right) = -\dfrac{nR}{V^2}$ $\dfrac{\partial}{\partial n}\left(\dfrac{\partial p}{\partial V}\right) = \dfrac{RT}{V^2}$ $\dfrac{\partial}{\partial n}\left(\dfrac{\partial p}{\partial T}\right) = \dfrac{R}{V}$

$\dfrac{\partial}{\partial V}\left(\dfrac{\partial p}{\partial T}\right) = -\dfrac{nR}{V^2}$ $\dfrac{\partial}{\partial V}\left(\dfrac{\partial p}{\partial n}\right) = -\dfrac{RT}{V^2}$ $\dfrac{\partial}{\partial T}\left(\dfrac{\partial p}{\partial n}\right) = \dfrac{R}{V}$

It should be clear how these six differentials readily fall naturally into three pairs. The results show that double partial differentiation *is* associative, and the Euler reciprocity relation *does* hold.

Chapter 23
Integration I: Reversing the process of differentiation

Self-test 23.1

23.1.1 $y = \dfrac{x^8}{8} + c$

23.1.2 $y = \dfrac{2x^6}{6} + c = \dfrac{x^6}{3} + c$

23.1.3 $y = \dfrac{x^{2.3}}{2.3} + c$

23.1.4 $y = \dfrac{3.7\, x^{9.2}}{9.2} + c$

23.1.5 $y = \dfrac{6}{-2x^2} = \dfrac{3}{x^2} + c$

23.1.6   We start by noting how $\dfrac{1}{x^{4.2}} = x^{-4.2}$. Integration then yields $y = \dfrac{-1}{3.2\, x^{3.2}} + 5x + c$

23.1.7 $y = \dfrac{x^3}{3} + \dfrac{3}{2}x^2 - 6x + c$

23.1.8 $\dfrac{4}{x^5} = 4x^{-5}$ and $\dfrac{3.5}{x^2} = 3.5x^{-2}$ so $y = \dfrac{-4}{4x^4} - \dfrac{-3.5}{x} = -\dfrac{1}{x^4} + \dfrac{3.5}{x} + c$

23.1.9 $\sqrt{x^7} = x^{7/2}$ so $y = \dfrac{x^{9/2}}{9/2} + c = \dfrac{2\, x^{9/2}}{9} + c.$

23.1.10 $\sqrt{x^6} = x^{6/2} = x^3$ so $y = \dfrac{x^4}{4} - \dfrac{2x^3}{3} + c$

Self-test 23.2

23.2.1 $y = \dfrac{\sin 4x}{4} + c$

23.2.2 $y = \dfrac{6}{4}\cos 4x + c = \dfrac{3}{2}\cos 4x + c$

23.2.3 $y = -\dfrac{5}{2x^2} + c$.

23.2.4 $y = 3\ln x + c$

23.2.5 $y = \dfrac{e^{5x}}{10} + c$

23.2.6 $y = 96.4\, x + c$

23.2.7 $y = \dfrac{x^5}{5} + \dfrac{\cos 2x}{2} + c$

23.2.8 $y = \dfrac{e^{dx}}{d} - \dfrac{e^{ex}}{e} + c$

23.2.9 $y = 5\ln x + \dfrac{e^{3.2x}}{3.2} + c$

23.2.10 $y = \dfrac{b}{a}\sin ax + \dfrac{a}{b}\cos bx = \dfrac{b^2 \sin ax + a^2 \cos bx}{ab} + c$

Self-test 23.3

23.3.1   The equation of the line is $y = \dfrac{x^4}{4} + c$.   Substituting yields $c = 9$.

23.3.2   The equation of the line is $y = \dfrac{e^{3x}}{3} + c$.   Substituting yields $c = -4.70$.

23.3.3   The equation of the line is $y = \dfrac{x^{1.5}}{1.5} + c$.   Substituting yields $c = -9.35$.

23.3.4   The equation of the line is $y = \ln x + c$.   Substituting yields $c = 3.61$.

23.3.5   The equation of the line is $y = \dfrac{x^4}{4} - \dfrac{x^3}{3} + c$.   Substituting yields $c = 0.75$.

23.3.6   The equation of the line is $y = 3 \ln x + c$.   Substituting yields $c = 0.92$.

Chapter 24
Integration II: Separating the variables and integration with limits

Self-test 24.1

24.1.1   Separating the variables: $dy = zx^3\, dx$
Adding the appropriate integration symbolism: $\int dy = \int zx^3\, dx$

Integration: $y = \dfrac{zx^4}{4} + \text{constant}$

24.1.2   Slight rearrangement: $\dfrac{1}{y^2} \dfrac{dy}{dx} = x^2$

Separating the variables and adding the appropriate integration symbolism:

$$\int \frac{1}{y^2} dy = \int x^2\, dx$$

Integration: $-\dfrac{1}{y} = \dfrac{x^3}{3} + \text{constant}$

24.1.3   Slight rearrangement: $\dfrac{1}{c^3} \dfrac{dc}{de} = \dfrac{1}{e}$

Separating the variables: $\dfrac{1}{c^3}\, dc = \dfrac{1}{e}\, de$

Adding the appropriate integration symbolism: $\int \dfrac{1}{c^3} dc = \int \dfrac{1}{e} de$

Integration: $-\dfrac{1}{2c^2} = \ln e + \text{constant}$

24.1.4   Slight rearrangement: $\dfrac{1}{h^3} \dfrac{dh}{dg} = 4 \sin g$

Separating the variables: $\dfrac{1}{h^3}\, dh = 4 \sin g\, dg$

Adding the appropriate integration symbolism: $\int \dfrac{1}{h^3} dh = \int 4 \sin g\, dg$

Integration: $-\dfrac{1}{2h^2} = -4 \cos g + \text{constant}$

Slight rearrangement: $\dfrac{1}{h^2} = 8 \cos g + \text{constant}'$

The prime on the constant alerts us to the way its value changes between the last two lines.

Self-test 24.2

24.2.1 $y = \left[\dfrac{x^5}{5} - \dfrac{x^3}{3}\right]_1^2 = \left(\dfrac{2^5}{5} - \dfrac{2^3}{3}\right) - \left(\dfrac{1}{5} - \dfrac{1}{3}\right) = (6.400 - 2.667) - (0.200 - 0.333)$

$= 3.733 - (-0.133) = 3.87\,(\text{to } 2\text{ d.p.})$

24.2.2 $y = \left[\dfrac{e^{3x}}{3}\right]_1^3 = \left(\dfrac{e^{3\times 3}}{3}\right) - \left(\dfrac{e^{3\times 1}}{3}\right) = 2701 - 6.7 = 2694\,(\text{to } 0\text{ d.p.})$

24.2.3 $y = \left[\dfrac{\sin 2x}{2}\right]_{\pi/3}^{\pi} = \left(\dfrac{\sin 2\pi}{2}\right) - \left(\dfrac{\sin(2\pi/3)}{2}\right) = 0 - \dfrac{\sqrt{3}}{4} - 0.433\,(\text{to } 3\text{ d.p.})$

24.2.4 $y = \left[\dfrac{x^4}{4} - \dfrac{x^3}{3} + \dfrac{x^2}{2} - 6x\right]_0^5 = \left(\dfrac{5^4}{4} - \dfrac{5^3}{3} + \dfrac{5^2}{2} - 6\times 5\right) - \left(\dfrac{0^4}{4} - \dfrac{0^3}{3} + \dfrac{0^2}{2} - 6\times 0\right) = 97.1$

24.2.5 $y = [2\ln x]_{9.5}^{10} = 2(\ln 10 - \ln 9.5) = 2\ln\left(\dfrac{10}{9.5}\right) = 2\ln(1.053) = 2\times 0.052 = 0.103$

24.2.6 $y = \left[-\dfrac{\cos 6x}{18}\right]_0^{\pi} = \left(-\dfrac{\cos 6\pi}{18}\right) - \left(-\dfrac{\cos 0}{18}\right) = -\dfrac{1}{18} + \dfrac{1}{18} = 0.$

Chapter 25
Integration III: Integration by parts, by substitution and integration tables

Self-test 25.1

25.1.1 Let $u = x$ and $dv/dx = \sin x$. Therefore, $du/dx = 1$ and $v = -\cos x$. Inserting terms into eqn (25.2) yields: $(-x\times\cos x) - \int(-\cos x)\,dx = -x\cos x + \sin x + c.$

25.1.2 Let $u = \ln x$ and $dv/dx = x$. Therefore, $du/dx = \dfrac{1}{x}$ and $v = \dfrac{x^2}{2}$. Inserting terms into

eqn (25.2) yields: $\left(\ln x \times \dfrac{x^2}{2}\right) - \int \dfrac{x^2}{2}\times\dfrac{1}{x}\,dx = \dfrac{x^2}{2}\ln x - \ln x - \dfrac{x^2}{4} + c.$

25.1.3 Let $u = x$ and $dv/dx = e^{ax}$. Therefore, $du/dx = 1$ and $v = \dfrac{e^{ax}}{a}$. Inserting terms into

eqn (25.2) yields: $x\times\dfrac{1}{a}e^{ax} - \int\dfrac{1}{a}\,e^{ax}\times 1\,dx = \dfrac{x}{a}e^{ax} - \dfrac{1}{a^2}e^{ax} + c.$

25.1.4 Let $u = x^2$ and $dv/dx = \sin x$. Therefore, $du/dx = 2x$ and $v = -\cos x$. Inserting terms into eqn (25.2) yields:

$\int x^2\,\sin x\,dx = -x^2\cos x - 2\int x\,(-\cos x)\,dx.$

The integral here requires an additional process of integrating by parts. The result (from Worked Example 25.1) is $x\sin x + \cos x$.

The overall result is therefore, $\int x^2\,\sin x\,dx = -x^2\cos x + 2(x\sin x + \cos x) + c$, where the brackets merely indicate where the terms come from.

25.1.5 Let $u = \ln x$ and $dv/dx = x^2$. Therefore, $du/dx = 1/x$ and $v = \dfrac{x^3}{3}$. Inserting terms into

eqn (25.2) yields: $\dfrac{x^3}{3}\ln x - \int\dfrac{x^3}{3}\left(\dfrac{1}{x}\right)dx = \dfrac{x^3}{3}\ln x - \dfrac{x^3}{9} + c.$

Self-test 25.2

25.2.1 Let $u = x^6 + 7$ and $\dfrac{du}{dx} = 6x^5$. Substituting these values into the original integral

yields: $\dfrac{1}{6}\int u^4 \dfrac{du}{dx}\,dx = \dfrac{1}{6}\int u^4\,du$, which is a simple integral equal to $\dfrac{u^5}{30} + c$.

Finally, we can substitute back into this expression for u in terms of x to give:

$\dfrac{1}{30}\left(x^6 + 7\right)^5 + c\,(x^6 + 7) + c$.

25.2.2 Let $u = \sin x$ and $\dfrac{du}{dx} = \cos x$. Substituting these values into the original integral

yields: $\int e^u \dfrac{du}{dx}\,dx = \int e^u\,du = e^u + c$. Finally, we can substitute back into this

expression for u to give: $e^{\sin x} + c$.

25.2.3 Let $u = \cos x$ and $\dfrac{du}{dx} = -\sin x$. If $x = 0$, $u = 1$ and if $x = \pi/2$, $u = 0$. Substituting these

values into the original integral yields: $\int_1^0 -u^3 \dfrac{du}{dx}\,dx = \int_1^0 -u^3\,du$.

Evaluating the integral: $\left[-\dfrac{u^4}{4}\right]_1^0 = \left(0 - \left(-\dfrac{1^4}{4}\right)\right) = \dfrac{1}{4}$.

Self-test 25.3

25.3.1 $\psi = A\left(1 + \dfrac{\left(-\dfrac{Zr}{a_0}\right)^1}{1!} + \dfrac{\left(-\dfrac{Zr}{a_0}\right)^2}{2!} + \dfrac{\left(-\dfrac{Zr}{a_0}\right)^3}{3!} + \ldots\right)$. The power of '1' is superfluous, but

does help demonstrate the trend.

It can be helpful to expand the terms, yielding: $A - \dfrac{AZr}{a_0} + \dfrac{AZ^2r^2}{2a_0^2} - \dfrac{AZ^3r^3}{6a_0^3} + \ldots$

Self-test 25.4

25.4.1 $e^{4x} = 1 + 4x + \dfrac{(4x)^2}{2!} + \dfrac{(4x)^3}{3!} + \dfrac{(4x)^4}{4!} + \ldots$.

so $\int e^{4x}\,dx = x + 4 \times \dfrac{x^2}{2} + \dfrac{4^2}{2!} \times \dfrac{x^3}{3} + \dfrac{4^3}{3!} \times \dfrac{x^4}{4} + \dfrac{4^4}{4!} \times \dfrac{x^5}{5} + \ldots + c$, where $c = \frac{1}{4}$.

Tidying the denominators yields,

$\int e^{4x}\,dx = \dfrac{1}{4} + x + \dfrac{4x^2}{2!} + \dfrac{4^2x^3}{3!} + \dfrac{4^3x^4}{4!} + \dfrac{4^4x^5}{5!} + \ldots$

We then multiply through by $4/4 = 1$:

$\int e^{4x}\,dx = \dfrac{1}{4} \times 1 + \dfrac{1}{4} \times \dfrac{4x}{1!} + \dfrac{1}{4} \times \dfrac{4^2x^2}{2!} + \dfrac{1}{4} \times \dfrac{4^3x^3}{3!} + \dfrac{1}{4} \times \dfrac{4^4x^4}{4!} + \dfrac{1}{4} \times \dfrac{4^5x^5}{5!} + \ldots$

Finally, we factorize out the ¼ terms, and simplify the numerators:

$\int e^{4x}\,dx = \dfrac{1}{4}\left(1 + 4x + \dfrac{(4x)^2}{2!} + \dfrac{(4x)^3}{3!} + \dfrac{(4x)^4}{4!} + \dfrac{(4x)^5}{5!} + \ldots\right) = \dfrac{1}{4}e^{4x}$

25.4.2 $xe^{8x} = x\left(1 + 8x + \dfrac{(8x)^2}{2!} + \dfrac{(8x)^3}{3!} + \dfrac{(8x)^4}{4!} + \ldots\right)$

$= x + 8x^2 + \dfrac{8^2x^3}{2!} + \dfrac{8^3x^4}{3!} + \dfrac{8^4x^5}{4!} + \ldots$

So $\int xe^{8x}\,dx = \dfrac{x^2}{2} + \dfrac{8x^3}{3} + \dfrac{8^2}{2!} \times \dfrac{x^4}{4} + \dfrac{8^3}{3!} \times \dfrac{x^5}{5} + \dfrac{8^4}{4!} \times \dfrac{x^6}{6} + \ldots$

This expression does not factorize readily, but will converge following the insertion of values of x.

25.4.3 $3x^2 e^x = 3x^2 \left(1 + x + \dfrac{x^2}{2!} + \dfrac{x^3}{3!} + \dfrac{x^4}{4!} + ...\right) = 3x^2 + 3x^3 + \dfrac{3x^4}{2!} + \dfrac{3x^5}{3!} + \dfrac{3x^6}{4!} + ...$

so $\displaystyle\int 3x^2 \exp x \, dx = \dfrac{3 \times x^3}{3} + \dfrac{3 \times x^4}{4} + \dfrac{3}{2!} \times \dfrac{x^5}{5} + \dfrac{3}{3!} \times \dfrac{x^6}{6} + \dfrac{3}{4!} \times \dfrac{x^7}{7} + ...$

This expression does not factorize readily, but will converge following the insertion of values of x.

25.4.4 $x^2 \exp(zx/a_0) = x^2 \left(1 + (zx/a_0) + \dfrac{(zx/a_0)^2}{2!} + \dfrac{(zx/a_0)^3}{3!} + \dfrac{(zx/a_0)^4}{4!} + ...\right)$

$= x^2 + (z/a_0)x^3 + \dfrac{(z/a_0)^2 x^4}{2!} + \dfrac{(z/a_0)^3 x^5}{3!} + \dfrac{(z/a_0)^4 x^6}{4!} + ...$

So $\displaystyle\int x^2 \exp(zx/a_0) \, dx = \dfrac{x^3}{3} + \dfrac{(z/a_0)x^4}{4} + \dfrac{(z/a_0)^2 x^5}{2! \times 5} + \dfrac{(z/a_0)^3 x^6}{3! \times 6} + \dfrac{(z/a_0)^4 x^7}{4! \times 7} + ...$

Again, this expression does not factorize readily, but will converge following the insertion of values of x.

Chapter 26:
Integration IV: Area and volume determination

Self-test 26.1

26.1.1 $\text{Area} = \displaystyle\int_3^5 (x^2 + 4)dx = \left[\dfrac{x^3}{3} + 4x\right]_3^5 = \left(\dfrac{125}{3} + 20\right) - \left(\dfrac{27}{3} + 12\right) = (41\tfrac{2}{3} + 20) - (9 + 12) = 40\tfrac{2}{3}.$

26.1.2 $\text{Area} = \displaystyle\int_2^6 \dfrac{1}{x^2} \, dx = -\left[\dfrac{1}{x}\right]_2^6 = -\left(\dfrac{1}{6} - \dfrac{1}{2}\right) = \tfrac{1}{3}.$

26.1.3 $\displaystyle\int_{T_1}^{T_2} \dfrac{C_p}{T} \, dT = \int_{T_1}^{T_2}\left(\dfrac{25.8 + 1.2 \times 10^{-2}T}{T}\right)dT = \int_{T_1}^{T_2}\left(\dfrac{25.8}{T} + 1.2 \times 10^{-2}\right)dT$

Integrating yields, $\Delta S = \left[25.8\ln T + 1.2 \times 10^{-2}T\right]_{T_1}^{T_2}.$

Inserting limits yields, $\Delta S\left(25.8\ln\left(\dfrac{T_2}{T_1}\right) + 1.2 \times 10^{-2}(T_2 - T_1)\right)$

Inserting values yields, $\Delta S = 25.8 \ln\left(\dfrac{350\,\text{K}}{300\,\text{K}}\right) + \{1.2 \times 10^{-2}(350\text{K} - 300\text{ K})\}$

so $\Delta S = 4.58 \text{ J K}^{-1} \text{ mol}^{-1}.$

Self-test 26.2

26.2.1   The line factorizes to $4x (x - 1)(x - 2)$, so the line crosses the x-axis at $x = 0$, 1 and 2. The limits will therefore be 0 and 1, and 1 and 2.

Integration yields, $\displaystyle\int y \, dy = [x^4 - 4x^3 + 4x^2].$

The first area $= [x^4 - 4x^3 + 4x^2]_0^1 = (1 - 4 + 4) - (0) = 1$ area unit.

The second area

$= [x^4 - 4x^3 + 4x^2]_1^2 = (16 - 32 + 16) - (1 - 4 + 4) = (0 - 1) = -1$ area unit.

So the total area $= 1 + 1 = 2$ area units.

26.2.2   The graph cuts the axis at $\pi/2$, so the limits will be 0 and $\pi/2$, and $\pi/2$ and π. $y = \cos x$, between the limits of 0 and π.

Integration yields, $\int_{x_1}^{x_2} y\,dx = \int_{x_1}^{x_2} \cos x\,dx = [\sin x]_{x_1}^{x_2}$

The first area	$= [\sin x]_0^{\pi/2} = (\sin \pi/2 - \sin 0) = 1 - 0 = 1$
The second area	$= [\sin x]_0^{\pi/2} = (\sin \pi - \sin \pi/2) = 0 - 1 = -1$
So the total area	$= 1 + 1 = 2$ area units.

Self-test 26.3.3

26.3.1   Using the result from Worked Example 26.4, the volume of this segment of a sphere is $V = 2\pi \left[r^2 x - \dfrac{x^3}{3} \right]_1^2$.

Inserting limits gives: $2\pi \left(9\times2 - \tfrac{1}{3}\times8\right) - \left(9\times1 - \tfrac{1}{3}\times1^3\right) = 2\pi\times(6\tfrac{2}{3}) = 41.9$. volume units.

The complete sphere of this radius has a volume of 113 volume units, so this segment comprises about 37% of the sphere's volume.

26.3.2   By definition, the volume V of this object is $\pi\int_3^5 y^2\,dx = \pi\int_3^5 3x\,dx$.

The integral is, $\pi \left[\dfrac{3x^2}{2} \right]_3^5 = \dfrac{3\pi}{2}(5^2 - 3^2) = 24\pi = 75.4$ volume units.

26.3.3   To find coordinates where the two lines overlap, we equate the two equations: $x^2 = -2x + 8$, so $0 = x^2 + 2x - 8$, so $0 = (x - 2)(x + 4)$. The two lines intersect at values of $x = -4$ and $+2$.

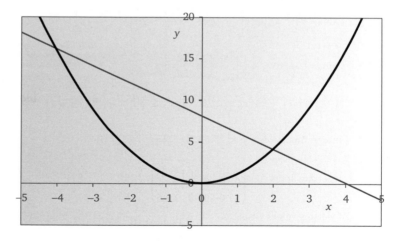

Area beneath the parabolic curve $= \int_{-4}^{2} x^2\,dx \left[\dfrac{x^3}{3} \right]_{-4}^{2} = 24$ area units.

Area beneath the line: $\int_{-4}^{2} -2x + 8\,dx = [-x^2 + 8x]_{-4}^{2} = 60$ area units. So the area of overlap $= 60 - 24 = 36$ area units.

Self-test 26.4

26.4.1   We first integrate with respect to z, treating x and y as constants:

$$\int_1^2 (x^2 + yz)\,dz = \left[x^2 z + \frac{yz^2}{2}\right]_1^2 = x^2 + \frac{3y}{2}$$

Note how z has completely disappeared from the answer. Next, we integrate the middle integral with respect to y; we treat x as a constant:

$$\int_1^2 \left(x^2 + \frac{3y}{2}\right) dy = \left[x^2 y + \frac{3y^2}{4}\right]_1^2 = x^2 + \frac{9}{4}$$

Note that y has disappeared from the expression on the right. Finally, we integrate the outer integral with respect to x:

$$\int_0^1 \left(x^2 + \frac{9}{4}\right) dx = \left[\frac{x^3}{3} + \frac{9x}{4}\right]_0^1 = \frac{31}{12}$$

It is worth checking, to show we obtain the same answer whatever the order in which we perform the integration.

26.4.2   Rewriting the data as an integral yields, $\int_0^4 \int_0^1 \int_0^{yz} dx\,dy\,dz$.

We then evaluate each integral in succession:

The x-plane: $\int_0^{yz} dx = [1]_0^{yz} = yz$

The y-plane: $\int_0^1 yz\,dy = \left[\frac{y^2 z}{2}\right]_0^1 = \frac{z}{2}$

The z-plane: $\int_0^4 \frac{z}{2}\,dz = \left[\frac{z^2}{4}\right]_0^4 = 4$

so the volume $= 4$ units.

Note: in this example, we must perform the integration with respect to x *before* the other integrations, because the limits depend on y and z.

Chapter 27:
Matrices

Self-test 27.1

27.1.1   For cyclohexane (**II**)

$$\begin{pmatrix} 6 & 1 & 0 & 0 & 0 & 1 \\ 1 & 6 & 1 & 0 & 0 & 0 \\ 0 & 1 & 6 & 1 & 0 & 0 \\ 0 & 0 & 1 & 6 & 1 & 0 \\ 0 & 0 & 0 & 1 & 6 & 1 \\ 1 & 0 & 0 & 0 & 1 & 6 \end{pmatrix}$$

27.1.2   For 2-methylpropanoic acid (**III**)

$$\begin{pmatrix} 1 & 1 & 0 & 0 & 0 & 0 & 0 \\ 1 & 8 & 1 & 0 & 0 & 0 & 0 \\ 0 & 1 & 6 & 1 & 0 & 0 & 2 \\ 0 & 0 & 1 & 6 & 1 & 1 & 0 \\ 0 & 0 & 0 & 1 & 6 & 0 & 0 \\ 0 & 0 & 0 & 1 & 0 & 6 & 0 \\ 0 & 0 & 2 & 0 & 0 & 0 & 8 \end{pmatrix}$$

Self-test 27.2

27.2.1   For methylethylamine, (**V**)

$$\begin{pmatrix} \alpha & 1.75\beta & 0 & 0 \\ 1.75\beta & 1.5\alpha & 1.75\beta & 0 \\ 0 & 1.75\beta & \alpha & \beta \\ 0 & 0 & \beta & \alpha \end{pmatrix}$$

$$
\begin{array}{c}
\quad 1 \quad\; 2 \quad\; 3 \;\; 4 \;\; 5 \quad\; 6 \\
\begin{array}{c} 1 \\ 2 \\ 3 \\ 4 \\ 5 \\ 6 \end{array}
\begin{pmatrix}
1.4\alpha & 1.6\beta & 0 & 0 & 1.6\beta & 0 \\
1.6\beta & \alpha & \beta & 0 & 0 & 0 \\
0 & \beta & \alpha & \beta & 0 & \beta \\
0 & 0 & \beta & \alpha & \beta & 0 \\
1.6\beta & 0 & 0 & \beta & \alpha & 0 \\
0 & 0 & \beta & 0 & 0 & \alpha
\end{pmatrix}
\end{array}
$$

27.2.2   For thiophene, (**VI**)

Self-test 27.3

27.3.1 $\begin{pmatrix} 1 & 2 \\ 4 & 6 \end{pmatrix} + \begin{pmatrix} 2 & -3 \\ 0 & 2 \end{pmatrix} = \begin{pmatrix} 1+2 & 2+(-3) \\ 4+0 & 6+2 \end{pmatrix} = \begin{pmatrix} 3 & -1 \\ 4 & 8 \end{pmatrix}$

27.3.2 $(3a\ 4b\ 8c\ 5d) - (a\ 3b\ 9c\ 5d) = ((3a-a)\ (4b-3b)\ (8c-9c)\ (5d-5d))$

$$= (2a\ \ b\ \ -c\ \ 0)$$

27.3.3 $\begin{pmatrix} 4 & -3 & 3 \\ 2 & 7 & -2 \\ -10 & 8 & 5 \end{pmatrix} - \begin{pmatrix} 2 & 9 & 0 \\ 3 & 8 & -2 \\ 2 & 7 & 2 \end{pmatrix} = \begin{pmatrix} 4-2 & -3-9 & 3-0 \\ 2-3 & 7-8 & -2-(-2) \\ -10-2 & 8-7 & 5-2 \end{pmatrix} = \begin{pmatrix} 2 & -12 & 3 \\ -1 & -1 & 0 \\ -12 & 1 & 3 \end{pmatrix}$

Self-test 27.4

27.4.1 $2\begin{pmatrix} 2 & 1 & -5 \\ 5 & 8 & 6 \\ -3 & 2 & 0 \end{pmatrix} = \begin{pmatrix} 2\times2 & 2\times1 & 2\times(-5) \\ 2\times5 & 2\times8 & 2\times6 \\ 2\times(-3) & 2\times2 & 2\times0 \end{pmatrix} = \begin{pmatrix} 4 & 2 & -10 \\ 10 & 16 & 12 \\ -6 & 4 & 0 \end{pmatrix}$

27.4.2 $4.2\begin{pmatrix} 0 & 3 \\ 5 & 1 \end{pmatrix} = \begin{pmatrix} 4.2\times0 & 4.2\times3 \\ 4.2\times5 & 4.2\times1 \end{pmatrix} = \begin{pmatrix} 0 & 12.6 \\ 21 & 4.2 \end{pmatrix}$

27.4.3 $a\begin{pmatrix} a & 1 & c^2 & 2ac \\ 0 & b & b^2 & ab \end{pmatrix} = \begin{pmatrix} a\times a & a\times1 & a\times c^2 & a\times2ac \\ a\times0 & a\times b & a\times b^2 & a\times ab \end{pmatrix} = \begin{pmatrix} a^2 & a & ac^2 & 2a^2c \\ 0 & ab & ab^2 & a^2b \end{pmatrix}$

27.4.4 $3x\begin{pmatrix} 2 & 5xy \\ 2x^{2.2} & x^2 \\ 4.3/x & 0 \end{pmatrix} = \begin{pmatrix} 3x\times2 & 3x\times5xy \\ 3x\times2x^{2.2} & 3x\times x^2 \\ 3x\times(4.3/x) & 3x\times0 \end{pmatrix} = \begin{pmatrix} 6x & 15x^2y \\ 6x^{3.2} & 3x^3 \\ 12.9 & 0 \end{pmatrix}$

Self-test 27.5

27.5.1 $\begin{pmatrix} 1 & 4 \\ 3 & 0 \end{pmatrix}\begin{pmatrix} 3 & 4 \\ 0 & 2 \end{pmatrix} = \begin{pmatrix} 1\times3+4\times0 & 1\times4+4\times2 \\ 3\times3+0\times0 & 3\times4+0\times2 \end{pmatrix} = \begin{pmatrix} 3 & 12 \\ 9 & 12 \end{pmatrix}$

27.5.2 $\begin{pmatrix} 2 & 0 & 3 \\ 1 & 2 & -1 \end{pmatrix}\begin{pmatrix} 1 \\ 2 \\ 4 \end{pmatrix} = \begin{pmatrix} 2\times1+0\times2+3\times4 \\ 1\times1+2\times2+(-1)\times4 \end{pmatrix} = \begin{pmatrix} 14 \\ 1 \end{pmatrix}$

27.5.3 $\begin{pmatrix} a & 0 \\ a^2 & 1 \end{pmatrix}\begin{pmatrix} a \\ 2a \end{pmatrix} = \begin{pmatrix} a\times a+0\times2a \\ a^2\times a+1\times2a \end{pmatrix} = \begin{pmatrix} a^2 \\ a^3+2a \end{pmatrix}$

Self-test 27.6

27.6.1   If a general point, (x, y, z) is reflected in an x–y mirror plane, then its image would lie at $(x, y, -z)$. Therefore, the matrix for this transformation is:

$$
R_{xy} = \begin{pmatrix} 1 & 0 & 0 \\ 0 & 1 & 0 \\ 0 & 0 & -1 \end{pmatrix}
$$

If we also look at the images of the points (1,0,0), (0,1,0) and (0,0,1):

$$\begin{pmatrix}1\\0\\0\end{pmatrix}\rightarrow\begin{pmatrix}1\\0\\0\end{pmatrix},\quad\begin{pmatrix}0\\1\\0\end{pmatrix}\rightarrow\begin{pmatrix}0\\1\\0\end{pmatrix},\quad\begin{pmatrix}0\\0\\1\end{pmatrix}\rightarrow\begin{pmatrix}0\\0\\-1\end{pmatrix}$$

then we can see that the column matrices representing the image of each of these points is also a column of the transformation matrix.

27.6.2　If a general point, (x, y, z) is rotated through 90° (in the anticlockwise direction) about the z-axis, then its image would lie at $(-y, x, z)$. Therefore, the matrix for this transformation is:

$$R_{90°}=\begin{pmatrix}0 & -1 & 0\\1 & 0 & 0\\0 & 0 & 1\end{pmatrix}.$$

If we also look at the images of the points (1,0,0), (0,1,0) and (0,0,1):

$$\begin{pmatrix}1\\0\\0\end{pmatrix}\rightarrow\begin{pmatrix}0\\1\\0\end{pmatrix},\quad\begin{pmatrix}0\\1\\0\end{pmatrix}\rightarrow\begin{pmatrix}-1\\0\\0\end{pmatrix},\quad\begin{pmatrix}0\\0\\1\end{pmatrix}\rightarrow\begin{pmatrix}0\\0\\1\end{pmatrix}$$

then we can see that the column matrices representing the image of each of these points is also a column of the transformation matrix.

27.6.3　The general matrix for the two-step transformation is: rotation through 90° (in the anticlockwise direction) followed by reflection in an x–y mirror plane.

$$\mathbf{R}=\mathbf{R}_{xy}\mathbf{R}_{90°}=\begin{pmatrix}1 & 0 & 0\\0 & 1 & 0\\0 & 0 & -1\end{pmatrix}\begin{pmatrix}0 & -1 & 0\\1 & 0 & 0\\0 & 0 & 1\end{pmatrix}=\begin{pmatrix}0 & -1 & 0\\1 & 0 & 0\\0 & 0 & -1\end{pmatrix}.$$

We can see that after the application of these two matrices, a general point (x, y, z) will move to $(-y, x, -z)$. This is not equivalent to a simple rotation or reflection.

Self-test 27.7

27.7.1　$\begin{vmatrix}1 & 2\\6 & 4\end{vmatrix}=1\times4-2\times6=-8$

27.7.2　$\begin{vmatrix}1 & -2 & 7\\3 & 0 & 2\\2 & 1 & 6\end{vmatrix}=1\begin{vmatrix}0 & 2\\1 & 6\end{vmatrix}-(-2)\begin{vmatrix}3 & 2\\2 & 6\end{vmatrix}+7\begin{vmatrix}3 & 0\\2 & 1\end{vmatrix}$

$\begin{vmatrix}0 & 2\\1 & 6\end{vmatrix}=0\times6-2\times1=-2;\quad\begin{vmatrix}3 & 2\\2 & 6\end{vmatrix}=3\times6-2\times2=14;\quad\begin{vmatrix}3 & 0\\2 & 1\end{vmatrix}=3\times1-0\times2=3$

therefore, $\begin{vmatrix}1 & -2 & 7\\3 & 0 & 2\\2 & 1 & 6\end{vmatrix}=1\times(-2)-(-2)\times14+7\times3=47$

27.7.3　$\begin{vmatrix}a & 2s & 4s\\-a & a & 4as\\0 & 3b & 0\end{vmatrix}=a\begin{vmatrix}a & 4as\\3b & 0\end{vmatrix}-2s\begin{vmatrix}-a & 4as\\0 & 0\end{vmatrix}+4s\begin{vmatrix}-a & a\\0 & 3b\end{vmatrix}$

$\begin{vmatrix}a & 4as\\3b & 0\end{vmatrix}=a\times0-4as\times3b=-12abs$

$\begin{vmatrix}-a & 4as\\0 & 0\end{vmatrix}=-a\times0-4as\times0=0$, i.e. this minor has no determinant

$\begin{vmatrix}-a & a\\0 & 3b\end{vmatrix}=(-a)\times3b-a\times0=-3ab$

Therefore, $\begin{vmatrix} a & 2s & 4s \\ -a & a & 4as \\ 0 & 3b & 0 \end{vmatrix} = a \times (-12abs) - 2s \times 0 + 4s \times (-3ab)$

$$= -12a^2bs - 12abs = -12abs(a+1)$$

27.7.4 As before, we must first reduce the determinant into cofactors and minors:

$$\begin{pmatrix} 1 & 3 & 2 & 8 \\ 3 & 1 & 6 & 3 \\ 2 & 6 & 1 & 0 \\ 8 & 3 & 0 & 1 \end{pmatrix} = 1\begin{vmatrix} 1 & 6 & 3 \\ 6 & 1 & 0 \\ 3 & 0 & 1 \end{vmatrix} - 3\begin{vmatrix} 3 & 6 & 3 \\ 2 & 1 & 0 \\ 8 & 0 & 1 \end{vmatrix} + 2\begin{vmatrix} 3 & 1 & 3 \\ 2 & 6 & 0 \\ 8 & 3 & 1 \end{vmatrix} - 8\begin{vmatrix} 3 & 1 & 6 \\ 2 & 6 & 1 \\ 8 & 3 & 0 \end{vmatrix}$$

Unfortunately, each of the four 3×3 determinants will also need to be reduced to form the corresponding cofactors, as follows:

$$\begin{vmatrix} 1 & 6 & 3 \\ 6 & 1 & 0 \\ 3 & 0 & 1 \end{vmatrix} = 1\begin{vmatrix} 1 & 0 \\ 0 & 1 \end{vmatrix} - 6\begin{vmatrix} 6 & 0 \\ 3 & 1 \end{vmatrix} + 3\begin{vmatrix} 6 & 1 \\ 3 & 0 \end{vmatrix} = 1 \times (1) - 6 \times (6) + 3 \times (-3) = 1 - 36 - 9 = -44.$$

$$\begin{vmatrix} 3 & 6 & 3 \\ 2 & 1 & 0 \\ 8 & 0 & 1 \end{vmatrix} = 3\begin{vmatrix} 1 & 0 \\ 0 & 1 \end{vmatrix} - 6\begin{vmatrix} 2 & 0 \\ 8 & 1 \end{vmatrix} + 3\begin{vmatrix} 2 & 1 \\ 8 & 0 \end{vmatrix} = 3 \times (1) - 6 \times (2) + 3 \times (-8) = 3 - 12 + (-24) = -33.$$

$$\begin{vmatrix} 3 & 1 & 3 \\ 2 & 6 & 0 \\ 8 & 3 & 1 \end{vmatrix} = 3\begin{vmatrix} 6 & 0 \\ 3 & 1 \end{vmatrix} - 1\begin{vmatrix} 2 & 0 \\ 8 & 1 \end{vmatrix} + 3\begin{vmatrix} 2 & 6 \\ 8 & 3 \end{vmatrix} = 3 \times (6) - 1 \times (2) + 3 \times (6 - 48) = 18 - 2 + (-126) = -110.$$

$$\begin{vmatrix} 3 & 1 & 6 \\ 2 & 6 & 1 \\ 8 & 3 & 0 \end{vmatrix} = 3\begin{vmatrix} 6 & 1 \\ 3 & 0 \end{vmatrix} - 1\begin{vmatrix} 2 & 1 \\ 8 & 0 \end{vmatrix} + 6\begin{vmatrix} 2 & 6 \\ 8 & 3 \end{vmatrix} = 3 \times (-3) - 1 \times (-8) + 6 \times (6 - 48) = -9 - (-8) + (-252) = -253.$$

So $\begin{vmatrix} 1 & 3 & 2 & 8 \\ 3 & 1 & 6 & 3 \\ 2 & 6 & 1 & 0 \\ 8 & 3 & 0 & 1 \end{vmatrix} = 1 \times (-44) - 3 \times (-33) + 2 \times (-110) - 8 \times (-253) = 1859.$

Self-test 27.8

The effective Hamiltonian matrix is $\begin{pmatrix} \alpha - E & \beta & 0 \\ \beta & \alpha - E & \beta \\ 0 & \beta & \alpha - E \end{pmatrix}$

The determinant is, $\det A = (\alpha - E)\begin{vmatrix} \alpha - E & \beta \\ \beta & \alpha - E \end{vmatrix} - \beta\begin{vmatrix} \beta & \beta \\ 0 & \alpha - E \end{vmatrix} + 0\begin{vmatrix} \beta & \alpha - E \\ 0 & \beta \end{vmatrix}$

$$0 = (\alpha - E)\{(\alpha - E)^2 - \beta^2\} - \beta^2(\alpha - E)$$

This expression readily simplifies to $(\alpha - E)\{(\alpha - E)^2 - 2\beta^2\} = 0$. This equation factorizes as $(\alpha - E)(\alpha - E + \sqrt{2}\beta)(\alpha - E - \sqrt{2}\beta) = 0$. Therefore, the three roots are: $E_1 = \alpha, E_2 = \alpha + \sqrt{2}\beta$ and $E_3 = \alpha - \sqrt{2}\beta$.

The molecular wavefunctions have the form: $\Psi = c_1\phi_1 + c_2\phi_2 + c_3\phi_3$

To find the eigenvectors corresponding to each eigenvalue we must substitute for the eigenvalue back into the eigenvalue equation.

The normalization factor can be found by solving $c_1^2 + c_2^2 + c_3^2 = 1$.

When $E_1 = \alpha$, the eigenvalue equation is:

$$\begin{pmatrix} 0 & \beta & 0 \\ \beta & 0 & \beta \\ 0 & \beta & 0 \end{pmatrix}\begin{pmatrix} c_1 \\ c_2 \\ c_3 \end{pmatrix} = \begin{pmatrix} 0 \\ 0 \\ 0 \end{pmatrix},$$

where c_1, c_2 and c_3 are the molecular orbital coefficients.
If we multiply the matrices, we find:

$c_2 = 0$ and $c_1 = -c_3$.

$\Psi' = c_1\phi_1 - c_1\phi_3 = c_1\left(\phi_1 - \phi_3\right)$

Since $c_1^2 + c_2^2 + c_3^2 = c_1^2 + (-c_1)^2 = 2\,c_1^2 = 1,\ c_1 = \dfrac{1}{\sqrt{2}}$.

The normalized wavefunction is: $\Psi = \dfrac{1}{\sqrt{2}}\left(\phi_1 - \phi_3\right)$.

When $E_2 = \alpha + \sqrt{2}\beta$, the eigenvalue equation is:

$$\begin{pmatrix} -\sqrt{2}\beta & \beta & 0 \\ \beta & -\sqrt{2}\beta & \beta \\ 0 & \beta & -\sqrt{2}\beta \end{pmatrix} \begin{pmatrix} c_1 \\ c_2 \\ c_3 \end{pmatrix} = \begin{pmatrix} 0 \\ 0 \\ 0 \end{pmatrix},$$

where c_1, c_2 and c_3 are the molecular orbital coefficients.
If we multiply the matrices, we find:

$-\sqrt{2}\beta c_1 + \beta c_2 = 0$ So, $c_2 = \sqrt{2}c_1$

$\beta c_2 - \sqrt{2}\beta c_3 = 0$ So, $c_2 = \sqrt{2}c_3$

Overall: $\sqrt{2}c_1 = c_2 = \sqrt{2}c_3\ \ c_1 = c_3$

$\Psi = c_1\phi_1 + \sqrt{2}c_1\phi_2 + c_1\phi_3 = c_1\left(\phi_1 + \sqrt{2}\phi_2 + \phi_3\right)$

Since $c_1^2 + c_2^2 + c_3^2 = c_1^2 + (\sqrt{2}c_1)^2 + c_1^2 = 4\,c_1^2 = 1, c_1 = \dfrac{1}{2}$.

The normalized wavefunction is: $\Psi = \dfrac{1}{2}\left(\phi_1 + \sqrt{2}\phi_2 + \phi_3\right)$

When $E_3 = \alpha - \sqrt{2}\beta$, the eigenvalue equation is:

$$\begin{pmatrix} \sqrt{2}\beta & \beta & 0 \\ \beta & \sqrt{2}\beta & \beta \\ 0 & \beta & \sqrt{2}\beta \end{pmatrix} \begin{pmatrix} c_1 \\ c_2 \\ c_3 \end{pmatrix} = \begin{pmatrix} 0 \\ 0 \\ 0 \end{pmatrix},$$

where c_1, c_2 and c_3 are the molecular orbital coefficients.
If we multiply the matrices, we find:

$\sqrt{2}\beta c_1 + \beta c_2 = 0$ So, $c_2 = -\sqrt{2}c_1$

$\beta c_2 + \sqrt{2}\beta c_3 = 0$ So, $c_2 = -\sqrt{2}c_3$

Overall: $-\sqrt{2}c_1 = c_2 = -\sqrt{2}c_3\ \ c_1 = c_3$

$\Psi = c_1\phi_1 - \sqrt{2}c_1\phi_2 + c_1\phi_3 = c_1\left(\phi_1 - \sqrt{2}\phi_2 + \phi_3\right)$

Since $c_1^2 + c_2^2 + c_3^2 = c_1^2 + (-\sqrt{2}\,c_1)^2 + c_1^2 = 4\,c_1^2 = 1,\ c_1 = \dfrac{1}{2}$.

The normalized wavefunction is: $\Psi = \dfrac{1}{2}\left(\phi_1 - \sqrt{2}\phi_2 + \phi_3\right)$

Self-test 27.9

27.9.1 $\mathbf{A} = \begin{pmatrix} 1 & 2 \\ 6 & 4 \end{pmatrix}$ so $\mathbf{A}^{\mathrm{T}} = \begin{pmatrix} 1 & 6 \\ 2 & 4 \end{pmatrix}$

27.9.2 $\quad \mathbf{A} = \begin{pmatrix} 1 & -2 & 7 \\ 3 & 0 & 2 \\ 2 & 1 & 6 \end{pmatrix} \quad \mathbf{A}^T = \begin{pmatrix} 1 & 3 & 2 \\ -2 & 0 & 1 \\ 7 & 2 & 6 \end{pmatrix}$

27.9.3 $\quad \mathbf{A} = \begin{pmatrix} a & 2s & 4s \\ -a & a & 4as \\ 0 & 3b & 0 \end{pmatrix} \quad \mathbf{A}^T = \begin{pmatrix} a & -a & 0 \\ 2s & a & 3b \\ 4s & 4as & 0 \end{pmatrix}$

27.9.4 $\quad \mathbf{A} = \begin{pmatrix} 1 & 3 & 2 & 8 \\ 3 & 1 & 6 & 3 \\ 2 & 6 & 1 & 0 \\ 8 & 3 & 0 & 1 \end{pmatrix} \quad \mathbf{A}^T = \begin{pmatrix} 1 & 3 & 2 & 8 \\ 3 & 1 & 6 & 3 \\ 2 & 6 & 1 & 0 \\ 8 & 3 & 0 & 1 \end{pmatrix}$

In this case, $\mathbf{A} = \mathbf{A}^T$, so $\mathbf{A}$ is symmetrical.

Self-test 27.10

In each case the determinant is non-zero, indicating that the inverse of $\mathbf{A}$ exists.

27.10.1 Using eqn (27.11), $A^{-1} = \dfrac{1}{-8} \begin{pmatrix} 4 & -2 \\ -6 & 1 \end{pmatrix} = \begin{pmatrix} \frac{1}{2} & \frac{1}{4} \\ \frac{3}{4} & -\frac{1}{8} \end{pmatrix}$

27.10.2 The determinant is 6.

$$A^{-1} = \frac{1}{6} \begin{pmatrix} 3 & -3 & -1 \\ 0 & 6 & -4 \\ 0 & 0 & 2 \end{pmatrix} = \begin{pmatrix} \frac{1}{2} & -\frac{1}{2} & -\frac{1}{6} \\ 0 & 1 & -\frac{2}{3} \\ 0 & 0 & \frac{1}{3} \end{pmatrix}$$

27.10.3 The determinant of the matrix is:

$$\det \mathbf{A} = 1 \begin{vmatrix} 3 & 2 \\ 2 & 1 \end{vmatrix} - 2 \begin{vmatrix} 4 & 2 \\ 0 & 1 \end{vmatrix} + 3 \begin{vmatrix} 4 & 3 \\ 0 & 2 \end{vmatrix} = (3-4) - 2 \times (4) + 3 \times (8) = 15$$

The minors associated with the *top* row elements are: $\begin{vmatrix} 3 & 2 \\ 2 & 1 \end{vmatrix}, \begin{vmatrix} 4 & 2 \\ 0 & 1 \end{vmatrix}, \begin{vmatrix} 4 & 3 \\ 0 & 2 \end{vmatrix}$

The minors associated with the *middle* row elements are: $\begin{vmatrix} 2 & 3 \\ 2 & 1 \end{vmatrix}, \begin{vmatrix} 1 & 3 \\ 0 & 1 \end{vmatrix}, \begin{vmatrix} 1 & 2 \\ 0 & 2 \end{vmatrix}$

The minors associated with the *bottom* row elements are: $\begin{vmatrix} 2 & 3 \\ 3 & 2 \end{vmatrix}, \begin{vmatrix} 1 & 3 \\ 4 & 2 \end{vmatrix}, \begin{vmatrix} 1 & 2 \\ 4 & 3 \end{vmatrix}$

Inserting the cofactors of each element into a new 3×3 matrix yields:

$$\begin{pmatrix} -1 & -(4) & 8 \\ -(-4) & 1 & -(2) \\ -5 & -(-10) & -5 \end{pmatrix} = \begin{pmatrix} -1 & -4 & 8 \\ 4 & 1 & -2 \\ -5 & 10 & -5 \end{pmatrix}$$

The inverse of $\mathbf{A}$ is $\mathbf{A}^{-1} = \dfrac{1}{15} \begin{pmatrix} -1 & 4 & -5 \\ -4 & 1 & 10 \\ 8 & -2 & -5 \end{pmatrix}$

27.10.4 The determinant of the matrix is:

$$\det \mathbf{A} = a \begin{vmatrix} c & b^2 \\ ac & c \end{vmatrix} - b \begin{vmatrix} a^2 & b^2 \\ ab & c \end{vmatrix} + bc \begin{vmatrix} a^2 & c \\ ab & ac \end{vmatrix} = a(c^2 - ab^2c) - b \times (a^2c - ab^3) + bc \times (a^3c - abc)$$

It is hard to simplify the determinant further.

The minors associated with the *top* row are: $\begin{vmatrix} c & b^2 \\ ac & c \end{vmatrix}, \begin{vmatrix} a^2 & b^2 \\ ab & c \end{vmatrix}, \begin{vmatrix} a^2 & c \\ ab & ac \end{vmatrix}$

The minors associated with the *middle* row are: $\begin{vmatrix} b & bc \\ ac & c \end{vmatrix}, \begin{vmatrix} a & bc \\ ab & c \end{vmatrix}, \begin{vmatrix} a & b \\ ab & ac \end{vmatrix}$

The minors associated with the *bottom* row are: $\begin{vmatrix} b & bc \\ c & b^2 \end{vmatrix}, \begin{vmatrix} a & bc \\ a^2 & b^2 \end{vmatrix}, \begin{vmatrix} a & b \\ a^2 & c \end{vmatrix}$

Inserting the cofactors yields:

$$\begin{pmatrix} c(c-ab^2) & -(a(ac-b^3)) & ac(a^2-b) \\ -(bc(1-ac)) & ac(1-b^2) & -(a(ac-b^2)) \\ b(b^2-c^2) & -(ab(b-ac)) & a(c-ab) \end{pmatrix}$$

The transpose of this matrix then yields the adjoint as,

$$\widehat{\mathbf{A}} = \begin{pmatrix} c(c-ab^2) & -bc(1-ac) & b(b^2-c^2) \\ -a(ac-b^3) & ac(1-b^2) & -ab(b-ac) \\ ac(a^2-b) & -a(ac-b^2) & a(c-ab) \end{pmatrix}$$

Self-test 27.11

We start by rearranging the equations:

$$x - 2y + 8z = 5$$
$$2x - 3y + 15z = 6$$
$$-4x + 8y - 30z = -22$$

These coefficient yield the matrix: $\begin{pmatrix} 1 & -2 & 8 \\ 2 & -3 & 15 \\ -4 & 8 & -30 \end{pmatrix} \begin{pmatrix} x \\ y \\ z \end{pmatrix} = \begin{pmatrix} 5 \\ 6 \\ -22 \end{pmatrix}$.

....

It might have been more logical (and easier!) to first rewrite the third equation, multiplying it throughout by -1, to yield '$4x - 8y + 30z = 22$'.

....

The inverse matrix is $\mathbf{A}^{-1} = \begin{pmatrix} -15 & 2 & -3 \\ 0 & 1 & 0.5 \\ 2 & 0 & 0.5 \end{pmatrix}$.

Matrix multiplication yields, $x = 3, y = -5$ and $z = -1$.

Chapter 28:
Vectors

Self-test 28.1

28.1.1 Using eqn (28.4), $|V| = \sqrt{2^2 + 3^2 + 5^2} = \sqrt{38} = 6.16$ (to 2 dp.)

28.1.2 Using eqn (28.4), $|V| = \sqrt{6^2 + 0^2 + 12.6^2} = \sqrt{194.76} = 13.96$ (to 2 dp.)

28.1.3 Length in the x-direction $= 3 \cos 40° = 2.298$
Length in the y-direction $= 3 \sin 40° = 1.928$

28.1.4 Length in the x-direction $= |\, 6.2 \cos (-150°)\,| = 5.37$. (The computed result (prior to taking the modulus) is actually negative, but a length is actually a scalar quantity; the minus sign merely indicates the *direction* is backwards along the x-axis).
Length in the y-direction $= |\, 6.2 \sin (-150°)\,| = 3.1$. Again, the sign (prior to taking the modulus) indicates that this vector will be along the negative part of the y-axis.

....

Remember to set the calculator to *degrees*.

....

Self-test 28.2

28.2.1 $1.5 \times \begin{pmatrix} 2 \\ 4 \end{pmatrix} = \begin{pmatrix} 1.5 \times 2 \\ 1.5 \times 4 \end{pmatrix} = \begin{pmatrix} 3 \\ 6 \end{pmatrix}$ 28.2.2 $2.73 \times \begin{pmatrix} 5 \\ 2.2 \end{pmatrix} = \begin{pmatrix} 2.73 \times 5 \\ 2.73 \times 2.2 \end{pmatrix} = \begin{pmatrix} 13.65 \\ 6.01 \end{pmatrix}$

Self-test 28.3

In both cases, we will perform the necessary vector algebra in two ways.

28.3.1 $4\mathbf{a} + 2\mathbf{b} = 4(3,2,1) + 2(5,0,4) = (12{+}10,\ 8{+}0,\ 4{+}8) = (22,8,12).$

Alternatively, $4\mathbf{a}+2\mathbf{b}=4\begin{pmatrix}3\\2\\1\end{pmatrix}+2\begin{pmatrix}5\\0\\4\end{pmatrix}=\begin{pmatrix}12\\8\\4\end{pmatrix}+\begin{pmatrix}10\\0\\8\end{pmatrix}=\begin{pmatrix}22\\8\\12\end{pmatrix}$

Magnitude $=\sqrt{22^2+8^2+12^2}=\sqrt{692}=26.31$ (to 2 dp).

28.3.2 $3\,\mathbf{a}-2\,\mathbf{b}+4\,\mathbf{c}=3(3,2,1)-2(5,0,4)+4(3,9,1)=(9{-}10{+}12,\ 6{-}0{+}36,\ 3{-}8{+}4)$
$\qquad\qquad\qquad = (11,42,-1).$

Alternatively, $3\mathbf{a}-2\,\mathbf{b}+4\,\mathbf{c}=3\begin{pmatrix}3\\2\\1\end{pmatrix}-2\begin{pmatrix}5\\0\\4\end{pmatrix}+4\begin{pmatrix}3\\9\\1\end{pmatrix}=\begin{pmatrix}9\\6\\3\end{pmatrix}-\begin{pmatrix}10\\0\\8\end{pmatrix}+\begin{pmatrix}12\\36\\4\end{pmatrix}=\begin{pmatrix}11\\42\\-1\end{pmatrix}$

Magnitude $=\sqrt{11^2+42^2+(-1)^2}=\sqrt{1886}=43.43$ (to 2 dp).

Self-test 28.4

28.4.1 x-direction: $2-(-5)=7$
y-direction: $7-3=4$
so the vector describing the relative motion is $(7, 4)$.

28.4.2 x-direction: $4.2-(-6.2)=10.4$
y-direction: $3.1-(-2.2)=5.3$
so the vector describing the relative motion is $(10.4, 5.3)$

Self-test 28.5

28.5.1 $\mathbf{r}=6\,\hat{\imath}+5\,\hat{\jmath}$ and $\mathbf{s}=2\,\hat{\imath}-8\,\hat{\jmath}$. From eqn (28.7), $\mathbf{r}\cdot\mathbf{s}=(6\times2)+(5\times(-8))=12-40=-28.$

28.5.2 $\mathbf{F}_1\cdot\mathbf{F}_2=4\times6\times\cos40°=18.39\ \text{N}^2.$

28.5.3 From eqn (28.4), $|\mathbf{p}|=\sqrt{2^2+1^2+(-1^2)}=2.449$ (to 3 d.p.) and,

$|\mathbf{q}|=\sqrt{1^2+0^2+1^2}=1.414$ to 3 d.p.

Also, from eqn (28.7) we have $\mathbf{p}\cdot\mathbf{q}=\begin{pmatrix}2\\1\\-1\end{pmatrix}\cdot\begin{pmatrix}1\\0\\-1\end{pmatrix}$

$$= (2\times1)+(1\times0)+(-1\times-1)=3$$

Since $\mathbf{p}\cdot\mathbf{q}=|\mathbf{p}||\mathbf{q}|\cos\theta$, we now have $3=2.449\times1.414\times\cos\theta$, so

$\cos\theta=\dfrac{3}{2.449\times1.414}=0.866,$ so $\theta=30°.$

28.5.4 From eqn (28.4), $|\mathbf{a}|=\sqrt{4+9+1}=3.742$ (to 3 d.p.) and $|\mathbf{b}|=\sqrt{16+1+9}=5.099$
to 3 d.p.
Also, from eqn (28.7) we have $\mathbf{a}\cdot\mathbf{b}=(2\times4)+(-3\times1)+(1\times-3)=2.$
Since $\mathbf{a}\cdot\mathbf{b}=|\mathbf{a}||\mathbf{b}|\cos\theta$, we now have $2=3.742\times5.099\times\cos\theta$, so

$\cos\theta=\dfrac{2}{3.742\times5.099}=0.1048,$ so $\theta=84°.$

28.5.5 Kinetic energy KE is given by the equation, $E=\frac{1}{2}mv^2$. The molecular mass of
adrenaline $C_9H_{13}NO_3$ is 183 amu $=3.038\times10^{-22}$ g.

Component of the KE in the x-direction $=\frac{1}{2}\times3.04\times10^{-22}\times3^2=2.73\times10^{-20}$ J

Component of the KE in the y-direction $=\frac{1}{2}\times3.04\times10^{-22}\times2^2=6.08\times10^{-20}$ J

Self-test 28.6

28.6.1   We can solve this problem using matrices. We start with the matrix:

$$\begin{pmatrix} \hat{\imath} & \hat{\jmath} & \hat{k} \\ 2 & 1 & -3 \\ 0 & 4 & 5 \end{pmatrix}$$

To solve the problem, we reduce the matrix:

$$\begin{pmatrix} \hat{\imath} & \hat{\jmath} & \hat{k} \\ 2 & 1 & -3 \\ 0 & 4 & 5 \end{pmatrix} = \begin{vmatrix} 1 & -3 \\ 4 & 5 \end{vmatrix}\hat{\imath} - \begin{vmatrix} 2 & -3 \\ 0 & 5 \end{vmatrix}\hat{\jmath} + \begin{vmatrix} 2 & 1 \\ 0 & 4 \end{vmatrix}\hat{k}$$

which reduces to, $(1\times5-4\times(-3))\hat{\imath} - (2\times5-0\times(-3))\hat{\jmath} + (2\times4-0\times1)\hat{k}$

so $(5-(-12))\hat{\imath} - (10-0)\hat{\jmath} + (8-0)\hat{k} = 17\hat{\imath} - 10\hat{\jmath} + 8\hat{k}$

28.6.2 $a\times b = (2-3)\hat{\imath} - (1-3)\hat{\jmath} + (1-2)\hat{k} = -\hat{\imath} + 2\hat{\jmath} - \hat{k}$

In matrix notation, $a\times b = \begin{pmatrix} 1 \\ 2 \\ 3 \end{pmatrix} \times \begin{pmatrix} 1 \\ 1 \\ 1 \end{pmatrix} = \begin{pmatrix} -1 \\ 2 \\ -1 \end{pmatrix}$.

Self-test 28.7

28.7.1   The component differentials are,

$$\left(\frac{\partial F}{\partial x}\right) = 3y^2z - 8xyz^3 \quad \left(\frac{\partial F}{\partial y}\right) = 6xyz - 4x^2z^3 \quad \left(\frac{\partial F}{\partial z}\right) = 3xy^2 - 12x^2yz^2$$

So $\nabla F = (3y^2z - 8xyz^3)\hat{\imath} + (6xyz - 4x^2z^3)\hat{\jmath} + (3xy^2 - 12x^2yz^2)\hat{k}$

Inserting the coordinates of the point (3,1,2) yields:

$\nabla F = (3(1)^2(2) - 8(3)(1)(2)^3)\hat{\imath} + (6(3)(1)(2) - 4(3)^2(2)^3)\hat{\jmath} + (3(3)(1)^2 - 12(3)^2(1)(2)^2)\hat{k}$

$\nabla F = (6-192)\hat{\imath} + (36-288)\hat{\jmath} + (9-432)\hat{k}$

$\nabla F = -186\hat{\imath} - 252\hat{\jmath} - 423\hat{k}$

28.7.2   The component differentials are,

$$\left(\frac{\partial \chi}{\partial x}\right) = 2z^4 - 2xy \quad \left(\frac{\partial \chi}{\partial y}\right) = -x^2 \quad \left(\frac{\partial \chi}{\partial y}\right) = 8xz^3$$

So $\nabla\chi = (2z^4 - 2xy)\hat{\imath} + (-x^2)\hat{\jmath} + (8xz^3)\hat{k}$

Inserting the coordinates of the point (2,3,4) yields:

$\nabla\chi = (2(4)^4 - 2(2)(3))\hat{\imath} + (-(2)^2)\hat{\jmath} + (8(2)(4)^3)\hat{k}$

$\nabla\chi = 500\hat{\imath} - 4\hat{\jmath} + 1024\hat{k}$

Self-test 28.8

28.8.1   The separate differentials are:

$$\frac{\partial G_z}{\partial y} = \frac{\partial(-4y^3z)}{\partial y} = -12y^2z \qquad \frac{\partial G_y}{\partial z} = \frac{\partial(3x^2z^3)}{\partial z} = 9x^2z^2$$

$$\frac{\partial G_x}{\partial z} = \frac{\partial(2xyz)}{\partial z} = 2xy \qquad \frac{\partial G_z}{\partial x} = \frac{\partial(-4y^3z)}{\partial x} = 0$$

$$\frac{\partial G_y}{\partial x} = \frac{\partial(3x^2z^3)}{\partial x} = 6xz^3 \qquad \frac{\partial G_x}{\partial y} = \frac{\partial(2xyz)}{\partial y} = 2xz$$

Inserting terms into eqn (28.12) yields,

curl G $= (-12y^2z - 9x^2z^2)\,\hat{\mathbf{i}} + (2xy - 0)\,\hat{\mathbf{j}} + (6xz^3 - 2xz)\,\hat{\mathbf{k}}$

Inserting the coordinates of the point (3,4,1) yields,

curl G $= (-12(4)^2(1) - 9(3)^2(1)^2)\,\hat{\mathbf{i}} + (2(3)(4) - 0)\,\hat{\mathbf{j}} + (6(3)\,(1)^3 - 2(3)\,(1))\hat{\mathbf{k}}$

curl G $= (-192 - 81)\hat{\mathbf{i}} + (24 - 0)\,\hat{\mathbf{j}} + (18 - 6)\hat{\mathbf{k}}$

so   **curl G** $= \nabla \times G = -273\,\hat{\mathbf{i}} + 24\,\hat{\mathbf{j}} + 12\hat{\mathbf{k}}$

28.8.2   The separate differentials are:

$$\frac{\partial \beta_z}{\partial y} = \frac{\partial(4xy^2)}{\partial y} = 8xy \qquad \frac{\partial \beta_y}{\partial z} = \frac{\partial(-yz^3)}{\partial z} = -3yz^2$$

$$\frac{\partial \beta_x}{\partial z} = \frac{\partial(2x^2y^3z)}{\partial z} = 2x^2y^3 \qquad \frac{\partial \beta_z}{\partial x} = \frac{\partial(4xy^2)}{\partial x} = 4y^2$$

$$\frac{\partial \beta_y}{\partial x} = \frac{\partial(-yz^3)}{\partial x} = 0 \qquad \frac{\partial \beta_x}{\partial y} = \frac{\partial(2x^2y^3z)}{\partial y} = 6x^2y^2z$$

Inserting terms into eqn (28.12) yields,
   curl $\beta = (8xy - 3yz^2)\,\hat{\mathbf{i}} + (2x^2y^3 - 4y^2)\,\hat{\mathbf{j}} + (0 - 6x^2y^2z)\hat{\mathbf{k}}$
Inserting the coordinates of the point (0,2,5) yields,
   curl $\beta = (8(0)(2) - 3(2)(5)^2)\,\hat{\mathbf{i}} + (2(0)^2(2)^3 - 4(2)^2)\,\hat{\mathbf{j}} + (0 - 6(0)^2(2)^2(5))\hat{\mathbf{k}}$
   curl $\beta = (0 - 150)\,\hat{\mathbf{i}} + (0 - 16)\,\hat{\mathbf{j}} + (0 - 0)\hat{\mathbf{k}}$
so   curl $\beta = \nabla \times \beta = -150\,\hat{\mathbf{i}} - 16\,\hat{\mathbf{j}}$

Self-test 28.9

28.9.1   The separate differentials are:

$$\frac{\partial G_x}{\partial x} = \frac{\partial(2xyz)}{\partial x} = 2yz \qquad \frac{\partial G_y}{\partial y} = \frac{\partial(3x^2z^3)}{\partial y} = 0$$

$$\frac{\partial G_z}{\partial z} = \frac{\partial(-4y^3z)}{\partial z} = -4y^3$$

The divergence of **G** is:
$\nabla \cdot \mathbf{G} = 2yz + 0 - 4y^3 = 2yz + -4y^3$
At (3,4,1),
$\nabla \cdot \mathbf{G} = 2 \times 4 \times 1 - 4 \times 4^3 = -248$ (sink at this point)

28.9.2   The separate differentials are:

$$\frac{\partial \beta_x}{\partial x} = \frac{\partial(2x^2y^3z)}{\partial x} = 4xy^3z \qquad \frac{\partial \beta_y}{\partial y} = \frac{\partial(-yz^3)}{\partial y} = -z^3$$

$$\frac{\partial \beta_z}{\partial z} = \frac{\partial(4xy^2)}{\partial z} = 0$$

The divergence of β is:
   $\nabla \cdot \boldsymbol{\beta} = 4xy^3z - z^3 + 0 = 4xy^3z - z^3$
At (0,2,5),
   $\nabla \cdot \mathbf{G} = 0 - 5^3 = -125$ (sink at this point)

Chapter 29:
Complex numbers

Self-test 29.1

29.1.1	real		29.1.2	imaginary
29.1.3	complex		29.1.4	complex
29.1.5	imaginary		29.1.6	real

Self-test 29.2

29.2.1 3i 29.2.2 6i

29.2.3 11−10i 29.2.4 14 − 6 i

29.2.5 $12 = 4 \times 3$, so $\sqrt{4 \times 3} = \sqrt{4} \times \sqrt{3} = 2 \times \sqrt{3}$. Therefore, $8 - 2\sqrt{3}i$

29.2.6 $50 = 25 \times 2$, so $\sqrt{25 \times 2} = \sqrt{25} \times \sqrt{2} = 5 \times \sqrt{2}$. Therefore, $2 - 5\sqrt{2}\,i$

Self-test 29.3

29.3.1 $9 + 3i$ 29.3.2 $2.6 - 0.9\,i$

29.3.3 $45.53 - 5.27i$ 29.3.4 $6a + 26bi$

29.3.5 $13a + 24b + 4c + 17di$

29.3.6 $-48 = -(16 \times 3)$, so $\sqrt{16 \times -3} = \sqrt{16} \times \sqrt{-3} = 4 \times \sqrt{-3} = 4 \times \sqrt{3}\,i$.

Also, $-75 = -(25 \times 3)$, so $\sqrt{-25 \times 3} = \sqrt{25} \times \sqrt{-3} = 5 \times \sqrt{-3} = 5 \times \sqrt{3}\,i$.

Therefore, $\sqrt{-48} + \left(-1 - \sqrt{-75}\right) = \left(4\sqrt{3}i\right) + \left(-1 - 5\sqrt{3}i\right) = -\left(1 + \sqrt{3}i\right)$

29.3.7 $-72 = -(36 \times 2)$, so $\sqrt{-36 \times 2} = \sqrt{36} \times \sqrt{-2} = 6 \times \sqrt{-2} = 6 \times \sqrt{2}\,i$.

Also, $-98 = -(49 \times 2)$, so $\sqrt{-49 \times 2} = \sqrt{49} \times \sqrt{-2} = 7 \times \sqrt{-2} = 7 \times \sqrt{2}\,i$.

Therefore, $19 + \sqrt{2}\,i$.

Self-test 29.4

29.4.1 $(-33 + 9\,i)$ 29.4.2 $(2 + 14\,i)$

29.4.3 (13) 29.4.4 $(123.32 + 18.44\,i)$

29.4.5 $(a^2 + b^2)$ 29.4.6 $((6a^2 + 21b^2) - 5abi)$

29.4.7 $((3a^5 + 8b^3) + (6a^3b^2 - 4a^2b)\,i\,)$ 29.4.8 $((2a + 20b) + (6\sqrt{ab}i))$

Self-test 29.5

29.5.1 $12\,(\cos(\theta + 2b) + i\sin(\theta + 2b))$

29.5.2 $a^3\left(\cos\dfrac{3\pi}{4} + i\sin\dfrac{3\pi}{4}\right) = a^3\left(-\dfrac{1}{\sqrt{2}} + i\dfrac{1}{\sqrt{2}}\right)$

Self-test 29.6

29.6.1 $15 - 3i$ 29.6.2 $-1 - 3i$

29.6.3 $3 - 2i$ 29.6.4 $4.2 - 7.3i$

29.6.5 $a - b\,i$ 29.6.6 $a^2 - 2b^2i$

29.6.7 $\sqrt{a} - 5\sqrt{b}i$

Self-test 29.7

29.7.1 $\dfrac{(2 + 3i)}{i} \times \dfrac{-i}{-i} = -2i - 3i^2 = -2i + 3 = 3 - 2i$

29.7.2 $\dfrac{(4+5i)}{2i} \times \dfrac{-2i}{-2i} = \dfrac{-8i+10}{4} = \dfrac{5}{2} - \dfrac{2i}{}$

29.7.3 $\dfrac{2+7i}{6+i} \times \dfrac{6-i}{6-i} = \dfrac{12-2i+42i-7i^2}{37} = \dfrac{19+40i}{37}$

29.7.4 $\dfrac{4-9i}{5-2i} \times \dfrac{5+2i}{5+2i} = \dfrac{20+8i-45i-18i^2}{29} = \dfrac{38-37i}{29}$

Self-test 29.8

29.8.1 $4(\cos\theta + i\sin\theta)$ 29.8.2 $b\,e^{d\varphi i}$

29.8.3 $(\cos\varphi + i\sin\varphi)^4 = e^{4\varphi i}$ 29.8.4 $c^2(\cos fx + i\sin fx) = c^2(\cos x + i\sin x)^f$

Self-test 29.9

29.9.1 $\dfrac{1}{z(\text{total})} = \dfrac{1}{R} + \dfrac{1}{1/i\omega C} = \dfrac{1}{R} + \dfrac{i\omega C}{1} = \dfrac{1+i\omega CR}{R}$ Therefore, $z(\text{total}) = \dfrac{R}{1+i\omega CR}$

Chapter 30:
Dimensional analysis

Self-test 30.1

30.1.1   rate $= k$ [reactant].
Inserting units: $[\text{mol m}^{-3}\,\text{s}^{-1}] = ? \times [\text{mol m}^{-3}]$.
Cancelling yields $[\text{s}^{-1}] = ?$ So the units of k_1 are s^{-1}.

30.1.2   rate $= k\,[\text{O}_2] \times [\text{hydrocarbon}]$.
Inserting units: $[\text{mol m}^{-3}\,\text{s}^{-1}] = ? \times [\text{mol m}^{-3}] \times [\text{mol m}^{-3}]$. Cancelling yields,
$[\text{s}^{-1}] = ? \times [\text{mol m}^{-3}]$.
Dividing throughout by the right-hand square bracket yields, $? = \text{mol}^{-1}\,\text{m}^3\,\text{s}^{-1}$. So the
units of k_2 are $\text{mol}^{-1}\,\text{m}^3\,\text{s}^{-1}$. We often write these units as $(\text{mol m}^{-3})^{-1}\,\text{s}^{-1}$.

Self-test 30.2

30.2.1 $\Delta H_2 = \Delta H_1 + C_p\,(T_2 - T_1)$. Inserting units: $[\text{J mol}^{-1}] = [\text{J mol}^{-1}] + [\text{J K}^{-1}\,\text{mol}^{-1}] \times [\text{K}]$.
The K^{-1} and K terms cancel, leaving $[\text{J mol}^{-1}] = [\text{J mol}^{-1}] + [\text{J mol}^{-1}]$. The dimensions
of each term are the same, so the equation *is* dimensionally correct.

30.2.2 $\Delta T = K_{(\text{cryoscopic})} \times \left(\dfrac{\text{mass of solute}}{\text{molar mass of solute}}\right) \times \dfrac{1000}{\text{mass of solvent}}$.

....
We ignore the factor of
1000, because it is not a
dimension.
....

Inserting units: $[\text{K}] = [\text{K kg mol}^{-1}] \times \left[\dfrac{\text{g}}{\text{g mol}^{-1}}\right] \times \left[\dfrac{1}{\text{kg}}\right]$.

Cancelling the kg and the mol^{-1} terms on the right-hand side yields $[\text{K}] = [\text{K}]$. The
dominions of each side are the same, so the equation *is* dimensionally correct.

30.2.3 $\Delta G^{\ominus} = \Delta H^{\ominus} - \Delta S^{\ominus}$: $[\text{J mol}^{-1}] = [\text{J mol}^{-1}] + [\text{J K}^{-1}\,\text{mol}^{-1}]$. The third term is clearly
anomalous. The third term has been multiplied by K^{-1}, that is, when compared to
the first two terms. This K^{-1} term needs to be cancelled somehow. The simplest way
to remove the K^{-1} term is to multiply it by K. So we suggest the $\Delta S^{\ominus}$ term should be
multiplied by a temperature term. The equation should be $\Delta G^{\ominus} = \Delta H^{\ominus} - T\Delta S^{\ominus}$.

30.2.4   Inserting units: $\dfrac{[\text{J mol}^{-1}]}{[\text{K}]} = [\text{J mol}^{-1}] + [\text{J mol}^{-1}]\left(\dfrac{1}{[\text{K}]} - \dfrac{1}{[\text{K}]}\right)$.

The right-hand bracket becomes $1/[\text{K}]$ so, dimensionally the equation is:
$[\text{J K}^{-1}\,\text{mol}^{-1}] = [\text{J mol}^{-1}] + [\text{J K}^{-1}\,\text{mol}^{-1}]$.

The equation is incorrect because the three terms differ. Because the units of the first and third terms agree, we will assume it is the middle term that is in error. The middle term is J mol^{-1}, and the others have an additional K^{-1} term. The simplest way to accommodate the difference is for the middle term to be divided by temperature. This kind of analysis cannot tell us what form that temperature should take.

In fact, the correct form of the Gibbs–Helmholtz equation is:

$$\frac{\Delta G_2}{T_2} = \frac{\Delta G_1}{T_1} + \Delta H\left(\frac{1}{T_2} - \frac{1}{T_1}\right)$$

Self-test 30.3

30.3.1 Molecular velocity [v], so [v] = [m] ÷ [s] = [m s^{-1}]

30.3.2 Acceleration [a], so [a] = [m s^{-1}] ÷ [s] = [m s^{-1}] × [s^{-1}] = [m s^{-2}]

30.3.3 Wave number [$\bar{v}$], so[$\bar{v}$] = 1/[m] = [m^{-1}]

Self-test 30.4

30.4.1 [F] = [ma] = [kg] [m s^{-2}] = kg m s^{-2}

30.4.2 [p] = [$m \times a/A$] = [kg] [m s^{-2}] ÷ [m^2] = kg m m^{-2} s^{-2} = kg m^{-1} s^{-2}.

30.4.3 We first rearrange $pV = nRT$: $R = pV/nT$.
Inserting units yields: ([kg m^{-1} s^{-2}] [m^3])/([mol] [K]) = kg m^2 s^{-2} mol^{-1} K^{-1}.

> The units for pressure come from Table 30.2 or from Self-test 30.4.2.

Self-test 30.5

30.5.1 1 °C = $\frac{5}{9}$ °F. We rearrange using the laws of BODMAS. First, 9 °C = 5 °F.
Then, 1 °F = $\frac{9}{5}$ °C.

30.5.2 (1 inch)^3 = (2.5 cm)^3. Multiplying the bracket, 2.5^3 = 15.625, so 1 inch^3 = 15.625 cm^3.
Alternatively, we could rearrange to say 1 cm^3 = 0.064 inch^3.

Self-test 30.6

30.6.1 ΔS/J K^{-1} mol^{-1} = 32

30.6.2 Energy released/kJ mol^{-1} = 15, or 10^{-3} E/J mol^{-1} = 15

30.6.3 Mass, m /g = 12

30.6.4 Temperature increase ΔT/K = 1.2

30.6.5 The simplest answer is, Pressure, p/Pa = 3.2 × 10^5.
But we should go further, writing 10^{-5} Pressure, p/Pa = 3.2, or
10^{-5} Pressure, p/10^5 Pa = 3.2

30.6.6 The simplest answer is, Amount, n/mmol = 2.67.
Going further, we write, Amount, n = 2.67 × 10^{-3} mol, so 10^3 Amount, n = 2.67 mol,
and thence = 10^3 Amount, n/mol = 2.67.

Glossary

Mathematical symbols

+   add.

−   subtract.

×   multiply.

÷   divide.

±   plus or minus.

=   equal to.

≠   not equal to.

≈   approximately equal to.

≡   equivalent to.

∝   Proportional to.

Mathematical operators

∞   Infinity.

∫   Integral.

√   Square root (a small numeral n nestling to the left of the root sign implies the nth root).

<   The term to the right of this symbol is greater than the term to its left.

≤   The term to the right of this symbol is greater than or equal to the term to its left.

>   The term to the right of this symbol is less than the term to its left.

≥   The term to the right of this symbol is less than or equal to the term to its left.

∂   Partial differential.

%   Percentage: a rate or proportion per hundred; relating to one hundredth part of a whole, as defined by eqn (5.8) on p. 69.

e An irregular number occurring often in nature, having the approximate value of 2.178 (to 4 s.f.).

τ Tau: an irregular number occurring often in nature, having the approximate value of 1.618 (to 4 s.f.).

π pi: the ratio of the circumference of a circle to its diameter, having the approximate value of 3.142 (to 4 s.f.).

∇   The Laplacian operator (also called del), as defined by eqn (22.5) on p. 334.

Δ Delta: the symbol Δ, sometimes with subsidiary terms to its right, indicates that the quantity displayed is a difference. Δ is defined by eqn (2.1) on page 17.

dy/dx The differential of x with respect to y: the instantaneous gradient of a function of x in a graph of y (as ordinate) against x (as abscissa).

H The Hamiltonian operator that is essential to quantum mechanics.

Π The Pi notation is used to represent the product of a sequence of terms, often written with subsidiary limits written above and below, and defined by eqn (2.3) on p. 27. Not to be confused with the constant π.

Σ The sigma notation is used to represent the summation of a series of terms, often written with subsidiary limits above and below. Σ is defined by eqn (2.2) on page 19.

Abscissa   The horizontal axis when representing data with Cartesian coordinates.

Adjacent   The shorter of the two sides in a right-angled triangle that abuts the angle of focus.

Adjoint   of a matrix A (symbolized as Â). We obtain Â in a multistep process: first obtaining the minors of the source matrix, then obtaining the determinant of each. Next, we rewrite the matrix by replacing the cofactors with these determinants. Finally, we transpose the matrix.

Algebra   The branch of mathematics dealing with general statements of relationships. It utilizes numbers, letters and other symbols to represent specific sets of numbers, values, vectors, etc., in the description of such relations.

Algebraic phrase   Part of an equation.

Angle of focus   The angle formed when two straight lines meet. The angle is usually symbolized as θ.

Area integral   – see integral, Riemann.

Area   The spatial extent of a thing in two dimensions.

Argument   The independent variable of a function: in the expressions sin (ax), exp $(2x)$ or $(6\pi r)^2$, each portion in brackets is the respective argument of the functions sine, exponential, and square.

Associative   An operation on a set of elements that gives an equivalent expression when elements are grouped without change of order, as $(a+b)+c = a+(b+c)$. See also *commutative*.

Average   – see *mean, mode, or median*.

Azimuthal angle   The angle from a reference vector in a reference plane to a second vector in the same plane, pointing toward a thing of interest. In the cylindrical and spherical polar coordinate systems it is the angle.

Base   The number that serves as a starting point for a logarithmic or other numerical system or scale.

Bracket   Parentheses of various forms indicating that the enclosed quantity is to be treated as a whole.

Calculus   A method of calculation; one of several highly systematic methods of treating problems by a special system of algebraic notations: the topic of differentiation or integration.

Cancel/cancelling   A process employed when simplifying a fraction, in which we eliminate by striking out the common factors in both the denominator and numerator. Similarly, we can eliminate equivalent terms by striking out on opposing sides of an equation.

Cardinal number   Any of the numbers that express amount; they are generally taken to mean whole numbers, i.e. integers.

Cartesian   Two- or n-dimension analysis, in which the coordinate axes are orthogonal.

Characteristic   Description of a matrix determinant (det $(\mathbf{A} - \lambda\mathbf{I})$) if its roots yield the eigenvalues of the matrix $\mathbf{A}$.

Characteristic determinant   A polynomial derived, from (det $(\mathbf{A} - \lambda\mathbf{I})$) = 0, whose roots are the eigenvalues of the matrix $\mathbf{A}$.

Chord   The line segment between two points on a given curve.

Cofactor $(-1)^{i+j}$ times the corresponding minor, which has been found when reducing a matrix.

Commutative   The ability of series of operators to operate in any order. More generally, if two procedures give the same result when carried out in arbitrary order, they are commutative (e.g. $a \times b = b \times a$). Many exceptions occur, e.g. in vector multiplication.

Conservative   A vector field having a zero curl.

Constant   A number or algebraic quantity, the value of which does not change.

Controlled variable   A variable, the value of which is pre-chosen before a measurement is to be obtained, and is usually plotted on the x-axis. The parameter to be measured is the *observed* variable.

Coordinates   A system devised to define the position of a point, line, etc. The position is defined by reference to a fixed figure, system of lines, etc.

Coordinates   The means by which we define the position of a point, line, etc. in two-or n-dimensional space.

Cosecant (cosec)   In a right-angled triangle, the ratio of the hypotenuse to the side opposite a given angle. The cosecant is the reciprocal of the sine.

Cosine   In a right-angled triangle, the ratio of the adjacent to the hypotenuse. The cosine is defined by eqn (16.2) on page 247, and symbolized as 'cos'.

Cotangent (cot)   In a right-angled triangle, the ratio of the side adjacent to a given angle to the side opposite. The cotangent is the reciprocal of the tangent.

Cross-product   The product of two vectors computed according to eqn (28.9). The product is also a vector.

Curl   A vector operator that shows a vector field's rate of rotation about a point. The curl is a form of cross-product.

Cylindrical coordinates   A system of coordinates for locating a point in space by two polar coordinates, and its perpendicular distance to the polar plane.

Cylindrical   Having the form of a cylinder, or attributes that relate to a cylinder.

Data   (The plural of datum): a series of numbers, terms, statistics, or items of information.

Data set   A group of related data, e.g. obtained by changing a single variable.

Datum   (The singular of data): a single number, term, statistic, or item of information.

Decimal   Pertaining to tenths or to the number 10. Decimal numbers are written in base 10.

Degenerate   Description of a square matrix that has a determinant of zero, and is therefore without an inverse.

Del   In vector calculus, a vector differential operator represented by the nabla symbol, ∇.

Delta Δ The symbol Δ, sometimes with subsidiary terms to its right, indicates that the quantity displayed is a difference. Δ is defined by eqn (2.1) on page 17.

Denominator   The portion of a fraction on the bottom of a fraction, or lowered and to the right of the solidus.

Determinant   A rather abstract quantity but, in essence, it quantifies the scale factor for volume when a matrix is regarded as a linear transformation.

Determinate (error)   An error that can (in principle, at least) be corrected by careful experimental design.

Differential coefficient (or 'quotient')   The limit of the ratio of the increment of a function to the increment of a variable in it, as the latter tends to 0; the instantaneous change of one quantity with respect to another, for example velocity, is the instantaneous change of distance with respect to time.

Differentiation   The process of computing a differential coefficient.

Divergence   A measure of the flux density of a vector field.

Domain   The possible values of a variable that may be used in a function's argument.

Dot product   The product of two vectors computed according to eqn (28.6). The product is a scalar and rather than a vector.

Eigenvalue(s)   The permitted solution(s) to a problem. The term is particularly useful in matrix algebra; thus the eigenvalues (λ) of a matrix $\mathbf{A}$ are its permitted solutions and satisfy the eigenvalue equation: $\mathbf{Ax} = \lambda\mathbf{x}$, where $\mathbf{x}$ is the corresponding eigenvector.

Eigenvector(s)   This term is used in matrix algebra. Eigenvectors, $\mathbf{x}$, are column vectors that have a corresponding eigenvalue, λ, and satisfy the eigenvalue equation: $\mathbf{Ax} = \lambda\mathbf{x}$.

Element   of a matrix: each of the $n \times m$ algebraic terms or numbers in a matrix.

Equal   Like or alike in quantity, degree, value; as great as or the same as; balanced.

Equality   The statement that two quantities are equal.

Equation   An expression expressed in the form of algebra, that asserts the equality of two quantities.

Error   The difference between the observed or approximately determined value and the true value of a quantity; *see also determinate (error) and indeterminate (error).*

Exponent   The power to which a variable, function or constant is raised, e.g. the x in 3^x, e^x, etc.

Exponential   Pertaining to the natural constant e (an irregular number having the approximate value of 2.178 …); see p. 177.

Extrapolant   Having extrapolated a series of data on graph, this is the line lying beyond the actual data.

Extrapolate   Draw lines through the available data to determine the line of best fit. Having defined the line this way, the line of best fit is extended. The portion beyond the actual data is an *extrapolant.* The value of the variable obtained by reading from an axis where an *extrapolant* strikes it is a prediction and not a direct observation.

Factor   One of two or more numbers, algebraic expressions, etc. that, when multiplied together, produce a given product; a divisor: 2, 5 and 10 are factors of 20. A factor is related to, but differs from, a *cofactor.*

Factorial, $n!$ The Pi product, $1 \times 2 \times … \times n$, where n is an integer.

Factorize   To resolve into factors, to ascertain the factors in a number or algebraic expression.

Finite   Capable of being completely counted; not zero, infinite or infinitesimal; of a set of elements. *See also infinite.*

Fourier transform   A mathematical tool that enables the mapping of a function, as a signal (that is defined in one domain, such as space or time), into another domain (such as wavelength or frequency). The function is represented in terms of sines and cosines.

Fraction   A number usually expressed in the form a/b. See also *proper fraction*, *improper fraction*, *vulgar fraction*.

Gradient   The steepness of a slope, defined as the change of height divided by distance travelled.

Gradient of a vector field   The magnitude of a vector's greatest rate of change. In vector calculus, the gradient of a scalar field is a 'vector field.'

Graph   A pictorial representation of data; a series of points, discrete or continuous, forming a curve or surface, each of which represents a value of a given function.

Hypotenuse   The longest side in a right-angled triangle.

Improper fraction   A fraction, the value of which has a modulus greater than one (i.e. the numerator is greater than the denominator).

Indeterminate (error)   An error that is random; an error that cannot be corrected for by change of experimental design; the opposite of a determinate error.

Infinite   Not finite; being so large as to be beyond the possibility of counting. *See also finite.*

Infinitesimal   Indefinitely or exceedingly small; so immeasurably small the magnitude cannot be assigned.

Inflection, point of   The coordinates of the part of a graph where (instantaneously) the gradient ceases to change.

Integer   A whole number.

Integral, definite   The difference between the values of the integral of a given function $f(x)$ for an upper value b and a lower value a of the independent variable x.

Integral, indefinite   Any function whose derivative is a given function.

Integral, Riemann   A definite integral used to determine an area. This integral is defined as the magnitude of the integral bounded (i) by the line of a given mathematical function, (ii) by the x-axis, and (iii) on either side by ordinates drawn at the limits.

Integration   The operation of finding the integral of a function or equation, especially when solving a differential equation; the opposite of differentiation.

Inverse   An element of an algebraic system, corresponding to a given element, such that its product or sum with the given element is the identity element.

Inverse matrix B^{-1}   A matrix B has an inverse matrix B^{-1} if the determinant *is* non-zero. The inverse matrix is defined by eqns (27.11) and (27.12) on p. 410.

Invertible   Description of a matrix that has a non-zero determinant, and therefore does have an inverse.

Irrotational   *another name for conservative.*

Laplacian   A differential operator that is central in electrostatics and fluid mechanics, and derives from Laplace's equation and Poisson's equation. In quantum mechanics, it is used in the kinetic energy term of the Schrödinger equation. It is denoted by Δ or ∇^2 symbols.

Latent roots   The eigenvalues of a matrix.

Legendrian   (Λ^2) A differential operator. In the context of quantum mechanics, it is the part of the Laplacian operator in spherical polar coordinates that involves the differentiation of the function in terms of the angular coordinates θ and ϕ.

Limits   The numbers affixed to the integration symbol (when a definite integral), indicating the interval or region over which the integration is taking place.

ln   Natural logarithm, as defined by eqn (12.1) on page 181.

log   logarithm in base 10, as defined by eqn (12.2) on page 182.

Logarithm   The inverse function of taking a to the power x, where b is the base number. Common examples include log in base 10 and in base e, as defined by eqn (12.1) on p. 181 and eqn (12.2) on p. 182.

Maclaurin series   A Taylor series in which the reference point is zero; *see also Taylor series.*

Magnitude   The size, extent, dimensions of a thing or variable.

Matrices   The plural of matrix.

Matrix   A rectangular array of numbers, algebraic symbols, or mathematical functions.

Maximum   The largest of a group or data set.

Mean   A variable having a value intermediate between the values of other quantities; an average, especially the arithmetic mean.

Median   The middle number in a sequence of numbers; when a sequence has an even number of members, we obtain the median as the average of the two middle members.

Minimum   The smallest of a group or data set.

Mode   The value of a variable that occurs the most often in the frequency distribution of the variable.

Modulus   The magnitude of a number, term, phrase or thing, that is without sign. A modulus is always expressed as a positive entity.

Nabla   The symbol ∇, as used to denote the mathematical del operator.

Non-singular Description of a matrix that has a non-zero determinant, and therefore does have an inverse.

Normalize The process of determining a constant multiplier for a wavefunction such that it satisfies the equation, $\int \psi * \psi \, dV = 1$.

Notation The process or method of noting or setting down by means of a special system of signs or symbols.

Number A numeral or collection of numerals, comprising some or all of the ten digits, 0, 1, 2, 3, 4, 5, 6, 7, 8, and 9.

Numerator The portion of a fraction on the top of a fraction, or raised and to the left of the solidus.

Observed variable The variable obtained experimentally as a result of performing the measurement with a pre-chosen value of the other variables; usually plotted on the y-axis; the opposite is the *controlled* variable.

Operation The action of applying a mathematical process to a quantity or quantities.

Operator A symbol or function denoting a mathematical operation. The simplest examples are $+$, $-$, $\times$, and $\div$.

Opposite The side of a right-angled triangle farthest from the angle of focus.

Ordered pair Two quantities written in such a way they indicate that one quantity precedes or is to be considered before the other, as (2,4) indicates the Cartesian coordinates of a point in the plane.

Ordinate The vertical axis when representing data with Cartesian coordinates.

Orthogonal

(1) (of lines) at right angles or perpendiculars.

(2) (of a system of real functions) defined so that the integral of the product of any two different functions is zero.

(3) (of a system of complex functions) defined so that the integral of the product of a function times the complex conjugate of any other function equals zero.

(4) (of two vectors) having a dot product equal to zero.

Parabola the shape of graphs of the type, $y = x^2$ or $y^2 = x$.

Parallel Two or more straight lines are parallel if their gradients m are the same but their intercepts c differ. Parallel lines never meet.

Percentage An amount of something, often expressed as a number out of 100.

Perpendicular At right angles. See also *orthogonal*.

Phrase, algebraic *– see algebraic phrase.*

pi (π) An irregular number occurring often in nature, having the approximate value of 3.142 (to 4 s.f.). We obtain π as the ratio of a circle's circumference to its diameter. π is not to be confused with the operator Π.

Pi product, Π A notation used to represent the product of a sequence of terms, often written with subsidiary limits written above and below, and defined by eqn (2.3) on page 27. Not to be confused with the constant π.

Polar coordinates A coordinate system that is partially orthogonal–Cartesian (i.e. cylindrical polar coordinates) or based on angles and a sphere (spherical polar coordinates).

Polygon A closed shape comprising straight sides. Usually, the sides are of equal length, e.g. a square is a polygon with four sides.

Power

(i) The number of times a variable, function or constant is multiplied by itself: $3 \times 3 \times 3 \times 3 = 3^4$ is the fourth power of 3.

(ii) The exponent in an expression, e.g. x^a.

(iii) Negative powers denote a *reciprocal*.

(iv) Fractional powers indicate a *root*.

Proper fraction A fraction, the value of which has a modulus smaller than one.

Proper roots The eigenvalues of a matrix.

Qualitative measurements indicate the qualities of thing(s) to be investigated, but impart no explicit detail concerning magnitude.

Quantitative measurements indicate the quantity or magnitude of thing(s) to be investigated, but impart no explicit detail concerning its qualities.

Radian The measure of a central angle subtending an arc, equal to 57.3°, as explained on p. 253.

Reciprocal The quantity obtained by dividing the number one by a given quantity.

Reduce

(i) The conversion of a matrix of order n into smaller, more manageable matrices.

(ii) The process of simplifying a linear

equation of the form $ay = mx + c$, to ensure there are no multiples of y. The reduced equation is $y = (m/a)x + (c/a)$.

Relative motion The motion of one particle or body when viewed from another.

Riemann Integral – *see integral, Riemann*

Root A quantity that, when multiplied by itself a certain number of times, produces a given quantity: 2 is the square root of 4, the cube root of 8, and the fourth root of 16. Roots, when expressed as a power, are fractional.

Satisfy Data satisfy an equation if they lie on the line defined by the equation.

Scalar A quantity, variable or measurement that has magnitude but not direction. See also *modulus*.

Schematic Stylized or generalized.

Secant (sec) In a right-angled triangle, the ratio of the hypotenuse to the side adjacent of given angle. The secant is the reciprocal of the cosine.

Secular – *see characteristic*

Sigma, Σ A notation used to represent the summation of a series of terms, often written with subsidiary limits above and below. Σ is defined by eqn (2.2) on page 19.

Series *See Maclaurin series; Taylor series.*

Significant figure(s) Any digit of a number that is known with certainty.

Sine In a right-angled triangle, the ratio of the opposite to the hypotenuse. The sine is defined by eqn (16.1) on page 247, and symbolized as 'sin'.

Singular Description of a square matrix that has a determinant of zero, and is therefore without an inverse. *See also degenerate.*

Slope The steepness of a line or curve.

Solidus The mathematician's preferred term for the slash '/', the straight line separating the two halves of a fraction. It is more horizontal than a virgule.

Sphere A solid geometric body whose surface is at all points equidistant from the centre. In Cartesian coordinates, its equation is, $x^2 + y^2 + z^2 = r^2$.

Standard deviation A measure of dispersion in a frequency distribution, equal to the square root of the mean of the squares of the deviations from the arithmetic mean of the distribution.

Strategy A plan, method, or series of manoeuvres for obtaining a specific goal or result.

Substitution The replacement of a number or algebraic expression by other numbers or algebraic expressions, which are equivalent (i.e. the equality stated in the original equation remains true).

Tangent In a right-angled triangle, the ratio of the opposite to the adjacent. The tangent is defined by eqn (16.3) on page 247, and symbolized as 'tan'.

Tau (τ) An irregular number occurring often in nature, having the approximate value of 1.618 (to 4 s.f.).

Taylor series An approximation of a given function f at a particular point x, in terms of values of the function and its derivatives at a neighbouring point x_0, by a power series in which the terms are given by $f^{(n)}(x_0) (x - x_0)^n/n!$, where $f^{(n)}(x_0)$ is the derivative of order n evaluated at point x_0. *See also Maclaurin series.*

Transpose (A^T) The matrix obtained by simply exchanging matrix rows for columns, so if a matrix element is a_{nm}, the elements in the transpose are a_{mn}.

Triangle A polygon with three sides.

Turning point The coordinates of the part of a graph where (instantaneously) the gradient is zero. The term usually applies to Cartesian graphs.

Unit vector A vector having a length of one unit. Unit vectors in the x-direction are symbolized as $\hat{i}$; unit vectors in the y-direction are symbolized as $\hat{J}$; unit vectors in the z-direction are symbolized as $\hat{k}$.

Unity The value of unity is 1.

Variable An algebraic quantity, the value of which can change. See also *observed variable* and *controlled variable*.

Vector A quantity possessing both magnitude and direction. *See also unit vector.*

Virgule – *see solidus.*

Vulgar fraction A fraction, the value of which cannot be described adequately by a non-recurring or non-infinite decimal.

Bibliography

The opinions expressed below are those of the authors, and do not necessarily reflect the opinions of Oxford University Press.

Books of mathematics for chemists

Essential algebra for chemistry students, David W. Bell, Thomson, Belmont CA, 2006. A useful little book: it contains plenty of good examples, and its explanations are generally well phrased. It contains no calculus, though, and concentrates really on notation and units, elementary rearrangements, and graphs.

Mathematics for physical chemistry: A guide to calculation in physical and general chemistry, Robert G. Mortimer, Elsevier, Amsterdam, 2005 (3rd edn). This book's major strength concerns its wide array of Worked Examples. The level is quite high, and the amount of explanation is occasionally too slim. It also describes the use of computer programmes such as BASIC. But it *is* a good book.

Chemical calculations, Paul Yates, Chapman and Hall, London, 1997. This is Yates' first maths book. It is relatively short, and aimed at those with little or no maths. Yates' style is gentle and persuasive. It *encourages* the reader. He also employs a margin icon of a calculator, demonstrating its correct use at appropriate junctures.

Yates is a British physical chemist; the chapters of his book relate to a traditional course in physical chemistry (level I and some level II) so, within each chapter, Yates discusses the mathematics needed by a typical student. For example, Chapter 3 is entitled 'Solution Chemistry' and Chapter 4 is 'Kinetics', and so on. This intelligent approach works well for students who struggle during a course in physical chemistry, but may not work so well for students on courses explicitly aimed at teaching mathematics. Yates himself must have felt the approach failed somewhat, because he includes an appendix of pure mathematics. Also, there is no logical progression from the mathematical fundamentals via symbols through to algebra, powers, etc.

Chemical calculations at a glance, Paul Yates, Blackwells, Oxford, 2005. This book represents a 'restatement' of Yates' earlier book *Chemical calculations* (immediately above) since it covers the same material but is themed mathematically rather than according to the syllabus of a traditional course in physical chemistry.

The book is beautifully produced in A4 format, and shows a logical layout throughout, starting with notation and units, and moving eventually to calculus. The proportion of chemical examples is high. The number of Worked Examples is higher than the average, although occasionally more detail would have helped. Nevertheless, it's a helpful tool.

The chemistry maths book, Erich Steiner, (2nd edn), OUP, Oxford, 2008. The second edition improves what was already the ultimate title on the market. Steiner is an inspiring and gifted teacher, and this book displays that gift on every page. It is written beautifully and well produced. The text is authoritative and comprehensive. It's the only 'maths for chemistry' text with a truly satisfying index. Furthermore, Steiner litters the text with clever quotes; most pages glisten with fascinating historical detail in a series of well-designed footnotes.

Maths for chemists, Volume 1: Numbers, functions and calculus, Martin C. R. Cockett and Graham Doggett, RSC, Cambridge, 2003. This book is well produced, as are all titles in the Royal Society of Chemistry's 'Tutorial Chemistry Texts' series. It is surely astonishing that its authors think the chapter on algebra should be a mere 10 pages in length. There are no chemical structures, although to be fair the proportion of the book mentioning chemistry is pleasingly high. The index is short but better than average.

Maths for chemists, Volume 2: Power series, complex numbers and linear algebra, Martin C. R. Cockett and Graham Doggett, RSC, Cambridge, 2003. This book is the companion volume to the text immediately above. By 'linear algebra,' the authors mean vectors, arrays, determinants and the like. The book is far too advanced for the average student, although its treatment of quite difficult topics is generally good.

Beginning mathematics for chemists, Stephen K. Scott, OUP, Oxford, 1995. Chemistry staff like this book. Its approach is logical and gentle on the student. Its principal titles such as 'warming up' and 'relaxing down' also help. But the book is too short. Calculus occupies the centre stage, with about 50 of its 136 pages. Algebra occupies a tiny 21 pages.

Essential mathematics for chemists, John Gormally, Prentice Hall, Harlow, 2000. The book is visually appealing with its two-colour printing. Gormally arranges the seven short chapters into sensible and predictable groups, starting with 'handling numbers,' then 'handling algebra.' It has separate chapters to describe the differential and integral calculus. Its index is poor, but does include chemical terms as well as mathematics. The exercises are generally chemical, although almost all of these chemical terms only relate to physical chemistry.

Unfortunately, once more this book looks like a vehicle for mathematicians rather than for chemists. To the eye, its pages seem crammed with equations. The number of figures is minimal, and there are no chemical structures.

Basic mathematics for chemists, Peter Tebbutt, Wiley, Chichester, 1994. The text is very traditional in style and content, with monochrome pages and narrow margins. It is slightly long. The overwhelming majority of the book represents mathematical content rather than chemistry.

Tebbutt's book contains more figures than those above, the majority of which are clearly of chemical origin but, yet again, there are no chemical structures. The very short index includes chemical terms as well as mathematical terms. Most of these chemical terms relate exclusively to physical chemistry.

Applying maths in the chemical and biomolecular sciences, Godfrey Beddard, Oxford University Press, Oxford, 2009. We did not use this book during writing because it only came out in August 2009. It employs an example-based approach.

Calculations for A-Level chemistry E. N. Ramsden, (3rd edn), Stanley Thornes, Cheltenham, 1994.

This book was in print until quite recently, and some shops may still have a copy. A revamped edition is expected soon.

The book contains a few chemical structures, and *all* the examples derive from chemistry—and from all branches of chemistry, not just physical chemistry. No student could complain that this book was written for mathematicians.

The style is clear, concise and constructive. Like Yates, Ramsden chooses to teach by chemical discipline rather than mathematical topic, which will preclude its use from most courses of 'Mathematics for chemistry.' The intended audiences are students reading for pre-university qualifications, so its standard is only slightly above that of a modern HND.

Books describing the mathematics

The 'teach yourself' series of books are generally excellent. For example, each of the following is highly recommended:

- *Teach yourself mathematics*, Trevor Johnson and Hugh Neill, Hodder and Stoughton, London, 2003
- *Teach yourself algebra*, Paul Abbott and revised by Hugh Neill, Hodder and Stoughton, London, 2003
- *Teach yourself trigonometry*, Paul Abbott and revised by Hugh Neill, Hodder and Stoughton, London, 2003
- *Teach yourself calculus*, Paul Abbott and revised by Hugh Neill, Hodder and Stoughton, London, 2003.

The *Dummies*© series from Wiley Interscience is also a good resource for the novice. The following are useful:

- *Algebra for dummies*, Mary Jane Sterling, Wiley Interscience, New York, 2001.
- *Algebra II for dummies*, Mary Jane Sterling, Wiley Interscience, New York, 2006.

Web-based support

Several good websites support a degree course of mathematics for chemists. All the websites we cite were last accessed on 1 August 2009, unless stated otherwise.

The site *http://mathworld.wolfram.com* has short, informative and authoritative articles on most aspects of mathematics, and is always worth a glance. Its sister site *http://demonstrations.wolfram.com* has many short videos that help demonstrate rearrangements and act as tutorials.

The similar site *http://www.intmath.com* also hosts a large number of excellent, short summaries that supplement most aspects of the mathematics described in this book.

http://www.mathscentre.ac.uk is a different sort of good site. It hosts leaflets, video clips, Worked Examples, etc., for both staff and students. Highly recommended.

Another good site is the 'Mathematics Support Materials' at the University of Plymouth: *http://www.tech.plym.ac.uk/maths/resources/ PDFLaTeX/mathaid.html* that hosts a great many .pdf files of very high quality. It does not cover all the topics in this book, but its treatment of vectors, matrices and calculus is good.

Books describing physical chemistry

Most of the chemistry in these pages comes from the physical branch. Some may see this weighting as unfortunate, but physical chemistry is generally mathematical.

The physical chemistry in these pages is adequately described in straightforward books on physical chemistry. For example, the best selling textbook of physical chemistry in the world is undoubtedly Atkins' *Physical chemistry*. The edition cited below is the seventh by P. W. Atkins and Julio de Paula, Oxford University Press, Oxford University Press, Oxford, 2002. Many students find it rather too mathematical; and its treatment is deliberately thorough—which again means mathematical.

The 'little brother' of this book is *Elements of physical chemistry* P.W. Atkins, (fourth edition), Oxford University Press, Oxford, 2005, which is intended to overcome these perceived difficulties by limiting the scope and level of its parent text.

Several texts approach the topic by means of Worked Examples. *Physical chemistry*, C. R. Metz, (2nd edition), McGraw Hill, New York, 1989, is a member of the *Schaum* 'out-lines' series of texts, and *Physical chemistry*, H. E. Avery and D. J. Shaw, Macmillan, Basingstoke, 1989, is part of the 'College Work-out Series'. Both books are crammed with Worked Examples, self-assessment questions, and hints at how to approach typical questions. Avery and Shaw is one of the few general textbooks on physical chemistry that a non-mathematician can read with ease.

Monk has tried to move away from the traditional approach in his, *Physical chemistry*: *Understanding our chemical world*, Wiley, Chichester, 2005. Here, Worked Examples take centre stage, with the theory being deduced from the examples rather than derived mathematically from first principles.

Books describing analytical chemistry and statistics

The following texts contain sufficient material to explain the analytical chemistry described in these pages: *Statistics and chemometrics for analytical chemistry*, J. N. Miller and J. C. Miller, (5th edn), Pearson, Harlow, 2005, is well written and offers a well-balanced approach. While the book *is* mathematical, the maths is not overbearing. Alternatively, *Analytical chemistry* (5th edn), G. F. Christian, Wiley, New York, 1994, is fairly comprehensive and will cover some of the material discussed here.

Books describing electrochemistry

Electrochemistry is a notoriously mathematical disciple. Without doubt, the best 'all round' book for the electrochemist, and already regarded as a modern classic, is *Electrochemical methods: Fundamentals and applications* (2nd edn), A. J. Bard. and L. R. Faulkner, Wiley, New York, 2001. Its treatment is always authoritative and *very* mathematical.

Alternatively, *Fundamentals of electroanalysis*, P. M. S. Monk, Wiley, Chichester, 2001, is part of the distance-learning series, 'Analytical Texts in the Sciences' *AnTS*. It therefore brims with Worked Examples and illustrations, but without the accompanying mathematical detail.

Books describing quantum mechanics

Quantum mechanics is essential when trying to understand how and why bonds form. For a comprehensive treatment, the necessary mathematics often lies beyond the scope of even mature undergraduates. The following texts will help when confronting quantum-mechanical problems: *Quantum mechanics,* Yoav Peleg, Reuven Pnini and Elyahu Zaarur, McGraw Hill, New York, 1998 is part of the *Schaum* 'outlines' series. It is very good, but most chemists would want significantly more prose between the equations. It also fails to include a list of symbols and frequently lapses into jargon.

In some respects, *Physical chemistry: A molecular approach*, Donald A. MacQuarrie and John D. Smith, University Science Books, Sausalito, CA, 1997, is a better book for the budding quantum mechanicist: without ever lacking rigour, it contains more explanation and prose.

Erich Steiner, whose book *The chemistry maths book* is cited above, is a quantum mechanicist, so his book brims with pertinent detail and relevant examples.

Sources of the examples cited in the text

Chapter 2 (Algebra I: Introducing notation, nomenclature, symbols and operators)

Worked Example 2.8: for full details of the reaction, see: 'Synthesis of mono-substituted 2,2′-bipyridines,' J. Cordaro, J. McClusker and R. Bergman, *J. Chem. Soc., Chem. Commun.*, 2002, 1496.

Chapter 6 (Algebra V: Rearranging equations according to the rules of algebra)

Self-test 6.2: The Nernst diffusion layer is best described in *Electrochemical methods*, Bard and Faulkner, p. 339 and appears as eqn (9.3.25).

Chapter 7 (Algebra VI: Simplifying equations: brackets and factorizing)

Worked Example 7.12: the fascinating relationship between the Golden Ratio and τ is described in the web page, *www.austms.org.au/Modules/Fib/fib.pss* (site last accessed 10 June 2004). Also well worth a look are the websites *http://maven.smith.edu/~phyllo/About/FiboPhyllo.html, http://www.geocities.com/capecanaveral/lab/5833/cycas.html* and *http://www.swintons.net/deodands/archives/000088.html*.

Chapter 10 (Algebra VI: Solving simultaneous linear equations)

The original paper describing the Arnon assay for chlorophylls, is D.I. Arnon, *Plant Physiol.*, 1949, **24**, 1–15. For newer methods, and for accurate values of ε for both chlorophyll a and b, see: 'The chequered history of the development and use of simultaneous equations for the accurate determination of chlorophylls a and b', Robert J. Porra, *Photosynth. Res.*, 2002, **73**, 149–156.

Chapter 14 (Statistics I)

For a good refresher course on elementary statistics, the *Dummies* series is quite good. For example, *Intermediate statistics for dummies*, Deborah Ramsey, Wiley, New York, 2007, has good discussions of the various means calculable.

For the physical scientist, the classic text is, *Data reduction and error analysis for the physical sciences* (3rd edn) Philip R. Bevington and D. Keith Robinson, McGraw-Hill, Boston Burr Ridge, IL, 2003, which is authoritative, mathematical but actually very readable.

The various types of standard deviation can cause much confusion, in part because some workers use the same terms to mean different statistical parameters. The web page *http://mathworld.wolfram.com/StandardDeviation.html* has a concise but useful summary of the different usages.

Chapter 15 (Statistics II)

Data reduction and error analysis for the physical sciences, Bevington and Robinson is probably the best text to supplement this chapter. Steiner's *The chemistry maths book* (2nd edn). Chapter 20, is also well worth a careful read.

Worked Example 15.7: The coherent neutron-scattering length came from the web page, *http://www.ill.eu/news-events/news/new-results-for-the-neutron-scattering-lengths-of-carbon-13*.

Chapter 16 (Trigonometry)

Worked Example 16.4: concerning the adsorption of methyl viologen on platinum, see the paper: 'An analysis of the voltammetric adsorption waves of methyl viologen,' K. Kobayashi, F. Fujisaki, T. Yoshimina and K. Nik, *Bull. Chem. Soc. Jpn.*, 1986, **59**, 3715.

Worked Example 16.5: for an accessible introduction to magic-angle NMR, see: *Fundamentals of molecular spectroscopy* C. N. Banwell and E. M. McCash, (4th edn), McGraw–Hill, Maidenhead, 1994, pp. 274–276.

Chapter 17 (Differentiation I: Rates of change, tangents, and differentiation)

Additional problem 17.2: our data come from: *Physical chemistry*, MacQuarrie and Smith, p. 954.

Additional problem 17.6: the temperature dependence cited relates to the electron mobility due to ionized-impurity scattering; see: *The physics and chemistry of solids*, Stephen Elliott, Wiley, Chichester, 1998, p. 510.

Additional problem 17.7: the effective potential is described in, *Physical chemistry*, Atkins, p. 369.

Additional problem 17.8: the expression is given in, *Physical chemistry*, Atkins, p. 738. Light scattering in general is discussed from p. 738.

Additional problem 17.9: virial equations and other alternatives to the ideal gas equation are discussed at length in, *Physical Chemistry*, Atkins, p. 17.

Additional problem 17.10: the form given to b here may be found in *Physical chemistry*, Atkins, p. 842. (Atkins symbolizes this quantity as B.)

Chapter 18 (Differentiation II: Differentiating other function)

Additional problem 18.10: the equation for ψ comes from, *Physical chemistry*, MacQuarrie and Smith, p. 413 (where the quantity is symbolized as ϕ_{1s}^{STO}).

Chapter 19 (Differentiation III: Differentiating functions of functions)

Additional problem 19.2: the expression comes from, *Physical chemistry*, Atkins, p. 738.

Additional problem 19.3: these laser equations come from, *Physical chemistry*, Atkins, p. 553 ff. Non-linear effects and frequency doubling are discussed on p. 559.

Additional problem 19.4: for an accessible introduction to magic-angle NMR, see, *Fundamentals of molecular spectroscopy*, Banwell and McCash, pp. 274–276.

Additional problem 19.7: the equation for ψ comes from, *Physical chemistry*, MacQuarrie and Smith, p. 413 (in which the ψ is symbolized as ϕ_{1s}^{GF}).

Additional problem 19.8: the example here has been amended slightly from, *Physical chemistry*, Metz, p. 296.

Additional problem 19.9: the expression linking U and r is given in, *Physical chemistry*, MacQuarrie and Smith, p. 540. We have amended it somewhat.

Additional problem 19.10: the expression linking θ and λ is explored in, *Physical chemistry*, MacQuarrie and Smith, 1197.

Chapter 20 (Differentiation IV: The product and quotient rules)

Worked Example 20.4: the equation comes from, *Fundamentals of molecular spectroscopy*, Banwell and McCash, p. 157.

Additional problem 20.4: shielding and screening are discussed in, *Physical chemistry*, Atkins, p. 261.

Additional problem 20.7: the derivation of these expressions is available in, *Physical chemistry*, MacQuarrie and Smith, p. 972, 973.

Additional problem 20.8: the coiling of polymer chains is discussed in, *Physical chemistry*, Atkins, p. 721.

Additional problem 20.10: many physical chemistry texts offer a poor treatment of the Eyring equation. Indeed, in many, the equation is different from that cited here. For a thorough justification for our version, see Monk, *Physical chemistry*, p. 416–420.

Chapter 23 (Integration I: Reversing the process of differentiation)

For more information on Newton and Leibniz, see the web site, *http://www.bbc.co.uk/ history/historic_figures/newton_isaac.shtml.*

Concerning old writing styles, see the interesting detail at *http://www.waldenfont. com/public/dhmanual.pdf.*

Additional problem 23.2: for a discussion of this relationship between C_p and T, see: *Physical chemistry*, Atkins, pp. 50 and 105.

Additional problem 23.3: the dependence between C_p and T at very low temperatures is mentioned in, *Physical chemistry*, Atkins, p. 51.

Additional problem 23.4: the dependence of this phase boundary comes from, *Physical chemistry*, MacQuarrie and Smith, p. 951.

Additional problem 23.5: for more information, including a full derivation of A, see: *Physical chemistry* (3rd edn), Gilbert Castellan, Addison Wesley, Reading, Ma., 1983, 671.

Chapter 24 (Integration II: Separating the variables and integration with limits)

Additional problem 24.1: the derivation of an expression for ΔG from the appropriate Maxwell relation is discussed in *Physical chemistry*, Atkins, 128 ff., and *Physical chemistry*, Monk, p. 154 ff.

Chapter 25 (Integration III: Integration by parts, by substitution, and integration tables)

In the *Teach yourself* series, *Calculus* by Abbott and Neill, p. 167 ff. is a useful resource, and has a few suitably pitched examples.

The full citation for the latest edition of the 'Rubber Handbook' is, *CRC Handbook of chemistry and physics*, CRC Press, Boca Raton FL., 2010. The integrals in Table 25.1 come from the 66th edn, R. C. Weast (ed.), CRC Press, Boca Raton, FL., 1985, pp. A-20–65

An excellent compendium of Maclaurin series can be found at the site, *http://mathworld.wolfram.com/MaclaurinSeries.html.* Concerning Taylor series, the relevant page is, *http://mathworld.wolfram.com/TaylorSeries.html.*

Chapter 26 (Integration IV: Area and volume determination)

Worked Example 26.6: for a fuller discussion of this particular probability integral, see Steiner, *The chemistry maths book*, (2nd edn), p. 302.

Additional problem 26.7: for a fuller treatment, see Steiner, *The chemistry maths book*, pp. 302, 303.

Chapter 27 (Matrices)

For an introduction to matrices, first visit the websites *www.maths.surrey.ac.uk/explore/ emmaspages/option1.html* and *http://mathworld.wolfram.com/Matrix.html* and its online bibliography. Next, try Steiner, *The chemistry maths book*. Chapters 17 (Determinants), 18 (Matrices and linear transformations), and 19 (Matrices and the eigenvector problem). Finally, read the relevant sections of *Matrix operations*, Richard Bronson, McGraw Hill, New York, 1989 (part of the excellent *Schaum* 'outlines' series). Bronson is hugely mathematical, but does have many Worked Examples.

Worked Examples 27.1 and 27.2: the website *http://www.maths.surrey.ac.uk/exlore/ mmaspages/option1.html* is informative, and several of its Worked Examples are useful. Alternatively, the website *http://www.stolaf.edu/depts/chemistry/courses/toolkits/247/ js/huckel* has several dozen examples of Hamiltonian matrices of cyclic compounds; none contains a heteroatom.

Chapter 28 (vectors)

For an introduction to vectors, first visit the websites *http://www.netcomuk. co.uk/~jenolive/homevec.html* and *http://www.mathsrevision.net/gcse/pages.php? page=2*. Next, try the book *Vector analysis: And an introduction to tensor analysis*, Murray R. Spiegel, McGraw Hill, New York, 1959, is a part of the excellent *Schaum* 'outlines' series. It is good, thorough, but very highly mathematical.

Worked Example 28.11: based on the example in Spiegel, *Vector analysis,* p. 67.

Chapter 29 (Complex numbers)

Chapter 14 of *Algebra II for dummies*, Mary Jane Sterling, is a good starting point when reading complex numbers. Its Worked Examples are easily followed yet thorough.

Chapter 30 (Dimensional analysis)

This is not a topic that is treated well in most textbooks. Try first the website *http://www.efm. leeds.ac.uk/CIVE/CIVE1400/Section5/dimensional_analysis.htm*. Alternatively, *Dimensional analysis and intelligent experimentation*, Andrew C. Palmer, World Publishing, Singapore, 2008, has some nice examples.

Other websites of interest

Concerning the fascinating and usually fruitful relationship between the etymology of a word and its present meaning(s), try the website *http://www.etymonline.com*.

Index

Chemicals and substances in the text

Scientists cited in the text

Main index